D0577623

ENVIRONMENTAL MANAGEMENT STRATEGIES

THE 21ST CENTURY PERSPECTIVE

ISBN 0-13-739889-1

9 780137 398898

PRENTICE HALL PTR
ENVIRONMENTAL
MANAGEMENT SERIES

ENVIRONMENTAL

MANAGEMENT STRATEGIES

THE 21ST CENTURY PERSPECTIVE

Gabriele Crognale, P.E.

Independent Consultant

Prentice Hall PTR
Upper Saddle River, New Jersey 07458
http://www.phptr.com

Library of Congress Cataloging-in-Publication Data

Crognale, Gabriele.
    Environmental management strategies : the 21st century perspective
    / Gabriele Crognale.
        p.   cm.
    Includes Index.
    ISBN 0-13-739889-1
    1. Environmental management.   2. Environmental management—United
States.   I. Title.
    GE300.C76   1999
    363.7'05'0973—dc21   98-34643
                                                                        CIP

Acquisitions editor: Bernard M. Goodwin
Cover design director: Jerry Votta
Manufacturing manager: Alan Fischer
Marketing manager: Kaylie Smith

© 1999 by Prentice Hall PTR
Prentice-Hall, Inc.
Upper Saddle River, New Jersey 07458

Prentice Hall books are widely used by corporations and government agencies for
training, marketing, and resale.

The publisher offers discounts on this book when ordered in bulk quantities. For more information contact:

> Corporate Sales Department
> Phone: 800-382-3419
> Fax: 201-236-7141
> E-mail: corpsales@prenhall.com

> Or write:

> Prentice Hall PTR
> Corp. Sales Dept.
> One Lake Street
> Upper Saddle River, New Jersey 07458

All rights reserved. No part of this book may be reproduced, in any form or by any means,
without permission in writing from the publisher.

Printed in the United States of America
10 9 8 7 6 5 4 3 2 1

ISBN: 0-13-739889-1

Prentice-Hall International (UK) Limited, *London*
Prentice-Hall of Australia Pty. Limited, *Sydney*
Prentice-Hall Canada Inc., *Toronto*
Prentice-Hall Hispanoamericana, S.A., *Mexico*
Prentice-Hall of India Private Limited, *New Delhi*
Prentice-Hall of Japan, Inc., *Tokyo*
Simon & Schuster Asia Pte. Ltd., *Singapore*
Editora Prentice-Hall do Brasil, Ltda., *Rio de Janeiro*

I dedicate this book to my mother Adelia, whose courage through her long period of agonizing pain and suffering has been a spiritual inspiration to me and holds no bounds. Her encouraging words and deeds through these tough physical and emotional times will always remain in my heart and soul. Angels finally escorted her into heaven on November 4, 1998. No one can take her place in our hearts. Ma, Dio ti benedice per tutto l'eternita.

CONTENTS

PART 3

PUTTING INNOVATION INTO PRACTICE: A CLOSER LOOK   255

CHAPTER 10

CHAPTER 11

EFFECTIVE RISK MANAGEMENT AND COMMUNICATION: TIPS ON WORKING
   WITH THE PUBLIC   **294**

PART 4

WHERE ENVIRONMENTAL MANAGEMENT MAY BE HEADED   369

CHAPTER 13
UTILIZING SOFTWARE TO ENHANCE ENVIRONMENTAL MANAGEMENT
   RESPONSIBILITIES   370

CHAPTER 14
SOURCES OF ENVIRONMENTAL INFORMATION   388

PREFACE

Environmental management, as seen from the eyes of a regulated organization's environmental manager and the regulator, has undergone a metamorphosis from its inception over twenty-five years ago with the creation of the U.S. Environmental Protection Agency (EPA). Since that time, environmental management, as we know it today, has matured into a viable business enterprise that bears little resemblance to what it used to be—a compilation of engineers and scientists focused on developing reactionary compliance responses to comply with EPA and state regulatory agencies' regulatory programs and regulations.

In general, managers at many facilities during the early stages of environmental management viewed this area as a nuisance business function created to manage a company's wastes, discharges, and emissions, and hence, it was seen as a detraction in time, effort, and monies to a company's bottom line. In addition, company environmental managers and their support staff was not only burdened with managing all the regulatory requirements, but also had to contend with the fact that they did not have as much top management support within the organization as they have today in addressing these same environmental issues.

With the changing times, and in the evolutionary awareness process within many forward-thinking organizations, several new management venues were introduced into corporations to help champion a more proactive corporate approach. These venues include: the practice of developing and having an environmental policy in place that is usually signed by an organization's chief executive officer; the development of a similarly focused corporate mission statement; more and more emphasis on life-cycle analyses and design for the environmental considerations in research and development departments for launching new products; and the increasing issuance of environmental reports by publicly held corporations to share with the public their respective environmental standing. These are provided as examples of what some regulated organizations have accomplished over the past several years to highlight their environmentally progressive achievements.

Stepping back again to the early years of companies' environmental efforts, it was common to see top management at some regulated companies perceive maintaining environmental compliance as a low priority. For some, subsequent penalties imposed by EPA or state regulators on their facilities for noncompliance were viewed as part of the cost of doing business. Progressive initiatives such as environmental audit programs, regulatory compliance and worker awareness training and pollution prevention opportunities, and other "tools" in the environmental manager's toolbox, were not yet as commonplace in the regulated community as they are today.

Fast forward to the present, environmental management has matured to a level where it has been assimilated into an organization's mainstream as a recognized business entity. That status level has helped push environmental management issues and concerns to the forefront of companies' strategic planning, and has helped channel top-management's efforts at many organizations to champion environmental stewardship as a goal for which to strive. The evolution of environmental management has taken organizations from their initial goal to maintain compliance to a new set of goals where the exploring of innovative ways to manage raw materials, processes and waste streams in a way that is both environmentally sound and cost-effective, is the wave of the next generation of environmental management.

On the regulatory side, while organizations continue exploring such innovative ways to maintain and look beyond maintaining compliance, the EPA continues exploring innovative ways to increase the regulatory compliance levels of organizations by continuing its message to organizations regarding their environmental responsibilities in relation to their business operations. While this concept may still feel like "command-and-control" that many organizations equate to the "old" EPA, we need to keep in perspective the intent behind EPA's compliance and enforcement efforts. These efforts still reflect EPA's general tenet, which is the protection of human health and the environment, both within the boundaries of an organization, and the general populace and the environs as well.

As a point of reference to some of the regulatory milestones achieved in EPA's push to have organizations return to and maintain regulatory compliance, the following EPA documents are worth noting as pivotal instruments: the Clean Air Act Amendments (1990), the Pollution Prosecution Act (1990), EPA's Revised Civil Penalty Policy (1992), and EPA Office of Enforcement's Revised Enforcement Four-Year Strategic Plan (1992). Of these activities, the issuance of EPA's penalty policy and the setting in motion of the Four-Year Strategic Plan (a discussion of this enforcement vehicle is provided in Chapter 2) may have contributed to sending a strong message to some nonbelievers in regulated industries who still held the misconception that penalties for regulatory noncompliance were part of the cost of doing business. During that period, it was not unusual to see EPA issue enforcement actions against organizations with penalties in excess of $10 million for violations of various federal regulations, including the Resource Conservation and Recovery Act (RCRA), the Clean Water Act (CWA), the Clean Air Act (CAA) and the Toxic Substances Control Act (TSCA). If the message had not yet been clear to the regulated community that payments of enforcement action penalties in lieu of maintaining compliance would no longer be viewed by EPA as a viable course of action, each subsequent and larger EPA enforcement action and penalty helped bring the point home to more organizations that such activities would soon be a thing of the past.

At some turning point, more companies began to see that it indeed paid to be in compliance, not only for the sake of compliance, but because it made good business sense as well. In order to sustain such efforts,[1] many organizations also saw that it made good business sense to have a dedicated group of professionals and support staff to stay on top of environmental compliance matters and to have clear lines of communication with top management to allow for strategic planning efforts.

Since then, considerations in which the next generation of environmental managers have been involved include bridging environmental management with other business enterprises within organizations to provide opportunities for growth and identifying areas for improvement and potential areas to decrease or eliminate waste streams altogether. In certain situations, the possibility exists for coordinated efforts between environmental professionals and their counterparts in research and development, purchasing, inventory/warehousing, human resources and MIS to achieve such waste reductions. The key, as in any new undertaking, lies in several elements: communication, coordination, and dedication. In addition, new information sectors that incorporate software innovations, such as environmental management information systems (EMISs) and enterprise document management (EDM), have emerged that can be very powerful MIS tools the environmental manager can tap into, in ways more effective than the traditional approach to information management used by environmental or MIS professionals. In having such information at their fingertips, environmental professionals can achieve specific goals to address any number of management concerns. The ability of the environmental professional to compile, assimilate, and synthesize key data for top management is, and will continue to be, a powerful management tool. Its potential has still to be tapped to its fullest.

What does this signal to the well-heeled environmental professional and the other business enterprise professionals? To the author, environmental management is no longer an isolated entity within a corporation in which its main focus is to maintain compliance. Rather, environmental management today overlaps with just about every other business enterprise within a company, and if, among other goals, a company wants to achieve environmental excellence and stewardship, it is almost mandatory that environmental professionals begin to look "out-of-the-box" to identify those areas that can help move a company forward. These individuals also need to be out in the forefront with top management to help champion environmental considerations that take a holistic view regarding environmental responsibilities while also keeping stakeholders' concerns in focus.

This paradigm shift in the mindset of many high-profile organizations has not gone without notice by the EPA, and in an attempt to initiate a radically different approach to the typical regulatory response of "command-and-control," EPA began exploring possibilities to work with organizations in a pro-business manner to achieve the agency's goal of increased regulatory compliance. Part of this strategy was outlined in a series of EPA documents and memoranda, beginning with the Five-Year Strategic Plan, issued in 1994 to outline EPA's grand scheme for the remainder of the twentieth century.[2] One of the major tenets of this plan focused on EPA's efforts to lead the nation in reorienting efforts to reduce and eliminate pollution at the source. Among the strategies identified as part of private sector and general public partnerships are increased emphasis on nontraditional programs, such as the 33/50 Program and the Common Sense Initiative, and expanding new approaches to build upon these programs.

As had been previously seen in EPA's first attempts to usher in a series of industry voluntary programs to promote greater voluntary compliance and a willingness to work with the agency's compliance programs, companies were willing to participate to show their sincerity to work with EPA. For example, the relative success with the participants of the Industrial Toxics Project, or the 33/50 Program, initiated by former EPA Administrator William K. Reilly, paved the way for a continuing series of EPA-led industry voluntary initiatives under present EPA Administrator Carol Browner. This new era began with the *Federal Register* notice launching the first Environmental Leadership Program (ELP) in 1993; followed by the subsequent *Federal Register* notice in 1994 launching the ELP Pilot Program, in its final stage at the time this book went to press; the Common Sense Initiative launched in 1994; and finally Project XL, launched in 1995, which is the latest of the EPA-led industry voluntary programs. As this book went to press, Project XL was in its early pilot phase.

Meanwhile, in regulatory enforcement relief, EPA continued its efforts to make its audit disclosure policy more "user-friendly," and first issued a draft policy in 1995, and a final policy in 1996 to address industry concerns regarding the use of environmental audits by regulated industries and possible issues related to audit disclosures. This topic has been the subject of much discussion on both sides, and while full disclosure to EPA may bring regulatory enforcement relief to organizations, there may still be a considerable way to go before full audit disclosure by EPA-regulated organizations becomes a *de facto* way of doing business for them.

Given this dynamic state of regulatory and voluntary flux in which we currently find ourselves, one may wonder where EPA sees the regulated community at this juncture—will EPA continue to move in the direction of industry-friendly programs, or will EPA reverse direction and revert to the old "command-and-control" enforcement focus? Time will tell whether we are at a crossroads in enforcement and compliance, or whether regulated industries can continue to expect a pro-business EPA with increased opportunities for regulatory relief. Some of the answers for the regulated community may lie in how EPA interprets the regulated community's response to a series of initiatives currently on the negotiating table, such as EPA's Audit Policy and self-disclosure; where regulatory compliance benefits may be seen by EPA with respect to organizations' conforming to the environmental management standards of the International Organization for Standardization (ISO), more commonly referred to as ISO 14001; and whether more organizations will commit to any one of EPA's voluntary programs and implement innovative solutions to address a regulatory concern or issue;[3] and other considerations that EPA may be currently tracking.

Again, one needs to keep in perspective EPA's main goal for being in business: seeing the regulated community return to compliance and then staying in compliance to ensure the protection of human health and the environment. Whatever innovative solutions organizations may want to explore to achieve EPA's primary goal is their prerogative—whether through quality management methodologies, such as TQM and ISO 9000; or through more comprehensive environmental management systems linked to ISO 14001, or EPA-led voluntary initiatives—provided that EPA feels confident the regulated community is maintaining and sustaining environmental regulatory compliance to ensure EPA's objectives. The extent that the regulated community can sustain this compliance vision of EPA to some degree of satisfaction may be reflected in the path, whether enforcement driven or not, that EPA decides to take in the future.

Taking this perception one step further, the intent of our book is to identify opportunities, management practices, methodologies, and innovative strategies that organizations can implement to improve upon their existing environmental management programs and other business enterprises from a holistic business overview. The logic is that such progressive undertakings by regulated organizations can also help diffuse any possible return to the old "command-and-control" approach by EPA and the states' enforcement programs, especially if the intent seems sincere and a strong sense of commitment is portrayed by organizations.

By way of positive example, several of the chapters in the book were written by senior environmental managers from Fortune 500 companies, who graciously shared some of their day-to-day experiences to provide added insight into the environmental workings at their companies from a real-world perspective. The case studies they provide exemplify, in both a quantitative and qualitative manner, the points expressed in each of their respective chapters, from a discussion of environmental audit strategies, the effectiveness of certification to ISO 14001, the use of sophisticated software programs and insight into a few actual EMIS and EDM programs in place, and other similar considerations; to the anecdotal insight from dealing with EPA's environmental enforcement, health and safety requirements, risk management, the Internet, and what to consider from various innovative environmental strategies that can be put in practice.

The key to the examples provided is that regulatory compliance is not the only driving force—while it is prominent, related forces are also noted. These include: market forces, stakeholders, and public perceptions, among others, that are all interwoven into the fabric of the regulated community, interacting with each other in a dynamic and ever-changing setting.

Also, as part of the maturing process of environmental management, this sector's increasing interaction with other business areas (enterprises) has created new opportunities for environmental management that may not have been previously possible. Such opportunities have been generated from increased internal cooperation and teamwork between business units leading to more-effective compliance and opportunities for preventing pollution. Without the gradual paradigm shift toward a more proactive position regarding an organization's environmental responsibilities, and understanding that such positions can also be good for the corporate "bottom line," such strides by many organizations may not have been achieved so quickly.

By way of example, the 1990s business upheaval forces that began with the emergence of re-engineering, and continued with downsizing, outsourcing, and other management-driven initiatives to cut costs from an employee-based standpoint, also may have helped environmental management to move forward into its next generation. With the advent of environmental management as a recognized business entity, with its own accountability system for cost, profit and loss, and its increased interaction with other company business enterprise units, this new openness between unit managers and environmental managers may have prompted a new work ethic allowing these groups to collaborate more often. The relative success of environmental managers learning to converse in "business-speak" rather than "enviro-speak" with other business managers to convey the results of environmental expenditures related to permitting, auditing, pollution prevention, ISO 14000, etc., has had some far-reaching effects. To highlight this further, contributing authors in several chapters describe the benefits achieved within their companies by maintaining a more comprehensive and effective environmental management system that incorporates other business units considerations into the management of their environmental responsibilities.

Furthermore, with more and more emphasis being placed on computer software systems and the Internet to help manage this deluge of environmental data being synthesized for top management's review, what used to be a difficult, if not impossible task, is being reduced to the click of a computer mouse and a few keystrokes of a PC in order to have this information in the hands of the environmental manager, who sees this data as a valuable management tool. Three chapters of the book are devoted in various degrees to this topic and provide in-depth information regarding the use of software as one of the tools at the environmental manager's disposal.

The contributing authors' experiences and case studies provide a certain amount of insight into the environmental activities within their respective organizations that can offer some answers or solutions to the readers to issues they may face in their day-to-day work activities.

The author wishes to thank the following people, without whom this book would not be possible: Dan Rush, who championed the material to Bernard Goodwin, who had faith in me and in publishing the book; Diane Spina, Bernard's assistant; and of course, all my contributors who gave of their time and energies, despite their own heavy work loads, and who believed in me from the get-go: Bill Parker, Janet Peargin, Ed Spaulding, Joe Hess, Maria Kaouris, Paul Dadak, Pat Davies, Margie Aleo, Stu Nicholson, Bob Ruczek, Mike Hill, and Ted Firetog, and a very special thanks to Roland Schumann, a former Acquisitions Editor for Government Institutes, who holds a special place: He saw my writing potential and offered me my first book opportunity—*The Greening of American Business.* Without Roland to steer me in this direction, who knows if I would have come this far. Thanks, Roland. You are a true friend.

Of course, a special thanks goes out to Marie, my very patient wife and new mom, who gave me the courage to go on despite the overwhelming odds; my parents, Cleto and Adelia, who instilled in me a drive to achieve to the best of my ability and never give up; to my sister, Francine, for being there for me, and to the newest addition of the Crognale clan, little "Doey," who has given me new meaning to looking at life from a holistic perspective. To all of you I say "Thank You."

Most of all, this book is dedicated to my mom, Adelia, who instilled in me a drive to seek answers where sometimes there are only questions. I admire her courage.

ENDNOTES

[1]In response to environmental regulatory changes at the time, the oil industry initiated a study that projected a minimum expenditure of $166 billion over a twenty-year period to comply with existing and anticipated environmental regulations, according to an article written by A. Sullivan, "Oil Industry Projects a Surge in Outlays to Meet US Environmental Standards" that appeared in *The Wall Street Journal*, August 31, 1993, p. A2.

[2]EPA document, *The New Generation of Environmental Protection: EPA's Five-Year Strategic Plan*, July 1994.

[3]Of EPA's voluntary programs, the Common Sense Initiative recently lost two industry-sector participants—the automotive and petroleum refining industries—due to a number of considerations specific to these industry sectors. The remaining four industry-sector participants, including metal finishing, computers and electronics, and iron and steel are still active participants. For further information, see V. Leclair, "Common Sense reform initiative falters," *Environmental Science & Technology*, May 1997, p. 222 A.

ABOUT THE EDITOR

Gabriele Crognale, P.E., an independent consultant, provides client assistance in developing more efficient environmental management systems and health and safety programs. Some of this assistance comes from experience gained in courses he taught for various professional and trade groups and universities, such as Government Institutes and Environmental Education Enterprises, and in conducting environmental and management system audits, and management system verifications (MSVs) for CMA-member companies. He bases his expertise on more than 23 years of experience in the environmental field, with concentration in RCRA, TSCA, OSHA, and ISO 14000. Among his current duties, he is a regular contributor to *Safety Compliance Letter,* and *Maintenance Management,* where he is on the Advisory Panel, two publications of the Bureau of Business Practice; teaches ISO 14001 Lead Auditor's courses for IQUES; is an adjunct at Quincy College; serves as a mentor for environmental graduate students of the Gordon Institute at Tufts University; and provides ISO 14001 auditing services for Scott Quality Systems Registrars, Inc. He has also been active as a presenter at several environmental conferences and trade association meetings in the U.S. and Europe. His previous works include contributions to the *Greening of American Business* and *Auditing for Total Environmental Quality.* He is a member of one U.S. Technical Advisory Groups to ISO's Technical Committee 207 and is also a RAB-accredited EMS Auditor. He earned his BS in Civil Engineering from the Northeastern University.

ABOUT THE AUTHORS

Janet Peargin is a policy analyst with Chevron Corporation, working on federal Clean Air Act legislation and regulations.

Peargin's experience includes environmental management and engineering positions with Chevron's pipe line and production companies, along with environmental consultation and training for Chevron operations in China, Nigeria, Papua New Guinea, Indonesia and South America.

She was Chevron's pollution prevention technical consultant for six years, and continues to teach pollution prevention classes at the University of California Berkeley Extension.

Peargin has a degree in environmental engineering from Vanderbilt University.

Edward R. Spaulding is public and government affairs manager for central California and Nevada for Chevron Corporation. He represents Chevron's oil and gas production, pipeline and gasoline marketing activities.

Previous assignments include policy development for federal and state environmental issues, environmental positions at Chevron's El Segundo Refinery in Los Angeles and Salt Lake Refinery in Utah, refinery operations, refinery engineering and project design.

He is a team leader in Chevron's quality fitness review process, where Chevron facilities are reviewed against the Malcolm Baldridge National Quality Award criteria. In 1994, Spaulding chaired a team that produced *Measuring Progress—A Report on Chevron's Environmental Performance,* Chevron's second corporation environmental report. In 1991, Spaulding was on the team that created *Protecting People and the Environment,* Chevron's environment and safety program.

Spaulding is a graduate of the University of California Berkeley with a degree in chemical engineering.

Robert (Bob) A. Rusczek is currently the president of Envirocomp Inc., located in Springfield, Massachusetts. Rusczek has extensive experience in safety and occupational health, and is a certified industrial hygienist and certified safety professional.

Rusczek started his safety and industrial hygiene career as an OSHA compliance safety and health officer in the Hartford, Connecticut, area office. He next received a Master's of Science degree in industrial hygiene, from the University of Michigan.

Rusczek embarked on his manufacturing experience with Monsanto in Nitro, West Virginia. He was responsible for plant safety and health at the facility. He was transferred to the Monsanto Springfield, Massachusetts, facility in 1987, and was promoted to safety and industrial hygiene superintendent. In addition to safety and industrial hygiene, responsibilities included site security, emergency response, TSCA compliance, and being the facility SARA coordinator.

As a safety and health consultant since 1989, Rusczek provides services to a variety of industries, concentrating on OSHA compliance.

Renee R. Bobal is currently senior manager, corporate environmental and safety affairs at Hoffmann-La Roche Inc. She is responsible for leading and managing environmental, health, and safety legislative and regulatory affairs and emerging issues for Roche facilities across the United States and Puerto Rico. Bobal has greater than seventeen years of environmental, health, and safety experience with manufacturing and service industries, a scientific research organization, and government. Bobal has held management responsibility for various corporate environmental programs, including environmental audits, policies and guidelines, site assessments for acquisitions/divestitures, and technical advice for Superfund projects. She also managed various compliance projects and programs, including conceptual environmental design for a grassroots manufacturing facility, permits and agency negotiations, involving the Clean Water Act, Safe Drinking Water Act, Resource Conservation and Recovery Act, Spill Prevention Control and Countermeasures, and site remediation at the federal, state, and local levels. Bobal is currently chairperson of the environmental, health, and safety steering committee of the Pharmaceutical Research & Manufacturers' Association (PhRMA) and is past chairperson of the PhRMA Water Working Group. She has been active as a presenter at various environmental conferences and seminars.

Bobal holds a Bachelor of Science and Master of Science in environmental sciences from Cook College/Rutgers University. Her M.S. research involved evaluation of the biodegradation of PCBs via mesophilic and thermophilic composting.

Joe Hess joined SGS-Thomson's Rancho Bernardo Site in 1996. He is the site's environmental champion (SEC) responsible for compliance and ISO 14000/EMAS conformance.

Hess has seven years of experience in the environmental and health field. He has a Master's of Public Health in Occupational and Environmental Health and is a Registered Environmental Assessor (REA) in the state of California. He is also a Certified Hazardous Materials Manager (CHMM) at the master's level.

Prior to joining ST, he was a project manager at Dames & Moore. During this time he assisted with the development of Rancho Bernardo's Environmental Manual and participated in the EMAs validation process.

While at Dames & Moore, Hess managed and conducted numerous environmental- and health-related compliance audits, assisted clients with waste minimization/pollution prevention studies, conducted environmental assessments for real estate transactions, prepared environmental manuals, and conducted employee training.

Prior to joining Dames & Moore, Hess was an environmental engineer with General Dynamics' space systems division. In this position, he was responsible for the air quality and hazardous waste program and the management of environmental data collected from the division's four sites.

Mitchell Gertz has a Master of Science in environmental engineering from the New Jersey Institute of Technology. He is a certified hazardous materials manager. He has twenty-one years of experience in the environmental field, including three years in consulting, five years as an environmental engineer in the nuclear energy industry, and thirteen years in the chemical industry. The last nine years and currently he has been the manager of environmental affairs for The PQ Corporation. Gertz has been responsible for developing and implementing internal and external audit programs throughout his career. Gertz is an active member of the pollution prevention task force of the Chemical Manufacturers Association and is a member of the Air and Waste Management Association, Water Environment Federation, and the National Association of Environmental Managers.

Paul L. Dadak is an environmental engineer who has more than twenty five years experience working with consultants, regulatory agencies, and industry. He has worked for a consultant engineering firm, the federal USEPA Region I, and industry. He holds both a Bachelors of Science in Civil Engineering and MSCE in environmental engineering and is a registered professional engineer in Massachusetts, New Hampshire, and Rhode Island. His professional experience has included assignments that allowed him to view the environment as a regulator, as affected by regulations, and as an environmentalist.

William H. Parker III serves as both director of environment, safety and health (ES&H) programs and director of chemical programs for EG&G. His ES&H duties include design and development of worldwide policies and programs that govern the company's compliance with federal, state, and local laws pertaining to ES&H standards and providing advice to management on processes and procedures to prevent or abate pollution and to ensure the safety and health of EG&G employees and associates. His chemical duties include assurance of proper operations at EG&G-operated facilities and support of the chemical demilitarization program. In both positions, he is responsible for developing new business opportunities for the company based on the resources within the company.

Prior to joining EG&G, Parker served under both President Reagan and President Bush in the Office of the Secretary of Defense as Deputy Assistant Secretary of Defense (Environment). In this position, he had responsibility for all DOD's environmental programs worldwide, including the Defense Environmental Restoration Program, compliance programs, natural resources programs for over 25 million acres of federal land, and the armed forces pest management control program. He also provided technical support to the Under Secretary of Defense (Acquisitions) for global Chem Demil programs and various advance systems.

Parker is a civil engineer by profession, having obtained his bachelor's degree in civil engineering from the University of Maine in 1960 and Master of Science degree from Northeastern University in 1966, both with an emphasis on environmental engineering. He has also earned Master's of Engineering Management (MEM) in 1981 and Business Administration (MBA) in 1982. Currently, he is a doctoral candidate in business administration at the NOVA University in Ft. Lauderdale, Florida.

Recently, Parker received the Secretary of Defense Medal for Outstanding Public Service. Honorary societies in which he is a member include Tau Beta Pi, Chi Epsilon, Sigma XI, Alpha Kappa Psi, Phi Kappa Phi, and Beta Gamma Sigma. He is also a Diplomate in the American Academy of Environmental Engineers and is listed in Who's Who in Technology. In 1995, he became an honorary member of the National Security Industries Association. Parker is the author of over seventy papers, articles, and testimony before Congress on technology and management.

Diana J. Bendz is currently the Endicott Senior Location Executive and Director of Environmentally Conscious Products, Corporate Operations and Environmental Affairs.

Bendz earned a Bachelor of Science degree in Polymer Chemistry from Syracuse University. She joined IBM as an engineer in electronic packaging development, working on advanced package designs, new materials and package reliability. She has managed package development organizations, manufacturing operations, site planning, site assurance, and cost performance. She became director of product safety and chemical management, corporate staff in 1990, and was named director of integrated safety technology, technical operations staff, IBM corporate headquarters in 1992.

She received the company's patent award, has published numerous technical articles, and is a member of the IBM Academy of Technology.

Bendz is a Fellow of IEEE, cited "for leadership and contribution to electronics manufacturing and to environmental impact and policy." In 1984 she received the IEEE Centennial Key to the Future Award and in 1990 received the IEEE/Components, Hybrids and Manufacturing Technology Society (IEEE/CHMT) Special Electronics Manufacturing Technology Award.

In 1996, Bendz received the Environmental Progress Individual Award from the Electronic Industries Association Environmental Issues Council for her significant contribution to the environmental achievement of the electronics industry.

Theodore W. Firetog is an Environmental Attorney in private practice in Long Island, New York, where he provides a wide range of legal counseling on all aspects of environmental law, including representing clients with respect to regulatory compliance, environmental litigation, and environmental liability in a variety of federal, state, and local law contexts. He also provides clients (as well as other attorneys) with legal advise on securing environmental permits and variances, coordinating environmental audits, hazardous waste matters, and the impact of environmental laws on real estate and other business transactions. He also is recognized as one of the leading authorities on environmental reporting and recordkeeping requirements and strategies for industry.

Firetog received his Bachelor of Science Degree from the University of Michigan, School of Natural Resources, his Master of Science (Natural Resources) Degree from the University of Michigan, and his Juris Doctor Degree from the State University of New York School of Law at Buffalo, where he was also a Sea Grant Law Fellow.

His environmental law expertise has been built on the practical experience that he received as a Staff Attorney with the Environmental Law Institute, an Attorney-Advisor with the United States Environmental Protection Agency in Washington, D.C., and as environmental counsel to the former New York City law firm of Shea & Gould. Mr. Firetog has authored various articles on environmental law and has served as a faculty member and speaker for numerous environmental law symposiums and conferences. He is the editor of *Environmental Reporting and Recordkeeping Requirements (3rd edition)*, published by Government Institutes, Inc.

Firetog is a member of the New York State Bar Association, Environmental Law Section; American Bar Association, Section of Natural Resources Law; Associate Member, Environmental Law Institute; and is listed in *Who's Who in American Law.*

Daniel McDonnell is manager of Product Safety, Asset Protection and Environmentally Conscious Products in IBM Corporate Operations in Endicott, New York. McDonnell is responsible to coordinate efforts across all IBM divisions and geographics for product safety, asset protection, and ECP programs. The position provides a closed loop process linking corporate end of life management, design for the environment and asset recovery.

Mr. McDonnell earned a BS from IUP and an MS from the University of Pittsburgh. He joined IBM as a Product Industrial Engineer. He has managed test engineering, site planning and operations, manufacturing operations, distribution and, most recently, successfully starting and managing the Endicott Asset Recovery Center. He assumed his current position in 1996.

Stuart A. Nicholson's background spans several areas relevant to business risk and environmental management. Originally trained in biological sciences (ecology/environmental science emphasis—B.S., M.S., University at Albany (SUNY), Ph.D., Georgia). As a professor, then consultant and research manager with government, he branched out into research and consulting on applied environmental problems, including extensive work on pesticide and other anthropogenic stressor risks and impacts, pest management, ecosystem recovery and remediation, and biodiversity and natural values protection. Following his interests in environmental policy and responsibility, he went on to receive his J.D. (North Dakota) and practiced several years in nationwide high stakes litigation. Subsequently Dr. Nicholson returned to academia, consulting, and related activities focusing on environmental responsibility, management and other business issues. While continuing in these areas he has most recently expanded his interests and activities into information technology, including its application to business management, processes and risks.

Thomas Harding is a member of the Lucent Technologies Global Environment, Health and Safety Organization. Harding conducts environmental, health and safety compliance and management system audits at Lucent Technologies facilities worldwide. Harding contributed toward the development and implementation of the Lucent Technologies environmental, health and safety global compliance audit program for the past ten years and participated in implementation of the environmental management system (EMS) internal audit program. Harding has over twenty nine years experience as an Environmental Engineer. He is a member of the Environmental Auditors Registration Association (EARA) and the American Institute of Chemical Engineers (AIChE). He holds a BS in Chemical Engineering from Newark College of Engineering.

David B. Jeffries is a member of the Lucent Technologies Global Environment, Health and Safety Organization. Jeffries conducts environmental, health and safety compliance and management systems audits at Lucent Technologies manufacturing facilities worldwide. Jeffries contributed to the development and implementation of the Lucent Technologies environmental, health and safety compliance audit and EMS internal audit programs. Jeffries has over twenty eight years experience as a quality and environmental auditor. He is a member of EARA and the International Register of Certificated Auditors (IRCA). He holds a BS in Forestry from Northern Arizona University.

John Williams, an environmental consultant, has worked closely with Lucent Technologies for over five years in the development and implementation of the compliance audit and EMS internal audit programs. He has led EMS implementation and internal audit training programs and performed internal systems audits at Lucent manufacturing and corporate office locations. He has over thirteen years experience as an environmental consultant, and holds an MS in Natural Resource Planning from the University of Vermont and a BS in Geology from St. Lawrence University.

Maria Kaouris is the Manager for Remediation and Technical Support at the Lucent Technologies Global, Environment, Health and Safety Organization. Kaouris was formerly the manager for the Compliance Assessment Department and had responsibility for compliance and management system audits at Lucent facilities worldwide. Kaouris has been instrumental in further developing and expanding the environmental, health and safety compliance audit program. In addition, she worked with members of the Compliance Assessment Department to develop and implement the management systems audit program. Kaouris has over fifteen years of experience in the environmental field and holds a BA in Biology from Cornell University, a BS degree in Chemical Engineering from City College of New York and an MS degree in Environmental Engineering from Stevens Institute of Technology.

PART 1

STARTING OUR JOURNEYS

Part 1 sets the tone for topics discussed in the subsequent chapters of the book. Chapter 1 talks about where environmental management is at the present, and what environmental managers should shift their focus to for keeping abreast of emerging trends. In similar fashion, Chapter 2 provides a glimpse into EPA's main driver—enforcement—and how this fits into the overall scheme of environmental considerations of regulated organizations. Chapter 3 provides a discussion into related health and safety issues that complement environmental issues, for a total EH&S or ES&H or HS&E perspective. Chapter 4 focuses on SEC requirements relating to environmental disclosure, which has slowly gained additional prominence within the regulated community. Chapter 5 takes an in-depth view of EPA inspections and how regulated companies can prepare for them, followed-through in Part 2. Chapter 6 takes a look at innovative strategies a company's environmental manager can initiate and provides entree into Part 2.

CHAPTER 1

ENVIRONMENTAL MANAGEMENT AT A CROSSROADS: TIME FOR A RADICAL BREAKTHROUGH

Gabriele Crognale, P.E.

> *"It is easier to resist at the beginning than at the end."*
>
> *Leonardo da Vinci*

INTRODUCTION

To the general public, it would seem that the strides many industrial companies have taken today in having achieved a certain level of environmental stewardship and awareness is a comforting feeling that reflects well on both the top management and environmental management of these organizations. A peek into some of these forward-thinking companies is provided courtesy of the environmental professionals who share with readers their particular experiences in dealing with day-to-day environmental matters. They are part of a core group of individuals within their respective companies charting an environmental course for their organizations to greater achievements as they proceed into the twenty-first century, from which all of us should benefit.

As we will see in each of the chapters of this book, this enlightened environmental awareness level has spread through the ranks of environmental professionals and top-level management within regulated business entities, regulatory agencies that administer environmental compliance programs, and a number of support service providers. In similar fashion, the paradigm has also shifted in other business areas or enterprises within organizations that previously viewed environmental concerns and issues as something not within their responsibilities, who now are beginning to realize the tremendous cost savings potential environmental consciousness brings to the bottom line. With such a "shtick," these business units are being brought into organizations' awareness circles as a means to identify and correct potential environmental issues, and more importantly, to identify areas for overall improvement within these organizations providing a positive return on their investment. Gone are the days, at least at the more progressive companies, when company managers believed that environmental and health and safety (EH&S) regulatory compliance and management matters were just a drain on the company's bottom line. A good number of organizations have made the case that an effective EH&S program and management system that looks to streamline operations, save resources, keep workers well-trained and proficient in their duties, and minimize the amount of waste exiting a facility, can serve a dual purpose: keep the regulators satisfied, and save the organization unnecessary expenditures.

For example, forward-thinking company managers at regulated organizations look to incorporate environmental improvements at various levels within the organization in a new, methodical and coordinated fashion, such as in: research and development, purchasing, marketing, human resources, and other business areas where environmental improvements can be identified and achieved. Taking a closer look at research and development, a growing number of organizations are looking to incorporate Design for the Environment (DfE) principles and Life Cycle Assessments (LCAs) into the development and design stage of new products and their potential impact upon the environment,[1] whether as an aspect of raw materials used, energy consumed, the associated waste generated in making the product, or the potential of used product components to be recycled rather than being disposed of as a waste.

One company in particular that is championing product recycling or takeback is IBM, which made strides to execute this effort as a sound business decision—*"Waste = $ off the bottom line"* [2]—rather than as a reactionary response in anticipation of regulatory pressures on the horizon, such as mandated product takeback legislation aimed at the electronics industry. Such legislation has emerged in a number of European Union member countries, and may also soon expand into the Asia-Pacific Region:[3] As such, some U.S. multinational companies may soon face regulatory hurdles from the European Union regarding the disposal of product packaging waste. For example, the European Union's directive on packaging and packaging waste, which are expected to come into effect by July 2001, will require countries to recover over 50 percent of their used packaging, and recycle 25 percent of the material.[4] The benefit to U.S. multinationals in preparing for such sweeping directives can come as redesigning packaging and shipping systems to adhere to these directives ahead of schedule, and possibly save considerable monies in so doing, according to one study.[5]

Incorporating DfE and LCA methodologies into packaging rethink and redesign before products are shipped, whether domestically, or into EC member countries, could result in substantial savings for those multinational organizations that may choose to follow this lead. Of note, the Hewlett-Packard Company was reusing its product containers and pallets as early as 1992 during the distribution of its products in Germany, as a result of that country's ban on solid waste. Since that time, a number of companies have taken a closer look at the concept of responsible packaging, such as Georgia-Pacific Corporation and Dixon Paper, to name a few.[6]

In other considerations, efforts begun by the Coalition for Environmentally Responsible Economies, better known as CERES, to raise the public's and corporations' socially conscious ethics awareness levels, have also spun off a side business enterprise as a result of CERES's efforts to have organizations adhere to the CERES Principles[7]—a series of ten environmentally conscious business principles tied to organizations' business practices that highlight organizations' commitments to conduct their businesses in an environmentally sound and accountable manner (for a closer look at the CERES Principles, refer to Appendix D). The CERES Principles have struck a strong chord with socially conscious groups and have generated a market for mutual fund organizations that invest solely in socially conscious publicly-held organizations that do not manufacture certain products viewed as harmful to human health and the environment. These mutual funds are managed by a growing number of organizations including: the PAX Fund, Domini Social Index, and the Calvert Group.

In a more global context, major industry groups, such as the Chemical Manufacturers Association (CMA); the American Petroleum Institute (API); the non-industry specific professional association, the Public Environmental Reporting Initiative (PERI); the Global Environmental Management Initiative (GEMI), and others, developed and put in place their own principles and initiatives to highlight their commitment to a higher level of environmental awareness and action. For example, the CMA launched Responsible Care® in 1988[8] and requires each of its member companies to adhere to the principles of Responsible Care®. In similar fashion, the API launched its Strategies for Today's Environmental Program (STEP) in 1990[9] which follows the general tenet of the CMA's principles and also requires each of its member companies to adhere to the program's elements.

With respect to PERI, its founding member companies launched the PERI guidelines[10] as a guidance document for companies to follow in preparing their voluntary environmental reports for the public. The impetus to develop such guidelines has been a viable driving force to allow environmental reports to be developed in a consistent and clear fashion for the benefit of concerned stakeholders. By utilizing the PERI guidelines, companies desiring to prepare environmental reports can do so by following the guidelines that provide a level playing field of sorts allowing their environmental reporting information to be developed in similar fashion and consistent with other companies that follow the same guidelines. Among the PERI initiative signatories are: AMOCO, IBM, DOW, Polaroid, and Phillips Petroleum. Further information on CMA's Responsible Care® principles, and the PERI guidelines is provided in Appendix D.

Viewing this dynamic landscape within the bigger picture, or holistic view, a good number of organizations have reached a relative plateau with respect to environmental achievement. In retrospect, the success achieved by many of these regulated organizations is admirable and there is a strong likelihood that these successes will continue, given the current environmental steward mindset that places greater emphasis on a more linked environmental management system (EMS) to streamline the management of environmental responsibilities versus the old corporate mindset that still sees as an objective the maintaining of compliance with environmental regulations without any additional considerations to improve upon existing conditions.

As an example, top management at a number of regulated industries has seen the benefits of and supports the implementation of various environmental, health and safety programs, and continue to look ever more closely at ISO 14000 as another management tool to help enhance their environmental management systems. Certifying conformance to ISO 14001, as well as looking at innovative ways to combine ISO 9001 with ISO 14001 certification, may soon become a strategic agenda item for discussion at a number of organizations' top management meetings.

As a brief introduction into the ISO environmental management standards—ISO 14001 through ISO Guide 64—these standards continue to move toward individual approval status[11] via a consensus development review and approval process of the members of Technical Committee (TC) 207 of the International Organization for Standardization. TC 207 members consist mainly of environmental professionals from industry, regulatory agencies, other stakeholders and service providers charged with developing the ISO 14000 suite of standards.

The transformation of looking further into environmental responsibilities from a holistic view has allowed regulated organizations an opportunity to see the regulatory landscape evolve from a "command-and-control" compliance response mindset into a business approach that has begun to look at environmental considerations in an ever-expanding circle of influence to provide beneficial solutions to the organization and allow for an increased pro-business work relationship with the regulators.[12]

Such a sea change in attitude was fueled in part by an increasing number of regulated organizations whose goals include attaining a level of continual improvement and environmental stewardship as a means to attain higher protection levels than required by environmental regulations. This corporate attitude change for the better probably can trace its beginning to EPA's 33/50 Program, which was launched in 1991.[13] As part of this revolutionary and innovative program, those organizations willing to "test the waters" in committing to an EPA-led voluntary emissions reduction program may have come away from the table with EPA experiencing a somewhat different regulator than they may have previously experienced. This was the begining of an era where organizations began to experience a regulatory agency—the EPA—more willing to listen to and work with the regulated community to attain a common goal—protecting human health and the environment and doing so in ways to foster pollution prevention initiatives and financial incentives to follow in the path first taken by EPA Administrator William K. Reilly.[14]

The relative success of EPA's 33/50 Program, which attracted a total of almost thirteen hundred companies by the time the program ended in December 1995,[15] was viewed with sufficient encouragement from EPA to have the agency launch a series of subsequent industry voluntary programs designed to *work with* and *not against* regulated companies. The first of these subsequent voluntary programs, the Environmental Leadership Program, or ELP, was first described in a *Federal Register* notice in 1993, outlining EPA's goal for this experimental program.[16] As a result of the favorable responses received after the first *Federal Register* notice of the ELP, a second *Federal Register* notice was issued announcing a call to regulated organizations to submit proposals to EPA for participating in the ELP pilot program.[17] Of note, readers are encouraged to visit EPA's ELP site to learn more about the project participants, including the Gillette Company, Aba-Geigy and Motorola, as this program evolves. Up-to-date information can be found at http://es.inel.gov/elp, or www.epa.gov/reinvest/elp.htm. The Gillette Company was one of the participants in this pilot program. Further discussion of the pilot program is contained in Gillette's case study in Chapter 10. As this book was going to press, the ELP was in its final stages of the first-year pilot program and final results were not yet available.

In addition to the ELP pilot program, EPA rolled out the Common Sense Initiative (CSI) on July 20, 1994[18] and Project XL, which stands for eXcellence and Leadership, created by the Clinton administration on March 16, 1995 in the environmental regulation initiative.[19] Web sites from which supplemental information can be obtained on EPA's 33/50 Program, the ELP and ELP Pilot Program, CSI and Project XL are listed in Appendix G.

In addition, the EPA, as a member of the ISO's Technical Committee (TC) 207, continues to play a role as an active participant of the various subcommittees within the U.S. Technical Advisory Committee (TAG) of TC 207. As an example, within the body of the text contained within ISO 14001, the cornerstone document of ISO's environmental management standards is the phrase, "... *prevention of pollution.*" This phrase was included in ISO 14001 at EPA's request as a guide in helping industries focus more closely on another of EPA's long-term goals—preventing pollution—and as a target for organizations to achieve in doing more with less, while saving both materials and monies in practicing pollution prevention.

Whether industries come together as participants in one of these voluntary EPA projects, or work side-by-side with EPA personnel in drafting and consensus-building the remaining ISO 14000 standards, it would seem that a greater number of environmental professionals from the regulated community have begun to see regulatory agencies in a more positive way. The lessons learned from one such example, Amoco's Yorktown Project, may have served as a benchmark for those companies that may have previously perceived EPA as the usual "command-and-control" adversary[20] rather than as an agency willing to work with the regulated community toward a common goal: protecting human health and the environment.

The benefits to be derived from such organizational changes include the opportunity for organizations to achieve sustainable development, aptly described by the Xerox Company as "... *a standard that scientific realities and social demands will require of all economic activity. Those companies that can recognize the nature of the transformation required and can better anticipate the escalating demands they will face—from customers, competitors, government, and the communities in which they do business—can shortcut their learning curve. They can plan their future, not be forced into it. The call to prevent pollution, not just to clean it up, is now standard among environmentally-enlightened companies. So is the effort to reduce, reuse and recycle materials...*" [21]
In addition, the Washington, D.C.–based Investor Responsibility Research Center (IRRC) conducted a survey in 1996 of Standard and Poors 500 companies—the S&P 500—highlighting and quantifying the results of several key indicators: the development of environmental policies; the importance of environmental performance as a factor to determine various levels of management compensation; the adherence to voluntary initiatives, such as ISO 9000 and 14000, Responsible

Care®, and other similar nongovernmental organization (NGO) initiatives; the establishment of a board of directors' committee to address environmental issues; the issuance of environmental reports; and whether these organizations had some form of an environmental auditing program in place.[22]

Clearly then, the stage has been set for more changes to occur within regulated organizations that can provide long-term benefits from a variety of viewpoints and satisfy a multitude of stakeholders. As organizations aspire to reach the ever-higher hanging fruit of their labors, a new order, or radical breakthrough in thinking, may need to be explored by multiple levels of management and workers alike in achieving such noteworthy goals. As one example of this ideological shift, a number of organizations have come to embrace the Public Environmental Reporting Initiative (PERI) guidelines previously described in developing their environmental reports in a consistent and open fashion and readily available to the public. Since the issuance of the first environmental reports about ten years ago (1988), the PERI guidelines have helped to create a *de facto* voluntary standard for companies to follow in preparing these environmental reports. As in any process that changes and improves over several iterations—PCs and their operating software for example—some of the earlier[23] environmental reports issued by companies were not as full of verifiable and substantial environmental data, and hence were less credible as company public documents as these are today.

To a well-heeled individual, environmental professional or regulator, there are a number of valuable nuggets of information that can be gleaned from these environmental reports—glimpses into the environmental activities of publicly-held U.S. corporations. These nuggets of information can provide insight in a number of ways to a number of individuals. For example, to the environmental management specialist or analyst at another regulated organization, the information could help explain how to address similar situations at his or her particular facility as a "lessons learned" opportunity; or provide examples for steering a company into other areas where the ever-higher hanging fruit of cost savings could be picked; help the organization brainstorm and experiment with a new idea, or look at an existing environmental issue or concern from a different angle. Any of these scenarios could be useful in bringing to light alternative ideas that could foster feasible and workable solutions for an organization's environmental management teams.

In addition to such benefits, such reports could provide information that may obviate the need for further "command-and-control" scrutiny, similar to the intent behind EPA's Audit Self-Disclosure Policy (this subject is provided in further detail in Chapter 9). A member of the general public could also gain considerable insight into reading about the facility down the street, or in a nearby community that makes the products he or she may use in day-to-day activities. As we can see, these reports can be of value to a number of individuals and interests, especially if the information depicted therein is factually accurate.

For example, in some of the reports we reviewed for inclusion in this book referenced at the end of this chapter, we found cases that depicted environmental audit programs (Chevron and IBM are two representative samples highlighted[24]) developed as a means to measure the organization's environmental performance against internal and external requirements; or to highlight the company's success in pollution prevention efforts and participating in voluntary programs (IBM). In addition, Chevron provides a case study on P2 efforts within the company, described further in Chapter 10.

In general, organizations' environmental reports provide a good amount of in-depth information that depicts the activities of each organization within a certain window of time, similar to the "snapshot in time" aspect of environmental audits. The individuals responsible for developing and editing different portions of these reports usually have a very full plate in providing a report that is factually accurate and depicts the company's progress to date. The reports we reviewed are no exception. Another aspect of these environmental reports, as we view it, is that they can allow a certain amount of employee wellness and empowerment to feel good about themselves in their environmental achievements as depicted in these reports for all to see. Such feel-good attitudes can spell the difference in having an organization's environmental policy objectives filter down from top management to line workers to be put into practice versus having the policy fall flat on its face.

For example, situations may have occurred at facilities where top or line management may have felt that one way to move employees into a progressive and more sensitive mode to their environmental or even work surroundings would have been by using motivational or theme posters. Some of these may highlight environmental awareness, promote pollution prevention, reinforce employee health and safety awareness, or promote general team building and empowerment concepts. There may be situations where our readers may have come across these ubiquitous displays in their visits or audits of facilities' production area corridors, human resources offices, lunch rooms, and other common workplace areas as a constant reminder to employees to "do the right thing" and/or other motivational ideas. One case in particular regarding these motivational theme posters and similar appurtenances was highlighted in *BusinessEthics* and exemplifies what such "tools" can really do for employees in certain situations. While the stories can be amusing, there is a lesson to be learned—posters alone can sometimes fall short of management expectations, while action and positive feedback may provide greater results. [25]

As good as these intentions may seem to be, however, advertisements alone may just not be enough[26] to convey these messages to all affected employees. Rather, what may be needed for organizations to aspire to and emulate some of the companies that release environmental reports of their accomplishments, milestones, and even areas for improvement, could be a combination of sincere commitment and follow-through from top management to the production line workers that can allow two-way communication from employees to management, thus fostering workplace creativity that allows a mechanism for offering suggestions or improvements of identified issues. Dynamic forces that encompass greater worker input, the use of horizontal organizations, and a larger and more diverse workforce, helped to set in motion work situations that continue to change the work landscape as we know it today, and with each succeeding day, another force alters the work landscape that was previously in place.[27] In some organizations, changes such as these are not out of the ordinary; rather, these may also be equated to business survival as discussed further in the following paragraphs.

To highlight the constantly changing business environment, we reference two books worth noting that provide some insight into the prevailing forces within the business world. These forces should be reviewed in greater detail by an environmental manager since they may also impact upon environmental, and health and safety considerations from a different perspective and may provide valuable additional information to an environmental specialist in helping to solve a particular problem, or at least better understand a particular business or employee situation.

The first of these books is titled *Transforming Company Culture*, by David Drennan, published by McGraw-Hill, Inc., in which Drennan points out that in order to initiate a visible management commitment, all that really counts is *action*. He also makes the observation that transforming the culture of any organization requires a clear vision of the future, and a set of goals that are readily identifiable by employees. The key to making these goals a reality lies with top management and their commitment. This commitment needs to be consistent and visible. The other part of this "equation" is that employees also need to see some tangible "evidence" in their own work areas, otherwise it just may not work. For these reasons, Drennan feels that ". . . action is all that counts."[28]

It is interesting to note that while this book was written from the perspective of a people-management professional and consultant in 1992, many of his observations, areas of consideration, and recommendations fit within the general tenet of environmental management, especially with respect to top management commitment and diligence to both follow-through and motivating employees by deeds instead of words alone (such as in the motivational posters described previously, or in a finely crafted environmental policy that rings hollow on the production shop floor). Many of his fundamental ideas have universal value, whether the company is a service provider, manufacturer, or raw material or intermediate materials provider/producer. The previous reference to the environmental reports also points to environmental and/or business situations that can be enhanced by such management commitment as a further argument to substantiate the elements of ISO 14001; that standard, too, embodies such commitment principles. Chapter 12 provides in-depth discussion into ISO 14001.

The second book worth noting is *The Living Company: Habits for Survival in a Turbulent Business Environment*, by Arie de Geus, published by the Harvard Business School Press. In his book, de Geus refers to successful organizations as "living companies" and writes about the success of these companies and how they have survived in the business world from earlier times. He writes in part, "*. . . What do extraordinarily successful companies have in common? . . . Our group saw four shared personality traits that could explain their longevity—Of significance to our book, two specific traits are worth noting: sensitivity to the world around them . . . (and) . . . tolerance to new ideas . . .*"[29] Expanding upon the first trait—*sensitivity to the world around them*—the living companies are able to adapt to the changes around them.

One example of this trait is the ever-increasing interest regulated organizations are giving to ISO 14001 certification and to those organizations that have certification. This ISO standard is market driven at present, since there is neither a formal EPA regulatory "carrot" nor "stick." Time will tell what positive impacts

ISO 14001 may have on environmental issues currently being managed by organizations. In addition, whether a facility's environmental management system can be in conformance with and certify to ISO 14001 *and* play a pivotal role in a facility's long-term success will continue to be tracked closely by those companies waiting on the sidelines before they "test the waters" regarding certification to 14001 and the benefits that may be derived from certification. As in the "living company" case, acceptance of ISO 14001 as a business decision may be viewed as a "tolerance to new ideas" approach that may need to be explored as a market incentive for organizations.

Looking more closely at the second trait identified in de Geus's book—*tolerance to new ideas*—the living companies tolerate activities, such as experiments and eccentricities, that stretch their understanding. In another context, multinationals must be willing to change in order to succeed. For example, in the area of life cycle assessment, the reuse and recycling of materials, and the conservation of energy sources, organizations are looking at innovative ways to do more with less, and to incorporate at the research and development stage the use of alternative designs and materials to create newer and better products. In some organizations, business lines may be spun off to change the focus of the organization, or companies may be acquired to strengthen the organization's product lines. In any of these instances, multinationals could be implementing changes necessary to achieve one or more of the organization's objectives in order to survive in a competitive business climate.[30]

As another example of the "tolerance to new ideas" concept, and to reinforce the transforming company culture concept, a young manager of a Chrysler plant in Ontario, Canada, undertook as one of his objectives, the complete upheaval of van assembly manufacturing to achieve greater throughput and output. His methods and results, at first perceived as questionable, worked wonderfully by the time he was promoted to manage a new plant being built in Latin America.[31] The lessons learned from this example point to a new plant manager who had the determination to stick through with what he believed, and also relied on his subordinates for ideas regarding how to proceed in the face of obstacles, and also putting his own job and reputation on the line. The key to this case study is that when the manager was transferred, his successor was able to pick up where he had left off, simply because he had "...changed the culture." This example brings to mind the often-used saying, "Say what you do, and do what you say."

As a brief look forward, several of the ensuing chapters in the book also take a closer look regarding specific business or environmental considerations and provide an in-depth glimpse into the operations of the subject organizations at various facilities. The authors' distinct and unique perspectives, provide a unique insight into their particular areas of expertise, and how their actual experiences provide valuable "lessons learned" not only for their own improvement but for providing examples for others to learn from their experiences. For example, in Chapter 10, Bill Parker of EG&G shares his insight as EG&G's director of environment, safety and health programs, in his description of EG&G's environmental programs, which includes their Waste Reduction Pays (WARP) Program, and EG&G's commitment to environmental excellence. EG&G is among the first of U.S. organizations that have certified to the British Standards (BS 7750), the

Irish Standards (IS 310), and the European Union's Environmental Management Auditing Scheme (EMAS). In addition, Parker's summary includes a discussion of valuable lessons that were learned, such as evaluating internal environmental training.

As you, the reader, will note, each author provides his or her own particular insight and viewpoint into a different facet of an environmental topic, issue, or item of consideration that is unique to his or her organization, and provides the readers an opportunity to glimpse into their organizations' EH&S programs and how their particular environmental management systems are managed. In each case, the authors provide a detailed understanding into how they manage specific environmental issues and the steps they take as an organization to improve upon their particular environmental program or management system.

ENVIRONMENTAL MANAGEMENT: THE PRESENT

There is no doubt that the field of environmental management has made considerable progress over the past several years, and by coincidence, the impetus behind much of this organizational mindset change seems to parallel the efforts of many forward-thinking environmental professionals who decided, in 1991, to pool their collective resources and knowledge and become part of the Strategic Advisory Group to the Environment (SAGE), the predecessor to the present ISO Technical Committee 207. This is not to say this is the only movement in place helping the paradigm shift prevalent at many forward-thinking organizations; rather, a combination of the paradigm shift, a gradual mindset change within regulated organizations, a universal understanding that environmental compliance just for the sake of compliance would not work, and that environmental issues should be studied in a different light, all helped play a pivotal role in fostering a move for organizations to look inward more closely. With that perspective, each organization's environmental management system gained more relevance, and with that understanding, it could be seen whether such systems, as structured, could positively or adversely affect their employees, the environment, and their businesses as well from various indicators: suppliers, customers, regulators, investors, the public, and all other affected stakeholders. We take a closer look at a number of these organizations through the eyes of several third-party groups that conducted benchmark surveys of a number of high-profile companies to look more closely at issues affecting EH&S considerations that we discuss here. That information is provided in greater detail in Chapter 7.

As such, with the addition of various factors described in the benchmarking surveys in Chapter 7 into the environmental management equation, environmental management today has reached a more mature stage where it is recognized as a viable business unit. What used to be business as usual—looking at environmental compliance and management considerations from a regulatory framework only—has been replaced by a new paradigm: holistic environmental management that takes into account and is incorporated into all facets of an organization's business units.

As a further item of consideration, a number of regulated organizations have begun systematically compiling and analyzing data to verify cost-saving and environmental benefits of having an environmental management system (EMS) or program in place as a structured mechanism for managing an organization's environmental responsibilities. Some of these processes have been able to show substantial cost savings and improvements in EH&S functions, such as the Millipore example in Chapter 8 or the Chevron example in Chapter 10. The EMS, by its structure, consists of people at various levels of organizational responsibilities whose daily task is to ensure that the system is functional, and daily, weekly, monthly, and other periodic organizational milestones are mandatory. In a smooth-running and efficient EMS, the EH&S staffers are on top of things and coordinate with other business units within the organization, while the EMS also helps to uncover system deficiencies and provide for corrections or improvements to such systems.

Environmental, or EMS audits, are among the tools used by environmental professionals to help single out system deficiencies, or findings, that are ultimately utilized in audit reports as a basis for audit recommendations, that over iterations, should provide opportunities for improvements within a facility's or organization's environmental management system. Besides maintaining regulatory compliance, such improvements can also allow an organization to become more cost effective, help uncover areas where unnecessary costs may have been expended, and identify opportunities for different or innovative approaches to produce goods and services at a lower cost while generating fewer wastes.

The bar has been raised a little higher, and while maintaining compliance is still one objective the regulatory agencies would like organizations to sustain, preventing pollution in the first place has a greater significance in the greater scheme of things, which makes sense.

ENVIRONMENTAL AUDITS: EFFECTIVE MANAGEMENT TOOLS

The term environmental audit, as defined by the EPA, is:

> "a systematic, documented, periodic and objective review by regulated entities of facility operations and practices related to meeting environmental requirements."[32]

Similarly, the term as defined in ISO 14010 is:

> "systematic, documented verification process of objectively obtaining and evaluating audit evidence to determine whether specified environmental activities, events, conditions, management systems, or information about these matters conform with audit criteria, and communicating the results of this process to the client." [33]

In either instance, environmental audits can be viewed as an efficient means or tool to gauge the adequacy and soundness of a facility's environmental management system (EMS), viewed in this context as a cross-sectional "snapshot" of the facility's EMS. Moving beyond this point, the ultimate goal of an organization's top management for the environmental audit program is to help identify those areas within the operational and management systems that need some type of improvement to help the organization manage its environmental responsibilities in a more efficient and effective manner. In this capacity, environmental audits can display their full potential as a management tool in identifying areas for improvement. This is partially due to the nature of audits—to reflect a "snapshot" of specific areas of a facility's operations within a predetermined time frame—with team members documenting their observations and recommendations. Since audits are also conducted with a specific scope, and with a specific mix of audit team members with a certain skill mix, it is not uncommon for audits to skip portions of a facility's operations, which may later be reviewed by another team, or by the same team during a follow-up audit. For these and other more subtle reasons, an audit program's effectiveness is usually seen over a certain period of time. In many instances, follow-up audits at a facility usually provide additional details about that facility that can be utilized as beneficial information to both facility management and corporate management in pointing out areas that need some sort of improvement. Again, the key is that the areas for improvement manifest themselves in an evolutionary and reiterative fashion to allow improvements to occur progressively with the affected employees, rather than via a lock-step approach that may not be as effective or popular with some employees.

To clarify the management tool aspect of audits as these relate to improvement opportunities, let us take, as an example, environmental audits that note some sort of process or management system excursions that may lead to repeat or similar regulatory or management system findings. Delving deeper into the findings, one may infer that such situations may be indicative of a number of management system deficiencies that may need to be studied more closely by both the audited facility and top management to ensure greater success or improvement of existing environmental conditions. Clear and concise audit findings and recommendations that accurately portray existing deficiencies that also identify root cause issues can provide added value to the audited facility's management, especially in those cases where repeat or similar findings are being routinely uncovered at each subsequent audit event.

This is one of the aspects of environmental auditing that provides an organization's top management the tools it needs to identify and correct environmental management system and compliance flaws that, left unchecked, may lead to regulatory noncompliance, or repeat instances where unnecessary expenditures are being made in a number of situations that are reflected as costs to the organization. Areas that come to mind in this regard include: the classic mislabeling of hazardous waste drums where a less hazardous substance is mislabeled as a more hazardous substance; the co-mingling of hazardous and nonhazardous wastes; the storing of reactive chemicals near sources of water, or the storing of incompatibles in close proximity to each other without some form of separation barrier; the inappropriate use or nonuse of a worker's personal protective equipment (PPE); lack of oversight of a forklift operator's training and safety records; and many more—all of which point to areas that can be improved once management has been appraised.

In summary, while an environmental audit may be viewed as a "snapshot in time" of a facility's operations, it can be an invaluable tool to an organization's management when utilized to its full potential to help improve environmental performance and reduce overall operational costs. Having the right audit team mix and "chemistry" with the audited facility can also assist in this process.

To highlight a few of the additional benefits that environmental auditing can have upon audited facilities, the following bullets are provided, courtesy of a U.S. General Accounting Office report:[34]

- Increasing environmental awareness and capability among employees;
- Potential relaxed regulatory scrutiny;
- A record of a facility's environmental performance is established;
- Planning and budgeting for environmental projects are facilitated.

We have only scratched the surface of the complex subject of environmental audits and audit programs in this opening chapter as a matter of reference. Additional discussion of the subject is provided in Chapter 9: The Environmental Audit as an Effective Management Tool. Here, environmental auditing is covered in greater detail by Mitch Gertz, Paul Dadak, Dave Jeffries, and Tom Hauding, where the authors provide additional insight into how their respective facilities conduct environmental audits, and the lessons each has learned through participating in their respective audit programs.

One final note with respect to environmental audits as compared to EMS (ISO 14001) audits. They are both management tools, similar in concept to management system verifications (MSVs) performed as part of Responsible Care®. Among the differences, though, between environmental (compliance) audits and EMS audits, is that environmental audits can be more prescriptive, referring to compliance with specific regulations and/or company policies, while EMS audits refer to elements within ISO 14001 that describe rather than prescribe conformance to an organization's environmental policy and the requirements of the environmental management system.

There are also similarities between environmental audits and EMS audits. For example, in environmental audit recommendations there are usually references to "best management practices," or BMPs, identifying a particular area or function that should be corrected that goes beyond regulatory compliance parameters. This is analogous to the commitment to continual improvement ingrained within the structure of ISO 14001.

Therefore, if the end result of these audit programs is to recommend improvements, either audit can accomplish this goal. How effective either audit can be may rest with how well an organization's management utilizes their management tools.

The contributing authors in Chapter 9 touch upon this subject area in greater detail.

POLLUTION PREVENTION: A GOAL WORTH CONSIDERING

In any forward-looking environmental program or management system, among the elements that work hand-in-hand with an environmental audit program are the opportunities for prevention of pollution, or elimination of waste at the front of the pipe. In some organizations, specifically tailored pollution prevention audits are conducted to help identify those areas where waste streams in process or manufacturing areas may be reduced or eliminated altogether. In this fashion, astute facility or plant managers can minimize the need for managing various hazardous waste streams, aqueous discharges requiring some form of wastewater treatment, or inventory of either fugitive or permitted air emissions, to name a few.

In any of these exercises to identify and execute pollution prevention, or P2 opportunities, a concerted effort is required by all responsible and affected workers, from shift line workers, wastewater operators, hazardous waste handlers, to environmental management and plant management staff, in order to be truly worthwhile. Even the "low-hanging fruit" that many organizations may uncover after a first pass to uncover P2 opportunities could be short-lived if a structured and regular P2 program is not part of the employees' regular routine and mind-set. For example, let us say that a facility has a well-planned and written EMS on paper, including well-documented internal procedures and various documents to ensure that all applicable regulatory requirements are met. However, the facility still needs to implement a mechanism or management system to track new employees requiring training, either within the preview of EH&S or human resources departments. In addition, the facility does not maintain an open line of communication between the EH&S department and say, procurement or research and development. In the right conditions, what may seem minor glitches could eventually derail even the best-laid plans and subsequently negate any P2 cost savings that may have been initially realized at this hypothetical facility.

Given this hypothetical situation, let us further suppose that the facility also imports exotic or toxic chemicals for use in R&D projects or used for cleaning purposes without first going through various internal EH&S screens and regulatory procedures. Besides the fact that such scenarios may pose regulatory issues with federal or state environmental regulations, these may pose adverse impacts with respect to P2 opportunities. With such a disconnect in the inventory system for new chemicals coming into a facility, a number of P2 issues could occur. For example, the importing of chemicals that are inappropriate for specific uses within a facility, or importing of chemicals that could have been substituted by less toxic ones that still meet process specifications, could lead to one of a number of non-P2 situations: chemicals that may be more costly to dispose; pose more serious health hazards during use and may require greater ventilation; or, could have been replaced by a less toxic or volatile chemical in a specific application with similar, or perhaps, better results.

In another example, appropriate training of all employees, affected in some way by environmental considerations, such as in the business areas described previously, are also key to helping to maintain some control over the generation of unnecessary wastes. Effective and specialized environmental awareness-type training can hold an abundance of P2 opportunities, and given the right mix of trainers and trainees can allow these opportunities to flourish. However, if the P2 opportunities that appropriate and effective training can provide to individuals within a facility are not acted upon in a timely fashion, or are acted upon randomly thus negating any long-term beneficial effects, or if training programs are not effectively tracked, such items can lead to missed opportunities for pollution prevention and the subsequent cost savings that could have been realized. Put in another way, without a mechanism in place within an organization, either at the corporate level or at the facility level, to ensure adequate and appropriate employee training to increase their awareness and competence levels regarding all applicable environmental considerations and issues, unnecessary spills, releases, misuse of chemicals, or other inappropriate chemical or waste handling situations may occur that may have the potential to generate costly, or worse, disastrous outcomes.

On a grander scale, even the EMS standards of ISO 14001 identify P2 opportunities within its core elements, such as in the environmental policy. This management-developed document contains the genetic code, if you will, that defines the main objectives and targets for an organization to achieve with respect to its activities, products, and services *and* how these will mesh with an organization's commitment to continual improvement and pollution prevention.

An example of some questions that may be asked regarding an organization's commitment to adhere to its environmental objectives and targets include:

- What does the organization really want to achieve as a goal for environmental performance, and how do P2 and cost savings enter into this equation?
- What contingencies will the organization develop to track and maintain such innovative opportunities as part of the EMS and continual improvement aspects?
- What innovations in designs and management practices can be introduced to assist in this regard?

In similar fashion, within the elements of checking and corrective action and management review, two main cornerstone elements of ISO 14001, lie the actual P2 opportunities that can be realized within an organization. Within corrective action, for example, specifically trained professionals, or auditors, can monitor and measure noted key characteristics of internal activities that may adversely affect the EMS and possibly human health and the environment. This could be in the form of ensuring that process equipment is operating at peak efficiency and all valves, piping, and seals are functioning properly and not leaking; ensuring that the target facility is in compliance with all applicable environmental rules and regulations; ensuring that all necessary permits are kept up-to-date; and that all pertinent regulatory required documents are properly filed and maintained on-site with proper document control procedures in place.

Finally, the ultimate responsibility with respect to the effectiveness of the EMS as it relates to a facility in meeting its P2 goals rests with top management. One measure of the effectiveness of management's review of a facility's EMS centers around any findings that may have been developed by an EMS audit team that pinpoint such opportunities and whether management was successful in implementing any audit recommendations, either as part of a just-completed audit, or as an item for follow-on action during the next go-round of audits.[35]

In short, identifying pollution prevention opportunities are not difficult to find—they just take a concerted, dedicated, and sustained effort by facility personnel in order to be successful.[36]

The preceding discussion focusing on P2 is only a brief introduction of the subject and should be viewed as such. Additional detailed discussion of P2 opportunities is provided in various chapters of this book, specifically, Chapters 6, 8 through 10, and 12 offer additional in-depth coverage on this timely subject from the perspective of several contributing authors. The underlying commonality among the contributors' works with respect to P2 all point to noticeable cost savings to their companies as a result. The graphical presentation of actual dollars saved by Chevron and IBM in their P2 efforts portray clear examples of how P2 can be tremendously beneficial to facility operations in minimizing waste streams and producing cost savings for the companies as well.

DESIGN FOR THE ENVIRONMENT (DfE): OUT-OF-THE-BOX THINKING

If a starting point is needed to pinpoint the beginning of DfE considerations, it can be argued that the regulators may have had a prominent role through their use of "command-and-control" methodologies prevalent throughout the last fifteen years or so. One of the positive end results of this regulatory control mindset was to foster from the more cutting-edge regulated industries innovative approaches to address regulatory responsibilities, such as exploring opportunities to do more with less, and focusing less on end-of-pipe only compliance strategies to satisfy regulators. As a result, there are a number of organizations that have spent considerable energies to look further upstream in a manufacturing process to meet and exceed regulatory requirements, or completely rethink the processes as a means to conserve both raw materials and energy sources. At the heart of this next generation process thinking is a fundamental drive by manufacturers to consume less raw materials in producing goods and thus generate less waste, seeking to reuse materials and base components, while making products that can be beneficially reutilized, such as with a modular approach to manufacturing techniques. In the electronics industry, for example, many of the companies in this industry constantly study ways to reuse basic components of manufactured goods, such as plastic and base metal components.

Among the DfE tools some of the more progressive organizations are utilizing to achieve these far-reaching goals are life cycle management and design to help promote and champion from within, and various design, manufacturing and recycling strategies. One such life cycle design was part of a demonstration project undertaken by AT&T and AlliedSignal that became the basis for a well-developed report[37] as part of EPA's Pollution Prevention Research Program. In brief, AT&T, now Lucent Technologies, focused on achieving greater material and energy efficiency, improving recyclability, and using and releasing fewer toxic constituents in their design of a telephone terminal. AlliedSignal developed design criteria to guide their design engineers to develop improved next-generation oil filters.

Within the framework of life cycle design as described in this EPA report, input from various stakeholder sources is also important to attain a successful design improvement. The framework can be broken down into its basic components:

- Product life cycle system
- Goals
- Principles
- Life cycle management system
- Development process

Within product life cycle, the process follows a general pattern: raw material acquisition; bulk and specialty processing; manufacturing and assembly; use and service; retirement; and disposal. Also to be considered in a life cycle design are accurate measurements of both environmental and economic performance of the goods being designed. From the design side of the equation, engineers should also have an open mind to help tackle new challenges to their creativity in working both within environmental parameters, future market considerations, and in keeping abreast of their competitors.[38]

For example, take the life cycle of computers and related peripherals—printers, faxes, modems, scanners, and possibly soon, Web-TVs. The intermediate products or components that are used in the manufacturing of these electronic goods each have their own environmental loop or cycle that takes into account their specific environmental considerations related to design, manufacture, and recyclability. In addition, given the dynamic and competitive nature of such high-end electronic goods, each of the manufacturers associated with these goods are also acutely aware of the economic constraints placed upon their products, and constantly have to adjust to such considerations while maintaining ever-decreasing and paper-thin profits.

It is clear then, that each area within the supply chain plays an integral part in the life cycle of these goods and is also a consideration in DfE strategies. From the useful life of the goods to beneficial reuse and recycling of components at the end of the goods' life cycle, to the innovative strategies for next generation products still in research and development departments in the organizations within the electronics industry, such considerations are key in sustaining environmental and product stewardship. The considerations to be explored further in this market include: the recycling of PC and peripheral components,[39] and how this can be effectively accomplished to be favorably received by consumers; P2 opportunities and materials reuse in recycling—from CO_2 emission decreases to the reuse of semiconductor components and base metals, from printed wiring board reclamation and recycling of plastic components to the re-design of these goods that address market concerns, like modular construction; and other considerations.

In similar fashion, Hewlett-Packard Corporation (HP) stated in its 1996 Commitment to the Environment report[40] that the company is commited to providing its customers with products and services that are environmentally sound throughout their life cycles and to conducting their business worldwide in a responsible manner. As part of the company's product stewardship program, which the company initiated in 1992, HP strives to prevent or minimize any negative impacts to human health or safety, or to the ecosystem, that may occur at any point in the life of a HP product. A copy of HP's environmental policy is included in Appendix E. A revised version can also be obtained by accessing HP's home page at http://www.hp.com. We provide additional discussion about Hewlett-Packard in Chapter 10.

Clearly then, as described in the previous illustrative examples, DfE, or life cycle design, considerations have found a secure spot in the design of subsequent goods and services that address the main tenets of DfE—the integration of environmental considerations into process design and rethink at the outset, and finding a perfect harmony between environmental considerations and performance, cost, market acceptance, social, legal, and other stakeholder criteria.

SOFTWARE: A ROLE PLAYER IN EH&S/EMS MANAGEMENT

With the advent of Windows™, Lotus Notes®, and other PC-based software applications, EH&S-specific software programs were brought out by various developers into the marketplace to help EH&S professionals and staff alike get a better handle on managing the mountain of data that environmental programs generate and that are also required by various EPA regulatory programs in air, water, waste management, and soil.

The first generation of these off-the-shelf software packages were MS-DOS-®based and were somewhat limited in their applicability, effectiveness, and user-friendliness. With an understood industry shift to Windows™ and Lotus® more developers were able to design software packages that could do more and were also much more user-friendly. Applications in which software has gained some popularity include: in MSDS and manifest generation and tracking; in EPA reporting, such as Toxic Release Inventory (TRI) data, for which EPA developed and supplies regulated industries with the software; in federal and state regulations on CD-ROM; in environmental auditing protocol and checklist development and findings tracking; and in ISO 9001 and 14001 elements tracking and document management; in predictive emissions monitoring systems (PEMS); and in a widening array of other uses, from PC to UNIX® and Oracle-based workstation applications, to more powerful intelligent systems, man-machine interface (MMI) and supervisory control and data acquisition (SCADA) applications that operate in real-time simulation mode and are used almost exclusively in chemical, petrochemical, discrete component, and aerospace applications, among others, to help prevent process system failures, predict containment vessel flows and monitor such variables as temperature, pressure, liquid levels, and other situations that may adversely affect the processes if not addressed .

In the area of EH&S management, some success has been attained by various software system developers that focuses on the management of data within environmental management, otherwise known as an environmental management information system, or EMIS. EMIS can be described in software terminology as more of a systems integrator, in that EMIS takes into account the various software applications within a facility, local-area network (LAN), or a wide-area network

(WAN) within several facilities of an organization to help manage the information being compiled for any number of regulatory requirements. The advantage an EMIS provides an organization is that it can assimilate information being generated from other business units within the facility to help coordinate information that is vital to the facility into a central location.

The business of software applications used in environmental, health, and safety (EH&S) considerations is only briefly discussed in this chapter. EH&S software, which has mushroomed into a very profitable business area for many of its software developers, cannot be done justice in these few pages. Given the vast field of software packages currently available, and the gamut of services the software can provide, a conscious decision was made to devote additional space to this vast and intriguing subject area. As such, we provide additional discussion of software applications in Chapter 8 (a case study), and Chapter 13.

ISO 14000 CHAMPIONS A NEW ERA FOR ENVIRONMENTAL MANAGEMENT

Since the introduction of ISO 14000 into the business world following its go-ahead recommendation to proceed from the Strategic Advisory Group on the Environment (SAGE),[41] events have moved rapidly forward toward the truly global acceptance of the ISO 14000 family of standards[42] as a means to enhance environmental management for large and small regulated organizations alike, from Bangor to Bangkok, and every point in between. As more organizations become more inquisitive with respect to the standards, their managers may begin to look more closely at ISO 14000-related events. This may include reading about organizations that have certified to ISO 14001 in various trade publications. From such research, they may begin to get a better understanding for what can be accomplished by an organization with a more standardized and cohesive plan for managing environmental responsibilities, and in the process, become more efficient and effective. It is quite possible that as organizations learn more about the standards, either through researching data or benchmarking those organizations that have already certified to ISO 14001, they may reach a certain comfort level to make an educated decision regarding these ISO standards.

As a further insight into the ISO 14000 suite of standards, it is important to stress that ISO 14001, as the main driver of these standards, has been developed as one of the tools organizations worldwide can use that maps out the main elements of an *effective* environmental management system, or EMS. It can, by its nature and similarity with ISO 9001, be integrated with other management requirements within the organization's other business enterprises and assist organizations in aspiring to achieve both environmental and financial goals.

Or, put in another context, consider the ISO 14000 standards as the mechanism to raise a bar for an athletic event. The athletes now have to try even harder to surpass that bar to succeed in the event—those that do not try, do not achieve the potential rewards associated with those that surpass the bar. Similarly, ISO 14000, as developed, seeks to provide a mechanism for all organizations—those with complex environmental management programs in highly regulated countries, such as the United States, as well as organizations with simple or no environmental management programs in countries with lax or no regulatory programs—to have an equal chance to develop an EMS that addresses an organization's environmental responsibilities, regardless of the host country's regulatory climate. The elements of the standards can facilitate the development of an organization's EMS that can be consistent and more effective than currently may exist within an organization, and that can be applied equally in any facility around the globe. In addition, since the standards are market driven, and not a regulatory requirement of any nation, this business aspect can be a more powerful persuasive tool for an organization's top management to implement than the regulatory "stick" approach.

This topic also is covered in greater detail in Chapter 12, in which Joe Hess of SGS-Thompson provides very good insight into his facility's experience with being certified to ISO 14001. There is also an introductory comment provided by Jack Bailey of Achushnet Rubber Company, one of the first U.S. facilities to be certified to ISO 14001.

Further information on the ISO 14000 standards can be found in a number of books in print by various publishers. Of the numerous books on the subject matter with which we are familiar are: *ISO Certification—Environmental Management Systems*, W. Lee Kuhre, © 1995 Prentice Hall PTR; *ISO 14001 and Beyond*, Christopher Sheldon, ed. © 1997 Greenleaf Publishing; *ISO 14000 Guide*, Joseph Cascio, Gayle Woodside, and Philip Mitchell, © 1996 McGraw-Hill; and *A Guide to the Implementation of the ISO 14000 Series on Environmental Management*, Ingrid Ritchie and William Hayes, © 1997, Prentice Hall PTR.

WHAT THIS SIGNALS FOR THE MATURING FIELD OF ENVIRONMENTAL MANAGEMENT

During the past several years, a movement has begun where a greater number of organizations continue to pursue innovative strategies with respect to environmental management considerations. As a barometer of such forward-thinking, there has been an increase in the number of conferences and seminars being targeted to facility environmental and quality managers, corporate and outside counsel, consultants and other related service providers by various professional organizations and conference providers. For example, the author attended a recent conference that focused on environmental management information systems (EMIS) applications, in which environmental professionals from a number of Fortune 100 companies shared their successes in the utilizing of electronic data to help manage their companies' EH&S documents.

In addition, greater emphasis is being placed on management information systems (MIS) that can bridge information between the various business enterprises within an organization, which has led to the successful marketing of enterprise document management (EDM) solutions by various software developers. One of the pioneers in this field is IBM, and in Chapter 13, a case study is provided on the benefits of EDM solutions, especially as it relates to increasing an EMS's efficiency, and how such solutions can work in concert with EMISs and other software systems as another mechanism to continually improve upon a facility's and organization's environmental responsibilities from the viewpoint of the entire organization, not just EH&S personnel who have historically carried the responsibility, the burden, and the frustrations that are part of EH&S programs at all organizations that create goods and services that have an impact upon human health and the environment.

What does this all mean to environmental management as a business, and what can we expect from all these external and internal forces coming at us, the environmental professionals and service providers, from all directions? The answer, surprisingly enough, may be better things. The world has slowly changed with respect to business issues—downsizing, re-engineering, quality driven goods, profit margins, and environmental accountability, have all had a part in revamping the business climate—while the quest for greater access to data as an integral component of more effective business decisions has led to the development of more powerful and sophisticated software to meet this business challenge and stay competitive; and, at the other end of the spectrum, regulatory agencies, led by the USEPA, have initiated a different approach for regulated organizations to be able to conduct their business in a less "hostile" regulatory climate, and to provide these organizations with mechanisms to show EPA how they can achieve regulatory compliance in a voluntary fashion, without having to revert to the time-honored "command-and-control" approach to have the regulated community return to environmental compliance.

WHAT'S ON THE HORIZON FOR REGULATED ORGANIZATIONS, AND WHERE MIGHT WE BE GOING?

To address that question, we may need to look back for just a moment to see how far we have come along, and need to ask ourselves whether we feel comfortable with the progress we have achieved to date. Anyone who may think not may need to take a closer look at a few of the milestones that have been achieved. For example, the Cuyohoga River that flows through Cleveland, Ohio, no longer catches on fire as it once did, and many successful businesses have established themselves along the banks of the river. More companies are taking a second look at recycling various waste products and streams as a means to cut costs; more companies are looking at ways to become more environmentally conscious of their activities and being viewed as good corporate citizens; and a greater number of larger companies are expending great effort to do the right thing when it comes to environmental considerations. We're getting there, and while we have overcome the inertia to remain status quo as business as usual, we still need to keep our sights high to continue in our efforts if we truly want to succeed and aspire to higher environmental and health and safety goals.

WHAT SHOULD WE LOOK FOR?

We should look more closely, among the items to be considered, at opportunities that may have been previously overlooked, or not looked at in greater detail, for benefits that may be derived without that much sustained effort on the part of the employees involved. For example:

Opportunities do exist that lend themselves to EMS improvements. The "low-hanging fruit" beneficial opportunities are there in various locations of organizations' facilities. What is required is a concerted effort by all employees and managers involved to work together in this effort, which is not just a "one-shot deal." Take for example, the beneficial opportunities that can result from implementing waste segregation as one means to reduce waste streams and cut costs; or, evaluating which chemicals or materials could be substituted by less toxic and harmful materials that work equally as well as another means to accomplish the same end result. Another opportunity could lie in a process's reuse of wash waters and wastes, such as in the pulp and paper industry; or, depending upon the processes in place at a facility, the implementation of process changes as a means to reduce raw material consumption, energy use, and waste generation, as other beneficial means. In a more global fashion, providing opportunities for employees engaged in chemical handling from inventory to waste management to receive in-depth and appropriate training in best management practices in handling these chemicals from a safety management *and* environmental management perspective as a means to minimize spills, not just to understand how to manage spills *after* they occur. After all, improvements are designed to act as lessons learned from previous events, so as to prevent their reoccurrence, not as a training mechanism to see how fast a spill can be remediated after it occurs the next time around.

Actual examples culled from various sources include:

- One paint company switched primers in its formula, saving costs and waste expenses;
- Another company looked closely at its processes to eliminate the root causes of all its waste streams;
- While another company reduced paint use by 30 percent and recycled thinners;
- In another example,[43] one manufacturing company worked with its paint and solvent suppliers as a means to accomplish their environmental objectives, improve product quality and reduce manufacturing costs. The changes in processes and supplies significantly reduced the generation of waste paint and solvents, paint sludge, water pollution, air pollution and Toxic Release Inventory (TRI) figures required under SARA Section 313.

Managers may need to expand opportunity horizons to other business enterprises. In our constantly changing business environment where more is expected to be done with less, whether the commodity is materials or people, among the ways that such expectations may be met include looking beyond one's own area of responsibility to anticipate and plan contingencies for unfavorable conditions or business issues that may arise. For example, a manufacturing company that utilizes various chemicals may want to ensure that its EH&S Department is kept in the loop with its research and development department and inventory department with respect to new chemicals coming into a facility. Of potential concern, may be any new chemicals that are imported into the facility that would need to undergo one or more Toxic Substances Control Act (TSCA) requirements, in addition to any Resource Conservation and Recovery Act (RCRA) requirements.

Or, to refer back to the chemical handling example, environmental managers may want to keep a close watch on fork-lift operators and their safety records to ensure that they have taken the necessary safety training, as well as have appropriate staffers check both chemical inventory storage areas and hazardous waste storage areas on a routine basis to ensure conditions are under control at all times.

In a different slant, some environmental managers have begun bringing environmental concerns and issues to other business enterprise managers and top management in a more business-friendly way, in other words, using more "business-speak" rather than "enviro-speak" in discussing specific environmental issues for beneficial resolution. Those managers that have been successful in this transition have experienced various degrees of success. Additional detailed discussion about this specific topic is contained in Chapter 10.

In similar fashion, the ability of environmental and quality managers to be able to cross-educate their staffers, as well as cross-educate and include the employees of other affected business units within the facility, can go a long way in team-building as a means to bring everyone to the table and to be on the same "sheet of music." If a facility lacks such fundamental team-building business concepts, any efforts for them to implement a series of more comprehensive and complex worker programs may be precluded by employee limitations that do not allow them to achieve their full potential. On the other hand, within those companies that have taken steps to allow their EH&S staffers to expand into the other business areas of a facility, and allow for cross-training opportunities, the potential can be unlimited.[44] In another example, Dow Chemical Company had sent an employee on a two-year stint to one of its customer/suppliers, Nalco Chemical Company, to show Nalco a few things about environmental and safety operations. In exchange, Dow got a Nalco employee to share his expertise in sales and marketing. In such an instance, both companies were able to borrow from each other's strengths.[45]

Finally, the professional training managers, featured in preeminent business training publications, all point to one or more training methodologies as effective in work settings. Such terms as: empowering, team building, motivational exercises, out-of-the-box thinking, etc., may be trade terms, but they zero in on a key element—uplift employees, give them praise, be sensitive, and use common-sense—focusing on the employees as people who want to do their best, and who just may need a chance to show it.

WHAT ADDITIONAL BENEFITS CAN WE ACHIEVE?

The preceding pages of this chapter focused upon the changing environmental management landscape and where companies may be headed with respect to maintaining and sustaining their environmental duties in a responsible manner. This chapter also provided a glimpse into what some organizations are doing as a means to voluntarily commit to improve their EH&S programs, or initiate some other proactive means, in an effort to move away from a "command-and-control" regulatory climate. For such organizations, moving forward as progressive and environmentally-responsible organizations, and equally as important, being viewed as such by all concerned parties, can be of consequence to the organizations' bottom line. Actual cases of positive results experienced by facility managers of their environmental efforts in identifying and improving upon companies' business processes and products have begun to emerge in environmental conferences, trade publications, and EPA-led and -sponsored workshops that others can learn from and emulate as they see appropriate at their particular facilities. Benchmarking such positive examples, and initiating a "lessons learned" approach, can lead to additional opportunities at other organizations.

Continuing in this vein, some organizations have focused their long-term goals on horizons that promise more effective environmental management of their environmental programs, look toward greater regulatory relief through the USEPA and other federal and state regulatory agencies, and look to save costs along the way as well, since many of them have seen the beneficial results of such concerted management efforts in auditing, pollution prevention, training opportunities, better management of data, and increased cooperative efforts between the different business units within a facility, or organization, to name a few.

A LOOK FORWARD

Among the activities organizations should take into consideration as means to reap positive benefits from many of the factors previously described are:

- Organizations, beginning at the facility or other micro level, should evaluate areas (taking a benchmarking approach) where operations can be streamlined, improved, or otherwise changed, that can provide opportunities for maximizing the available work force's potential for their output in a nonthreatening, cooperative, and responsible manner that addresses their work, corporate commitment, and environmental and health and safety responsibilities in a truly holistic fashion. Such achievements help keep the company competitive, maintain market share, and be favorably viewed by its customers, the public and other concerned stakeholders.
- In addition, organizations may need to look at their long-term business goals and how their current environmental management responsibilities are reflected in these goals. They may also need to ask themselves whether their existing environmental management systems are functioning in a manner to reflect their environmental policy, and could things be improved upon to better the existing systems. In a different view, organizations may need to reevaluate their processes as a means to introduce process revision-rethink or redesign to move forward, and also take into account additional business-oriented considerations that are associated with value migration and a narrow competitive field of vision and customer field of vision.[46] Parallels to these business concepts were

previously described in the high-tech companies case studies and "living company" description examples.

- In human resources, seen as an organizational raw material resource, organizations should not overlook the tremendous potential that exists within their ranks with respect toward having companies do more with less. Numbers alone do not necessarily equate to greater employee production of widgets, or the greater amount of services that can be provided; rather, the propensity, work ethic, and proper training, of individuals can lead to better prepared, more satisfied, more diligent employees. Such talent pools should be mined "early and often" as another means to extract beneficial elements that employees can utilize as resources to perform their duties. For example, in providing ISO 9000 quality management and/or ISO 14000 environmental management insight and related awareness-type training to employees without a clear plan for identifying, or retraining these employees, and not having in place a clear game plan for them to understand, may be futile. This may be especially critical to satisfying the organization's objectives, should the goal for such insight and training be to provide a foundation for a company's commitment to continual improvement. Continual improvement is an integral element within both ISO 9001 and 14001, and for an organization to be successful in achieving its stated environmental goals, the goals and objectives need to be clear from the start, so that everyone concerned is "on the same sheet of music."

- Furthermore, in employee-training considerations, whether within the environmental or quality area, or in related company areas where some improvements in these areas can be derived, it is important that organizations establish steps to ensure that affected employees are trained, and trained properly, and that a responsible management-type follow-through on the effectiveness of such training exists. Not all training is alike, neither are training providers, and a diligent staff follow-through may also require some research into the training provided.[47] Such invested up-front time may provide the company tangible benefits at some point.

A few points worth noting to provide additional insight into the training of workers engaged in various responsibilities dealing with environmental management in an organization and how this training can be an effective and integral component of an effective environmental management system, include:

(1) Effective employee training takes time and a continuous effort—It's not a one-shot deal. The importance of clear communications between workers, their trainers, and supervisors, is important to the success of the training program's objectives, and in any follow-up training.

(2) Make a concerted effort to delve into the effectiveness of the training program, to ensure that proper training is being provided and that trainees understand the subject matter and are putting it into practice—what we call the "Onion Concept" analogy. The "Onion Concept" refers to learning about a subject by delving further into it, like peeling an onion one layer at a time. Use any information gained from any follow-up exercises to gauge this effectiveness.

(3) Employees need some form of motivation, or other subjective factors to maintain the momentum of training objective(s), such as role-playing or other types of training games.[48]

(4) Real-life examples, or role-playing help to bring the points raised in training programs home. Creativity is a key.[49]

(5) The organization must stand behind the environmental management program governed by the EMS to help keep the momentum going. This is not something to be perceived by workers as mere "window dressing" on the part of the organizations. Organizations truly committed to sustaining effective environmental management programs must do as they say, or "walk the talk."

(6) Don't rest on your laurels—start over. This is ingrained within the text of ISO 14001 and within ISO 14031 as part of the continuous improvement aspect. Without such a concerted and diligent effort, true improvement cannot be achieved as envisioned.

- In similar fashion, organizations should not overlook the tremendous across-the-board learning opportunities that can be derived from such "lessons learned" as noteworthy or catastrophic incidents to point out extreme examples of some management system failure. To further define this point, many articles and books have been written about various events or incidents, where readers have been able to gain valuable insight from these examples.[50] In some situations, U.S. federal agencies have also responded with new or revised regulations aiming to eliminate the reoccurrence of such incidents, as the Department of Transportation (DOT) did in airline safety. [51]

In summary, regulated organizations do not operate in an isolated vacuum. Rather, dynamic forces constantly barrage managers and workers at these organizations, from environmental regulatory and other regulatory constraints, customer concerns and other market-driven forces, and the cornucopia of internal and external stakeholder concerns. How well such organizations function as a direct result of these forces may be attributable to how well they are prepared to manage all these forces.

The intent of this book, therefore, is to provide a few of the answers to the issues highlighted herein, and again in greater detail in each of the ensuing fourteen chapters. Environmental management responsibilities in the dynamic 1990s are not for the faint of heart. Good luck!

ENDNOTES

[1]Among the companies that may be the farthest along in recycling and reusing products and packaging are high-tech companies. Among them, Philips, NEC, Panasonic, and IBM, all have evaluated to some degree the reuse of electronic components. References: *From Necessity to Opportunity,* Philips 1997 Corporate Environmental Review; NEC and Panasonic, Care Innovation 96 (Frankfurt); and IBM's 1996 environmental report, *Environment: A Progress Report.*

[2]Excerpted from a presentation made by Daniel McDonnell of IBM at the Inverse Manufacturing Symposium, March 10, 1997, in Tokyo, Japan. Highlights of this presentation are provided in further detail in Chapter 15.

[3]See "Focus Report: Is Mandated Product Takeback Coming to a Country Near You?," *Product Stewardship Advisor* ™, Volume 1, No. 1, May 1997, p. 2.

[4]Diane Hairston, "Responsible Packaging," *Chemical Engineering,* May 1997, p. 61.

[5]Ibid.

[6]Ibid.

[7]Originally released as the Valdez Principles in 1989 by the Coalition for Environmentally Responsible Economies.

[8]Responsible Care® had its beginning through the efforts of the Canadian Chemical Producers' Association (CCPA) in 1988, and was soon adopted by the U.S. Chemical Manufacturers Association (CMA). Reference: CMA report, *Chemical Manufacturers Association, Who We Are, What We Do,* undated.

[9]For additional information on the API's STEP Program, see 5th Annual Report, Petroleum Industry Environmental Performance, May 1997.

[10]The PERI Guidelines were developed between 1992 and 1993 by a number of companies from various industry sectors, with additional input from stakeholders as well. Reference: PERI Guidelines.

[11]As of this writing, ISO 14001 and 14004 were published as International Standards, September 1996; ISO 14010, 14011, and 14012 were published as International Standards, October 1996; ISO 14040 was published as an International Standard, June 1997; and ISO 14050 was published as an International Standard, May 1998. The ISO 14020s and ISO 14031 were in Draft International Standard (DIS) status, with the remaining standards in draft status. *Source: International Environmental Systems Update,* July 1998, pp. 21–22.

[12]As one example of this cooperative effort, many states have initiated technical assistance programs to assist small to medium organizations with their compliance problems and in identifying P2 opportunities. Among the states with such programs are: Massachusetts, New York, and Ohio.

[13]The 33/50 Program was announced at an EPA press conference on February 7, 1991 and was begun by former EPA Administrator William K. Reilly.

[14]A related article outlining EPA Administrator Carol Browner's approach to environmental protection titled, "EPA's Browner to Take Holistic Approach to Environmental Protection," was featured in *Chemical & Engineering News,* March 1, 1993, p. 19.

[15]Additional information on the progress of the 33/50 Program is found in a series of EPA reports on the program, including: EPA's 33/50 Program First Progress Report (July 1991); Second Progress Report (Feb. 1992); Sixth Progress Update (September 1995); and Company Profile Reduction Highlights, Volume II (December 1995).

[16]The Environmental Leadership Program, *Federal Register,* Vol. 58, No. 10, January 15, 1993, 4802.

[17]On June 21, 1994, EPA published a second *Federal Register* notice requesting proposals for the ELP. As a follow-up on April 7, 1995, an EPA press release proclaimed the launching of the one-year pilot of the ELP.

[18]The Common Sense Initiative was announced by EPA Administrator Carol Browner on July 20, 1994 in an EPA press release, *EPA Environmental News.*

[19]Project XL was listed in the *Federal Register* on May 23, 1995, reference FRL-5197-9, Regulatory Reinvention (XL) Pilot Projects.

[20]One of the classic cases highlighting this perceived adversarial role is Amoco's Yorktown Project. A summary of that project and the ensuing follow-up "bridge building" to bring the parties back together was described in an article by C. Solomon, "What Really Pollutes? Study of a Refinery Proves an Eye-Opener," *The Wall Street Journal,* March 29, 1993, p. A1.

[21]Xerox Company advertisement, "Sustainable Development—What Does It Mean for Industry?," *Business Week,* July 26, 1993.

[22]Additional detail about the IRRC survey is provided in Chapter 7.

[23]Timothy Aeppel, "Firms Reveal More Details of Environmental Efforts But Still Don't Tell All," *The Wall Street Journal*, December 13, 1993, p. B1.

[24]Excerpted from Chevron's and IBM's 1996 Environmental Reports.

[25]See "Working Ideas—Bad Ideas in Management Awards," *BusinessEthics*, May/June 1997, p. 10.

[26]Bleakley, F., "Many Companies Try Management Fads, Only To See Them Flop," *The Wall Street Journal*, July 6, 1993, p. A1.

[27]Two items of interest that help provide additional insight into this topic include: J. Tannenbaum, "Worker Satisfaction found to be Higher at Small Companies," and J. Kaufman, "Gray Expectations: A Middle Manager Struggles to Adapt to the Times," *The Wall Street Journal*, May 5, 1997, p. A1.

[28]Drennan, D., *Transforming Company Culture*. New York: McGraw-Hill International (UK) Limited, 1992, pp. 46–69.

[29]For further information, see Arie de Geus, "The Living Company," *Harvard Business Review*, March–April 1997, pp. 51–59, an article summary of de Geus's book by the same name.

[30]See n. 1.

[31]Gabriella Stern, "Shifting Gears: How a Young Manager Shook Up the Culture at Old Chrysler Plant," *The Wall Street Journal*, April 21, 1997, p. A1.

[32]This definition was contained in EPA's Environmental Auditing Policy Statement, *Federal Register*, Vol. 51, No. 131, July 9, 1986, 25004. A reference to this EPA Policy is contained in Appendix F.

[33]International Standard, ISO 14010, First Edition 1996-10-01, "Guidelines for Environmental Auditing—General Principles," International Organization for Standardization, 1996.

[34]Reference: GAO Report, "ENVIRONMENTAL AUDITING, A Useful Tool That Can Improve Environmental Performance and Reduce Costs," GAO/RCED-95-37, April 1995, p. 23. We refer to this update again in Chapter 7.

[35]Crognale, G., "P2 Financial Incentives," *Environmental Protection*, May 1997, p. 36.

[36]A compilation of P2 success stories is contained in the EPA document, *EPA Pollution Prevention Success Stories*, EPA/742/96/002, April 1996. It is a handy reference document, and is worth obtaining. Another document worth obtaining is *EPA's Partnerships in Preventing Pollution—A Catalogue of the Agency's Partnership Programs*, Spring 1996.

[37]For further insight, readers are referred to the EPA document, *Life Cycle Design Framework and Demonstration Projects: Profiles of AT&T and AlliedSignal*, EPA/600/R-95/107, July 1995.

[38]See B. Filipczak, "It Takes All Kinds: Creativity in the Workforce," *Training*, May 1997, pp. 32–40.

[39]Refer to reference in n. 2.

[40]Excerpted in part from Hewlett-Packard's environmental report, *Hewlett-Packard's Commitment to the Environment*. Hewlett-Packard Company, Palo Alto, CA, 1996.

[41]SAGE was formed in August 1991, through the ISO, to assess the need for international environmental management standards and to recommend an overall strategic plan for developing such standards. SAGE consisted of a panel of environmental experts from member countries. In the fall of 1992, SAGE gave its recommendation to ISO's Technical Board, which soon thereafter authorized the creation of Technical Committee (TC) 207. With the creation of TC 207, ISO commissioned its second effort in the broad field of management standards, in hope of duplicating a successful quality formula aimed at the envi-

ronmental area. Based on the most recent minutes of the U.S. Technical Advisory Group's (TAG) meeting, the USTAG's membership as of July 30, 1998 was 356 members. (*Source:* ASTM, memo to USTRG members.)

[42]See n. 11.

[43]T. Weisman and B. Knight, "Improving Processes and Partnerships," *Environmental Protection*, October 1996.

[44]A noteworthy reference article is J. Rigdon's, "Using New Kinds of Corporate Alchemy, Some Firms turn Lesser Lights into Stars," *The Wall Street Journal*, May 5, 1993, p. B1. We also provide Boeing's Cross Training Program in Appendix G.

[45]A labor letter news item appeared in *The Wall Street Journal*, "TRADING PLACES: Chemical Companies Switch Employees to Learn New Tricks," April 12, 1994, p. A1.

[46]Described further, "value migration" is "... the flow of economic and shareholder value [that flows] away from obsolete business models to new, more effective designs [and] reflects changing customer needs that are beginning to be, and ultimately will be, satisfied by new competitive offerings." B. S. Shapiro, A. J. Slywotzky, and R. S. Tedlow, "How To Stop Bad Things From Happening to Good Companies," *Strategy & Business*, First Quarter 1997, pp. 25–41.

[47]G. Crognale, "How To Choose Environmental Management Courses: The 'Consumer Reports' Viewpoint," *Environmental Management Report*, April 1997, p. 17.

[48]The author maintains a data base of many such games used in practice.

[49]The author maintains a data base of such examples in his reference file.

[50]One noteworthy publication is, Trevor Kletz, *What Went Wrong? Case Histories of Process Plant Disasters*, 3rd ed., (Gulf Publishing Company), 1998, Houston, TX.

[51]As a means to increase safety aboard passenger airlines, the U.S. Department of Transportation (DOT) issued a press release on December 30, 1996 for a permanent ban on the transportation of chemical oxygen generators as cargo on passenger airplanes. On that same day, a *Federal Register* notice was issued describing the Final Rule and Prohibition of Oxidizers Aboard Aircraft; 68952, *Federal Register*, Vol. 61, No. 251. Ancillary information is provided in two General Accounting Office (GAO) reports dealing with aviation safety, GAO/RCED-96-193, and GAO/RCED-97-2. In a related story relating to airplane safety in the wake of several U.S. disasters in 1996, see "A Greater Threat Than Terrorism?" W. Stern, *Business Week*, September 9, 1996.

REFERENCES

Articles & Books

"EPA's Browner to Take Holistic Approach to Environmental Protection," *Chemical & Engineering News,* March 1, 1993, p. 19.

"Focus Report: Is Mandated Product Takeback Coming to a Country Near You?," *Product Stewardship Advisor*™, Vol. 1, No. 1, May 1997, p. 2.

"Working Ideas—Bad Ideas in Management Awards," *BusinessEthics* May/June 1997, p. 10.

Chemical Week, July 17, 1991, issue devoted to Responsible Care®.

Crognale, G. "How To Choose Environmental Management Courses: The Consumer Reports Viewpoint," *Environmental Management Report,* April 1997, p. 15.

Crognale, G. "P2 Financial Incentives," *Environmental Protection,* May 1997, p. 36.

Crognale, G. "What's Up in EPA's Interim Policy Statement on environmental Auditing," *Corporate Environmental Strategy*, Vol. 3, No. 2, Winter 1996, pp. 37–39.

de Geus, A. "The Living Company," *Harvard Business Review*, March–April 1997, pp. 51–59.

de Geus, A. *The Living Company: Habits for Survival in a Turbulent Business Environment*. Boston, MA: Harvard Business School Press, 1997.

Dreenan, D. *Transforming Company Culture*. London: McGraw-Hill International Limited, 1992.

Flipczak, B. It Takes All Kinds: Creativity in the Workforce," *Training*, May 1997, pp. 32–40.

Hairston, D. "Responsible packaging," *Chemical Engineering*, May 1997, p. 61.

International Environmental Systems Update, July 1998, pp. 21–22.

Kletz, T. *What Went Wrong? Case Histories of Process Plant Disasters*, 3rd ed. Houston: Gulf Publishing Company, 1997.

Plunkett, L. C. and Fournier, R. *Participative Management: Inplementing Empowerment*. New York: John Wiley & Sons, Inc., 1991.

Shapiro, B. and Tedlow, R. S. "How to Stop Bad Things from Happening to Good Companies," *Strategy & Business*, First Quarter 1997, pp. 25–41.

Sheldon, Christopher, Ed. *ISO 14001 and Beyond: Environmental Management Systems in the Real World*, Sheffield, United Kingdom: Greenleaf Publishing, 1997.

Sullivan, T. F. P. Ed. *The Greening of American Business: Making Bottom-Line Sense of Environmental Responsibilities*. Rockville, MD: Government Institutes, 1997.

Weisman, T. and Knight, B. "Improving Processes and Partnerships." *Environmental Protection*, October 1996.

EPA & Miscellaneous Industry & Company Reports

60 *Federal Register* 16875, April 3, 1995.

60 *Federal Register* 66706, December 22, 1995.

American Petroleum Institute's 5th Annual Report, *Petroleum Industry Environmental Performance*. Washington, DC: Author, 1997.

Chemical Manufacturers Association. *Who We Are, What We Do*, undated. Washington, DC: Author.

Chemical Manufacturers Association. *The Year in Review 1995–1996, Responsible Care® Progress Report: Members and Partners in Action*, Washington, DC: Chemical Manufacturers Association, 1996.

Environmental Protection Agency. *33/50 Program: First Progress Report*, July 1991; Second Progress Report, February 1992; Sixth Progress Report, September 1995; Company Profile Reduction Highlights, Volume II, December 1995.

Environmental Protection Agency. *Life Cycle Design Framework and Demonstration Projects: Profiles of AT&T and Allied Signal*. EPA/600-R-95/107, July 1995.

Environmental Protection Agency. *Pollution Prevention Success Stories*. EPA/742/96/002, April 1996.

Federal Register 25004, Vol. 51, No. 131, July 9, 1986.

Federal Register 4802, Vol. 58, No. 10, January 15, 1993.

Federal Register, FRL-519709, Regulatory Reinvention (XL) Pilot Projects, May 23, 1995.

Government Accounting Office. *Environmental Auditing, A Useful Tool That Can Improve Environmental Performance and Reduce Costs.* GAO/RCED-95-37, April 1995.

International Standard, ISO 14010, "Guidelines for environmental auditing." Geneva, Switzerland: International Organization for Standardization, 1996.

National Environmental Technology Strategy, *Bridge to a Sustainable Future,* 1995. Washington, DC: Author.

PERI Guidelines, undated.

Publicly-held companies environmental reports: Hewlett-Packard's *Commitment to the Environment,* 1996; IBM's *Environment: A Progress Report,* 1996; Phillips Petroleum's *1996 Health, Environmental and Safety Report* Philips Electronics' *From Necessity to Opportunity: Corporate Environmental Review,* 1997; Polaroid's *1995 Report on the Environment;* Chevron's *1996 Environmental Report;* Texaco's *Environment, Health and Safety Review 1996.*

CHAPTER 2

THE USEPA'S ENFORCEMENT STRATEGY: THE EVOLUTION OF ENVIRONMENTAL ENFORCEMENT

Gabriele Crognale, P.E.

> *"The American people want, expect and demand to be protected from pollution, and they want effective enforcement of environmental laws in order to guarantee that protection."*
>
> *Steven Herman,*
> *EPA Assistant Administrator*

INTRODUCTION

Adhering to environmental regulations is one way to ensure that industrial processes, whether in extracting raw materials, or manufacturing durable and consumer goods, do not adversely affect the health and safety of workers, the community in which these industries reside, and the environment that surrounds these industries beyond the fenceline. How well or how poorly industries adhere to these regulations, policies, and norms is part of the charge of the regulators who regulate these industries at the federal, state and local levels.

Another item for consideration is whether the regulations take into account various environmental and health and safety aspects to ensure these regulations effectively carry out the intent of the regulatory policymakers who wrote the laws in the first place.

In this chapter, we take a closer look at both issues and how they relate to each other with respect to regulatory compliance and enforcement. We also take a closer look at the steps the EPA has taken over the past several years and continues to take to dispel the regulated community's perception of EPA as a "command-and-control" agency that takes an enforcement approach to one that "partners" with the regulated community to attain the agency's primary goals—the protection of human health and the environment. We also need to remind ourselves that EPA's enforcement arm is primarily directed to those companies that continue to choose noncompliance as the way to do business; for those companies, "partnering" with EPA may not be an option. As insight into enforcement and compliance directives, we provide an overview of EPA's FY 1997 enforcement priorities.

In this overall context, we focus on some of the issues that are important to members of the regulated community and the regulatory agencies as one way to strive for "cleaner, cheaper, smarter" solutions to environmental issues, borrowing from EPA's Common Sense Initiative slogan.

EPA ENFORCEMENT AT A GLANCE: AN OVERVIEW

To put EPA's enforcement approach in perspective, it is important to provide some historical insight to this approach, beginning with the period from the late-1980s and early-1990s to the present.[1] One of the cornerstone documents during that era was the RCRA Civil Penalty Policy, or RCPP, that was issued in October 1990.[2] The RCPP was developed primarily as an enforcement enhancement tool by EPA after the agency began perceiving from the regulated community that many violators considered EPA penalties and fines nothing more than the cost of doing business. During that period, one of EPA's senior officials, Bruce Diamond, the Director of the Office of Solid Waste, had stated that both EPA and the states were seeking higher penalties in enforcement actions. A number of companies since that time have received penalties in the millions and tens of millions of dollars,[3] in marked contrast to the penalties previously assessed in the hundreds of thousand dollars range prior to the issuance of the RCPP.

To put the overarching tenet of the policy in perspective, some insight can be gained from the introductory paragraph to a chapter devoted to compliance and enforcement in a 1990 EPA study. It may be historical, but it is still timely. It reads as follows:

> An effective enforcement program must detect violations, compel their correction, ensure that compliance is achieved in a timely manner, and deter other violations. The RCRA enforcement program will obtain substantial voluntary compliance only if the regulated community perceives that there is a greater risk and cost in violating a requirement than in complying with it.[4]

In addition to learning to understand EPA's enforcement approach, it is also important to know something about EPA's enforcement strategy and where it begins. With this understanding, the regulated community can be better prepared to address regulatory considerations, and possibly move beyond regulatory compliance by initiating innovative approaches to meet (or better) compliance requirements.

EPA's enforcement strategy begins within EPA's main policymaking group for enforcement, the Office of Enforcement and Compliance Assistance (OECA),[5] at EPA headquarters in Washington, D.C. OECA is charged with managing EPA's civil and criminal enforcement program, along with other programs,[6] and includes the enforcement of violations of the Clean Air Act, Clean Water Act, the Resource Conservation and Recovery Act (hazardous waste violations), and the Toxic Substances Control Act (toxic substances and PCB violations), to name a few. OECA is also the generating point of various EPA policy memoranda and guidance documents, such as EPA's Audit Disclosure Policy[7] and the Exercise of

Investigative Discretion.[8] In addition, OECA sets certain enforcement priorities for each of EPA's ten regional offices. At the regional level, EPA oversees the administration of state regulatory programs and may either lead EPA enforcement activities or provide assistance to state enforcement activities, depending upon the relationship EPA may have established with each state enforcement agency as part of the state's implementation program.

Among OECA's other offices to assist in enforcement activities include the National Enforcement Investigations Center (NEIC) in Denver and the Office of Criminal Enforcement (OCE). NEIC and OCE often work in tandem in pursuit of criminal cases. The NEIC is designed to provide EPA with highly specialized technical and criminal and civil support when necessary. Each regional office houses NEIC specialists who work with regional inspectors and Office of Regional Counsel in pursuit of environmental enforcement actions. These NEIC representatives are also empowered with police powers as U.S. Marshals and can serve search warrants if necessary. More detailed information about each of OECA's offices is provided later in this chapter.

1. Some of the Enforcement Tools Used by EPA

EPA has been empowered by Congress with a number of enforcement tools that the agency can use to enforce the violations of various environmental laws that can result in administrative actions, criminal prosecutions, or if permits have been issued, the revocation of the permits. These enforcement actions can also have a penalty component.[9]

Enforcement actions can take the form of Notices of Violations (NOVs) for minor violations and Administrative Orders (AOs) for more serious violations, that may take the form of a Consent Order or a Unilateral Order. In those cases where EPA cannot achieve compliance with the violator, EPA can then present the case to an administrative law judge for resolution. If that referral does not provide resolution, EPA can then refer such cases to the Department of Justice and pursue legal action in federal court.

The most serious or egregious violators also face the possibility of criminal enforcement. These cases[10] usually result from overt or covert illegal activity such as the knowing and willful illegal disposal of hazardous waste,[11] altering or fabricating of permit or wastewater discharge values, tampering with monitoring equipment, falsifying hazardous waste manifests or falsifying hazardous waste labels, or other such wanton acts that could adversely impact human health and/or the environment. In addition, the federal sentencing guidelines provide additional guidance for imposing significant sanctions upon egregious environmental violators.

2. EPA's Evolving Compliance and Enforcement

Given this array of enforcement tools at its disposal, EPA has been ratcheting up its enforcement efforts over the past several years[12] with an increase in enforcement actions and penalties imposed upon violators. Aware of such occurrences, some companies during this period were urging other companies to initiate proactive moves to avoid sizable enforcement penalties.[13]

To provide additional insight into the evolution of compliance and enforcement, we provide an overview of several noteworthy EPA guidance documents, among them: (1) EPA's Four-Year Plan; (2) EPA's Five-Year Plan; and (3) EPA's FY 1997 Enforcement and Compliance Assurance Priorities. The focus, structure, and "architecture" of these three documents are very similar, telegraphing an agency that is consistent in its approach.

EPA's Four-Year Plan[14]

This plan was developed, in part, to provide additional long-term insight to EPA's environmental compliance and enforcement efforts under the guidance of EPA's Office of Enforcement and Compliance Assurance (OECA), which replaced the former Office of Enforcement. This plan encompassed ten main themes to guide EPA in its enforcement initiatives through FY 1997.

The ten themes of the Four-Year Plan are:

1. Strengthen the institutional voice;
2. Develop an integrated, multimedia enforcement strategic planning process;
3. Target initiatives for maximum enforcement results;
4. Screen violations for appropriate responses;
5. Use innovative enforcement tools and authorities;
6. Improve EPA enforcement relationships with states, Indian Tribes and local governments;
7. Improve EPA enforcement relationships with other federal agencies;
8. Communicate the impacts and results of enforcement actions;
9. Improve the infrastructure and training; and
10. Promote international environmental enforcement programs.

Of the ten themes, perhaps the ones with the greatest far-reaching impact include themes 3, 4, and 5 that focus on enforcement results, responses, and tools, respectively. Some of the by-products of these themes include: (1) enforcement targeting by industry, processes, specific pollutants, and geographic region, with a greater emphasis on TRI inventory data to help data compilation; (2) developing cases with the greatest potential for risk reduction and deterrence; and (3) placing greater emphasis on industry's use of pollution prevention initiatives and environmental auditing/self-disclosure practices, and utilizing the criminal sentencing guidelines[15] when and where appropriate as a deterrent mechanism. Of note, additional discussion about the guidelines is provided in Chapter 8 by Ed Spaulding of Chevron.

EPA's Five-Year Plan[16]

The plan, according to EPA, represents the combined insight, energy, and forward-thinking of EPA's senior leadership, employees, and its stakeholders as they define EPA's role and direction into the twenty-first century. EPA's Five-Year Plan, excerpted from the summary,[17] consists of a series of guiding principles the agency will emphasize to achieve sustainable environment and economy through: ecosystem protection, environmental justice, pollution prevention, strong science and data, partnerships, reinventing EPA management, and finally, environmental accountability (of the regulated community).

The plan is viewed by EPA as a blueprint for change at EPA and is intended to guide EPA in planning, resource allocation, and decision-making processes from 1995 to 1999. EPA's vision for the plan is to set direction for those changes that will shape EPA's environmental agenda into the next century. In the next section, we view EPA's agenda in a smaller time frame, and provide some insight into EPA's FY 1997 objectives.

EPA's FY 1997 Enforcement and Compliance Assurance Priorities[18]

As a follow-on to EPA's FY 1996 OECA priorities, Steve Herman, the Assistant Administrator for Enforcement and Compliance Assurance highlighted in the FY 1997 priorities, documents EPA's milestones achieved to implement the agency's compliance assurance program in FY 1996. These include: EPA's final policy on auditing self-policing/self-disclosure that is discussed in Chapter 9; the administration's efforts to reinvent environmental regulations and address environmental problems on a multimedia, industry-wide basis, as well as introduce pollution prevention opportunities, technologies, and techniques; and policies that provide incentives for small businesses and small communities to make good faith efforts to voluntarily identify and seek technical assistance to correct violations.

In addition, the FY 1997 priorities document also showcases EPA's enforcement efforts in FY 1996 that included a record 262 criminal cases that were referred to the Department of Justice (DOJ) and the work of EPA's civil enforcement program that referred 295 more cases to DOJ than in FY 1995. In dollars, EPA assessed a record $76.7 million in criminal fines and about $30 million more in civil penalties than in FY 1995. Part of EPA's focus in these cases was combining significant penalties, injunctive relief, and other innovative features that are intended to protect both public health and the environment. Among some of the actions EPA took that are highlighted in this priorities document are:

- **Iroquois Pipeline Operating Company.** The company and four of its top officials pled guilty to degrading scores of wetlands and streams while constructing a natural gas pipeline from Ontario, Canada to Long Island. In the largest environmental criminal enforcement settlement since Exxon Valdez, the company will clean up thirty streams and wetlands and pay $22 million in fines and penalties, $2.5 million of which will be used to create additional wetlands.
- **Georgia-Pacific Corporation.** The company will spend more than $35 million in injunctive relief, penalties, and supplemental projects to settle allegations that it illegally emitted tons of VOCs annually into the air at its wood product factories in eight southeastern states. The settlement will reduce ozone-forming emissions from these plants by at least 90 percent, that translates into a reduction of 10 million pounds of air pollutants per year.
- **General Motors.** The company paid an $11 million fine, spent more than $25 million to recall and retrofit polluting vehicles, and spent another $8.75 million on projects to offset emissions to settle charges that "defeat devices" on Cadil-

lacs resulted in illegal releases of approximately 100,000 tons of excess carbon monoxide pollution.

ADDITIONAL EPA FY 1997 COMPLIANCE ASSURANCE PRIORITIES

As EPA continues to complete its efforts to reorganize enforcement and compliance programs at headquarters and at the ten regional offices, the goal is to fully implement key enforcement and compliance assurance priorities and approaches that began in 1995. EPA sees a continued and aggressive implementation of its enforcement and compliance assurance program, emphasizing community-based, sector and media-specific priorities. EPA makes it clear in this priorities document that "OECA will adhere to the fundamental principle that formal law enforcement is the central and indispensable element of effective governmental efforts to ensure compliance."

EPA further expands upon this general tenet by making reference to two OECA policy statements to reemphasize that "a strong enforcement program is the foundation upon which (EPA's) complementary compliance incentive and compliance assistance efforts depend." The first of these policy statements refers to "core" federal functions that must be performed by EPA *to assure the protection of public health and the environment and to keep polluters from gaining a competitive advantage over regulated entities that do comply with the law.* This policy statement is part of the Core EPA Enforcement and Compliance Assurance Functions, issued February 21, 1996.

The other policy statement, Operating Principles for an Integrated Enforcement and Compliance Assurance Program, issued November 27, 1996, provides an overview of EPA's enforcement and compliance tools and authorities, as well as provides guidance for their most effective use. In addition, EPA makes it known that the priority actions the agency takes in FY 1997 reflects the principles, strategies, and approaches outlined in both statements. Whether EPA may effectively gauge such actions by using the regulated community as the study area could become a review item for the U.S. General Accounting Office (GAO) as part of its continued oversight of EPA's activities relating to reinventing environmental regulations or other related EPA activities.

A brief encapsulation of EPA's priorities lists the following general themes the agency plans to follow:

- **Employ** the full range of administrative, civil judicial, and criminal enforcement tools and remedies in the most efficient and appropriate way.
- **Continue** to refine and improve its compliance and monitoring capabilities.
- **Promote** the widespread use of compliance incentives embodied in the self-disclosure, small business, and small communities policies.
- **Continue** to expand efforts to provide regulatory and technical information to the private sector and municipalities.
- **Foster** environmental justice to ensure that environmental pollution does not disproportionately affect minorities and low-income groups.

- **Develop, pilot, and implement** improved measures of success for the full spectrum of enforcement and compliance assurance activities, and make those measures available to the public.
- **Continue** the major effort begun in FY 1996 to strengthen the agency's partnerships with the states and tribes.

These priorities, in turn, will be initiated by each of OECA's offices, which consist of the:

- Office of Regulatory Enforcement (ORE)
- Office of Criminal Enforcement, Forensics and Training (OCEFT)
- Office of Site Remediation Enforcement (OSRE)
- Office of Federal Activities (OFA)
- Office of Compliance (OC)
- Office of Environmental Justice (OEJ)
- Federal Facilities Enforcement Office (FFEO)
- Enforcement Capacity and Outreach Office (ECOO)

In summary, EPA's FY 1997 Enforcement and Compliance Assurance Priorities document provides a snapshot view of EPA's goals and objectives for ensuring the protection of human health and the environment as best it can given these tools. How well or how effective such tools can work to achieve EPA's goals into FY 1998 and beyond still remains to be seen and measured.

In the following sections, we take a closer look at some EPA initiatives and reflect upon how well (or not-so-well) they have fared, and include some comments to date on their progress.

EPA AT A CROSSROADS

In the previous section, we provided some insight into current enforcement and compliance thinking at the highest levels of EPA. Aware of the continued controversy surrounding enforcement, increased regulations, and other aspects of maintaining compliance by the regulated community, EPA acknowledges that there will continue to be honest differences of opinion regarding the content and direction of some EPA programs and policies. In EPA's view, this can even be healthy as a means to flush out any one of a number of misdirections. Several documents, one of them an EPA report on EPA's reinvention efforts, and several GAO reports, are highlighted later in the chapter to provide additional insight into this particular issue and what recommendations can be made to help guide EPA in the coming years.

In juxtaposition to the continued enforcement and compliance scenario that OECA envisions to some degree, EPA has been at a crossroads since the early-1990s with the introduction of voluntary programs aimed at *working with* instead of *against* the regulated community. The veritable success of the flagship of these voluntary initiatives, the 33/50 Program, ushered in a new era for EPA that still continues to roll out progressive directives and voluntary initiatives. Some of those worth noting include: (1) The Environmental Leadership Program (ELP);

(2) The Common Sense Initiative (CSI); and (3) Project XL. We take a closer look at some participants in these initiatives and provide some insight into their accomplishments to date by summarizing their most recent project milestones.

1. THE 33/50 PROGRAM

One of the cornerstone projects credited with initiating EPA's move away from a strict command-and-control approach was the Industrial Toxics Project, more commonly known as the 33/50 Program, announced at a press conference on February 7, 1991 by the former EPA Administrator William K. Reilly. The 33/50 Program was an innovative concept in which EPA had identified seventeen chemicals, all of which still are: (1) serious health and environmental concerns; (2) high-volume industrial chemicals; and (3) can be reduced through pollution prevention. EPA's goal at that time was to reduce the aggregate release of the seventeen target chemicals (see Table 2.1) to 33 percent of 1988 levels by the end of 1992 and to 50 percent by the end of 1995—hence, the name 33/50.

EPA had also hoped that those companies who would later pledge to commit to these goals[19] would utilize pollution prevention practices rather than end-of-pipe treatment to achieve these goals. More importantly, EPA had hoped that the program would instill a pollution prevention ethic in American business where companies would analyze their operations as a means to reduce or eliminate pollution before it would even begin.

This program, the first of its kind, was a successful endeavor with regulated industries. The 33/50 Program's success with companies could have been attributed to the flexibility each company had in taking its own initiative to reduce or eliminate the seventeen targeted chemicals from their waste streams. A summary fact sheet on the program, called 33/50 Program *Achievements,* can be downloaded from EPA's web page at www.epa.gov/reinvent/3350.htm. It was last revised on January 15, 1997.

www

Table 2.1.   17 Chemicals of the 33/50 Program

Benzene
Cadmium and Compounds
Carbon Tetrachloride
Chloroform
Chromium and Compounds
Cyanides
Lead and Compounds
Mercury and Compounds
Methyl Ethyl Ketone
Methyl Isobutyl Ketone
Methylene Chloride
Nickel and Compounds
Tetrachloroethylene
Toluene
Trichloroethane
Trichloroethylene
Xylenes

Confident from this initial endeavor with the 33/50 Program, EPA again decided to test the waters in the area of voluntary initiatives with the regulated community, and issued a public notice to describe its Environmental Leadership Program, or ELP, asking for public comments.[20] As a result of the one hundred or so comments EPA received on the ELP, the agency decided to issue a second public notice requesting proposals for the ELP.[21] That notice generated considerable comment from the regulated community, and culminated in the ELP pilot program.

At about this same time, while EPA was looking for additional innovative ways to work with the regulated community, the agency was still intent upon companies maintaining compliance with the regulations, and to ensure that compliance would remain a driving force, several plans were developed by EPA to help move compliance forward. These documents included the Four-Year Strategic Plan, and the Five-Year Strategic Plan that were previously discussed. Both documents were developed as guidance documents to help guide EPA's forward through 1999.

2. The Environmental Leadership Program[22]

The Environmental Leadership Program (ELP) was designed to allow EPA to recognize and provide incentives to facilities willing to develop and demonstrate accountability for compliance with existing laws. The program began in January 1993 with a *Federal Register* notice on the design of a proposal outlining a sector approach to compliance in which EPA had requested public input, and culminated with several follow-on *Federal Register* notices, including a request for pilot projects that jump-started the program. The intent of the request was to solicit proposals for pilot projects that would demonstrate: state-of-the-art compliance management programs, EMSs, independent audits and self-certification, public accountability and involvement, P2 approaches and mentoring. The ELP is one of the current administration's reinvention of regulation to achieve environmental results at the least cost. Among the main priorities[23] of the ELP are:

1. To examine the basic components of a state-of-the-art compliance management system;
2. To identify the verification procedures that ensure the ELP is working, for example, in third-party audits or (ISO 14001) self-certification;
3. To establish measures of accountability so that these management systems will be credible to the public; and
4. To promote community involvement in understanding and supporting innovative approaches to compliance.

An interested party may want to know exactly what benefits may be derived from such projects for his or her particular organization in exchange for providing EPA this commitment and tying up employee and other company resources. EPA's response to this inquiry includes the following salient points directed to ELP participants:

1. Public recognition for their ELP participation—that could be tied into corporate communication-type press releases or inclusion in their voluntary environmental reports as an important "atta boy."

2. A limited (grace) period in which to correct any violations disclosed to EPA and the states during the ELP pilot program, which is similar in structure to EPA's Audit Disclosure Policy of December 1995.

3. Participation in a joint EPA/states initiative to identify methods for reducing inspections and streamlining reporting requirements. Our translation—identifying innovative approaches to counter the former "command-and-control" mindset.

EPA's pilot program began in earnest soon after EPA's press release issued in April 1995 that identified twelve pilot projects: ten companies and two federal facilities. These twelve organizations agreed to participate in innovative, common-sense, and cost-effective ways to comply with existing environmental laws. Part of the added benefit of this pilot program is that the participants pledged to share their learning experiences in an effort to allow more organizations to follow this example in a "lessons learned" approach and thus join the ranks of "high achievers."

Specific elements of the program the participants tested include:

- Advancing the design of sophisticated environmental management systems (EMSs), such as EMSs in conformance with ISO 14001;
- Incorporating multimedia compliance assurance into facility management;
- Third-party verification to provide assurance of performance and self-certification for reporting, thus allowing the opportunity for a fresh set of eyes to review each project;
- Public measures of accountability, as one of the means to ensure complete openness;
- Community involvement in setting goals and reviewing results, a practice that is also one of the components of the Chemical Manufacturers Association Responsible Care® Program. (Additional discussion about Responsible Care® is provided in Chapter 10.)
- Mentoring programs to help small businesses find cost-effective ways to comply with the law.

The culmination of the efforts of the twelve pilot-phase participant organizations were highlighted at a National Stakeholders Conference in Washington, D.C., in November 1996 hosted by EPA. That conference drew approximately 160 attendees from sixteen states, environmental groups, and industry.

Since that time, EPA's Internet home page—www.epa.gov/reinvent/elp.htm—is devoted to the Environmental Leadership Program to keep interested parties up-to-date on ELP events. These events are listed on an EPA Reinvention Activity Fact Sheet that notes the next steps EPA would take in this program. These steps include:

WWW

- Write and make available a conclusions report, including proposed criteria and incentives for participation in a full-scale ELP;

- Work with a full range of stakeholders, including states, environmental groups, and others to develop and implement "full-scale" programs;
- Publish a notice in the *Federal Register* and provide for an open comment period for a "full-scale" ELP.

In addition, EPA also published its annual report on reinvention activities in March 1998, and among EPA's "List A" or high-priority projects for reinvention is the ELP as a vehicle to test third-party audits for industry compliance. As part of the overarching tenet of the "lessons learned" approach of the pilot participants we described earlier, the EPA will evaluate the results or findings of third-party audits conducted at several facilities as a basis for potentially developing a permanent performance-based reward and recognition program. The ELP pilot program as it evolves, with input from environmental groups, industry, and states, will evaluate criteria for third-party audits that will assure the public that environmental requirements are being met by facilities, and any violations noted are disclosed and promptly corrected.[24] One link that might add value to the criteria for third-party audits that EPA will be evaluating could be for EPA to review such audit criteria or information with any audit disclosure information companies may provide as part of EPA's Audit Disclosure Policy, and compare the results for any similarities or patterns, or possibly initiate some form of benchmark study to further evaluate the findings of any "best-in-class" companies. Further discussion about these two topics is provided in greater detail in Chapters 7 and 9.

Since the ELP is a work-in-progress and is still evolving, it is difficult to bring this program to closure at this time. However, at the time this book was going to press, the most recent update on this program from EPA stated that the implementation of a full-scale ELP "... is still on hold ... it is taking longer than initially expected to develop a strategy for coordinating EPA's voluntary programs."[25] Readers are strongly urged to consult this EPA web site and others for regular updates on any EPA activity. A complete listing of frequently accessed EPA web sites is included in Appendix G.

3. THE COMMON SENSE INITIATIVE[26]

The Common Sense Initiative, or CSI, was ushered in by EPA's Administrator Carol M. Browner during an announcement that highlighted the selection of the first six major U.S. industries to participate in a new effort to transform existing processes for developing environmental regulations into a comprehensive system for strengthened environmental protection. The key to new regulations under the CSI was to achieve increased environmental protection at a lower cost through the creation of pollution control and prevention strategies via an industry-by-industry basis versus the status quo pollutant-by-pollutant approach. The catchphrase, or main guiding principles, for the CSI was to develop "cleaner, cheaper, smarter" regulations.

The six industries that make up the CSI and have participated since the inception of the program are automobile assembly, computers and electronics, iron and steel, metal plating and finishing, petroleum refining, and printing. Of note, these six industry sectors had accounted for 12.4 percent of the toxic releases reported by all American industry in 1992.[27] Additional information regarding the CSI group's make-up can be found on EPA's web page for the CSI Program at www.epa.gov/reinvent/new/7-97/page3.htm.

www

As envisioned by the EPA, the CSI program was developed to examine every aspect of environmental regulation as it effects each industry and the environment. The premise is that each team within the CSI workgroups will be able to attain cleaner, cheaper, and smarter performance in six main areas encompassing: (1) *Regulations*—review existing regulations for opportunities to improve upon them for more effective environmental results at less costs; (2) *Pollution Prevention (P2)*—actively promote P2 as a regular business practice and a main driver for environmental protection; (3) *Reporting*—facilitating the dissemination of relevant environmental (compliance) information; (4) *Compliance*—assisting those companies wishing to maintain and exceed compliance while initiating enforcement activities to those companies that do not; (5) *Permitting*—Revise permitting procedures to enhance efficiency, foster innovation, and introduce public participation opportunities; and (6) *Environmental Technology*—give regulated industries the opportunity, incentives and flexibility to develop innovative technologies for meeting and exceeding regulatory requirements while providing for cost-cutting measures.

A sampling of recommendations by several of the CSI subcommittees found the following items for consideration:

- **The Metal Finishing Group**[28]—This group raised issues concerning the then-proposed metal products and machinery effluent guideline rule and how that might have affected POTWs; interactions between industry, states and POTWs over permits; initiating greater communication with individual facilities, such as through a helpline, brochure, or other vehicle to guide platers in understanding what they should be doing; and looking at new projects to promote compliance via a metal finishing resource center, the use of voluntary compliance audits with reduced penalties, developing a POTW training project to work with metal finishing operations; and conduct regional meetings to solicit increased input to the CSI process.

- **The Petroleum Refining Sector**[29]—Among the topics discussed at this meeting were a "One-Stop" Initiative for recordkeeping and reporting; public participation and risk management; regulatory overlap and reform; accident prevention; and how new project proposals are introduced.

- **The Computers and Electronics Sector**[30]—Several workgroups within this sector made recommendations to EPA. Among the recommendations were: (1) the EPA administrator establish a process to ensure that EPA regulatory interpretations and/or determinations intended or likely to affect environmental management practices of companies be compiled, made accessible, and publicized to interested stakeholders; and (2) a draft *Federal Register* notice to solicit pilot project proposals for: (a) the development and operation of alternative strategies to replace or modify specific regulatory requirements, (b) pilots that complement Project XL and the ELP, and (c) increased regulatory flexibility resulting in enhanced environmental, health and safety performance and increased accountability to communities and workers.

- A subsequent **Computers and Electronic Sector** Subcommittee meeting identified specific issues in the following areas: (1) overcoming barriers to P2, product stewardship, and recycling in the manufacturing process and in post-consumer disposal of electronic equipment; (2) reporting and information access; and (3) an alternative performance-based system for environmental protection.

From another viewpoint, while the CSI program shows much potential for achieving EPA's intended goals as one of the cornerstones of EPA's reinvention efforts, an independent study by the General Accounting Office (GAO) concluded that the progress of the CSI has been slow in view of the high expectation EPA set for it in the beginning, and that EPA underestimated the time required to gather and analyze the information needed for developing recommendations and establishing the necessary relationships among the various stakeholder groups needed to reach consensus on complex issues.[31] The GAO report recommends that EPA provide an improved operating framework that accomplishes the following: more clearly defines CSI's cleaner, cheaper, smarter goal to environmental protection; specifies how the council and its subcommittees and workgroups will accomplish their work, clarify consensus issues, examine how the CSI's goal should be interpreted and applied to individual projects and to what extent representatives of all stakeholder groups should be included in each level of projects and workgroups.[32] In response to the GAO report, and in defense of the CSI Program, J. Charles Fox, Associate Administrator, EPA Office of Reinvention states, in part, "...many times we learn more from the unexpected outcomes of activities than from those we had expected. While concrete environmental improvements are the ultimate goal of CSI, it takes time to identify, test and evaluate new, innovative approaches to achieving these environmental improvements."[33]

Possibly in response to the GAO reports, EPA issued the agency's annual report on reinvention that addresses many of the CSI issues highlighted in the GAO reports. Specifically, regarding the most recent activities to date in the CSI Program, EPA sees the CSI process among its "List A" projects for reinvention, that will be evaluating regulatory negotiation and consensus-based rulemaking. As described in EPA's report on reinvention, EPA views the CSI process as a vehicle for identifying regulations that might be developed through negotiations and consensus.[34]

One such example of regulations that might be developed through negotiations and consensus is a landmark agreement that EPA entered into with the metal finishing sector of the CSI. This agreement is the first of its kind—the first sector-wide CSI agreement reached by participating stakeholders including industry; environmental organizations; federal, state, and local government organizations; labor and community groups. The agreement contains industry-wide goals for: (1) full compliance, (2) improved economic payback, (3) brownfields prevention, and (4) enforcement of chronic noncompilers. The agreement also includes a comprehensive action plan for all stakeholder groups, and provides incentives, creates tools, and removes barriers for metal finishers to achieve their voluntary performance goals and objectives.[35]

In closing, since the CSI Program is another of EPA's works-in-progress, the information we provide herein will become dated by the time this book goes to press. As such, readers are strongly urged to consult EPA's CSI web site for any updates on this program, along with any other relevant information that may be of interest to readers.

4. PROJECT XL[36]

Project XL was one of the new initiatives designed to give regulated organizations and communities the flexibility to develop alternative strategies that will replace or modify specific regulatory requirements conditional to producing greater environmental benefits. The *Federal Register* notice that solicited project proposals by EPA were aimed at various facilities, including federal facilities, that would demonstrate *eXcellence* and *Leadership* (hence the term "XL") in environmental management while striving to reduce costs and achieve environmental performance beyond that required in existing regulations.[37]

Project XL[38] is the most recent of a number of EPA's voluntary programs that function as a national pilot program designed to test various innovative ways of achieving better and more cost-effective public health and environmental protection. Through the efforts of site-specific agreements with various site-specific sponsors, EPA is compiling data and project experiences that will become beneficial in helping EPA redesign current approaches to sustaining public health and environmental protection. With Project XL, the various project sponsors, whether private facilities, industry sectors, federal facilities, or communities, can implement innovative strategies that can produce superior environmental performance, replace specific regulatory requirements, and promote greater accountability to stakeholders.

Furthermore, XL projects are real-world tests of innovative strategies that aim to achieve cleaner and cheaper results than conventional regulatory approaches would achieve. One carrot that EPA provides to project participants is that they will receive regulatory flexibility from EPA in exchange for commitments to achieve better environmental results than would have been attained through full compliance with the regulations. EPA's goal for the program is to implement a total of fifty XL projects in four categories and it is important that each project evaluate new ideas that have the potential for wide application and broad environmental benefits. The four categories will consist of: (1) facilities, (2) industry sectors, (3) government agencies, and (4) communities.

As with the ELP and the CSI program, EPA has earmarked Project XL as a high priority and lists this among the agency's "List A" projects identified in its 1998 annual report on reinvention in two specific areas: "Project XL" and "alternative strategies for agencies."[39]

As of March 1998, a total of seven pilot projects were underway under Project XL and twenty more are being developed. The project sponsors include the EPA, states, co-regulators, and other interested stakeholders that participate in day-to-day activities and negotiations that culminate in Final Project Agreements (FPAs) outlining the details of the projects and each party's commitment.

As a brief overview, a random sampling of three of the pilot projects consist of:

- **Weyerhaeuser Flint River Operations (GA)**—The company's pulp manufacturing facility seeks to minimize the environmental impact of its manufacturing processes by pursuing a long-term minimum-impact mill. **Innovative approach**—The facility will test operations under a minimum-impact goal, and whether new approaches to meet ambitious environmental goals can be created. **Benefits**—*First year results show plant effluent decreased by as much as 32 percent, air emissions by 13 percent, and projections show that bleach plant effluent will be cut by 50 percent over a ten-year period, water usage will be reduced by 1 million gal/d, and hazardous waste constituents will also be reduced. In dollars, the company saved $176,000 in just one year in operating costs and projecting; it will save another $10 million in future capital spending.*

- **Merck Stonewall Plant (VA)**—The company's pharmaceutical manufacturing facility seeks to avoid production delays in drug manufacturing and reduce emission levels for SO_2 and NO_x to protect visibility and reduce acid deposition in nearby Shenandoah National Park and the local community. **Innovative approach**—Merck's project will focus on whether: (1) a cap on criteria air pollutants for the site provides better air quality while providing greater operational flexibility than the current permitting system; (2) a cap can create better incentives to minimize emissions; and (3) a system that requires increased monitoring, recordkeeping, and reporting as emissions approach the cap ensures compliance and generates additional emission reduction incentives. **Benefits**—*The facility projects to achieve reductions in certain air pollutant emissions by 20 percent (~300 ton/yr), decreases in SO_2 and NO_x emissions by 60 percent (900 ton/yr), and reductions in hazardous air pollutants (HAPs) by 65 percent (47 ton/yr). Merck also benefits in as long as facility emissions remain below the caps: The company will not need to undergo time-consuming and costly EPA and state permit reviews for changes that increase emissions. Merck also has the option to install new control technology or decrease the facility's emissions through emission reductions that would have been achieved with the new technology by some other means.*

- **OSi Specialties, Inc (WV)**—The company, a subsidiary of Witco Corporation is a specialty chemical manufacturer. The company's plant agreed to: (1) install air pollution controls on a production unit well ahead of anticipated EPA requirements; (2) reuse/recycle methanol; and (3) study the feasibility of reducing its waste streams. OSi 's project will extend until 2002 unless additional environmental benefits warrant a continuation of the regulatory flexibility. **Innovative approach**—The project will determine whether: (1) providing flexibility to control pollution more cost-effectively will produce tangible benefits for both the environment and industry; (2) deferring regulations for encouraging P2 activities is environmentally beneficial; and (3) quantifiable results can be obtained through a P2 study: This last time could also be used in comparison in a future benchmark study of other chemical companies to compare tangible "apples-to-apples" results. **Benefits**—*The project will result in the destruction of 98 percent of the organic compounds in the vent stream (~310,000 lb/yr) and the company will also recover and reuse an estimated 500,000 lb/yr of methanol that would otherwise be treated in its wastewater system. This, in turn, will result in a reduction of about 815,000 lb/yr of sludge from the facility's wastewater treatment system. In return, EPA and the West Virginia Department of Environmental Protection have deferred new organic air emission regulations (RCRA Subpart CC) applicable to OSi's two hazardous waste surface impoundments.*

Additional and updated information on these and the remaining XL projects can be found at www.epa.gov/ProjectXL/ or via Project XL's fax-on-demand line at 1-202-260-8590.

TRACKING EPA'S REINVENTION EFFORTS: ARE WE THERE YET?

While EPA has been moving forward in its efforts to reinvent itself as a forward-thinking agency charged with keeping the public and our environment safe from environmental harm and managing for environmental results, the GAO and the U.S. Chamber of Commerce (COC), have also been keeping tabs on how well EPA has fared in its efforts. To provide additional insight into this area, we reference several GAO reports, the COC report, and related studies, including EPA's own tracking efforts in these matters.

Among the drivers for GAO to evaluate how EPA has been managing itself over the past decade are a number of both internal and external studies recommending that EPA manage for environmental results as one way to improve and better account for its performance. One of these studies was conducted by the National Academy of Public Administration (NAPA) in 1995 in which NAPA recommended that EPA take a series of steps to manage for results, including actions to improve and integrate its processes for planning, budgeting, and ensuring accountability.[40] In response to NAPA's recommendations, EPA announced in March of 1996 its plans to create a new office that would develop and implement an integrated planning, budgeting, and accountability system for the agency.

1. GAO REPORTS

As a follow-on to EPA's efforts to date regarding NAPA's recommendations, the GAO was asked by Congress to review the status of EPA's efforts and issued its report in June 1997.[41] The highlights of that particular report include steps EPA needs to take in designing management systems as a means to fully implement an integrated planning, budgeting, and accountability system that would be useful in establishing direction, setting priorities, and (very important) measuring performance. GAO further recommended that the EPA administrator, after consulting with key stakeholders, establish benchmarks for how EPA's new management system is to operate, and furthermore, utilize these benchmarks to monitor its progress in implementing the system.

In response to the findings and recommendations of the GAO report, EPA officials agreed with GAO's recommendations and developed an outline for its new planning, budgeting, and accountability system, which includes the following: (1) establish strategic or long-term goals and objectives; (2) develop strategies to achieve these goals and objectives; (3) identify the desired outcomes for these strategies; (4) translate these goals, objectives, strategies, and desired outcomes into annual goals, planned activities, and performance measures; and (5) use these performance measures to assess EPA's and individual programs' performance in achieving the desired outcomes.[42] Two studies currently underway that may effect EPA's goals and objectives are EPA's National Environmental Goals Project, which will establish a set of long-range environmental goals with realistic milestones for the year 2005; and a study conducted by EPA's Science Advisory Board, the Integrated Risk Project, culminating in a draft report that ranks the relative risks of environmental problems and helps develop methodologies that EPA can use to rank risks in the future.[43] As of this writing, the final report was not yet available.

In a similar focus, a related GAO report[44] highlights challenges facing EPA's regulatory reinvention initiative efforts and the issues EPA needs to address to achieve the intended results of these initiatives. Based on the findings of this report, EPA has in the hopper at least thirty-five "high priority actions " and "other significant actions" listed, which is also listed in EPA's *Annual Report on Reinvention*, referenced previously. Among the more prominent of these projects are Project XL and the Common Sense Initiative that were also referenced.

The issues identified by the GAO include the need for EPA to: (1) develop a greater focus on key initiatives as one means to improve prospects of success; (2) realize that agreement among all stakeholders is difficult, if not impossible, to achieve; (3) realize that problem resolution requires a sustainable process to be successful—citing the 3M Company's failed XL project submittal due to perceptions about EPA headquarters and regional personnel; and (4) systematically evaluate the effectiveness of initiatives in progress.

Several EPA regional offices that were contacted as part of the GAO study identified several issues that would need to be addressed if EPA's reinvention efforts are to succeed.

These include:

- A large number of EPA initiatives may be diverting attention from high-priority efforts;
- Some key EPA initiatives have required a greater resource commitment than originally anticipated, specifically Project XL and CSI;
- Unclear objectives and guidance pose barriers for two key reinvention programs—Project XL and CSI;
- Stakeholders have questioned the extent of EPA's commitment to reinvention;
- Achieving full consensus has been challenging;
- Some environmental and local interest groups lack resources to participate fully in project negotiations;
- EPA is not systematically evaluating reinvention initiatives' effectiveness; and
- Environmental statutory framework limits potential to reinvent environmental regulation.

Among the conclusions that GAO found in developing this report for the U.S. Congress were that: (1) the management of a large number of complex initiatives has caused difficulties for EPA staff and stakeholders; (2) achieving and maintaining consensus among all stakeholders has proved to be an enormous challenge; (3) the development of a long-term, institutional process for quickly resolving reinvention problems could help EPA avoid a number of issues, such as miscommunication and disagreements, that undermined some of its previous reinvention projects; (4) EPA should develop an evaluation component for its initiatives to allow EPA to make informed decisions about which initiatives to continue and what improvements, if any, can be made to them; and finally, (5) considerable discussion has centered on the need for statutory changes as a precondition for reinventing environmental regulation. In addition, the GAO recognizes that without legislative changes, EPA will be limited in how far it can go to truly reinvent environmental regulation.[45]

2. U.S. CHAMBER OF COMMERCE SURVEY ON U.S. BUSINESSES

From another perspective, a U.S. Chamber of Commerce (COC) report[46] takes a closer look at various federal regulations, including environmental and natural resource regulations, and how companies are impacted by them. This report was compiled by Voter/Consumer Research, a public opinion research firm, that summarized the results of a questionnaire survey sent to four thousand chamber member companies. More than eight hundred responses were received, representing a 20 percent response rate.

Among the findings depicted in the report, a good number were not surprising and seem to reflect some of the issues raised by the studies we referred to earlier in the chapter and in several chapters later in the book.

Of note, the executive summary of the report found[47] that: (1) federal regulations generate huge compliance costs—the second- and third-highest costs are found in labor (38 percent) and environment (33 percent); (2) tracking new regulations is troublesome; (3) businesses want to participate more in the regulatory process; (4) pension and OSHA regulations are the most burdensome; and (5) the Clean Air and Clean Water Acts are the most burdensome of environmental and natural resource regulations.

Among the detailed findings of the report, several point to a recurring theme—allowing industries to participate. For example, a large majority of companies queried would like to provide input to the regulators during the regulatory process, such as: (1) participating in the regulatory drafting process to make it more user-friendly; (2) participate in some form of government-business dialogue; and (3) feeling that such input would be much more beneficial to the regulatory agencies than the status quo comment period after publication in the *Federal Register*. To bolster this widespread opinion among member companies, many businesses believe it would be beneficial to have input throughout the development of a regulation affecting their industry.

Two other factors that seem to recur within the companies that responded to the questionnaire include: Industry prefers compliance over enforcement, and voluntary compliance with environmental regulations is a particularly high priority. Statistics from the survey point to an impressive 86 percent of respondents who agreed that the current environmental enforcement system should be reformed to provide incentives for voluntary compliance, such as a voluntary compliance program (*EPA's audit disclosure policy is one such measure being tried*), industry standards (*ISO 14001 can fit into this area*), or other incentive mechanism. Another statistic shows that 65 percent of large manufacturers conduct environmental audits. Furthermore, the COC feels that if voluntary environmental audits become a weapon to be used against the company, environmental quality will decline and the cost of enforcement will increase.[48] From a different view, the COC also feels that an emphasis on voluntary compliance by regulated industries will lead to genuine improvement in the environment. The big *if* is that should EPA and allied federal agencies shift from an emphasis on using the "stick" to using the "carrot," environmental quality will benefit.

Finally, with respect to environmental standards, survey respondents showed overwhelming support for using best available science in the area of risk assessment, along with qualified and independent scientists to review any compiled risk assessment findings. The numbers show: 91 percent of respondents felt that agencies like EPA should be required to utilize the best available science in developing new regulations affecting workers' and the population's health and safety and the environment; and 71 percent felt that a review process involving independent scientists versus the standard practice to solicit comments in the *Federal Register* would be the most appropriate mechanism for reviewing the findings of EPA's risk assessment program.

3. EPA's Two-Year Anniversary Report[49]

One of the milestone accomplishments of 1995 was the establishment of an ambitious agenda to reinvent environmental protection as part of a bigger picture—a federal government that can work better and cost less. As part of this goal, EPA made an announcement about a new Office of Reinvention that would help oversee the charge of EPA continuing to pursue important environmental and public health protection improvements in a number of areas.

Within those areas, EPA actively promoted the use of innovative environmental technologies and management approaches, as well as provided the flexibility needed to put these tools to the test. Among some of the notable highlights of this report include:

A. Promoting Innovation and Flexibility
- Project XL—highlights EPA working with companies to develop alternative strategies for improving environmental performance beyond that which could be attained under traditional regulatory methods.
- The CSI program—highlights EPA working with multiple environmental stakeholders to create new industry-by-industry approaches to environmental regulations versus the standard pollutant-by-pollutant approach as was the previous norm.
- EPA partnerships—highlights partnering with industry, trade groups, and other outside parties, in which EPA established four new environmental technology verification programs to provide objective, reliable information regarding cost and performance of new technologies.
- By year-end 1996, EPA completed actions to improve environmental quality and economic opportunity around "brownfields."[50]

B. Increasing Community Participation and Partnerships

- EPA redesigned its Internet site to make it more user-friendly to specific audiences, such as researchers and scientists, authors, business and industry, and concerned citizens, among others.
- As of the date of the report, over five hundred companies pledged to participate in various voluntary environmental improvement programs, bringing the total number of companies to over seven thousand.

C. Improving Compliance

- EPA initiated the "Sector Facility Indexing Project" to allow interested parties to obtain site-specific environmental performance data for facilities within a specific industry.

As a summary, the goal of EPA's reinvention agenda, as was first proposed in 1995, is to provide improvements on two different levels: externally and internally. Externally, EPA's reinvention efforts focus on a series of high-priority projects, designed to promote innovation, increase community participation, improve regulatory compliance, and cut red tape. As this process evolves, it has become apparent to EPA that a change is taking place: Besides the fact that companies are beginning to see positive results in participating in this process, reinvention is creating a new mindset among EPA staffers at various levels. To foster this continued culture change, EPA created the Office of Reinvention to help provide consistent focus on reinvention throughout the agency, as well as help ensure steady progress in meeting EPA reinvention commitments. With that, EPA hopes these steps position the agency to build on the reinvention progress of the past several years since the reinvention of environmental protection was first proposed and pursue an even more aggressive agenda for the future.

ADDITIONAL EPA CONSIDERATIONS

Let us take this time to pause a moment and briefly summarize where this flurry of activity at EPA is leading with respect to the regulated community and the protection of human health and the environment. It is probably a good guess to say that EPA, as a whole, has come a long way from the old "command-and-control" regime and moved ever more increasingly toward more creative and, hopefully, less onerous environmental regulations as time goes on. As some examples that were previously described, we can point to: well-received alternatives to traditional regulation, such as Project XL and the CSI; EPA's audit disclosure policy; EPA's most recent *Federal Register* notice[51] dealing with ISO 14001 pilots; positions evolving state and federal brownfields initiatives; EPA's Hazardous Waste Identification Rule (HWIR); and the newly emerging regulation dealing with suspected endocrine disruptive chemicals. Is this a portend of a beneficial trend by EPA to shift toward "smart" regulations?

As Michael Hill points out from the examples he provided us for use in this chapter,[52] the majority of environmental requirements have become increasingly more flexible in the past several years. He attributes some of this flexibility to improved communications between the "warring" parties—the regulators and the regulatees. The closest analogy would be East-West relations warming up between Western and Eastern Europe within the past ten years following the tearing down of the Berlin Wall. Although, in actuality, some of the increased communications may have been a result of EPA's losing ground in enforcement powers in recent court battles, hence, prompting EPA officials to establish more meaningful dialogue to reach a middle ground. Hill provides several court cases to support his conclusion, including *United States v. Olin Corp.*, 927 F. Supp. 1502, S.D. Ala. 1966, and *Kenecott Utah Copper Corp. v. U.S.* No. 93-1700, D.C. Cir. July 16, 1996.

Additional considerations where he feels that EPA is moving in the right direction include the fact that EPA is looking more and more at using regulations that rely on results, as opposed to procedure, in regulating wastes; coming to realize that pursuing litigation against parties who contributed *de minimis* amounts of waste at a Superfund site does not make any practical sense, and as such, EPA Administrator Carol Browner announced in June 1996 that EPA would double the volumetric cut-off for such Superfund settlements and would take an active role in promoting them; realizing that it pays to keep small and medium-sized businesses informed of their environmental obligations before deadlines pass, to allow them to take the necessary steps beforehand to meet these imposed deadlines. One example involves EPA's notification to underground storage tank (UST) owners to remind them of several interim steps they must take beforehand to meet EPA's December 22, 1998 upgrade requirements. Finally, in borrowing from the U.S. Army Corps of Engineers (COE), EPA has begun to evaluate issuing "general permits" for RCRA in similar fashion to the COE's general permits for any dredging and filling under Section 404 of Clean Water Act. These are referred to as Section 404 permits for work in "waters of the U.S."

Hill also notes that while communication has improved, much still needs to be done to evolve further, such as an expanded forum similar to that of the Common Sense Initiative, and greater use of advisory committees. The GAO noted similar findings in their reports noted previously. Hill feels, and justifiably so, that such communication improvements are of particular importance in "new" areas of regulation, where science, technology, and effects are either unknown or relatively unknown, as in the area of suspected endocrine disruptive chemicals.

INDUSTRY PARTICIPATION IN DEVELOPING SCIENCE AND REGULATION: THE CASE OF ENDOCRINE DISRUPTERS

One example of the challenges facing the new areas of EPA regulations can be found in those areas where the science, technology, and effects of the regulations are not yet completely known, and thus, the far-reaching effects of such regulations cannot yet be determined or measured. One area in particular focuses on emerging regulations pertaining to suspected endocrine disruptive chemicals. Regulatory proponents identify endocrine-disrupting chemicals (EDCs) as probable causes of, among other things:

- Reduction in sperm count and other reproductive problems;
- Learning disabilities; and
- Behavioral abnormalities.

EDC developments could affect risk assessments, and as a result, affect other environmental statutes, such as: CWA and CAA permits; CERCLA and RCRA cleanup action levels; and OSHA. In addition, these could also affect testing equipment and cost, cleanup techniques and emission controls, labeling requirements, and toxic tort litigation by workers, neighbors and consumers.

This area of regulation is being pushed by August 1996 amendments to two environmental statutes: the Safe Drinking Water Act (SDWA) and the Federal Food, Drug and Cosmetic Act (FFDCA). Among the requirements of the statutory amendments of these two regulations are:

- *August 1998*—EPA must develop a screening and testing program to determine whether certain substances may have an adverse effect on the endocrine systems of humans or wildlife.
- *August 1999*—Industry must implement the program. Entities that import, manufacture, or register suspected EDCs must conduct the tests and report their results to EPA. Related proposals are in Congress and in the legislation of some states, such as Illinois and Massachusetts.

In addition, EPA must also take any other actions that may be necessary to ensure the protection of public health. This could involve:

- Reporting, including amended confidential business information provisions;
- Labeling;
- Restrictions in manufacturing/use; or
- Lower emission or cleanup action levels.

A chronology of federal regulatory developments depict the following:

- *August 1996*—The passage of statutory amendments;
- *Fall of 1996*—Two EPA-sponsored papers discuss research needs;
- *November 1996*—The White House's Office of Science and Technology's Committee on Environment and Natural Resources (CENR) generates a federal framework for research on endocrine disruption;
- *December 1996*—EPA convenes the first meeting of Endocrine Disruptor Screening and Testing Advisory Committee (EDSTAC);
- *March 1997*—EPA White Paper is remarkably candid in acknowledging the uncertainties;
- *March 1998*—EPA reports anticipated screening and testing program to Science Advisory Board;
- *August 1998*—EPA required to develop screening and testing program;
- *August 1999*—EPA required to implement screening and testing program.

What is the universe of chemicals that would fall under this screening and testing program? According to Hill, EPA has narrowed the list of potential targets for the testing program down to *seventeen thousand chemicals* out of *over seventy thousand* commercially available chemicals in the United States today. Interestingly, many are commonly used.

As a "heads-up," some possible steps Hill suggests that companies can take in anticipation of this upcoming EPA program could include:

- Internal assessment
  - What suspected EDCs does the company manufacture? Use?
  - What literature exists for those substances and their connection to human or animal hormone systems?
- Possible responses
  - A Commission peer-reviewed paper, based on good science, before August 1988;
  - Conducting tests on substances companies use/produce;
  - Risk management, including minimization of exposure to workers, neighbors and consumers;
  - Track laws and initiatives at the federal, state and international level;
  - Spearhead regulatory advocacy, whether individually or as a group, with Congress and with EPA.

What could the immediate impact be to environmental regulation, response from industry,[53] or other affected entity? EDCs could spawn the development of lab equipment and techniques that may have low detection limits or be able to determine additive or synergistic effects. There could also be the promulgation of more stringent rules and standards relating to: cleanup levels of groundwater, soil and sediment; treatment of drinking water; emission levels to air and water; and worker safety. Risk management could also be impacted, as well as new products and processes that may already be in the pipeline, or in research and development.

Since we are at the forefront of this "new" area of regulation that may be incorporating science and technology in the development process and taking into account the short- and long-term effects of such regulations on the populace, no definitive conclusions can be drawn. But if EPA's March 1997 White Paper is any indication, EPA may be a very receptive audience to work with to draw some definitive conclusions regarding endocrine disrupting chemicals and their effect on humans and/or wildlife.

A CLOSING THOUGHT: ARE WE THERE YET?

Since EPA was first created in 1970, the intent of our lawmakers in Congress and federal and state regulators was to ensure that regulated companies do their best to maintain and sustain environmental regulatory compliance, while undoing years of environmental neglect that spawned a self-igniting Cuyohoga River in Cleveland, Ohio, Love Canal in Buffalo, New York, and other environmental scars at the thousands of Superfund sites across the country.

We have come a long way from those early EPA years, seeing EPA and its state counterparts becoming ever more formidable in their efforts to combat environmental scofflaws and criminals. It would seem the handwriting is finally on the wall as seen by an ever-growing number of large and small companies alike that include environmental issues among their top priorities. EPA's voluntary initiatives have generated their own "cult followers," while EPA has also learned to be less combative to keep the environment safe, and somewhere the pendulum will swing until it reaches a happy medium, unless something happens that causes the Congress and EPA to revert to the old ways of "command-and-control". That would not help to propel us forward, but only time will tell whether we are truly there yet, are on our way there, or have fallen down, and we can't get up yet.

ENDNOTES

[1] For additional insight, see G. Crognale, "Corporate Environmental Liability: Criminal Sanctions, Government Civil Penalties, and Civil Suits," pp. 55–80; and "Allocating Corporate Resources for Environmental Compliance," pp. 169–195, *The Greening of American Business*, Rockville, MD: Government Institutes, 1992.

[2] For additional insight on this policy and the ensuing ramifications, see H. Bradford, "EPA's RCRA Program Gets Tough," *Engineering News Record*, July 22, 1991, p. 8.

[3] For a historical perspective, the reader may want to refer to highlights of enforcement settlements depicted in *Hazardous Materials Intelligence Reporter* during this period, such as: the issue of August 2, 1991, Vol. XII, No. 31 (GE); May 17, 1991, Vol. XII, No. 20 (United Technologies); and September 11, 1992, Vol. XIII, No. 37 (Dexter Corp.), among others.

[4] Excerpted from the EPA document, *The Nation's Hazardous Waste Management Program at a Crossroads: The RCRA Implementation Study*, EPA/530-SW-90-069, Chapter 6, p. 57.

[5] EPA established the Office of Enforcement and Compliance Assurance (OECA) on June 8, 1994.

[6] For additional information about the make-up of OECA, the reader is directed to the EPA booklet, *The Office of Compliance; An Introductory Guide*, EPA 300-F-95-002, January 1995.

[7] Published in the *Federal Register* on December 22, 1995. (Discussed in greater detail in Chapter 9.)

[8] EPA Memorandum from Earl Devaney, Director, Office of Criminal Enforcement, dated January 12, 1994.

[9] For example, penalty provisions are provided under various sections of the statutes governing the various regulatory programs, such as Section 3008(a) of RCRA, Section 106 of CERCLA, and Section 16 of TSCA, among others.

[10] A good source of such cases can be found in both environmental and health and safety trade publications; monthly newsletters, such as BBP's *Safety Compliance Letter*; and national papers, such as *The Wall Street Journal*.

[11] News item in *Hazardous Materials Intelligence Reporter*, September 18, 1992, Vol. XIII, No. 38 (Recticel Foam).

[12] One of the turning points in EPA's enforcement came in 1991 when EPA announced that it had beefed-up its enforcement strategy and had planned to hit RCRA violators even harder than before. Additional historical information can be found in an article by H. Bradford, "EPA's RCRA Program Gets Tough," *Engineering News Record*, July 22, 1991, p. 8 and BNA's *Environmental Reporter*, July 12, 1991, p. 602.

[13] Refer to the BNA's *Environmental Reporter*, July 19, 1991, p. 648.

[14] See EPA memorandum from Herbert H. Tate, Assistant Administrator, "Revised Four-Year Strategic Plan FY 1994—FY 1997," dated July 27, 1992.

[15]For additional information about the sentencing guidelines, the reader is referred to the following documents: M. Gelacak and H. Epstein, "Sentencing Guidelines for Organizations: A Commissioner's View," *Business Crimes Bulletin*, Vol. 1, No. 4/May 1994; J. Gibson, "Avoiding Criminal Investigations," *The National Environmental Journal*, May/June 1994; and R. Schmitt, "Plan for Tough Pollution Penalties Sparks Opposition from Business," *The Wall Street Journal*, March 14, 1994, p. B4.

[16]See the EPA document, *The New Generation of Environmental Protection: EPA's Five-Year Strategic Plan*, EPA 200-B-94-002, July 1994. Its complementary summary document can be downloaded from EPA's web site @ www.epa.gov/docs/strategic_plan/summary.txt.html.

[17]The summary is titled *The New Generation of Environmental Protection: A Summary of EPA's Five-Year Strategic Plan*, EPA 200-2-94-001, July 1994.

[18]See EPA's FY 1997 Enforcement and Compliance Assurance Priorities, Steven Herman, Assistant Administrator for Enforcement and Compliance Assurance. This document was downloaded from http://es.epa.gov/oeca/naag97.html.

[19]By the time the 33/50 Program had concluded in December 1995, over 1,300 companies had pledged their support to participate and had eliminated over 700 million pounds of pollution generated during the life of the program, and had exceeded EPA's goal of 50 percent. Reference: EPA Report, 33/50 Program Achievements, December 1995.

[20]See the *Federal Register*, 58 FR 4802, dated January 15, 1993.

[21]See the *Federal Register*, 59 FR 32062, dated June 21, 1994.

[22]Ibid.

[23]Supplemental information about the ELP was contained in *Environmental News*, an EPA press release dated April 7, 1995.

[24]This latest material came from the EPA document, *The Changing Nature of Environmental and Public Health Protection: An Annual Report on Reinvention*, EPA 100-R-98-003, March 1998, p. 45.

[25]EPA ELP Update # 12, last updated on March 18, 1998 at www.epa.gov/elp/.

[26]Announced in an EPA press release about the CSI program on July 20, 1994.

[27]Source: *The Common Sense Initiative: A New Generation of Environmental Protection*, an undated EPA fact sheet about the CSI Program.

[28]Memo from Bob Benson, EPA CSI/Metal Finishing Group, dated March 20, 1995.

[29]Preliminary Agenda for Petroleum Refining Sector Meeting on July 25, 1995.

[30]CSI Computers and Electronics Sector Subcommittee Meeting summaries for September 25, 1995 and November 13, 1995.

[31]GAO report, *Regulatory Reinvention: EPA's Common Sense Initiative Needs and Improved Operating Framework and Progress Measure*, GAO/RCED-97-164, July 1997, p. 27.

[32]Ibid., p. 28.

[33]Ibid., p. 48.

[34]Refer to the EPA document noted in n. 24, p. 42.

[35]Source: EPA's Common Sense Initiative Update at www.epa.gov/commonsense/jan98.htm.

[36]This project was initially highlighted in the President's Reinventing Environmental Regulation initiative on March 16, 1995. EPA later solicited Project XL proposals in a *Federal Register* notice dated May 23, 1995.

[37]Reference: *The XL Pipeline*, First Edition, EPA's Project XL, undated, p. 1.

[38]The following information about Project XL was excerpted from information obtained from the Project XL web page as described in *Project XL: Summary of Current Pilot Projects.*

[39]Refer to the EPA document in n. 24, pp. 44–45.

[40]The results of this study are found in the NAPA document, *Setting Priorities, Getting Results: A New Direction for EPA*, April 1995.

[41]This document is the GAO report, *Managing for Results: EPA's Efforts to Implement Needed Management Systems and Processes*, GAO/RCED-97-156, June 1997.

[42]Ibid., p. 18.

[43]Ibid., p. 21.

[44]See the GAO report, *Environmental Protection: Challenges Facing EPA's Efforts to Reinvent Environmental Regulation*, GAO/RCED-97-155, July 1997.

[45]Ibid., for further information, refer to pages 32–56 of the GAO report.

[46]See the U.S. Chamber of Commerce report, *Federal Regulation and Its Effect on Business*, June 25, 1996.

[47]Ibid., pp. 6–8.

[48]Ibid., pp. 25–28.

[49]See the EPA report, *Managing for Better Environmental Results*, EPA 100-R-97-004, March 1997.

[50]Brownfields refer to abandoned or underused industrial or commercial sites where real or suspected contamination discourages redevelopment.

[51]EPA's position statement regarding ISO 14001 and environmental management systems can be found in the *Federal Register* at 63 FR 12094, dated March 12, 1998.

[52]Excerpted with permission from the following sources: M. Hill, "A Trend toward 'Smart' Regulations," *Environmental Solutions*, January 1997, pp. 14–16; and a presentation by Michael Hill at the 1997 HazMat Conference titled, *Regulatory Update: A Federal Perspective*. Hill works for the Washington D.C. law firm, Collier, Shannon, Rill, and Scott.

[53]The Chemical Manufacturers Association has lauded the efforts of EPA's EDSTAC, according to an article in *CMA News*, October 1998.

REFERENCES

EPA Documents

The Changing Nature of Environmental and Public Health Protection: An Annual Report on Reinvention, EPA 100-R-98-003, March 1998, www.epa.gov/reinvent

Managing for Better Environmental Results, A Two-Year Anniversary Report on Reinventing Environmental Protection, EPA 100-R-97-004, March 1997.

The Nation's Hazardous Waste Management Program at a Crossroads: The RCRA Implementation Study, EPA/530-SW-90-069, July 1990.

The New Generation of Environmental Protection: A Summary of EPA's Five-Year Strategic Plan, EPA-200-2-94-001, July 1994.

The New Generation of Environmental Protection: EPA's Five-Year Strategic Plan, EPA-200-B-94-002, July 1994.

The Office of Compliance: An Introductory Guide, EPA 300-F-95-002, January 1995.

GAO Documents

Environmental Protection: Challenges Facing EPA's Efforts to Reinvent Environmental Regulation, GAO/RCED-97-155, July 1997.

Key Management Issues Facing EPA, GAO/RCED-98-153 R, April 1998.

Managing for Results: EPA's Efforts to Implement Needed Management Systems and Processes, GAO/RCED-97-156, June 1997.

Regulatory Reinvention: EPA's Common Sense Initiative Needs and Improved Operating Framework and Progress Measure, GAO/RCED-97-164, July 1997.

Other References

EPA progress reports dealing with the ELP, the CSI and Project XL and regular Internet updates noted in the chapter and Appendix G.

Hill, M. "Regulatory Update: A Federal Perspective" presented at the 1997 Hazmat Conference, Atlantic City, NJ.

Hill, M. "A trend toward 'Smart Regulations," *Environmental Solutions,* January 1997, pp. 14–16.

The US Chamber of Commerce and Voter/Consumer Research's report, *Federal Regulations and Its Effect on Business,* June 25th 1996.

Neale, K. "Consensus Report and Recommendations for Endocrine Screening and Testing." *CMA News,* October 1998, pp. 9–11.

CHAPTER 3

OSHA SAFETY AND HEALTH MANAGEMENT RESPONSIBILITIES

Robert Rusczek, CIH, President, EnviroComp, Inc.,
with introduction by Gabriele Crognale, P.E.

> "Environmental manage-
> ment is not the only as-
> pect ... quality and occu-
> pational health and safety
> are also core management
> concerns which must be
> addressed and communi-
> cated to interested par-
> ties."
>
> *Dick Hortensius*

INTRODUCTION

Oftentimes, when we environmental professionals talk about EH&S, we usually are referring only to the "E" for environmental, as opposed to environmental, health and safety. Yet, as many of us know, we cannot separate environmental from (occupational) health and safety considerations, whether in the workplace, or in our immediate surroundings. As a passing thought, there may yet come a time when health and safety standards will join the ISO 9000 quality management standards and ISO 14000 environmental management standards. While the jury may still be out in this endeavor at the ISO decision-making level, taking into account health and safety considerations along with environmental and quality management considerations from a management system overview perspective makes a lot of sense and can eliminate unnecessary duplication of effort. Peering ahead, we are already beginning to see some signs that environmental and health and safety considerations are coming closer together.

As an example to reinforce this point, let us look at OSHA's Process Safety Management (PSM) rule[1] and EPA's Risk Management Program (RMP) rule[2]. Both rules focus on operational facilities that use certain chemicals where implementing an RMP can help to prevent accidental releases. Similarly, under OSHA's Hazard Communication Standard[3] or Haz Com as it commonly known, OSHA requires an employer to conduct a hazard assessment, provide employees with right-to-know information, and educate them about on-the-job chemical hazards, among other considerations. Under RCRA, employers are required to provide employees responsible for generating and managing hazardous wastes with specific training to enable them to respond to emergencies, as provided under 40 CFR Part 265.16. The point being, employees subject to both require-ments can gain valuable added insight if they can piggy-back their Haz Com with RCRA training and gain two bites of the apple, for example. Employee right-to-know training using Material Safety Data Sheets (MSDS) that can be linked to de-partments that produce or import new chemicals, say, that employees may ulti-mately manage as wastes, can add a whole new dimension to their under-standing of their job responsibilities. In essence, relating workplace and chemical hazards, operating procedures, and worker safety to sepcific RCRA or other envi-ronmental compliance regulations can provide a more fertile ground for line workers to better understand their EH&S responsibilities and how they are actu-ally interlinked in their workplace.[4]

Another example comes to us by way of Dick Hortensius of the Netherlands Standardisation Institute (NNI). Dick feels, and rightly so, that while environ-mental management has proved to be the focal point of the 1990s, occupational health and safety management systems may soon receive similar attention. As an example, he notes that the British Standards Institute, or BSI, published the *British Standard, BS 8800, Guide to Occupational Health and Safety Management Systems* in April 1996. BSI's standard is based on the management systems model of ISO 14001. He also notes that a guideline for occupational health and safety manage-ment was being prepared in the Netherlands at the time his article was going to press.[5]

Therefore, we could not, in all honesty, prepare a book about environmental management strategies unless we included a discussion regarding health and safety considerations and how these considerations can and do play a pivotal role in effective EH&S and EMS systems. Environmental and health and safety issues and responsibilities should not to be viewed as a lock-step process, but rather as fluid components of a process that works best when harmonized, much like a symphony orchestra that is effectively managed by a seasoned maestro.

REGULATORY BACKGROUND

A solid, effective safety and industrial health program is essential for any type of facility, especially one that stores, handles, and/or processes hazardous materi-als. The rewards of keeping employees safe and healthy are many, with the key ones being improved productivity, reduced worker compensation costs, and minimizing potential citations from the Occupational Safety and Health Admin-istration (OSHA).

This chapter provides an overview of OSHA, their procedures, and key OSHA safety and health standards. It is not intended to provide detailed infor-mation on hazard recognition, evaluation, or control. In addition, it addresses key federal OSHA requirements that apply to the majority of general industry. The federal or state OSHA standards for your specific industry need to be consulted for additional information. The main purpose of this chapter is to introduce the reader to *key* OSHA requirements that apply to most general industry establish-ments. It is important to note that all of these standards require employee involvement and ongoing administration, therefore strong safety management is critical to their implementation and ongoing effectiveness.

In addition, various books and assistance are available on occupational health and safety management. The National Safety Council (NSF)[6] for example, is one of several organizations providing safety information. The NSF has local chapters throughout the United States. Additional books to refer to are found in the references at the end of the chapter. A more complete listing of general references is contained in Appendix H.

OSHA Regulatory Requirements

The Occupational Safety and Health Act (OSHAct) of 1970[7] created two federal Agencies: the Occupational Safety and Health Administration (OSHA), and the National Institute for Occupational Safety and Health (NIOSH). OSHA was created to be the regulatory arm to enforce the OSHAct. OSHA's main responsibilities are to develop safety and health standards and to enforce them. NIOSH, which is currently located within the Health and Human Services Administration under the Center for Disease Control, was created to provide research and training as stipulated in the OSHAct.

There are approximately twenty states that administer their own OSHA programs. The OSHAct requires them to have regulations at least as stringent as the federal ones. Federal oversight is performed to evaluate the states' programs ongoing effectiveness.

OSHA Standards Promulgation

OSHA has various mechanisms to promulgate new standards, which can be a detailed, time-intensive undertaking. OSHA can issue new standards on its own, or in response to petitions from other groups, including NIOSH, state and local governments, or any nationally recognized standards-producing organization, or employer or labor representatives.

Advisory Committees

If OSHA determines that a specific standard is needed, OSHA will normally call upon one of several advisory committees to facilitate the process and develop specific recommendations. There are two formal standard setting committees:

- The National Advisory Committee on Occupational Safety and Health (NACOSH) advises and makes recommendations to the Secretary of Health and Human Services (HHS) and to the Secretary of Labor concerning the OSHAct.
- The Advisory Committee on Constructive Safety and Health advises the Secretary of Labor on formulation of construction safety and health standards.

In addition to these standing committees, others may be formed ad hoc. All advisory committees, standing or ad hoc, must have members representing management, labor, and state agencies, as well as one or more designees of the Secretary of HHS. The occupational safety and health professions and the general public also may be represented.

NIOSH Recommendations

Recommendations for standards also may come from NIOSH, which conducts research on various safety and health issues, and provides technical assistance to OSHA. NIOSH also recommends standards for OSHA's adoption. While conducting its research, NIOSH may make workplace investigations, gather testimony from employers and employees and require that employers measure and report employee exposure to potentially hazardous materials. NIOSH also may require employers to provide medical examinations and tests to determine the incidence of occupational illness among employees. These tests, if required by NIOSH, would be paid for or provided by NIOSH.

Standards Adoption

Once OSHA has developed a new or modified standard, it publishes it in the *Federal Register* as an Advance Notice of Proposed Rulemaking (ANPR) or a Notice of Proposed Rulemaking (NPR). An ANPR is often used to solicit information that can be used by OSHA in drafting an NPR. The NPR will include the terms of the new rule and provide a specific time (at least thirty days from the date of publication, usually sixty days or more) for the public to respond.

After the close of the comment period and public hearing, OSHA must publish in the *Federal Register* the full, final text of any standard amended or adopted and the date it becomes effective, along with an explanation of the standard and the reasons for implementing it. The *Federal Register* is published daily by the Government Printing Office (GPO).

Emergency Temporary Standards

Under certain limited conditions, OSHA is authorized to set emergency temporary standards that take effect immediately and are in effect until superseded by a permanent standard. OSHA must determine that workers are in grave danger due to exposure to toxic substances or agents determined to be toxic or physically harmful, or to new hazards, and that an emergency temporary standard is necessary to protect them. This is published in the *Federal Register*, where it also serves as a proposed permanent standard. It is then subject to the usual procedure for adopting a permanent standard except that a final ruling should be made within six months. The validity of an emergency temporary standard may be challenged in an appropriate U.S. Court of Appeals. For various reasons, OSHA rarely issues emergency temporary standards.

Standard Promulgation Summary

The process of promulgating a new standard is resource intensive and time-consuming. For example, in OSHA published an ANPR 1975 concerning "Standard for Work in Confined Spaces." The final rule "Permit-Required Confined Spaces" was published approximately eighteen years later on January 23, 1993.

INJURY AND ILLNESS RECORDS

OSHA requires employers to maintain injury and illness records. For employers whose employees work in dispersed locations, the records must be kept at the place where the employees report for work. There are various low-hazard industries, such as restaurants, that OSHA does not require records be kept. In addition, employers with no more than ten total employees, regardless of the industry, are not required to keep records. However, notification of inclusion in a Bureau of Labor Statistics survey would require recordkeeping and reporting for these otherwise exempted companies for the time period specified. OSHA Injury and Illness Recordkeeping is a separate process from Worker's Compensation. A recordable injury or illness may or may not be compensable, and visa versa.

RECORDKEEPING FORMS

Recordkeeping forms are maintained on a calendar-year basis. They are not sent to OSHA or any other agency. They must be maintained for five years at the establishment and must be available for inspection by representatives of OSHA, HHS, BLS, or the designated state agency. Only two forms are needed for recordkeeping, which are discussed below:

- **OSHA No. 200, Log and Summary of Occupational Injuries and Illnesses—** Each recordable occupational injury and illness must be logged on this form within six working days from the time the employer has knowledge. It is recommended that questionable recordable injuries and illnesses be added to the log, and after additional information is gathered as to their recordability, be either left on or lined out on the log. The OSHA 200 log summary must be posted in the workplace for the entire month of February for the previous calendar year. Even if there were no injuries or illnesses during the year, zero must be entered on the totals line, and the form posted.
- **OSHA No. 101, Supplementary Record of Occupational Injuries and Illnesses—**The OSHA 101 form contains detailed information about each injury and illness. It also must be completed within six working days from the time the employer learns of the work-related injury or illness. A substitute for the OSHA 101, such as workers' compensation forms, may be used if they contain all of the required information.

OTHER RECORDKEEPING REQUIREMENTS

In addition to these general injury and illness recordkeeping requirements, many specific OSHA standards have additional recordkeeping and reporting requirements. Training and inspection records are required under numerous OSHA standards. The applicable standards need to be reviewed to determine the specific requirements.

All employees have the right to examine any records kept by their employers regarding exposure to hazardous materials, or the results of medical surveillance. The employer is obligated to inform employees of this right on initial hire and annually thereafter.

POSTING REQUIREMENTS

OSHA requires that employers post certain materials at a prominent location in the workplace. The normal posters include the Job Safety and Health Protection workplace poster (OSHA 2203 or state equivalent) informing employees of their rights and responsibilities under the OSHAct. Besides displaying the workplace poster, the employer must make copies of the OSHAct and copies of relevant OSHA rules and regulations available to employees upon request. Any official edition of the poster is acceptable.

If the employer falls within the requirements of OSHA's Occupational Noise standard, a copy of the standard (29 CFR 1910.95) needs to be posted. In addition, the following items are also posted:

- Summaries of petitions for variances from standards or recordkeeping procedures.
- Copies of all OSHA citations for violations of standards. These must remain posted at or near the location of alleged violations for three days, or until the violations are corrected, whichever is longer.
- Log and Summary of Occupational Injuries and Illnesses (OSHA No. 200). The summary page of the log must be posted no later than February 1, and must remain in place until March 1.

WORKPLACE INSPECTION

Every establishment covered by the OSHAct is subject to inspection by OSHA Compliance Safety and Health Officers (CSHO). Compliance officers are trained to recognize safety and health hazards and enforce the applicable OSHA standards. Under the OSHAct, an OSHA compliance officer, upon presenting credentials, is authorized to

> Enter without delay and at reasonable times any factory, plant, establishment, construction site or other areas, workplace, or environment where work is performed by an employee of an employer ... Inspect and investigate during regular working hours, and at other reasonable times, and within reasonable limits and in a reasonable manner, any such place of employment and all pertinent conditions, structures, machines, apparatus, devices, equipment and materials therein, and to question privately any such employer, owner, operator, agent or employee.

Inspections are conducted without advance notice. There are a few special circumstances under which OSHA may give notice to the employer, but even then it will be less than twenty-four hours.

If an employer refuses to admit an OSHA compliance officer, or if an employer attempts to interfere with the inspection, the OSHAct permits appropriate legal action.[8]

INSPECTION PRIORITIES

OSHA has developed an inspection priority mechanism to best enforce the OSHAct. This system is designed to target the most hazardous situations. The following information on inspection priorities is presented in the order of the top priority to low priority.

IMMINENT DANGER

Imminent danger situations are the number-one priority. An imminent danger is any condition where reasonable certainty exists that a danger can be expected to cause death or serious physical harm. For a health hazard to be considered an imminent danger, there must be a reasonable expectation that toxic substances such as dangerous dusts or gases are present, and that exposure to them will cause immediate and/or irreversible harm. The OSHA area director reviews the available information and immediately determines whether there is a reasonable basis for the allegation.

If it is decided the case has merit, the area director will assign a compliance officer to conduct an immediate inspection of the workplace. Upon inspection, if an imminent danger situation is found, the compliance officer will ask the employer to voluntarily abate the hazard and to remove endangered employees from exposure. The OSHA inspector will advise all affected employees of the hazard and post an imminent danger notice.

Should the employer fail to abate the hazard, OSHA may apply to the nearest federal district court for appropriate legal action to correct the situation. Such action can produce a temporary restraining order (immediate shutdown) of the operation or section of the workplace where the imminent danger exists.

CATASTROPHES AND FATAL ACCIDENTS

Second priority is given to investigating fatalities and catastrophes resulting in hospitalization of three or more employees. Such situations must be reported to OSHA by the employer within eight hours. Investigations are made to determine if OSHA standards were violated, and to ensure that the hazards involved are eliminated.

EMPLOYEE COMPLAINTS

Third priority is given to employee complaints of alleged violation of standards or of unsafe or unhealthful working conditions. The OSHAct provides employees the right to request an OSHA inspection when the employee feels that there is a violation of an OSHA standard that creates a safety or health hazard. OSHA will maintain confidentiality of the person filing the complaint. OSHA may inspect just the area(s) where the complaint alleges the hazards exist, or may decide, based on various factors, to perform a wall-to-wall inspection. This priority category is the one that is utilized the most by employees. Approximately one-half of all OSHA inspections are originated by employee complaints.[9]

PROGRAMMED INSPECTIONS

Next in priority are programmed inspections aimed at specific industries, or hazardous materials or conditions. Industries are routinely selected for inspection on the basis of factors such as their injury and illness rates, and employee exposure to toxic substances. The applicable Standard Industrial Classification (SIC) code(s) will be determined, and reports of companies with these SIC codes are then generated for targeted inspections. Programmed inspections may be either regional or national in scope. States with their own occupational safety and health programs may use somewhat different systems to identify high-hazard industries for inspection.

FOLLOW-UP INSPECTIONS

A follow-up inspection determines whether previously cited violations have been corrected. If an employer has failed to abate a violation, the employer may face additional proposed daily penalties while such failure or violation continues. Follow-up inspections are usually randomly selected, and may occur at any time after the original citation's abatement dates pass for individual items.

It is important to note that a repeat violation does not necessary have to take place on the specific item originally cited by OSHA. It concerns a repeat violation of the specific regulation at the company's facilities.

INSPECTION PROCESS

Prior to performing the actual inspection, the compliance officer reviews available enforcement information about the company and the particular standards likely to apply. The compliance officer will almost always have a camera, tape measure, and other standard equipment. Appropriate equipment for detecting and measuring hazardous materials and physical agents such as noise, etc., will also be selected.

OPENING CONFERENCE

In the opening conference, the compliance officer (CSHO) should explain why OSHA is inspecting the company. If it is a complaint-driven inspection, the CSHO should provide the company a copy of the employee complaint. An authorized employee representative also is given the opportunity to attend the opening conference and to accompany the compliance officer during the inspection. If employees are represented by a recognized bargaining representative, the union ordinarily will designate the employee representative to accompany the compliance officer. Similarly, if there is a plant safety committee, the employee members of that committee will designate the employee representative (in the absence of a recognized bargaining representative).

If neither a union or formal safety committee exist, the compliance officer must consult with a reasonable number of employees concerning safety and health matters in the workplace; such consultations may be held privately. In addition, since the inspection requires paperwork review, this information may be asked for during the opening conference and reviewed.

WALK-AROUND INSPECTION

After the opening conference, the compliance officer and accompanying employer and employee representatives perform the "walk-around" part of the inspection, proceeding through parts or all of the establishment, inspecting the applicable work areas for compliance with OSHA standards. During this time, the CSHO can observe conditions, consult with employees, and may take photos, videotape the walk-through, take instrument readings, and examine records, among other activities as necessary during the inspection. Of course, any trade secrets observed by the compliance officer must be kept confidential as stated in the OSHAct. It is important for an employer to inform the CSHO of any trade secrets beforehand, so that they may be classified accordingly.

Employees are consulted during the inspection tour. The compliance officer may stop and question workers in private about safety and health conditions and practices in their workplaces. Each employee is protected under Section 11C of the OSHAct from discrimination for exercising their safety and health rights.

During the course of the inspection, the CSHO should inform the employer of any unsafe or unhealthful working conditions observed, which at this phase of the process are called "apparent violations." It is in the best interest of the employer to correct apparent violations as soon as they are identified by the CSHO. This indicates good faith with the compliance process, and the CSHO will document that the items have been corrected. OSHA, however, will still issue citations with penalties for apparent violations even if they are immediately abated.

CLOSING CONFERENCE

After the walk-around inspection, a closing conference is held between the CSHO and the employer. This may be held on the same day as the inspection or later. One reason for a delay is when health hazards are being evaluated and the laboratory results are needed to determine compliance. The CSHO should inform the employer of every apparent violation identified, and discuss items such as abatement methods and time periods.

The CSHO then explains the employer's various rights to the company's representative, including an informal contest and appeal to the Occupational Safety and Health Review Commission. The CSHO at this point does not indicate any proposed penalties. The employee representative may be present with the employer, or a separate closing conference may be held with the representative to discuss matters of direct interest to employees.

SAFETY AND HEALTH PROGRAMS

A company needs to have a comprehensive safety program if it intends to prevent occupational injuries and illnesses, and keep in compliance with the applicable regulations. The program needs to be in writing, communicated to all employees, and given a high level of support by management. The following part of this chapter will summarize the key elements of an effective safety and health program.

RESPIRATORY PROTECTION: NEED FOR RESPIRATORS

Respiratory hazards can be present in many different industries in numerous forms: lack of oxygen, dusts, mists, fumes, gases, and vapors that may lead to cancer, temporary or permanent lung impairment, other diseases, or death, to name a few. The proper type and use of a respirator can prevent the entry of harmful substances into the wearer's body. Some respirators also provide a separate supply of breathable air so work can be performed where there is inadequate oxygen or where greater protection is needed.

ENGINEERING CONTROLS

The prevention, or at least adequate control, of atmospheric contamination at the worksite generally should be accomplished by engineering control measures. Examples of engineering controls include enclosing the contaminant-producing operation, providing local exhaust ventilation to exhaust the contaminant, or substituting a hazardous material with a less hazardous one. The employer should make every feasible effort to engineer out a potential respiratory health hazard. However, when effective engineering controls are not feasible, or while those controls are being designed and installed, or during certain operations, a comprehensive respiratory protection program must be instituted.

The following highlights the key elements of an effective respiratory protection program. These elements are:

- written standard operating procedures
- program evaluation
- selection
- training
- fit testing
- inspection, cleaning, maintenance and storage
- medical examinations
- work area surveillance
- air quality standards
- approved respirators

Each of these are discussed in the next section. It is important to note that the following is a general, brief overview of respiratory protection. Specific

OSHA expanded health standards have more stringent respiratory requirements and need to be closely reviewed.

Written Standard Operating Procedures In workplaces where respirators are used in potentially hazardous atmospheres present during normal operations or emergency situations, employers need to have written respiratory protection standard operating procedures. An employee who may wear a respirator as part of the program must be familiar with these procedures.

The standard operating procedures should include the following elements of the program: concentrating on hazards, proper selection, usage, and limitations. OSHA and NIOSH have numerous documents on respiratory protection, as do various consensus organizations and equipment manufacturers, that provide guidance on developing standard operating procedures.

Program Evaluation A facility's respirator program should be evaluated at least annually to determine its ongoing effectiveness. The written operating procedure should be modified as necessary to incorporate changes. It may be helpful that someone not inherently involved in the program perform this evaluation. For example, a safety and health professional from the company's insurance company or a qualified consultant may be a useful resource.

Selection of Respirators Before the proper respiratory equipment can be selected, the airborne hazard must be fully identified in both its composition and concentration. Chemical and physical properties of the contaminant(s), as well as the toxicity and concentration of the hazardous material, must be determined, along with the potential for an oxygen deficient atmosphere. The nature and extent of the hazard, task demands, physical area to be covered by employees, mobility, and other work requirements are also important selection factors. Then, only NIOSH-approved respirators should be selected and used as designed and intended.

There are two basic classes of respirators: air purifying and air supplying. Air-purifying respirators (APR) use filters and/or sorbent materials to remove harmful substances from the air.

It is critical to know that APRs do not supply oxygen and therefore must never be used in oxygen-deficient atmospheres. In addition, because of their inherent limitations, they also must not be used in an atmosphere that is considered to be immediately dangerous to life or health (IDLH).

Sorbent-type APRs must only be used with materials that have been shown to be effectively adsorbed, and that have good warning properties. An example of a good warning property is a noticeable odor well below the safe exposure limit.

Supplied-air respirators (SARs) are designed to provide quality breathable air from a reliable clean source. These may be in the form of self-contained breathing apparatus (SCBAs) that employees wear on their backs, to airline systems that have an attached airline to an air source. In all cases, the air source must be of breathing-grade quality air. The Compressed Gas Association has developed criteria for breathing-grade quality air [Grade D breathable air described in

Compressed Gas Association (CGA) Commodity Specification G-7.1-1989]. The CGA criteria is referenced in OSHA's respiratory protection standard 29 CFR 1910.134 (the 1966 version is specifically referenced). There are various types of SARs; it is important to realize that different types provide different levels of protection.

The compressor for supplying air must be equipped with the necessary safety devices and alarms. Compressors must be constructed and situated to avoid any entry of contaminated air into the system and must be equipped with suitable in-line, air-purifying sorbent beds and filters installed to ensure air quality. The system also must have ample storage capacity to enable the wearer to escape from a contaminated atmosphere in the event of compressor failure. If an oil-lubricated compressor is used, it must have a high temperature or carbon monoxide alarm or both. If only the high temperature alarm is used, the air from the compressor must be tested frequently for carbon monoxide. Airline couplings must be incompatible with outlets for other gas systems to prevent accidental servicing of airline respirators with nonrespirable gases or oxygen.

Air-purifying respirators present minimal interference with the worker's ability to move about. Atmosphere-supplying respirators, however, may restrict movement and present potential hazards. For example, SARs, with their trailing hoses, can limit the area the wearer can cover and may present a potential hazard if the trailing hose comes into contact with machinery. Similarly, an SCBA, which may weigh twenty-five to thirty-five pounds, creates both a size and weight penalty. This may limit climbing and movement, in addition to the physiological stress that it creates.

Corrective glasses and contacts present a problem when wearing respirators. The temple bars on glasses break the seal of a full-face respirator, thereby preventing an acceptable fit. There are mountings brackets inside full facepieces that allow glass frames to be held in place without temple bars. Contact lenses, on the other hand, may present problems to the wearer if they happen to slip or fall out due to static pressure differences. However, with soft contact lenses this is not as great a concern. OSHA has stated in a directive that it will only be considered a "deminimis" (less than citable) violation to wear contact lenses with full face respirators.

Training A respirator will only protect the worker if it is properly selected, used, and maintained. Both supervisors and their employees must be trained in these and other important respirator elements. The training should also allow the employees opportunities to ask questions, and to try on ("don" the respirator) respirators in a classroom setting, then in a test atmosphere.

Training needs to include information on respirator fit and how to check the facepiece-to-face seal. In order to have the employees fully understand how respirators function, their limitations, and how to wear them, the training should include an explanation of the following:

- the specifics of the respiratory hazards
- the need for the respirator when performing certain tasks
- why a particular type of respirator is needed

- limitations of the selected respirator
- methods of donning the respirator and checking its fit and operation
- when and how to change cartridges and filters
- cleaning, inspection, and storage
- what to do in emergency situations

The institution of a quality, comprehensive respiratory-protection training program will have many rewards—the main one being that employees will be safer when wearing respiratory protection.

Fit Testing In order for tight-fitting respirators to work properly and keep the employees safe, a tight seal is paramount in the entire area where the facepiece contacts the face. There are currently hundreds of respirators on the market, each one having different characteristics that affect the fit to the wearer. There are obviously an infinite variety of faces relative to size and shape that need to be properly fitted. Although respirator manufacturers have and will continue to try to fit everyone with improved facepiece design, no one respirator will fit everyone. Employers will need to purchase different sizes of one brand, or different brands of each type to ensure proper fit for all of the workers who must wear one.

Various facial characteristics can adversely affect the fit of a respirator. Beards and other facial hair that is in the sealing area of the facepiece can substantially reduce the effectiveness of a respirator. Therefore, employers should not let their employees have facial hair in the sealing area of the respirator. Key court decisions have upheld employers' respirator programs concerning prohibiting facial hair when the programs were properly developed and administered fairly.

Various facial modifications, such as the absence or modification of dentures, surgery, scars, or significant weight loss or gain, can also seriously affect the fit of a facepiece.

In order to objectively evaluate the fit of a respirator to a wearer's face, a fit test should be performed following an established protocol. The fit test of the facepiece can be tested two general ways: qualitative and quantitative. Qualitative fit testing involves introducing a fragrant solvent vapor known as isoamyl acetate ("banana oil"), which is a saccharin aerosol or irritating fume, into the breathing zone around the respirator being worn. If no odor, taste, or irritation (depending on the test being performed) is detected by the wearer, a proper fit is indicated. By following an established protocol, such as the qualitative fit test protocol found in Appendix D of OSHA's lead standard 29 CFR 1910.1025, additional reliability can be given to this relatively simple and inexpensive test method.

Quantitative fit testing offers more accurate, detailed information concerning respirator fit. In summary, some type of analytical instrument measures the level of a test agent or airborne dust inside and outside of the facepiece. The larger the differential between these two values, the better the fit. The wearer performs various simple exercises, such as deep breathing, talking, and moving the head side-to-side that could cause facepiece leakage.

In addition to these two types of periodic fit-testing methods, a respirator must be checked for fit and proper function each time it is put on. This is performed by doing a fit check in which the air intake part of the respirator (cartridge or filter openings, etc.) are covered over with the wearers' palms and the wearer breathes in. A vacuum-like seal to the face should occur (negative-pressure check).

Inspection All respirators need to be inspected before and after each use. All parts of the respirator, such as the facepiece, straps, gaskets, hoses, and connections must be in proper working order. Respirators intended for emergency purposes must be inspected at least monthly. Inspection records need to be kept. A convenient time to perform the inspection is during cleaning.

Cleaning A respirator that has been used must be cleaned and disinfected before it is reissued. Respirators should only be cleaned by following the manufacturer's recommendations. Certain cleansers or high temperatures may damage the rubber or other materials of construction. Effective bactericidal agents frequently used are various quaternary ammonium compounds. Before a cleaned and disinfected respirator is placed in storage, it is recommended that it be fully dry.

Storage Respirators must be stored to protect them from contamination, temperature extremes, or damage. Often times a respirator will come in a strong, resealable plastic bag that is intended for ongoing storage.

Medical Examinations Employees designated to wear or who wish to use a respirator must be physically able to do so. Even wearing a simple air-purifying respirator adds resistance to breathing and increases the amount of air the wearer must inhale to overcome the dead air space in the facepiece.

OSHA regulations require that a local physician determine what specific health and physical conditions apply as part of the medical evaluation. Specific expanded health standards usually detail specific medical tests and questionnaires. The respirator user's medical status must then be reviewed on an annual basis.

Work Area Monitoring and Assessment In order to ensure the ongoing protection of the employees and the effectiveness of the respirator program, periodic evaluation of the work area must be performed and documented.

Changes in operating conditions and procedures, ventilation, maintenance levels, and work practices may influence the airborne concentration of a hazardous material. These and other factors require periodic monitoring of the air contaminant concentration. The collected information then needs to be compared to the most recent exposure limits, fit-test results, warning properties of the hazardous materials, and other pertinent information.

Summary A respiratory protection program needs to be properly developed, instituted, and administered on an ongoing basis in order for it to protect employees. The equipment issued to the employee must be properly selected, used, and maintained for a particular work environment and contaminant. Periodic work area surveillance, medical exams, and training need to be performed and documented.

OCCUPATIONAL NOISE EXPOSURE

Noise is considered one of the most pervasive occupational health concerns. Noise, a term used for unwanted sound, is created by many types of operations and equipment. Noise is longitudinal pressure waves in air or other media. The intensity is determined by the delta range of the pressure waves, whereas frequency is determined by the length of the waves. Noise is commonly measured as a sound pressure level on the decibel (dB) scale. The decibel scale is logarithmic, with zero decibel (0 dB) being designated as the "threshold of hearing." Because humans do not perceive all frequencies at the same intensity, the "A Scale" has been developed to approximate the human ear's discrimination. The human ear can perceive sound in the frequency range of 20 to 20,000 hertz (Hz), or cycles per second.

Exposure to high levels of noise causes hearing loss and may cause other harmful health effects as well. The extent of damage depends on the intensity of the noise ("loudness") and the duration of the exposure. Noise-induced hearing loss can be temporary or permanent. Generally, prolonged exposure to high noise levels over a period of time gradually causes permanent damage.

OSHA has a comprehensive occupation noise regulation found in 29 CFR 1910.95 for general industry; other OSHA industries' regulations have adopted the same requirements. The following is a general overview of these requirements.

NOISE MONITORING

The OSHA noise standard requires employers to monitor noise exposure levels to determine if employees are exposed to noise at or above 85 dBA averaged over the entire shift. The most accurate way to measure full-shift noise exposure is with noise dosimeters—battery-operated instruments that continually measure and accumulate noise exposure information. Monitoring should be repeated when changes in production, process, or controls increase noise exposure.

A sound level meter is also a useful measurement instrument that allows the employer to identify specific noise sources and their intensity.

EMPLOYEE TRAINING

As with many OSHA regulations, employees need training if they work in noise areas as defined in this section. They must be trained at least annually in the effects of noise; the purpose, advantages, and disadvantages of various types of hearing protective devices including their proper selection, fit and care; and the purpose and procedures of audiometric testing. The training program may be structured in any format as long as the required topics are covered and properly documented.

HEARING PROTECTION

A variety of hearing protective devices need to be available to all workers exposed to 8-hr time-weighted average (TWA) noise levels of 85 dB or above. In addition, they need to be worn in the following three situations. First, workers need to wear them from their hiring date to when they have their baseline audiogram. Second, any employee that has a standard threshold shift must wear them. Third, when employees are exposed to levels of or greater than 90 dBA over an 8-h TWA, they must be worn.

Hearing protection must provide an adequate amount of noise reduction. The Environmental Protection Agency (EPA) has developed a rating system for hearing protection called the Noise Reduction Rating (NRR) system. All hearing protection is required to provide an NRR on the package it comes in. OSHA then requires a determination be made of the adequacy of hearing protection based on workplace noise levels. Different methods detailed in one of the Occupational Exposure to Noise standard appendices are available for this determination.

General Personal Protective Equipment

An area receiving renewed interest and regulatory attention is the proper selection and use of personal protective equipment, or PPE. This includes body, eye/face, head, and hand and foot protection (respiratory and hearing protection have been addressed in an earlier section of this chapter). Reasons for this interest are the ongoing high incidence rates for worker injuries of these body areas, and the numerous advances made in the materials of construction, specialized uses, and updates in the respective consensus standards.

This is especially true for body and hand protection from hazardous chemicals, as proper selection requires an understanding of chemical compatibility and review of permeation data. Each of these areas are discussed in more detail below.

OSHA PPE REGULATIONS

In April of 1994, OSHA amended its personal protective equipment regulations (29 CFR 1910.132-138). These regulatory changes require employers to perform certified hazard assessments of their workplaces with regards to PPE.

Based on the assessments, PPE designed to meet applicable American National Standards Institute (ANSI) standards need to be selected, and employees must have instruction in the proper usage, wearing, and limitations. The regulations include various appendices to aid in compliance.

HEAD PROTECTION

The workplace needs to be evaluated for the potential of objects falling onto worker's heads or contact with live electrical equipment. If these hazard potentials exist, hardhats meeting the appropriate ANSI Z89.1-1986 classification need to be selected.

EYE/FACE PROTECTION

Eye and face protection continue to be a major emphasis area, as injury to the eyes can result in the loss of one of our important senses. Eye protection must be carefully evaluated, including considerations such as hazard type, source, and trajectory. For example, side shields on approved safety glasses or goggles should be used when the potential exists for projectiles to hit the worker from the side. Goggles may be required over safety glasses for protection from hazardous material contact.

Face shields also provide eye and face protection, but are not classified as primary eye protection. ANSI Z87.1-1989–approved glasses or goggles will normally have to be worn under a face shield.

FOOT PROTECTION

Foot hazards include hazardous material contact, objects dropping onto the feet, rolling/crushing, contact with sharp objects, and electrical conductivity. Proper foot protection is important to protect the worker from these hazards. Safety-toed shoes are a common type of foot protection, but note that various types of them are available. ANSI Z41-1991 should be reviewed for proper foot protection selection.

HAND PROTECTION

Hand protection is considered the most difficult form of general PPE to select. The hands are often very near or in direct contact with hazards, such as chemicals, heat, and cutting and piercing sources. Hand protection must not interfere with the wearer's ability to perform tasks requiring a high degree of dexterity.

When selecting hand protection to prevent contact with hazardous materials, chemical-specific data needs to be reviewed so that the material of construction is compatible with the hazardous material(s). The most important information to review is permeation data, which is the phenomena of chemical molecules migrating through the chemical protective clothing (CPC). The speed and rate that this occurs is evaluated for many common chemicals against a specific type and brand of CPC. No permeation is considered the best result for CPC selection. Other common CPC test parameters include degradation and penetration.

BODY PROTECTION

Body protection includes chemical splash suits, heat-insulating suits, welding leathers, and many other specialized types of protective devices that cover only parts or all of the body. As with hand protection, body protection is also difficult to properly select as many variables exist: amount of body mass that needs protection, CPC materials of construction, heat stress, and other factors. Body protection needs to be carefully evaluated, selected, and employees trained in its proper use and its limitations.

HAZARD COMMUNICATION

OSHA's Hazard Communication Standard is based on the core philosophy that employees have a right to know the hazards of chemicals in their work area, and the steps necessary to prevent exposure. This is performed through training, labeling, material safety data sheets, and other mechanisms to convey hazard information.

A facility's Hazard Communication Program (HCP) is one of the most important safety and health programs to develop and keep up to date. Many other regulations dovetail into the HCP, including environmental and transportation requirements. The HCP must be in writing and describe, at a minimum, the following information on labeling and other forms of warning: material safety data sheets and employee information and training. The following discussion provides an overview of the key requirements of this standard.

MATERIAL SAFETY DATA SHEETS

Chemical manufacturers and importers must evaluate the hazards of the chemicals they produce or import. The compiled information must then be stated on material safety data sheets (MSDS).

The manufacturers, importers, and distributors are responsible for ensuring that their customers are provided a copy of these MSDSs with or before the initial shipment. There is no specific format required for the MSDS, although certain information must be present on it. OSHA has developed a nonmandatory format, OSHA form 174, which is used by numerous manufacturers and importers. The MSDS must be prepared in English.

TRAINING AND INFORMATION

Employees who may be exposed to hazardous chemicals need to be trained prior to initial assignment to a work area in which hazardous materials are present. Additional training is required when a new chemical hazard is introduced. It is important to document that the training has been performed, including details of the training program.

The training contents need to include the physical and health hazards of the chemicals in the work area, methods to detect the presence of chemicals, and details of the site's hazard communication program. Information also needs to be readily available to employees concerning the OSHA Hazard Communication standard, where hazardous materials are present in the facility, and location and availability of the hazard communication program and referenced documents.

CONTAINER LABELING

A container includes tanks, barrels, drums, boxes, cylinders, bottles, etc.—essentially everything that may contain a hazardous material except for piping systems (employees need to be trained on how to determine the contents of unlabeled pipes).

The label on the container needs to include the hazardous material's name (same name as on the MSDS and the list of hazardous materials so that they can be cross-referenced), hazards, and target organ effects. In addition, existing labels on containers, such as those from the manufacturer, are not to be removed.

HAZARDOUS MATERIALS LIST

A list of hazardous materials needs to be kept up-to-date and readily available to employees so that they can cross-reference between labels and MSDS. For this reason, it is important to have the same name on these sources. The list may be either kept by department or for the entire facility.

EMERGENCY PREPAREDNESS AND RESPONSE

Every facility needs to have prepared plans and employee training on what to do in the event of an emergency. Emergencies may result from a number of causes, such as fire, injury, weather, and hazardous material spills. This review will concentrate on the regulations concerning hazardous material spills.

EVACUATION PLANS

The core emergency plan for a facility is the evacuation plan. The evacuation plan is designed to instruct all employees, contractors, and visitors how to safely exit a facility in the event of any emergency. With few exceptions, the evacuation plan needs to be in writing and employees need to be trained on the plan at initial hire and annually thereafter. The plan may include any necessary procedures employees may need to follow while preparing to exit, such as emergency shutdown of equipment.

OSHA has a standard requiring emergency action plans which include evacuation requirements, 29 CFR 1910.38(a), and OSHA vigorously enforces this. OSHA also closely looks at facilities to determine if the number and location of exits are adequate, and to evaluate if any access to exits, exits, or discharges from exits are blocked, or not maintained properly.

29 CFR 1910 Subpart E has numerous means of egress requirements, as well as criteria for fire prevention plans, that need to be reviewed. The National Fire Prevention Association's (NFPA) 101 Life Safety Code provides additional means of egress information.

HAZARDOUS MATERIAL EMERGENCY RESPONSE PLANS

In 1989, OSHA issued its Hazardous Waste Operations and Emergency Response standard 29 CFR 1910.120, commonly referred to as HAZWOPER. This standard applies to work at hazardous waste sites, hazardous waste treatment, storage and disposal (TSD) facilities, and hazardous material (hazmat) incidents.

Hazmat incidents [1910.120(q)] will be discussed briefly as follows as they relate to OSHA's HAZWOPER standard. Note that other agencies and their regulations also have hazmat regulations. These are beyond the scope of this chapter, but the reader should refer to them at the federal,[10] state and local level for additional information and insight.

If a facility has hazardous materials, facility management needs to formally decide what it will do in the event of a hazmat release. If management's decision is to evacuate employees, the facility is only required to have an evacuation plan (as discussed above) and its employees trained in the execution of the plan.

Many facilities also make a decision to respond to hazmat incidents. With that decision, they also need to carefully determine at what level they will plan to respond since the HAZWOPER standard differentiates between "offensive" response (i.e., full hazmat team) and "defensive" response (performing mitigating tasks at a safe distance from the incident). Within each of these categories are various training levels.

In addition to training requirements, the amount of equipment necessary for employee safety can be quite expensive to purchase and maintain. Regardless of the response level, if a facility decides that it will respond in some fashion to hazmat incidents, it needs to have in place an emergency response plan that includes all of the elements found in 29 CFR 1910.120(q).

ELECTRICAL SAFETY

Electricity is the most common form of energy in workplaces; unfortunately, most employees do not realize that it can also be deadly. Many employees are shocked or electrocuted every year. OSHA was concerned about the relatively high number of injuries and fatalities from electricity, especially across the broad range of job classifications of the affected workers, as well as the variety of tasks that were being performed. OSHA issued new general industry regulations in 1990 concerning electrical safe work practices, found in 29 CFR 1910.331-335.

In summary, these regulations are designed to do the following:

- educate employees about the hazards of electricity
- inform employees of proper procedures, protective equipment, and precautions to take to prevent electrical contact
- require employers to develop safe work practices and train their employees on them

It is important to understand that OSHA's electrical safe work practices standards apply to a wide range of employees, not just electricians. For employees that are not "hands-on" (called "unqualified" in the standard), think of the training as "electrical hazard communication" in that the company is educating workers to understand the hazards of electricity and how to safely work around it so that they do not get shocked or electrocuted. "Qualified" workers need to have additional training so that they can safety work near or on electrical equipment.

The electrical safe work practices standard dovetails with the lockout/tagout standard. As such, the possibility exists to have one comprehensive worker program that meets the requirements of both standards, thus eliminating duplication of effort, overlap, and possibly some confusion on the part of the workers.

OSHA has numerous other electrical standards that apply to certain equipment, hazard classification areas, and industries. These are found in Subpart S of 29 CFR 1910. Although this OSHA regulation applies to general industry, it is important to know that OSHA has many other electrical requirements found in Subpart S.

LOCKOUT/TAGOUT

OSHA's Lockout/Tagout standard 29 CFR 1910.147 (technically called "The Control of Hazardous Energy") was promulgated in 1989 to protect employees from the release of various forms of energy during the servicing and maintenance of equipment and machinery. Forms of energy may include electrical, hydraulic, pneumatic, thermal, kinetic, potential, radiation, pressurized fluids, etc.

PROGRAM REQUIREMENTS

In summary, this regulation requires employers to develop a lockout/tagout program that includes written procedures and employee training to prevent the accidental energizing of equipment and machinery during servicing and maintenance. This is to be accomplished by using locks and/or tags, or other devices to isolate the energy at an isolating device, such as an electrical disconnect or valve. The procedures need to be in writing, and specific to the equipment or machinery. The program must address several lockout/tagout scenarios such as when contractors are working on equipment, when shift change occurs, or when more than one employee is working on the same equipment. As this is a performance standard, employers may develop the program to best meet the company's needs, as long as the critical elements in the standard are met.

EMPLOYEE TRAINING

Lockout/tagout training is necessary for employees who will be performing lockout/tagout (called "authorized" employees), and the employees whose machines or equipment may be locked or tagged out by someone else (called "affected" employees). "Other" employees in the facility need to know about the program as well so that they do not remove locks or tags, or otherwise cause an unsafe incident.

MACHINE GUARDING

Although updating the OSHA machine guarding standards for general industry has received little attention in the last decade, enforcing existing standards continues to be aggressive. The machine guarding standards are either industry specific (i.e., paper industry, mills and calendars in the rubber and plastics industries, etc.) or function specific (i.e., woodworking, forging, etc.), or may be general to apply to essentially all other machines. In summary, the areas to be guarded may be at the point of operation, or be related to power transmission apparatus.

POINT OF OPERATION

The point of operation is where the specific function of the machine is taking place, such as cutting, shearing, punching, gluing, or riveting.

The guarding of the point of operation needs to be properly designed and adjusted so that an employee *cannot* contact the hazard area with any part of his or her body or clothing. In addition, the guard must not create a hazard in itself. In OSHA's mechanical power press standard 29 CFR 1910.217, Table O-10 lists maximum guard-opening dimensions dependent on the distance the guard is from the point of operation. This table is based on anthropometric data of finger, hand, and arm dimensions. Informally, safety professionals utilize this table for other guarding situations, even though it is specific to mechanical power presses in the OSHA standard.

Guards may be fixed on the machine to prevent contact with the point of operation. They may be constructed out of a variety of materials, but they must be substantial in nature and in themselves not create any hazards. They should be easily removed with tools so that maintenance can be performed on the machine (following lockout/tagout procedures), but not so easily removed that operators may remove them during normal production operations.

In addition to fixed guards, there are a variety of other acceptable methods to prevent employees from accessing the point of operation. These may include light curtains, impedance-stopping devices, presence-sensing devices, two hand controls, and other devices. It is important to note that not all devices may properly protect employees.

For example, a properly designed and constructed fixed guard will more than likely stop projectiles from hitting an employee, whereas a light curtain will do nothing other than possibly stopping the machine.

POWER TRANSMISSION APPARATUS

Power transmission apparatus is the term used collectively for belts and pulleys, chains and sprockets, rotating shafts, and similar devices used to transmit energy from one point to another. OSHA has detailed regulations on power transmission apparatus-guarding requirements, with the majority of them found in 29 CFR 1910.219. In summary, all power transmission apparatus needs to be guarded so that employees cannot contact it with any part of their body or clothing.

CONFINED SPACES

A confined space is defined by OSHA as a space that: 1) is large enough to bodily enter; 2) has restricted means of getting in or out of it; and 3) is not designed for continuous occupancy. In January of 1993 OSHA issued its Permit-Required Confined Spaces standard, codified as 29 CFR 1910.146. This standard addresses confined spaces and the numerous hazards that already may be present or created when employees work in them. If any hazards are present, potentially present, or may be created by the work to be performed, the confined space is then called a permit-required confined space.

CONFINED SPACE HAZARDS

Confined space hazards may include hazardous atmospheres (such as lack of oxygen, health hazard and fire/explosion), a material that may engulf an entrant (such as grain), a configuration that may trap an entrant (such as a cone-shaped tank), or any other recognized hazard. Another hazard that OSHA addresses in this standard concerns rescue. The majority of fatalities in confined spaces have been associated with untrained rescuers. This standard requires every facility that has confined spaces in which employees will enter to have either an internal rescue team or have formal arrangements with an external rescue service.

CONFINED SPACE PROGRAM

Employers need to perform a documented survey of their facilities to determine if they have confined spaces. If they do, they then need to decide if their own employees are to enter them. If employees are not to enter the spaces, the employer needs to prevent access into them and put up signs that state "Danger—Permit Required Confined Space—Do Not Enter" or similar language. If employees are to enter them, the employer needs to develop a comprehensive confined space entry program.

A written procedure needs to be developed that addresses preventing unauthorized entry, hazard recognition and control (before entry), responsibilities, safety entry and control measures, equipment, testing (air, etc.), and training. The procedure also has to include a confined space entry permit. The permit needs to include detailed information on it. The OSHA standard lists the permit's required information, as well as having a few examples in one of its appendices.

If an employer is having a contractor enter a confined space, the employer needs to inform the contractor of the known hazards and related procedures, that a permit entry program is required, and debrief the contractor at the end of the entry. If both site employees and the contractor are to enter the space, the entry needs to be coordinated with regards to permits, attendants, rescue, etc.

The procedures need to include a mechanism to audit the program on at least an annual basis. All completed permits need to be retained for at least one year; they need to be reviewed as part of the audit.

TRAINING

The employees that enter the spaces are called entrants; the employees that continually oversee their entry are called attendants; and the employees responsible for all aspects of an entry are called entry supervisors.

Employees may also be on the confined space rescue team. Each one of these classifications has specific training requirements. The training is required to be performed before employees are first assigned entry duties, when there is a change in entry operations or hazards that their current training did not address, and when deficiencies in the program are identified.

SUMMARY

This chapter strived to inform the reader of information concerning OSHA and key OSHA standards. We only provided a brief overview of the selected standards; for additional insight and information readers need to review the actual standards and their interpretations for their specific industry and application. In addition to understanding applicable standards, strong safety management and involvement of employees is critical to improving safety performance and achieving ongoing occupational health and safety compliance.

In addition, having a thorough understanding of OSHA requirements, and recognizing their relationship to environmental considerations as they relate to worker situations in the workplace, can allow for a seamless environmental and health and safety worker program.[11] For example, in alluding to the introductory portion of this chapter, being aware of environmental considerations is a concept that all workers should strive to attain in their regular work functions. A shop worker that disregards his or her PPE, such as impervious and inert gloves, in handling hazardous chemicals, may inadvertently cause himself or herself harm and create a release in the event the chemical spills onto the shop floor, or worse, those actions may allow a 55-gal drum to tip over and fall. Anything is possible, and proper precaution doesn't take much additional effort.

The subsequent chapters look at environmental considerations in greater detail that should provide supplemental information to health and safety to provide a comprehensive environmental, health, and safety management overview for issues facing EH&S professionals, workers, their management, other business units within the organization, and other affected stakeholders.

ENDNOTES

[1] 29 Code of Federal Regulations (CFR) Part 1910.119, passed on February 24, 1992.

[2] Noted in 61 *Federal Register* 31667, passed on June 20, 1996.

[3] 29 Code of Federal Regulations (CFR) 1910.1200, passed on November 25, 1983.

[4] For additional insight, refer to G. Crognale, "Take Advantage of EPA and OSHA's Regulatory Synergy to Enhance Training," *Safety Compliance Letter*, November 25, 1997, p. 2.

[5] For additional information, see D. Hortensius, "Beyond ISO 14001; An Introduction to the ISO 14000 Series," in *ISO 14001 and Beyond,* " Sheffield, UK: Greenleaf Publishing, 1997.

[6] National Safety Council, 1121 Spring Lake Drive, Itasca, IL; Tel. (708) 285-1121.

[7] Public Law 91-596; 91st Congress, S. 2193, December 29, 1970.

[8] The landmark case that established precedent for OSHA (and EPA) inspectors to take legal action, in other words, seek a search warrant to gain entry to perform an inspection, is *Marshall v. Barlows's Inc*. The decision for that case can be found in *98 Supreme Court Reporter*, 436 U.S. 307, 56 L. Ed. 2d 305, No. 176-1143, pp. 1816–1834.

[9] A similar mechanism exists within EPA's civil and criminal enforcement program to allow employees and concerned citizens to register a complaint against a suspected violator of environmental laws and regulations.

[10] The EPA regulation analogous to hazmat as described herein is 40 CFR 265.50, which is Subpart D—Contingency Plan and Emergency procedures, under RCRA.

[11] For additional insight, see G. Crognale's "Hazardous Chemicals: Forging a Link Between Safety and Environmental Concerns," *Safety Compliance Letter*, Bureau of Business Practice, August 10, 1997, pp. 2–3.

REFERENCES

Crognale, G. "Take Advantage of EPA and OSHA's Regulatory Synergy to Enhance Training," *Safety Compliance Letter*, Bureau of Business Practice, November 25, 1997, p. 8.

Crognale, G. "Hazardous Chemicals: Forging a Link Between Safety and Environmental Concerns," *Safety Compliance Letter*, Bureau of Business Practice, August 10, 1997, pp. 2–3.

Hansen, Doan J., ed. *The Work Environment: Occupational Health Fundamentals*. Boca Raton, FL: Lewis Publishers, Inc., 1991.

Kaletsky, R. *OSHA Inspections: Preparation and Response*. New York: McGraw-Hill, Inc., 1997.

Thomen, J. *Leadership in Safety Management*. New York: John Wiley & Sons, Inc., 1991.

Shelton, C. ed. *ISO 14001 and Beyond*, Sheffield, UK: Greenleaf Publishing, 1997.

29 Code of Federal Regulations 1910, the OSHA regulations.

40 Code of Federal Regulations Part 262, 265, RCRA regulations.

61 *Federal Register* 31667, June 20, 1996.

Public Law 91-596, S. 2193, December 29, 1970.

98 Supreme Court Reporter, 436 U.S. 307, 56 L. Ed 2d 305, No. 176-1143.

CHAPTER 4

AN OVERVIEW OF SEC ENVIRONMENTAL REQUIREMENTS FOR PUBLICLY-HELD COMPANIES

Gabriele Crognale, P.E.

> *"Institutional investors, now in control of more than half the shares of U.S. corporations, demand more accountability."*
>
> Richard H. Koppes[1]

INTRODUCTION

In strategic corporate planning involving environmental considerations, a growing number of discussions within top management circles continue to focus on environmental liability and disclosure, and the relevance to the corporation of such items that include, in some capacity or another, the disclosure of such information as: EPA-required annual toxic release inventory (TRI) reporting release figures; the organization's commitment to adhere to the CERES principles (noted in Chapter 1 and included in Appendix D); the potential far-reaching effects of EPA's Audit Disclosure Policy on EPA-regulated organizations[2] superceded the growing support for more corporate openness by organizations as fostered by the ISO 14000 environmental management standards; and the disclosure requirements of the EMAS[3] regulations of the European Union, and other similar disclosure items that may or may not be actual liabilities to the organization, unless they are not acted on in an appropriate time frame.

Given the array of such dynamic and diverse considerations as described, a greater number of publicly-held companies have taken this discussion of environmental liabilities to a higher level than was formerly the norm—as a description in their annual reports and quarterly statements, as required by the U.S. Securities and Exchange Commission (SEC) as part of U.S. securities' registration procedures that are made available to shareholders and the public. Expanding upon this SEC requirement, a number of SEC-regulated companies began issuing more detailed environmental reports that disclosed information beyond the scope of the SEC's regulations as a voluntary effort aimed primarily at their shareholders and interested third parties. These detailed environmental reports provide an encapsulated disclosure of their environmental efforts, achievements, and goals, as a benchmark of sorts, to where they were previously. Among the companies that initiated this effort in the early years—between 1988 and 1990—are the major petrochemicals, including: Chevron, Texaco, and Phillips Petroleum; and other firms such as Polaroid. Of note, Chevron senior environmental engineers provide some insight into their company's pollution prevention achievements as described in Chevron's 1996 environmental report in Chapter 10; while Alfred DeCrane, Jr., Texaco's CEO, provides an overview on sustainable growth with respect to the petroleum industry and how some environmental regulations may affect sustainable growth.[4] A press release of that presentation is provided in Appendix G, with the permission of Texaco.

This voluntary effort has expanded to a greater number of publicly-held companies as a means for them to highlight their environmental stewardship efforts. As described in Chapter 1, a number of industry organizations created specific guidelines to ensure that voluntary environmental reports would be developed in such a manner to be of benefit to interested parties. As a result of the increasing number of environmental reports that have been issued since the first voluntary reports became public, and the increased public interest generated from these reports, several studies have been initiated as a means to track or benchmark these reports. One such study was conducted by members of Duke University, in which the intent of the study was to understand the nature and scope of corporate environmental reports and to identify leading environmental reporting practices of U.S. companies by identifying and analyzing all known reports by large U.S. multinationals.[5]

In addition, Price Waterhouse LLP conducted a series of surveys on environmental matters that dealt with corporate America and touched upon key environmental issues. One such survey, the third in a series by this management consulting firm, focused on: environmental disclosures, formal environmental accounting policies, environmental auditing, and environmental reports as useful tools. The survey highlights many of the areas that deal with environmental disclosure aspects and how some companies view their responsibilities in this regard. Of note, survey developers approached over thirteen hundred major U.S. companies for the survey, and received a 34 percent response rate. Such a high response rate bodes well for respondents in that it demonstrates a strong and genuine top management interest. Additional research the author conducted in evaluating the twelve or so environmental reports referenced in this book, and reflected in other sources of information he references herein, follow very closely many of the Price Waterhouse survey's findings and conclusions.[6] As such, environmental disclosure has become the rule, rather than the exception.

Similarly, albeit from a different perspective, sufficient interest was generated from the international community to have TC 207, ISO's committee charged with the development of the ISO 14000 standards, to include a set of standards[7] that deal with environmental labels, declarations, and self-declaration environmental claims. In the terminology of ISO 14021, an environmental report is considered by the standard to be an environmental claim as well, which gives these reports an additional dimension in this regard. In addition, a report issued jointly by the Federation of Netherlands Industry and the Netherlands Christian Federation of Employers in 1991, titled *Environmental Reporting by Companies,* highlighted the importance of companies' environmental management systems as a means to stimulate regulatory compliance, control of environmental responsibilities and risks, and the opportunity for positive interaction between companies and various stakeholders. Of specific interest with respect to environmental disclosure, the report stressed the need for companies to consider publishing a separate environmental report with the goal to have the report contain all relevant company information to get a clear picture of a company's environmental policy (and its progress to date).[8] Further information on environmental reports as an example of environmental stewardship and as a beneficial aspect of the ISO 14000 standards are included in Chapters 10 and 12, respectively.

Regarding SEC's role in viewing environmental disclosures as a requirement of publicly-held companies, a number of events were instrumental in the SEC's decision to issue a staff accounting bulletin in 1993 to "ante up the stakes" with publicly-held companies. Among the companies that were the first recipients of SEC administrative and judicial actions included Allied Chemical Corporation (now Allied Signal) and United States Steel Corporation (now USX) dealing with environmental risks and expenditure issues.[9] In generic terms, staff accounting bulletins reflect the commission staff's views regarding accounting-related disclosure practices. Furthermore, they represent interpretations and policies followed by the Division of Corporation Finance and the Office of the Chief Accountant in administering the disclosure requirements of the federal securities laws.[10]

As a further impetus for the SEC to lead publicly-held companies to begin to take a more holistic approach to environmental disclosure and liability accountability, on June 14, 1993, the SEC issued its Staff Accounting Bulletin (SAB) 92, or as it is commonly referred to by the SEC, "the staff," as a means for publicly-held companies to more appropriately report their environmental liabilities.[11] In a related news item, SEC Commissioner Richard Roberts told financial executives at a December 1993 meeting in New York that some companies "would be drawn and quartered" by the SEC's enforcement division "for inconsistencies and lack of disclosure" about environmental liabilities in their SEC filings. In such instances, the SEC can flex its enforcement muscles to force a company to provide more environmental liability data and can fine the company up to $500,000 for each violation.[12]

Part of this "muscle-flexing" zeal exhibited by the SEC may have its roots in a 1992 survey conducted by Price Waterhouse ". . . that found 62% of respondents with 'known exposures not yet recorded in their financial statements.' These companies may be in for a rude awakening. SEC officials have lost patience with this attitude and, in fact, are prowling around for test cases to prosecute."[13]

In addition, the EPA, in its efforts to increase its enforcement coordination activities to ensure that more companies return to compliance, has expanded its Memoranda of Understandings[14] (MOUs) to include the SEC, along with other federal agencies, such as OSHA, Department of Justice, and the Federal Bureau of Investigation. The intent of this MOU was to provide an informational exchange of sorts to help the SEC monitor the adequacy of companies' environmental disclosures. The items EPA could make available to the SEC under this memorandum include: the names of potentially responsible parties (PRPs); a list of facilities barred from government contracts under the Clean Water Act and Clean Air Act; a list of all filed civil cases under Superfund (CERCLA) and RCRA; a list of a recently concluded civil cases under federal environmental laws; a list of all filed criminal cases under federal environmental laws; and a list of all RCRA facilities subject to cleanup requirements.[15]

Given this scenario, it would seem that in addition to the other internal and external forces that impact upon regulated organizations within the EPA and OSHA-regulated universe, the SEC also intends to play a pivotal role with these organizations as well, at least from the perspective of those EPA-regulated companies that are also regulated by the SEC. It would also seem that the SEC has initiated its first strike at a far-reaching sweep that touches upon every publicly-held corporation that is traded on Wall Street—the heart of financial America—with SAB 92. That staff has begun to have a noticeable impact on publicly-held companies' disclosure policies, and will probably increase in some manner as we proceed into the twenty-first century. This SEC document has a direct relationship not only with EPA's voluntary disclosure overtones for regulated organizations, but with companies' investors, their customers, and the public alike. As one *Wall Street Journal* Business Bulletin item noted on the topic of environmental records of U.S. multinational corporations, ". . . *Key points in gaining credibility include revealing negatives as well as positives and showing compliance data.*"[16]

The intent of this chapter is to provide additional insight into the staff, as well as other dynamic factors, affecting companies each day that may also play a role in having publicly-held corporations satisfy not only SEC reporting and disclosure requirements, but those of environmental regulatory agencies and concerns of investors and affected stakeholders, such as local residents.

THE SEC ENVIRONMENTAL REPORTING REQUIREMENTS

The issuance of SAB 92 provided the mechanism for the SEC to gauge the full extent of the number of publicly-held companies, or (SEC) registrants, that would adhere to the intent of this particular staff. The initial reaction of registrants to provide greater environmental disclosure, from SEC's perspective, was that a number of registrants did not seem to take SEC's concern for appropriate reporting of environmental liabilities seriously.[17] That perception may have prompted former SEC Commissioner Roberts to make his strongly worded point to financial executives regarding environmental disclosures, as described previously.

What does this say about the SEC, and what does this hold in store for publicly-held U.S. national and multinational companies? To address this rhetorical question, let us fast-forward to 1997.

Since the issuance of SAB 92 and the MOU between EPA and the SEC, a growing number of publicly-held organizations have taken considerable effort and great strides to go beyond SEC disclosure, not only as a means to move beyond compliance, but also to help showcase and share their results with other companies and concerned and interested stakeholders as a means to learn from each others' experiences and accomplishments, as a meaningful "lessons learned" approach.

For some organizations, though, the standards for reporting environmental liabilities may have been somewhat unclear, according to some critics, which may have led to an uneven reporting practice with regulated industries. As a result of this uncertainty regarding regulated entities in estimating their particular environmental liabilities, the American Institute of Certified Public Accountants (AICPA) issued a statement of position, SOP 96-1, "Environmental Remediation Liabilities," which took effect December 15, 1996.[18] SOP 96-1 outlines when a company must recognize an environmental liability and how it should measure it. This SOP is also meant to clarify an element of U.S. accounting principles—Financial Accounting Standard (FAS) 5. According to the AICPA, one of the reasons for uncertainty on the part of registrants had to do with the language contained within FAS 5: "FAS 5 said you record a liability when it's probable (it has been incurred) and the amount is reasonably estimable," according to Fred Gill of AICPA.[19] The reasonably estimable aspect was narrowly interpreted by many companies, and since they could not estimate the whole environmental item, they choose the "do nothing" approach. SOP 96, on the other hand, offers guidance to a registrant for disclosing what it *can* estimate.

For example, in calculating liabilities related to a particular site (cleanup activities), a company needs to calculate its share of liability costs and its share of costs related to a site that will not be paid by other potentially responsible parties (PRPs) or the U.S. government. Also to be included in the calculations are: current laws and regulations; the remediation technologies currently available; and estimates of what it will cost to execute all the phases of the site remediation.[20] Explained in another fashion, a company may want to evaluate and benchmark what other companies have done in this area, and even ask for their viewpoints on the remediation firms' estimates for cleanup. Additional inquiries with respect to the accuracy and relevance of their estimates and whether the technologies were the most practical and feasible could also provide additional insight into this perceived "black box."

SOP 96-1 adds an additional dimension to the rules, interpretative releases, and prior professional guidance that provide SEC's environmental reporting requirements. These include: Regulation S-K, in items 101, 103 and 303; the Securities Act Release, or SAR 6835; and SAB 92, which provides the most substantive guidance on identifying and reporting contingent environmental losses. A major tenet of SAB 92 is to ensure that registrants provide investors with factually accurate disclosures and to allow the reader of such disclosures to understand fully the events and other contingencies affecting the registrant's organization.

Specific elements within Regulation S-K are worth describing further:

Item 101 requires registrants to describe the material effect that compliance with federal, state, and local environmental laws will have on a company's earnings, capital expenditures, and how this may affect a company's market share in the open market. Also to be factored into this calculation is the potential for fines and related enforcement actions. Under this regulation, should the omission of such information be viewed as valuable by an investor, this information would have to be disclosed.[21]

Item 103 requires registrants to describe any material pending legal proceedings unless they involve routine litigation that is incidental to the business.[22] The SEC's position regarding legal proceedings includes any administrative orders that involve environmental matters.

Item 303, management discussion and analysis, or MD&A, requires an organization's management to prepare narrative reports discussing a company's financials, and any other information necessary to provide investors with a clear understanding of the registrant's financial condition.

The intent of the MD&A is to allow a company's investors an opportunity to see what top management sees by providing them an historical and prospective analysis of the registrant's financial condition and results of operation.[23]

Even without such regulatory persuasion, the environmental consciousness level has sufficiently changed at a number of registrant organizations, that in some of these organizations, top management is moving forward in environmental disclosure and from a different aspect, as if following the tune of a "different drummer"—one that is both market-driven and positive results-oriented.[24,25] There are a growing number of case studies that point to improvements achieved at some companies due to increased employee awareness, internal efforts to reduce pollution wherever possible, and the success of environmental auditing programs to help uncover findings and improve upon existing environmental management and compliance systems.[26]

People at various levels within regulated organizations are becoming more convinced of the environmental benefits that can be achieved in tandem with savings that can be realized, but they are also realizing that it takes a willingness to share one's success with others to help sustain internal improvements and environmental stewardship. The pilot programs of EPA's Environmental Leadership Program[27] and Project XL are two good examples of companies willing to share their triumphs and shortfalls in attempting to improve upon their environmental responsibilities without wasting unnecessary materials, chemicals, worker efforts, and company resources.

What then is the SEC's goal for registrant companies, and are these companies meeting this goal in the eyes of the SEC? While this may be a rhetorical question to some, we should not lose sight of the similarity in principal between the EPA and the SEC. Where EPA's main focus is to see the regulated return to compliance as a means to protect human health and the environment, and is devoting considerable attention to developing environmental indicators or measures for use in assessing various programs' performance and better informing the public about environmental conditions and trends,[28] the SEC's main focus with respect to SAB 92 and related SEC requirements is to ensure that registrant companies provide full and accurate environmental disclosure, and to take that disclosure aspect seriously to remain in the SEC's "graces" (an analogy to EPA's "returning to compliance").

Where do such companies stand with respect to disclosure in SEC's perspective, and how does the SEC view many of the voluntary reports being issued in increasing frequency and number by registrants?

To attempt to answer these questions, one needs to delve further into the strides that some organizations have taken in providing expanded and more detailed environmental disclosures of their operating facilities, and the various other external and internal factors that have come into play since the issuance of SAB 92 in 1993. Some of these factors may, in time, create a need for an updated SAB to stay in step with the accomplishments of some companies and related social, trade group, and professional organizations moving ever-forward into new environmental conscious territory.

WHAT THIS MAY ENTAIL FOR PUBLICLY-HELD COMPANIES

SEC reporting requirements dealing with environmental disclosure of publicly-held companies are not that far from what is currently being practiced by many well-known Fortune 100 and 500 companies, and in a more comprehensive fashion, what some EPA-voluntary program participants are providing to the public well beyond the SEC's requirements.[29]

In similar fashion, schools of thought among various practitioners of environmental accounting standards champion making a company's exposure to remediation liability healthy for the investment community in helping determine the relative amount of risk in their investments. Companies that may be undervalued in this scenario include DuPont, 3M, and Procter & Gamble. Of these, DuPont is often referred to by environmental accounting proponents as a benchmark for the industry and is favorably looked at with respect to its internal accounting practices.[30]

Should a publicly-held company fall somewhere in between with respect to its position on environmental disclosure, and need some additional guidance on what to do to maintain compliance with SEC's requirements, the following is provided as a general guide:

- A company should develop a baseline and "tickler" file to help determine which environmental costs should be disclosed, in what document, and when these should be disclosed. Benchmarking with other organizations in sister industries or in the same trade group or association could also be a source for additional information.
- One way to streamline this process could be to tie such information requirements with the organization's EMS or perhaps, more appropriately, the organization's EMIS, to keep track of these costs and liabilities. Such a tracking system could also be linked to internal or third-party environmental audit findings and reports that identify potential liabilities; or, it could be linked to any enforcement actions and related penalties associated with violations of CERCLA, RCRA, TSCA, CAA, and/or the CWA.
- Another way to streamline the efficiency within an organization may be to link environmental liability disclosure requirements to other business units within the organization and to link these concurrently with the financial and accrual departments responsible for compiling and disseminating this information to the SEC and related accounting service providers.

With respect to SOP 96-1, investors of publicly-held companies may need to wait until early 1998, when the first of 1997's annual reports are issued by companies to see how seriously companies and the SEC view this reporting standard.

Rather than wait until then, it may be prudent to evaluate what some companies can begin to expect from other forces before that time, some of which may be more of a driver for these companies than the SEC requirements.

Environmental disclosure, whether it is in the context as fostered by the SEC, EPA, or the tenets of the CERES principles, Responsible Care©, ISO 14021, EMAS, the PERI initiatives, or a host of other "drivers" or "search engines" (to borrow from the language of the Internet), is basically the same. Whether the disclosure and the associated costs center around existing conditions related to site clean-up required by RCRA or CERCLA; items uncovered in an environmental audit; penalties associated with an enforcement action; excessive TRI release figures; or other emissions and releases that evoke the CAAA, CWA, TSCA, or RCRA, or a combination thereof, it all needs to be addressed in some fashion if the disclosure the host organization wishes to make is sincere, straightforward, and unbiased, and wants to be perceived by its employees, the regulators, local community, shareholders, customers, suppliers, and the other stakeholders, as true environmental stewards.

The argument to be made in favor of disclosure, not only to satisfy SEC requirements, is that it provides opportunities for the organization to learn more about itself, to improve upon existing conditions, and to generate a positive cash flow as well, given the right conditions.[31] In addition, should a company desire to pursue certification to ISO 14001, and eventually make a self-declaration environmental claim—in other words, issue an environmental report—the company will need to follow specific guidance as described in the requirements of ISO 14021, contained in **4.3 Specific Requirements,**[32] which includes:

- The information specifies environmental improvements;
- The environmental claim should not misrepresent the environmental benefits;
- The environmental claim needs to be relevant to improvements made.

Among the environmental reports culled by the author for this book, the bulleted highlights of DIS/ISO 14021, 4.3, are all addressed in the reports reviewed, which seem to validate the conclusion that a number of publicly-held organizations understand the benefits that environmental reporting and disclosure can provide, when viewed in a business context versus solely a regulatory compliance context. The public wishes to be informed, and organizations, both U.S.- and foreign-based, are doing their best to keep the public up-to-speed. They realize it does not pay to limit or suppress disclosure; the only issue for some is whether the regulators allow them the benefit of the doubt in providing regulatory relief in such cases as these companies work to self-police themselves through their own voluntary efforts, such as through environmental audit programs to uncover excursions in their EMSs.

In close concert with the various forms of environmental disclosure are the aspects of risk management and risk communication, and how these additional considerations play a role in disclosure, versus what could happen to an organization without taking these items into consideration at some point. That discussion is provided in greater detail in Chapter 11.

A LOOK AHEAD

The two-year period, beginning in 1996, brought forth a whole new dimension in environmental management considerations that has accelerated the paradigm shift that began somewhere around the 1992 to 1993 time frame. While the SEC's SAB 92 may have been only one of a number of new players brought to the environmental awareness table, it was one more item to raise the bar another notch for regulated organizations.

As organizations move forward and begin to digest and assimilate the deluge of the forces that have come into play with these new players, a higher level of consciousness will be achieved, as these organizations begin gauging the relative accomplishments of companies who are willing to share their experiences in their efforts to better manage their environmental responsibilities.[33] Further discussion of environmental management software as described in the preceding footnote is provided in greater detail in Chapter 13.

As a final consideration to companies that may be somewhat skeptical in adhering to the SEC disclosure requirements, given the arguments for it in the preceding pages, their top management should take into consideration the pros and cons of not adhering, or adhering in a token manner at best.

Does the organization really benefit by not providing adequate disclosure, or might this be a factor for shareholders and other stakeholders to raise as an issue in annual meetings or other public forums?

The intent of additional disclosure is to provide information about an organization that can be used to educate both the public and the community in which the organization operates, and to satisfy the regulators—who keep on demanding more and more information from their regulated entities. In addition, such relevant information is useful to interested third parties who just want to be kept up-to-date and informed. The world is rapidly shrinking, and interested parties are becoming more educated and savvy and have access to an increasing number of information sources, such as environmental information of the Standard and Poor's 500 compiled by the Investor Responsibility Research Center (IRRC). With such information readily available to well-heeled interested third-parties, of what benefit would fabricated, misleading or erroneous information be to regulated organizations if their reports are going to be scrutinized by such individuals or groups? The regulated industries that do take the initiative and time to compile this comprehensive information are going to do their best to ensure the information is factually accurate and relevant to their intended audience. After all, irrespective of all this hard work and other good intentions by EPA regulated corporations, the information provided can be verified or disputed as easily as the click of a computer mouse and entering the Internet via http://www.doubting-thomas.com. Therefore, since such information is readily obtainable and accessible to be downloaded in the electronic world in which commerce also resides, it may just as well be as factually accurate and nonmisleading as possible right from the start. With environmental disclosure, there are neither shortcuts nor quick fixes—just the unadulterated facts. That is something even the most hard-core "Doubting Thomases" may have trouble refuting.

ENDNOTES

[1]R. H. Koppes, "Corporate Governance," *The Wall Street Journal,* April 14, 1997, p. B5.

[2]EPA's Voluntary Environmental Self-Policing and Self-Disclosure Policy was first listed in the *Federal Register* on April 3, 1995 as an interim policy, and later listed in the *Federal Register* on December 22, 1995, formally as "Incentives for Self-Policing: Discovery, Disclosure, Correction and Prevention of Violations," 60 FR 66706. Among the policy's elements of relevance to organizations include a commitment from top management to conduct environmental audits and provide for corrective action follow-up; a structured audit program that outlines management objectives, scope, and frequency of audits; and mechanisms to implement the necessary corrective actions to address the regulatory issues to have the companies return to compliance. Of equal, if not greater importance, with the bigger picture in focus, is that environmental audit programs are really management tools for organizations to use to improve upon their existing environmental management programs or systems (EMSs) and self-policing or disclosure should come from within the organization, rather than as an item for potential regulatory scrutiny and enforcement action, unless the organization warrants such scrutiny because of "hidden agendas."

[3]The Environmental Management Auditing Scheme was developed and passed by The Commission of the European Communities, as COM(91) 459 final, Brussels, 5 March 1992: A proposal for a Council Regulation (EEC) allowing voluntary participation by companies in the industrial sector in a community eco-audit scheme.

[4]A presentation by Alfred C. DeCrane, Jr., chairman and Chief Executive Officer, Texaco, Inc., New York: Marketing Conference of the Conference Board, October 26, 1993.

[5]D. Lober et al. "The 100+ Corporate Environmental Report Study," Duke University Center for Business and the Environment, January 30, 1996.

[6]Price Waterhouse LLP, *Progress on the Environmental Challenge: A Survey of Corporate America's Environmental Accounting and Management, Third in a Series,* Copyright © 1994 Price Waterhouse LLP, all rights reserved.

[7]Draft International Standard, ISO/DIS 14021, Environmental labels and declarations—Self-declaration environmental claims—Guidelines and definition and usage of terms, © International Organization for Standardization, 1996.

[8]*Environmental Reporting by Companies,* VNO Federation of Netherlands Industry and NCW Netherlands Christian Federation of Employers, May 1991, p. 4.

[9]For additional information, see S. Kass and J. McCarroll "Environmental Disclosure in Security and Exchange Commission Findings" *Environment,* April 1997, p. 4.

[10]Excerpted from an SEC web site cover document, *Selected Staff Accounting Bulletins,* obtained from the Internet, http://www.sec.gov/rules/acctindx.html, last update: 10/25/96.

[11]SAB 92 was listed in the *Federal Register* at 58 FR 32843 on June 14, 1993.

[12]See L. Berton, "SEC Rule Forces More Disclosure," and T. Aeppel "Firms Reveal More Detail of Environmental Efforts But Still Don't Tell All," *The Wall Street Journal,* De-

cember 13, 1993, page B1; and B. Birchard, "The Right to Know," *CFO,* November 1993, p. 28.

[13]B. Birchard, "The Right to Know," *CFO,* November 1993, p. 28.

[14]See *Inside EPA,* November 11, 1994, as a general reference.

[15]See *Superfund Report,* March 28, 1990, as a general reference.

[16]*The Wall Street Journal,* Business Bulletin, March 28, 1996, p. A1.

[17]See footnote No. 11 and M. Crough, "SEC Reporting Requirements; Environmental Issues," *Environmental Claims Journal,* Winter 1994/95, pp. 41–55.

[18]See W. Freedman, "Green Guidelines Clarify Reporting," *Chemical Week,* February 5, 1997, page 57; and H. Epstein and A. Ten Veen, "Position Statement Clarifies Liability Disclosure," *The National Law Journal,* March 17, 1997, p. B18.

[19]See W. Freedman, "Green Guidelines Clarify Reporting," *Chemical Week,* February 5, 1997, p. 57.

[20]Ibid.

[21]17 CFR 229.101(c) (xii).

[22]Ibid., and see 17 CFR 229.103.

[23]Ibid.

[24]As an example of the strides CMA member companies have taken to becoming more recognized for their environmental stewardship, the February 1996 issue of *CMA News*'s headline was "Building Credibility Through Annual Environmental Reports," and featured the first of a two-part series of articles about Monsanto's environmental commitments titled, "Reporting on the Environment," by S. Archer, p. 7. In part two, S. Archer continues with "Readers Like Environmental Reports," *CMA News,* March 1996, p. 22.

[25]A related story was authored by D. Mager, "Bottom-Line Benefits of Environmental Responsibility," *At Work: Stories of Tomorrow's Workplace,* September/October 1996, p. 11.

[26]Further information about EPA and industry efforts is provided in an EPA document, *Managing for Better Environmental Results,* March 1997.

[27]The EPA's web site contains a wealth of information on these and other EPA items of interest. A more complete list of web sites of interest is found in Appendix G.

[28]See GAO report, *Managing for Results: EPA's Efforts to Implement Needed Management Systems and Processes,* April 1997, as a general reference.

[29]See n. 27.

[30]Freedman, "Green Guidelines Clarify Reporting (n. 19), p. 57.

[31]Mager, "Bottom-Line Benefits of Environmental Responsibility (n. 25), p. 11.

[32]Draft International Standard (n. 7), p. 4.

[33]Several stories of note that encapsulate this theme are: B. Quinn, "Creating a New Generation of Environmental Management," *Pollution Engineering,* June 1997, p. 50; T. R. Wiseman and B. J. Knight, "Improving Processes and Partnerships," *Environmental Protection,* October 1996, p. 25; and G. Crognale, "Pollution Prevention: Financial Incentives." Environmental Protection, May 1997, p. 36.

REFERENCES

Accountants SEC Practice Manual. Product or Environmental Loss Contingencies—Sab 92, 4362, 213 9-93, © 1993, Commerce Clearing House, Inc.

Archer, S. "Reporting on the Environment," *CMA News,* February 1996, p. 7 and "Readers Like Environmental Reports," *CMA News,* March 1996, p. 22.

Aeppel, T. "Firms Reveal More Detail of Environmental Efforts but Still Don't Tell All," *The Wall Street Journal,* December 13, 1993, p. B1.

Berton, L. "SEC Rule Forces More Disclosure," *The Wall Street Journal,* December 13, 1993, p. B1.

Birchard, B. "The Right to Know," *CFO,* November 1993, p. 28.

Bridge to a Sustainable Future, the National Environmental Technology Strategy, April 1995.

Crognale, G. "Enhanced Environmental Openness: The European Community's Eco-audit Scheme," *Environmental Management Review,* First Quarter 1993, p. 80.

Crognale, G. "What's Up in EPA's Interim Policy Statement on Environmental Auditing," *Corporate Environmental Strategy,* Winter 1996, p. 37.

Crognale, G. "P2: Financial Incentives," *Environmental Protection,* May 1997, p. 36.

Crough, M. "SEC Reporting Requirements: Environmental Issues," *Environmental Claims Journal,* Winter 1994/95, pp. 41–55.

DeCrane, A. E., an untitled presentation on sustainable development, *The Conference Board,* October 26, 1993.

Draft International Standard ISO/DIS 14021, Environmental Labels and Declarations. Geneva, Switzerland: International Organization for Standardization, 1996.

Duke University Center for Business and the Environment, *The 100+ Corporate Environmental Report Study,* January 30, 1996.

Elliott, E. et al., Fried Frank, Harris Shriever and Jacobson, "A Practical Guide to Writing Environmental Disclosures," January 1994.

Environmental Management Auditing Scheme (EMAS), The Commission of the European Communities, COM(91) 459 final, Brussels, 5 March 1992: A proposal for a Council Regulation (EEC) allowing voluntary participation by industrial companies in a community eco-audit scheme.

Environmental Reporting by Companies, VNO Federation of Netherlands Industry and NCW Netherlands Christian Federation of Employers, May 1991.

EPA's *Inside EPA,* November 11, 1994, as a general reference.

EPA's *Managing for Better Environmental Results,* EPA 100-R-97-004, March 1997, as a general reference.

EPA's *Superfund Report,* March 28, 1990, as a general reference.

EPA notice, "Incentives for Self-Policing; Discovery, Disclosure, Correction and Prevention of Violations," *Federal Register,* 60 FR 66706, December 22, 1995.

Epstein, H. and A. Ten Veen, "Position Statement Clarifies Liability Disclosures," *The National Law Journal,* March 17, 1997, page B18.

Freedman, W. "Green Guidelines Clarify Reporting," *Chemical Week,* February 5, 1997, p. 57.

Freeling, K. "Environmental Law," *The National Law Journal,* July 24, 1995, p. B5.

GAO Testimony Report, *Managing for Results: EPA's Efforts to Implement Needed Management Systems and Processes,* GAO/T-RCED-97-116, April 1997.

Humphreys, S. "Complying with SEC Requirements for Reporting Environmental Costs," *Environmental Business Advisor,* Vol. 1, Issue 1, Spring 1995.

Kass, S. and J. McCarroll, "Environmental Disclosure in Securities and Exchange Commission Findings," *Environment,* April 1997, p. 4.

Koppes, R.H. "Corporate Governance," *The Wall Street Journal,* April 14, 1997, p. B5.

Lober, D. "The 100+ Corporate Environmental Report Study," Duke University Center for Business and the Environment, January 30, 1996.

Mager, D. "Bottom-Line Benefits of Environmental Responsibility," *At Work: Stories of Tomorrow's Workplace,* September/October 1996, p. 11.

Pitt, H. and K. Groskaufmanis, "The New Securities Litigation Reform Act should prompt a reevaluation of the form and content of corporate disclosures," *The National Law Journal,* January 22, 1996, p. B4.

Price Waterhouse LLP, *Progress on the Environmental Challenge: A Survey of Corporate America's Environmental Accounting and Management,* 1994, Price Waterhouse.

Quinn, B. "Creating a New Generation of Environmental Management," *Polluting Engineering,* June 1997, p. 60.

Securities and Exchange Commission (SEC), 17 CFR Part 211, Release No. SAB 92, *Federal Register,* 58 FR 32843, June 14, 1993 and Code of Federal Regulations (CFR), 17 CFR Part 229.

SEC document, *Selected Staff Accounting Bulletins,* obtained from SEC's web site at www.sec.gov/rules/acctindx.htm; last update: 10/25/96.

United Nations Environment Programme (UNEP) Technical Report No. 11—*From Regulations to Industry Compliance: Building Institutional Capabilities,* 1992, UNEP; Technical Report Series No. 6—*Companies' Organization and Public Communication of Environmental Issues,* 1991, UNEP; and UNEP Global Survey: Environmental Policies and Practices of the Financial Services Sector, January 1995.

Wiseman, T. R. and B. J. Knight, "Improving Processes and Partnerships," *Environmental Protection,* October 1996, p. 25.

Wall Street Journal, Business Bulletin, March 28, 1996, p. A1.

CHAPTER 5

An Overview of Multi- and Single Media EPA Inspections

Theodore W. Firetog, Esq., and Gabriele Crognale, P.E.

INTRODUCTION

As we saw in Chapter 2, enforcement and compliance assurance are still among the key components of EPA's mandates and responsibilities for ensuring that companies and communities comply with national laws for protecting human health and the environment.[1]

One of the tools that EPA (and the states) can utilize to gauge the extent of regulatory compliance within regulated entities is the regulatory inspection. Of note, the basic structure of an EPA (or state) inspection is not that different from an OSHA health and safety inspection, as was briefly described in Chapter 3.

Understanding how regulatory inspections are conducted, can provide additional insight to environmental managers and their superiors of regulated entities in preparation for such inspections, whether environmental and/or health and safety. Such preparatory work is especially relevant in the planning and conducting of environmental (and health and safety) audits by these individuals, whether or not they are already targets of enforcement agencies and may be scheduled for one or more inspections. Let us not lose sight of the fact that the anticipation of such inspections and possible enforcement actions are always a concern of regulated entities or organizations. As an example to put this concern into perspective, here are a few statistics provided by EPA: at the end of Fiscal Year (FY) 1996, EPA had referred a record 262 criminal cases and 295 more civil cases than in FY 1995 to the Department of Justice (DOJ), and assessed a record $76.7 million in criminal fines[2]; while at the end of FY 1997, EPA had referred a combined 704 criminal and civil referrals to DOJ and had assessed $264.4 million in fines and penalties—the highest one-year totals in the agency's history.[3]

In addition to increases in EPA enforcement actions as depicted stemming from regulatory inspections that have many managers at regulated companies concerned, these same managers also need to be fully appraised of several pieces of legislation that Congress passed in response to their own concerns with the effects of inflation on penalties assessed by EPA to companies. These congressional acts were aimed at providing for an adjustment in inflation in any penalties that EPA may administer.

Among these are the Federal Civil Penalties Inflation Adjustment Act of 1990, and the Debt Collection Improvement Act (DCIA) of 1996. These two acts of Congress were later reflected in EPA's Civil Monetary Penalty Inflation Adjustment Rule released in 1996.[4] As a result of this rule, EPA increased its maximum penalties for violations of its environmental regulations by 10 percent. For example, as Table 5.1 shows, the maximum penalties for various EPA media programs include:

As one method to counteract spiraling enforcement costs associated with inspections, an increasing number of regulated companies have come to rely on environmental audit programs, whether in-house or via third parties. With such audit programs, many companies are able to more effectively identify and correct regulatory findings within a regulated facility before they can become violations if noted by a regulatory inspector. Label such audits "preinspection" audits.

Such preinspection audits can telegraph to the facility in question any deficiencies that may exist in its environmental compliance program before the agency actually conducts its inspection, thereby affording the entity time to correct such problems before becoming the target of an EPA administrative, civil judicial, or criminal enforcement action. Environmental audits, then, are key components of a company's strategy in preparing for and lessening the regulator's potential enforcement response following an inspection. This is an important subject to cover, and is discussed in greater detail in Chapter 9.

The compliance inspection is just one example of the wide array of formidable tools with which EPA has been empowered by Congress to use in the agency's enforcement efforts to protect human health and the environment. Under the provisions of each of the major environmental statutes, EPA has authority to conduct inspections (see Table 5.2). There are two principal reasons why EPA conducts inspections:

1. To promote compliance with environmental laws.
2. To investigate violations of environmental laws.

In promoting compliance with environmental laws, EPA is concerned with informing facility owners about the statutory and regulatory requirements that may be applicable to the facility, as well as providing guidance on how the facil-

Table 5.1.   EPA's Civil Monetary Penalties—Some Examples

EPA program	Former Maximum	Revised Maximum
TSCA	$25,000	$27,500
CAA	$25,000/$200,000*	$27,500/$220,000*
CWA	$125,000*	$137,500*

*Maximum Administrative Proceedings

Table 5.2.   Federal Statutes/Regulations For Multimedia Investigations

	Air CAA	Water CWA	Superfund CERCLA and EPCRA	Pesticides RFRA	Solid Waste RCRA	Drinking Water SDWA	Toxics TSCA
Inspection Authority	114,[a] 211[a] [80, 86[b]]	308, 402 [112.41]	104	8.9 [160.15, 169.3]	3007, 9005 [270.30]	1445 [142.34, 144.51]	11 [717.17, 792.15]
Recordkeeping Authority	114, 208 [51, 57, 58, 60, 61, 79, 85, 86]	308,402, [122.41, 122.48]	103 372 10	4.8 [160 63, 169, 160, 185-195]	3001, 3002, 3003, 3004, 9003 [262.40, 263.22, 264.74, 264 279, 264 309, 265 74, 264 309, 270 30]	1445 [141 31 33, 144 51 144 54]	8 [704, 710, 717 15, 720.78, 761 180, 762 80,763 114, 792 185 195]
Confidential Information	206, 307 [2.201-2.215, 2 301, 53, 57, 80]	308 [2.201-2.215, 2.302, 122.7]	104 [2.201-2.215] 322 [350]	7, 10 [2.201-2.215, 2.307]	3007, 9005 [2 201-2.215, 2 305, 260 2, 270 12]	1445 [2 201-2 215, 2 304, 144 5]	14 [2.201-2.215, 2.306, 704.7, 707.75, 710.7, 712.15, 717.19, 720 80 95, 750.16, 750.36, 762.60, 763.74]
Emergency Authority	303	504	104, 106 [300.53, 300.65]	27 [164, 166]	7003	1431	7
Employee Protection	322	507	110		7001	1450	23
Permits		[122, 125]	110		[270]	[144, 147]	
Basic requirements include applications, standard permit conditions, monitoring, reporting							
EPA procedures for permit issuance	[124]	[124]			[124]	[124]	
Technical requirements	[52]	[129, 133, 136,302] BMP[d] [125] SPCC [112] Waivers [125, 130]			[260-266]	[146, 264]	
Specific References	NSPS[f] NESHAP[g] [61] CEM[h] [60] SIP[i] [52] PSD[h] [50]	Effluent guidelines [400-460] BMP [125], SPCC [112] Pretreatment [125, 403] Toxic [129]			Generators [262], Transporters [263] TSD[j] [265] Stds. for TSD Permits [264], Interim Stds [265], Storage <90 days [262], Exemptions [261].		PCBs [761] Dioxin [775]

[a]Statute (e.g., Clean Air Act, Section 114 or 211)
[b][80, 86] 40 CFR, Parts 80 and 86, CFR refers to Code of Federal Regulations
[c]Reportable quantities
[d]BMP—Best Management Practices
[e]SPCC—Spill Prevention Control and Countermeasures Plan
[f]NSPS—New Source Performance Standards

[g]NESHAP—National Emission Standards for Hazardous Air Pollutants
[h]CEM—Continuous Emission Monitoring
[i]TSD—Treatment, Storage and Disposal
[i]SIP—State Implementation Plan
[k]PSD—Prevention of Significant Deterioration

ity can comply with the such requirements. In general, however, any regulated entity that is not familiar with such requirements is extremely foolish to wait until an EPA inspection to find out how to comply with those requirements. Actually, most compliance inspections promote compliance by establishing a credible presence of the power and interests of EPA in ensuring that violations of environmental laws will be detected. The purpose of a preinspection audit, therefore, is to correct any potential violations before they are discovered during an EPA inspection.[5]

In addition to promoting compliance, EPA uses compliance inspections to investigate violations. Certainly, any compliance inspection will provide EPA with facts about a facility's compliance status, and in cases of noncompliance, allow the agency to collect and preserve evidence of such noncompliance. Detected violations can result in substantial civil and criminal penalties. Indeed, the fear of detection (i.e., being caught in noncompliance by an agency inspection) and the thought of paying and suffering the consequences of such noncompliance is perhaps the principle motivating factor for using preinspection audits.

In preparing for a preinspection audit, it is important to understand how EPA inspections are planned. Such an understanding is essential for determining the scope of the preinspection audit (thus, focusing on the probable issues that will arise during the EPA inspection), and for being prepared on the types of information that the agency may collect or require the facility to provide.

HOW EPA INSPECTIONS ARE PLANNED

One of the major problems that EPA has in implementing its goals associated with compliance inspections (ie., to promote compliance with environmental laws and to investigate violations of such laws) is the virtually limitless resources that would be necessary in order for the agency to inspect all facilities that are subject to a plethora of environmental laws and implementing regulations that are currently in effect. Because some environmental laws and regulations affect literally hundreds of thousands of facilities nationwide it would be almost impossible for EPA (and the states) to inspect more than a small percentage of these facilities.

Understanding its limitations in its ability to inspect all these facilities, and hence, ensure that the facilities are in compliance, EPA also has at its disposal a number of tools that the agency uses to administer its compliance and enforcement program. These tools can enhance EPA's limited enforcement resources, and are useful in planning its inspection priorities prior to the start of each fiscal year.

As a starting point, EPA regional offices review EPA Implementation Policies, Policy Memoranda and other pertinent agency enforcement documents. From there, the regional offices make recommendations to assign inspection priorities, and may introduce region-specific enforcement issues, such as complaints, repeat violators, or other information that may be available in various EPA databases, such as: IDEA (Integrated Data for Enforcement Analysis) which links all of EPA's national enforcement databases; Dunn & Bradstreet financial information and data from Form Rs (listing toxic release inventory data); and METS (multi-media enforcement tickler system), a tickler system designed by Region I containing information about current enforcement actions, planned inspections, and enforcement actions at facilities in Region I.

These data bases also contain statistical data from previous regulatory inspections. This data can include: the type and number of violations cited, whether the violations are repeat, or whether the inspection was the result of a complaint, state or other federal agency referral, or a follow-up inspection.

In addition to the information found in the databases referenced, regional offices can also set inspection priorities to address the general tenets of EPA long-term goals, such as exemplified in EPA's five-year strategic plan, and EPA's most recent report dealing with the agency's reinvention progress.[6]

As another consideration, during the facility priority setting and evaluation phase, EPA regional management can also evaluate one or more of the following factors for environmental significance at any given facility in its database such as:

- known or suspected release at regulated facility
- migration potential of such releases
- waste characteristics

As a general rule of thumb, environmental managers also are up-to-speed on their facility's enforcement track record can reasonably assess whether they may expect to receive a follow-up inspection some time soon. Keeping tabs on the facility's enforcement file to gauge whether past violations were corrected and the facility's environmental audit file to guage whether regulatory findings and their root causes have been corrected can greatly enhance their regulatory compliance profile should a surprise inspection loom on the horizon.

For additional guidance, we provide a series of articles about preparing for EPA inspections in Appendix A.

CATEGORIES OF COMPLIANCE INSPECTIONS

In determining its overall inspection strategy, the agency considers four categories of compliance inspections: routine, for cause, case development support, and follow-up. Routine compliance inspections are those conducted at facilities chosen or targeted under a neutral administrative inspection scheme. Under a neutral inspection scheme, the regulated community is divided into classes or segments based upon certain criteria such as the type of industry, the size of the facility, the quantity of regulated substances that are handled at the facility, or the amount of pollutants discharged. Specific facilities with each class or segment are then targeted for an inspection in a nonbiased way. Therefore, the selection of a facility for a routine inspection is not based upon any evidence that such a facility is in violation of any environmental law or regulation, or that a compliance problem even exists at the facility.

By contrast, a for-cause inspection is one conducted because EPA has some reason to suspect that an actual violation exists at the facility. The reason may, for example, be based upon a tip, complaint, self-monitoring report or audit, or an inspection conducted by another agency (e.g., OSHA). During a for-cause inspection, the inspector knows in advance what he or she is looking for in order to establish noncompliance. The focus of the inspection, therefore, is limited to obtaining evidence necessary to confirm or dismiss the potential violation.

If a violation is found during a routine inspection (or confirmed through a for-cause inspection), additional inspections may be necessary for continued case development support. Such case-development support inspections are geared towards collecting evidence requested by the agency's case development or litigation team in furtherance of prosecuting an incident of noncompliance.

Follow-up inspections are usually conducted on a facility that was found to be in violation during a prior inspection. The purpose of the such follow-up inspections is to ensure that the facility is now in compliance with the regulations and any consent decree or administrative order that may have been issued as a result of the previous violation. For example, pursuant to RCRA, HSWA, UST, and CERCLA, EPA has the authority to issue specific administrative orders (either consent or unilateral) to compel a violator to conduct one or all of the following: study the impacts of contamination [the RCRA facility assessment (RFA)]; evaluate the corrective action alternatives [the RCRA facility investigation (RFI)]; or undergo corrective action as a result of one or more EPA corrective action orders.[7] As part and parcel of such administrative orders, EPA will conduct follow-up inspections to ensure compliance with the terms and conditions of such orders. If the facility is found to still be out of compliance or to be in violation of the terms and conditions of an applicable administrative order, EPA will usually take a stronger enforcement action.[8]

In some cases, EPA may conduct a fifth type of inspection similar to a for-cause inspection. Under most of the environmental statutes, EPA has the authority to pursue criminal investigations. Criminal activities such as falsified information in records and reports and illegal disposal are typical examples of such activities that EPA special agents of the Criminal Enforcement Division will investigate, usually in coordination with the U.S. Attorney's office and other federal, state, and local law enforcement agencies. As with for-cause inspections, the special agent will know in advance what evidence he or she is looking for to support a criminal prosecution. Unlike for-cause inspections, however, inspections conducted in connection with a criminal investigation seldom are conducted on a program-specific agenda and almost always involve the issuance of a warrant.[9,10]

DEVELOPING AN EPA INSPECTION SCHEME

Each program area develops an overall inspection scheme that allocates a proportion of the area's available resources for conducting inspections in each of the four general categories described above (i.e., routine, for cause, case development support, and follow-up). The allocation is based upon the agency's identified enforcement priorities and program's operating year guidance and annual implementation plans developed by the agency. These may be supplemented by agency strategic plans. For instance, the tone for inspections through FY 1997 has been set in EPA's Office of Enforcement's Four-Year Strategic Plan (FY 1994–1997), and the Office of Enforcement & Compliance Assurance's Five-Year Strategic Plan.[11] Within this document, the need for agency-wide integrated strategic planning and the institutional mechanisms that should be developed by EPA in order to successfully implement a multimedia enforcement approach are discussed.[12] Both documents discuss innovative tools, such as pollution prevention in enforcement settlements, the need for additional measures of enforcement performance and success (e.g., environmental indicators), continued expansion and integration of the criminal enforcement program, and the role and oversight of states in the enforcement process. The strategies rely on the use of EPA's inspection authority, and continue to fuel the impetus for multimedia inspections (i.e., compliance inspections intended to determine a facility's status of compliance with respect to applicable laws, regulations, and permits covering a variety of program areas).

While multimedia inspections are increasing, the use of program-specific compliance inspections still constitutes the bulk of inspections conducted by EPA, with each program using a somewhat different approach for establishing its inspection priorities.

The goals and objectives for assigning priorities to implement RCRA, for example, are set forth in the RCRA Implementation Plan (RIP). It is revised each year to address EPA's priorities. The FY '95 RIP's focus, for example, included: (the regulated community's) promoting waste minimization practices, highlighting environmentally sound waste management practices, and identifying opportunities to reduce the risks of releases of hazardous materials and wastes.[13] EPA developed the FY '95 RIP as an addendum to the FY '94 RIP, a precursor to subsequent compliance and enforcement directives, as we previously discussed in Chapter 2.

Although each program area has a different approach to establishing its objectives, all program areas consider three common factors in developing their goals: 1) the probability that a violation in a particular class or segment of the regulated community (as identified under a neutral inspection scheme) will present a significant risk to human health and the environment; 2) the probability that particular class or segment of the regulated community will violate environmental laws or regulations; 3) the probability that inspections of a particular class or segment of the regulated community will deter noncompliance by demonstrating a credible enforcement presence.

Facilities in a high-priority class or segment nearly always are targeted for inspection. In a lower-priority class, only a small percentage of the facilities may be inspected. Thus, the neutral inspection scheme (which sets out how the regulated community is divided into classes or segments) plus the overall enforcement priorities (determined by the program's operating year guidance and annual implementation plan) determines what facilities are to be inspected, at least for routine inspections.

The majority of inspections conducted by EPA are routine inspections. A smaller percentage of EPA's resources are allotted for follow-up inspections, since EPA feels that such inspections are essential to ensuring the integrity of the enforcement program. As we discussed previously in planning inspections, that percentage devoted to follow-up inspections may vary from year-to-year depending upon other enforcement priorities that may need to be addressed by EPA regional inspectors. As such, because of the limitations on EPA enforcement resources, not all facilities that were found to be in violation during a previous inspection can be targeted for follow-up inspections. Indeed, EPA reliance on the self-monitoring requirements that are usually contained in its consent decrees and administrative orders is meant to help address the limited amounts of available agency resources. Such limited resources mean that not all for-cause inspections can be conducted. The tips or complaints that trigger the for-cause criteria must be evaluated against the program priorities and goals to determine whether an on-site inspection is necessary, or whether some other follow-up action is appropriate (e.g., an informational request).

All of a program area's inspections must be allocated between the various categories described above. This means, however, that certain trade-offs between the types of inspections to be conducted must sometimes occur. For example, the more inspections conducted under the for-cause category, the less available become the resources that are needed to conduct routine and follow-up inspections. Consequently, if there is a sudden increase in for-cause inspections, lower priority routine inspections may actually be postponed or deleted from the inspection schedule.

As mentioned previously, a preinspection audit is designed to alert the regulated facility to any deficiencies that may exist in its compliance program, and thereby, allow the entity to correct such problems. Such audits are usually designed to duplicate routine inspections, that is, inspections where the facility does not know whether a specific violation or problem exists. Indeed, the primary purpose of such audits is to detect any potential violations that may be present at a facility. However, a preinspection audit could be fashioned to respond to the other types of inspections that EPA conducts. The facility may, for example, audit its performance with a specific consent decree or administrative order, or focus on one particular known problem.

How the facility came to be included in the EPA inspection scheme will help to determine the objectives and scope of the preinspection audit, as well as to provide an indication as to the level of intensity that should be used for the audit review.

CHOOSING THE LEVEL OF EPA INSPECTION

The intensity and scope of an EPA inspection (as well as a preinspection audit) can range from a quick walk-through inspection to a more intensive sampling type of inspection. A walk-through inspection is limited to a site visit with a visual observation of the facility. Such an inspection would, for example, involve checking for the existence of control equipment, observing work practices and general environmental housekeeping, or ensuring that a records repository has been established. These walk-through inspections help establish EPA's enforcement presence, and are used by the agency as a screening tool to identify those facilities that should be targeted for a more intensive inspection at a later time.

A compliance evaluation inspection is the most common form of routine inspection used by EPA. Such inspection would include those observations that would be made during a walk-through inspection, as well as an in-depth review and evaluation of facility records, interviews with facility personnel concerning source monitoring methods, production processes, and control devices, and perhaps some environmental sampling.

If environmental sampling is known to be required prior to an inspection, a sampling inspection will be scheduled. This type of inspection includes some or all the activities of a compliance evaluation inspection, as well as a pre-planned sample collection plan to be implemented. In either type of "routine" inspections, it may be prudent for the facility to maintain on-site some sort of a regulatory inspection "protocol" that can help prepare facility personnel—including the receptionist—for a planned or unplanned surprise inspection. Among the items that should be discussed beforehand, and for which a consensus should be arrived, is the sensitive aspect of requesting a search warrant from an inspector. Should the inspection not warrant such a defensive stance, and such situations have undergone internal mock scenarios, facility representatives should carefully weigh the potential outcome, whether with or without counsel.[14]

An additional consideration regarding search warrants . . . requesting an EPA inspector to obtain a search warrant prior to the start of an inspection is not a call that a facility representative should make without considerable thought. There may be situations where a warrant may be a prudent course of action for a facility to take, but without just cause, a search warrant is not a good strategy. Search warrants, whether justified or not, provide an inspector with the added impetus to work deeper and further at regulatory compliance situations that may lead to violations. This also telegraphs to an inspector that the facility has something to hide, whether real or perceived. Before making a final decision. Facility representatives should pose this question to themselves: "Is this added scrutiny worth it?" You make the call.

Of note, EPA's Four-Year Strategic Plan depicts EPA as an agency focusing more of its energies toward multimedia enforcement. In response, and in conjunction with such enforcement strategies, EPA soon afterwards embarked on a program of conducting multimedia compliance inspections.

One of the first cases that was extensively written about as an example to highlight the need for multimedia enforcement, was Avtex Fibers, of Front Royal, Virginia.[15] In this particular case, several inefficiencies related to existing enforcement systems in place at both the federal and state level were illustrated by ineffective state/federal efforts to enforce various regulatory standards dealing with air, water and hazardous waste disposal issues.

The enforcement history of Avtex Fibers dates back to 1981, when groundwater contamination was noted by the Virginia Solid Waste Control Board (SWCB). Subsequent studies found that groundwater contamination had been caused by the leaching of carbon disulfide (CS_2) process wastes from unlined surface impoundments. In addition, CS_2 contamination on site had leaked through enough of the surface groundwater to have the facility designated a Superfund site in 1984. Among the documents that regulatory personnel had reviewed, included those documents that attributed groundwater contamination to poor hazardous waste management on site by company officials. Yet, despite such documentation, the company had not been cited for RCRA violations during this time.

Among the presumptions to explain the company's lax attitude toward clean-up efforts was that EPA had chosen to use Superfund as the vehicle to correct the on-site contamination. As a result, the company delayed expenditures on needed repairs. In addition, any corrective action that would have to be taken also required coordination with Virginia's Department of Waste Management. This additional layer of regulatory control may have also inadvertently delayed the stabilization of the site.

During the Superfund process, several additional obstacles occurred, which compounded the cleanup process. Specifically, a transformer blew, and as a result, it released PCBs into the Shenandoah River. This event was not reported to the proper authorities, and in a subsequent EPA inspection, numerous TSCA violations were observed. Yet EPA took no enforcement action at that time, and EPA's TSCA office did not communicate their findings to enforcement personnel until three years later.

In addition, during the period of 1980 to 1988, over nineteen hundred Clean Water Act violations were noted, and regulatory personnel issued no penalties; air emissions of carbon disulfide were also significant, and air quality modeling performed determined that the evaporation of carbon disulfide was a greater threat than previously believed.

In summary, past practices and enforcement activities at this site allowed the following to occur current to the date of the article:

- PCBs contaminated thirty-two miles of the Shenandoah River and one hundred acres of land;
- The groundwater is contaminated with carbon disulfide arsenic, cadmium, and lead;
- Industrial process acid spilled in quantities that ate through concrete floors and bedrock to contaminate the groundwater;
- One worker (apparently) died from inhaling toxic fumes;
- Clean-up was estimated (1993) in excess of $20 million, and final estimates are projected to $300 million;

We selected this case because it illustrates part of the reason behind EPA's movement toward a more integrated enforcement system to prevent a similar recurrence to the Avtex Fibers case and the need for EPA to conduct coordinated multimedia inspections. Since then, EPA has pushed forward with a more comprehensive multimedia enforcement program, directed by OECA's office of regulatory enforcement. This program is to focus on the national screening of companies with multiple facilities to determine whether they are characteristically in noncompliance with environmental laws on a facility-wide and nationwide basis. As an example, ASARCO, Inc., settled with EPA to address hazardous waste violations at two of its facilities and to establish a court-enforced environmental management system at over thirty facilities nationwide.[16]

MULTIMEDIA INSPECTIONS

All routine inspections conducted by EPA can be grouped into four categories of increasing complexity, moving from Category A (a program-specific compliance inspection) to Category D (a complex multimedia investigation), depending upon the complexity of the facility and the objectives of the investigation. Factors used to determine which category is applicable include the complexity of pollution source, facility size, process operations, pollution controls, and the personnel and time resources required to conduct the compliance investigation. EPA describes the following four categories of inspections in its "Multi-Media Investigation Manual":[17]

Category A: Program-specific compliance inspections, conducted by one or more inspectors. The objective is to determine facility compliance status for program-specific regulations.

Category B: Program-specific compliance inspections (e.g., compliance with hazardous waste regulations), which are conducted by one or more inspectors; however, the inspector(s) screen for and report on obvious, key indicators of possible noncompliance in other environmental program areas.

> **Category B** multimedia inspections have limited, focused objectives and are most appropriate for smaller, less complex facilities that are subject to only a few environmental laws. The objective is to determine compliance for program-specific regulations and to refer information to other programs based on screening inspections.

Category C: Several concurrent and coordinate program-specific compliance investigations conducted by a team of investigators representing two or more program offices. The team, which is headed by a team leader, conducts a detailed compliance evaluation for each of the target programs.

> **Category C** multimedia investigations have more compliance issues to address than the Category B inspection and are more appropriate for intermediate to large facilities that are subject to a variety of environmental laws. The objective is to determine compliance for several targeted program-specific areas. Reports on obvious, key indicators of possible noncompliance in other environmental program areas are also made.

Category D: These comprehensive facility evaluations address not only compliance in targeted program-specific regulations, but also try to identify environmental problems that might otherwise be overlooked. The initial focus is normally on facility processes to identify activities (e.g., new chemical manufacturing) and by-products/waste streams potentially subject to regulation. The by-products/waste streams are traced to final disposition (on-site or off-site treatment, storage, and/or disposal). When regulated activities or waste streams are identified, a compliance evaluation is made with respect to applicable requirements.

> **Category D** multimedia investigations are thorough and, consequently, resource intensive. They are appropriate for intermediate to large, complex facilities that are subject to a variety of environmental laws. Compliance determinations are made for several targeted program-specific areas. Reports on possible noncompliance are made for other program areas.
>
> The investigation team, headed by a team leader, is comprised of staff thoroughly trained in different program areas. For example, a large industrial facility with multiple process operations may be regulated under numerous environmental statutes, such as the Clean Water Act (CWA), Clean Air Act (CAA), Resource Conservation and Recovery Act (RCRA), Toxic Substances Control Act (TSCA), Comprehensive Environmental Response, Compensation and Liability Act (CERCLA), and the Federal Insecticide and Rodenticide Act (FIFRA). The on-site investigation is conducted during one or more time periods, during which intense concurrent program-specific compliance evaluations are conducted, often by the same cross-trained personnel. (We include a sample multimedia checklist in Appendix A.)

Generally, each category of inspections use the same protocols for conducting the inspections, including preinspection planning, use of a project plan, sampling, inspection procedures, and final report. The major difference will be in the number of different regulations that the inspections in Categories C and D address.

According to EPA, the multimedia approach for conducting compliance inspections (described in Categories C and D, above) has several advantages over a program-specific inspection, including:

- A more comprehensive and reliable assessment of a facility's compliance with fewer missed violations;
- Improved enforcement support and better potential for enforcement;
- A higher probability to uncover/prevent problems before they occur or before they manifest an environmental or public health risk;
- Ability to respond more effectively to non–program-specific complaints, issues, or needs and develop a better understanding of cross-media problems and issues, such as waste minimization.

SAMPLE ENFORCEMENT ACTIONS

As a reminder to us, trade publications, newsletters and national papers, regularly depict EPA and OSHA enforcement actions taken against regulated companies. We called a representative sample of these companies from various sources to culminate our focus on EPA inspections and the agency's push behind inspections—the enforcement of environmental laws and creating a deterrent presence within the regulated community.

The companies called from various sources include:

- **Louisiana-Pacific**   In a federal consent decree, EPA fined Louisiana-Pacific $11.1 million and ordered the company to install pollution-control equipment at 11 plants around the country, at a projected cost of $70 million. EPA said that Louisiana-Pacific, which manufactures wood paneling, violated the Clean Air Act by failing to obtain permits for or fully identify emissions at 14 plants. EPA said new incineration technology that the company is required to install would reduce emissions of carbon monoxide, volatile organic compounds (VOCs), and other pollutants.[18]

- **United Technologies**   EPA and the DOJ fined United Technologies $5.3 million for a series of abuses in handling and discharging hazardous wastes in recent years. In agreeing to pay the fines, the company said the government probe prompted it to take a closer look at its manufacturing practices and the materials they use. As part of the settlement, the company also agreed to undergo extensive audits of its environmental practice until the end of the decade. Frank McAbee, United Technologies' senior vice president for environment and business practices, said that the company had put in place in 1990 a comprehensive system to evaluate manufacturing processes to reduce pollutants. "This settlement has given us impetus to move that process even faster," he said.[19]

- **EPA Environmental Research Laboratories**   Inspector General John Martin cited two EPA Environmental Research Laboratories in Athens, GA, and Narragansett, RI., for contractor mismanagement and abuse, including documents that turned up missing from the labs' files after the inspector general's office had contacted them. At Narragansett, a research director "verbally directed Narragansett contract management staff to remove and shred documents from contract files," Martin said in an audit report. Martin discussed his findings at a hearing of the Senate Government Affairs Committee.[20]

- **Wyman-Gordon Co.**   OSHA charged Wyman-Gordon Co. with 149 safety violations at its North Grafton, MA metal forging plant and proposed over $1 million in penalties. OSHA said the alleged violations were discovered during an *inspection* of a manufacturing facility initiated *in response to an employee complaint*. The alleged hazards include unmarked and blocked exits, unguarded machine parts and lack of protective equipment for workers near furnaces and acid tanks.[21]

- **Georgia-Pacific Resins, Inc.**   Georgia-Pacific Resins agreed to pay $432,500 and make significant safety improvements following a four-month OSHA investigation into a fatal explosion at a company plant. According to OSHA head Charles Jeffers, "George-Pacific's Columbus facility cooperated fully . . . now, the company wants to move forward . . . and has agreed to implement additional measures to help prevent future accidents."[22]

- **Darling International, Inc.**   The Minnesota Pollution Control Agency (MPCA) fined Darling International $4 million after the company pled guilty in December 1996 to five felony counts of violating the Clean Air Act in Minnesota. Among the charges against the company were the illegal discharge of wastewater, diluting wastewater samples and falsifying reports to the MPCA.[23]

- **BFI Services Group, a subsidiary of Browning-Ferris Industries**   The U.S. District Court for the Eastern District of Pennsylvania (Philadelphia) fined BFI Services Group $3 million and ordered the company to pay over $600,000 to four publicly owned treatment facilities. The company pled guilty to charges arising from the illegal disposal of wastewater treatment sludge at these publicly owned facilities. The court also ordered the company to pay $1.5 million in remedial restitution to programs that address environmental concerns in southwestern Pennsylvania.[24]

- **American Insulated Wire Corporation**   EPA and the DOJ announced that American Insulated Wire agreed to a $2.4 million settlement to resolve claims that it violated the Clean Air Act, the Resource Conservation and Recovery Act (RCRA) and the Emergency Planning and Community Right-to-Know Act (EPCRA) at its manufacturing plant in Attleboro, MA. Under the agreement, the company will pay $1.4 million in penalties and spend another $1 million on two supplemental environmental projects (SEPs) that will reduce air and water pollution and ensure regulatory compliance.[25]

CONCLUSION

In closing, we would like to refer to the cases cited previously to reiterate our previous points. Specifically, while we see that EPA continues to place emphasis on multimedia inspections, EPA still conducts a good number of program specific or single media inspections. How EPA determines each inspection type may be based on criteria similar to our previous discussion about planning an inspection, among other considerations.

Does this mean that a preinspection audit should likewise focus on a specific media program? Would you conduct an environmental audit in a specific program only? Not necessarily, and each situation should first be determined on a case-by-case basis. If the purpose of a preinspection audit is to alert the regulated entity to any deficiencies that may exist in its compliance program, the audit should be designed to be multimedia in nature, with an intensity and scope equivalent to a "compliance evaluation inspection" that EPA conducts (that is, unless the targeted facility suspects that a follow-up, case-development support, or for-cause inspection is scheduled). Only a multimedia preinspection type of audit will ensure that the regulated entity identifies all potential violations and problems that exist at its facility. Such multimedia preinspection audits will also allow the regulated entity the opportunity to consider the full financial impact of any corrective action that needs to be undertaken.

Nevertheless, knowing whether the agency's planned inspection is a multimedia variety or a program-specific type is important to help the regulated entity fully prepare its personnel on probable issues that may arise during an EPA inspection and the types of information that the agency may require the facility to provide during the inspection.

If a facility is not prepared for a regulatory inspection, whether EPA, OSHA or the state, or chooses to ignore audit recommendations, their managers should not be surprised if they receive an administrative action with a sizable penalty component, just like the eight cases cited. The old adage "forewarned is forearmed" could not be said any louder.

ENDNOTES

[1]For additional information on this topic, see the EPA document, *Managing for Better Environmental Results,* EPA 100-R-97-004, March 1997, p. 18 and S. Herman, "EPA's 1998 Enforcement and Compliance Assurance Priorities," *National Environmental Enforcement Journal,* February 1998, p. 3.

[2]See S. Herman, "EPA's FY 1997 Enforcement and Compliance Assurance Priorities," downloaded from EPA's web page at es.inel.gov/oeca/naag.97.html, July 1997, p. 1 of 20.

[3](No. 1) p. 3.

[4]*Federal Register,* Volume 61, 69359, December 31, 1996. The corrections to the final rule appeared in the *Federal Register,* Volume 62, Number 54, 13513. Additional information is also provided in "EPA Adjusts Civil Penalties for Inflation," *Environmental Manager,* March 1997, p. 4.

[5]In some instances, it is not possible to correct a past violation as to make such a violation nondetectable (e.g., a reporting violation). Any attempt to do so, such as altering documents, may incur even stiffer penalties. However, a preinspection audit can be used to uncover such problems and to take the necessary steps to mitigate the problem by evidencing the facilities' good faith efforts to comply.

[6]For additional information, see the EPA documents, *The New Generation of Environmental Protection: EPA's Five-Year Strategic Plan,* EPA 200-B-94-002, July 1994, p. 71; and *The Changing Nature of Environmental and Public Health Protection: An Annual Report on Reinvention,* EPA 100-R-98-003, March 1998, p. 31.

[7]Authorized under Section 3004(u), Section 3008(h), Section 7003(u) of the Resource Conservation and Recovery Act (RCRA) and Section 106 of the Comprehensive Environmental Response, Compensation, and Liability Act (CERCLA). In addition, EPA also has the authority to administer penalties under the Clean Water Act (CWA), the Clean Air Act (CAA), CAA Amendments (CAAA), and the Toxic Substances Control Act (TSCA), and other federal regulations as listed in Table 5.2

[8]In most cases, the corrective action order also contains a penalty component, which is allowed under Section 3008(a) of RCRA. In the event a facility is unwilling to undergo any of the above corrective action orders, which can be written either as a unilateral or consent order, EPA then has the authority, under Section 3008(h) of RCRA to refer such enforcement actions to the Department of Justice (DOJ). Prior to referral to DOJ, the next step in negotiations would be to present the case before an administrative law judge.

[9]Further information on how warrants came to be used in some types of regulatory inspections, the reader is referred to the landmark case for warrants, *Marshall v. Barlow's.* This case is found in the Supreme Court decision No. 76-1143, 436 US 307, 56 L.Ed.2d 305, listed in 98 *Supreme Court Reporter,* pp. 1816–1834.

[10]For additional insight from the perspective of the safety officer and the OSHA inspector, the reader is referred to, "On the Docket—Stopping OSHA at the Door," *Safety Management,* June 1997, Number 411, p. 4, Bureau of Business Practice.

[11]Memorandum from Herbert H. Tate, Jr., Assistant Administrator, Office of Enforcement, USEPA, July 27, 1992.

[12]For additional insight into this subject area, the reader is referred to the article by P. Fontaine, "EPA's Multimedia Enforcement Strategy: The Struggle to Close the Environmental Compliance Circle," *Columbia Journal of Environmental Law*, Vol. 18, No. 1, April 1993, pp. 31–101.

[13]For additional insight, refer to the EPA document, Solid Waste and Emergency Response (OS-305), FY '94 RCRA Implementation Plan, March 1993: OSWER DIRECTIVE #9420.00-09, and the EPA's, Solid Waste and Emergency Response (5101), FY '95 RCRA Implementation Plan Addendum, OSWER DIRECTIVE #9420.00-10.

[14]See n. 9 and n. 10.

[15]Fontaine, "EPA's Multimedia Enforcement Strategy: The Struggle to Close the Environmental Compliance Circle," (n. 12), pp. 31–101.

[16]S. Herman, "EPA's 1998 Enforcement and Compliance Assurance Priorities," National Environmental Enforcement Journal, p. 8.

[17]EPA's Office of Enforcement, *Multi-Media Investigation Manual,* Appendix B, p. B-1, 1992, by Government Institutes.

[18]T. Noah, "Louisiana-Pacific Gets Hefty Penalty in Clean Air Care," *Wall Street Journal*, May 25 1993, p. B2.

[19]A. K. Naj, "United Technologies Fined $5.3 Million for Series of Environmental Violations," *Wall Street Journal*, August 24, 1993, p. B6.

[20]Washington (AP)—"EPA investigator says officials shredded documents, coached employees," *Civil Engineering News,* August 1993, p. 10, Mavietta, GA.

[21]News Item—"OSHA asks safety penalties of more than $1 million," *Wall Street Journal,* August 13, 1993, p. B2

[22]OSHA Highlights—Safety Compliance Letter, April 10, 1998, Bureau of Business Practice, p. 1.

[23]State Roundup—*Safety Compliance Letter,* August 25, 1997, p. 8.

[24]State Roundup—*Safety Compliance Letter,* September 25, 1997, p. 8.

[25]Issues & Trends, EPA—*Safety Compliance Letter,* July 10, 1998, p. 7.

REFERENCES

Naj, A.K. "United Technologies Fined $5.3 Million For Series of Environmental Violations," *Wall Street Journal*, August 24, 1993, p. B6.

Washington (AP) "EPA Investigator Says Officials Shredded Documents, Coached Employees," *Civil Engineering News,* August 1993, p. 10.

EPA adjusts Civil Penalties for Inflation, *Environmental Manager,* March 1997, p. 4.

EPA document, *Managing for Better Environmental Results,* EPA 100-R-97-004, March 1997, p. 18.

EPA document, *The New Generation of Environmental Protection: EPA's Five-Year Strategic Plan,* EPA 200-B-94-002, July 1994, p. 76.

EPA document, *The Changing Nature of Environmental and Public Health Protection: An Annual Report on Reinvention,* EPA IVO-R-98-003, March 1998, p. 31.

EPA document, Solid Waste and Emergency Response (OS-30J), FY '94 RCRA Implementation Plan, March 1993: OSWER DIRECTIVE # 9420.00-09.

EPA document, Solid Waste and Emergency Response (5101), FY '95 RCRA Implementation Plan Addendum, OSWER DIRECTIVE # 9420.00-10.

EPA document, EPA's Office of Enforcement, *Multi-Media Investigation Manual,* Appendix B, p. B-1, 1992, Rockville, MD: Government Institutes. All rights reserved.

Fontaine, P. "EPA's Multimedia Enforcement Strategy: The Struggle to Close the Environmental Compliance Circle," *Columbia Journal of Environmental Law,* vol. 12, No. 1, April 1993, pp. 31-101.

Herman, S. "EPA's 1998 Enforcement and Compliance Assurance Priorities," *National Environmental Enforcement Journal,* February 1998, p. 3.

Herman, S. "EPA's FY 1997 Enforcement and Compliance Assurance Priorities," downloaded from EPA's web page at es.inel.gov/oeca/naag.97.html, July 1997.

Issues & Trends, EPA—*Safety Compliance Letter,* July 10, 1998, *Bureau of Business Practice,* p. 7.

Marshall v. Barlow's, *98 Supreme Court Reporter,* Supreme Court decision No. 76-1143, 436 US 307, 56 L. Ed. 2d 305, pages 1816 through 1834.

On the Docket, Stopping OSHA at the Door, *Safety Management,* June 1997, *Bureau of Business Practice,* No. 411, p. 4.

OSHA Highlights—*Safety Compliance Letter,* April 10, 1998, *Bureau of Business Practice,* p. 1.

State Roundup—*Safety Compliance Letter,* August 25, 1997, *Bureau of Business Practice,* p. 8.

State Roundup—*Safety Compliance Letter,* September 25, 1997, *Bureau of Business Practice,* p. 8.

*Noah, T. "Louisiana-Pacific Gets Hefty Penalty in Clean Air Case," *Wall Street Journal,* May 25, 1993, p. B2.

News Item, "OSHA Asks Safety Penalties of More than $1 Million," *Wall Street Journal,* August 13, 1993, p. B2.

CHAPTER 6

A LOOK AT INNOVATIVE STRATEGIES FOR THE ENVIRONMENTAL MANAGER

Gabriele Crognale, P.E. and Renee Bobal, Hoffmann-LaRoche, Inc.

> *"If you can't convince them, confuse them."*
>
> *Harry Truman*

INTRODUCTION

Within the structure of environmental management at regulated organizations, be it at the corporate or plant level, managing environmental issues is usually within the purview of senior and experienced environmental professionals and their specialized support services staff. In some organizations, environmental management responsibilities at facilities are under the direct control of corporate environmental and health and safety (EH&S) managers, while at other organizations, each facility is under the control of autonomous facility environmental managers who are independent from their corporate EH&S counterparts. In other organizations, some of this work has begun to be outsourced to an increasing number of third-party contractors, which we discuss in greater detail in Chapter 15.

While individual managers may wear different hats from their varied functions in the organization, they share several things in common: their title—environmental manager, or EH&S manager—and the fact that routine assignments can change due to unforeseen circumstances requiring immediate attention. For example, accidents, spills, or surprise regulatory inspections can lead to unanticipated changes or shifting of a manager's priorities. Whether a manager's duties focus upon addressing regulatory compliance issues, such as spills, training, or other regulatory concerns, delegating staff to resolve issues, addressing environmental audit recommendations, or dealing with top management to address other job-related items, there is always something new on the horizon. Now, due to changes brought on by trade pressures, their responsibilities are slowly melding into the organization's mainstream. Some are also finding new responsibilities from other business areas, such as procuring various goods and services. As such, it becomes increasingly important for environmental managers and support staff, whether at the corporate or at the plant level, to have at their disposal additional tools to enable them to perform their job duties more effectively. Included in this new order of responsibilities is dispelling any ill-conceived perceptions by other business unit managers that environmental responsibilities are a drain on company resources and profits. In reality, with such tools at their disposal, they can execute a number of strategies to help them achieve their regulatory compliance goals, while also saving their companies measurable sums of money.[1]

Their tools can focus upon: (1) evaluating EH&S responsibilities and delegating the appropriate individuals to execute these responsibilities; (2) championing "out-of-the-box" thinking from EH&S staff and shop staff to help promote a new way of looking at environmental considerations and responsibilities within the company, such as in producing products more efficiently while generating less waste as a result, or agreeing to become part of an EPA-led voluntary program, such as proposals submitted by regulated organizations to EPA in the Environmental Leadership Program (ELP) and Project XL; (3) exploring the use of specific EH&S and other software tools, such as chemical process real-time software, training and human resource tracking software, as a means to help manage various EH&S responsibilities and keep track of compliance and technical data; (4) initiating benchmarking studies with "best-in-class" companies within the regulated community who are championing forward-thinking environmental strategies, as discussed in greater detail in the next chapter; and (5) championing other innovative thinking concepts from employees to promote worker efficiency, greater cost savings without "downsizing," and instill creativity at all employee levels to fuel improvements.

Of importance with respect to success is that one needs to help foster a feeling of togetherness among employees to put into practice the tenet of an organization's environmental goals, policies, and principles, with the overarching goal to benefit the organization in the long run. As such, while this chapter's focus is to look at strategies that managers can undertake in executing their job duties, the strategies we discuss in greater detail are not the exclusive domain of the environmental managers alone. Rather, our premise is that such strategies can also apply to staff members whose responsibilities may require some strategic thinking and planning in order to carry out the environmental tasks required, or plant workers and other employees who interact with a company's EH&S group; in short, no one within an organization is excluded; everyone can play a part—from the engineer in research and development, to the parts manufacturer, to the shipping clerk. In such a setting, benefits can be derived within a company when everyone contributes.

With that introduction, let us look at these strategies in greater detail and also evaluate a typical company posture some regulated organizations take, that of the regulatory compliance *reactionary* responses to address environmental compliance issues. One suggestion for moving beyond the "perverse spiral"[2] that generates a steady stream of additional regulations to control adverse environmental situations not previously regulated, is for organizations to explore innovative environmental programs that address or evaluate emerging environmental considerations that are not yet on an organization's "radar screen." This preemptive approach could provide an organization a "heads up" on emerging concerns as a result of anticipated issues. These issues could be gleaned from a number of indicators, such as: changing sentiment, legislative overtones, business directions, innovative products, and a wide variety of other variables that factor into this "crystal ball" to unravel the "perverse spiral." There may also be other subtle overtones that may need to be considered in any strategic preemptive actions to help regulated organizations move forward beyond compliance considerations that sometimes seem to bog down organizations, stakeholder groups, and even the regulators themselves.[3]

For example, as described in part in Chapter 1, the paradigm shift has changed from maintaining compliance ("end-of-pipe" solutions mindset) to going beyond compliance, to saving resources, worker effort, and monies ("beginning-of-pipe" solutions breakthrough thinking) as the main driver in environmental management responsibilities. In such fashion, a forward-moving philosophy fostered by an environmental policy that has top management support can provide organizations opportunities to maintain and sustain regulatory compliance while looking for opportunities to improve upon existing situations that can benefit all concerned parties. The key is that top management needs to show its commitment and support to all levels of employees to make them feel that management is behind them. In one example of management support, Dow Chemical Company allowed one of its EH&S specialists to provide his specific EH&S expertise to Nalco Chemical Company, while Dow received in return, a sales and marketing specialist from Nalco Chemical to help in one of its business lines. In this manner, an innovative concept was explored by both organizations as a means for each of them to obtain valuable assistance while providing a vehicle for the individual employees to see something new in their jobs.

In similar fashion, the Boeing Company's Cross Functional Assignment Program allows company opportunities to be placed from one organization or group into another for the mutual benefit of the participant and the organization. This program resides with Boeing's Manufacturing Research and Development Organization and was developed as one way for employees to sharpen their skills and broaden their horizons in various other work disciplines within the company. This concept is noteworthy because it provides employees opportunities to try new assignments within one's own company without the stress of looking for a new job with another company, and possibly in another location. Additional insight into Boeing's Cross Functional Assignment Program is provided in further detail in Appendix G, with permission from the Boeing Company.

A SAMPLING OF INNOVATIVE STRATEGIES

KEEPING UP WITH REGULATORY REQUIREMENTS

Part of the environmental manager's day-to-day responsibilities includes keeping abreast of current and emerging federal and state environmental and related regulations, such as those regulated by the DOT, OSHA, the SEC, and in some industries, the FDA. A let-up in this required reading exercise, which includes *Fed-*

eral Register notices, EPA policy memoranda, or other interpretations and guidance, and applicable state regulations and policies, can develop into issues for managers and their organization that may need to be resolved at some point.

While the volume of regulatory information to be managed never seems to ease, the task of collecting, collating, and archiving this data into a manageable compilation for internal use and synthesis need not be a formidable task. For example, some companies have developed CD/ROMs and $3\frac{1}{2}''$ diskettes that allow a subscriber to view EPA's and OSHA's federal regulations and corresponding state regulations on a personal computer. A subscriber to one of these services can also elect to receive regular updates, such as *Federal Register* notices, and other bits of information as part of the service provided. Other service providers allow subscribers to retrieve information about facilities in any geographic area that are stored in one or more EPA or state databases, such as RCRA or CERCLA databases. This particular information could be useful to facility managers as a check to ensure their facility has copies of all pertinent information relative to their facility that is maintained by the respective regulatory agencies; or in the event facility representatives may want to know additional information about a nearby facility or another facility they may be viewing in some potential merger and acquisition, or other company-specific use. On a grander scale, EPA first launched Envirofacts in 1994 as a tool for internal EPA use only, but with the advent and growing popularity of the Internet, this web site was subsequently made available to the public as well. This site currently provides information on a regulated company's compliance reports and similar environmental information and, as of November 1997, had garnered mixed reviews from the regulated community.[4]

Another method for environmental managers and their staff to keep up with current regulations and emerging trends can be through professional meetings, conferences, or symposiums. They can benefit from attending these events by hearing presentations from experts in the field on current issues and trends and how to solve specific EH&S issues; these events also provide them opportunities to network with fellow professionals, or explore other opportunities. A sampling of several noteworthy conferences that were held in 1996 to 1997 is provided in Chapter 7.

Should a company's travel budget not allow an environmental professional to attend one or more of such conferences or seminars, another method that environmental, or health and safety managers can use to keep up with this constant barrage of regulations is by accessing any of the pertinent trade publications, newsletters, and other periodicals that are readily available. To facilitate this task, managers could designate staff members to compile and catalogue these publications for internal use, in such categories as reference for specific ES&H issues, research data on some internal memo or report, or other practical application. This library does not have to be a major undertaking, and should be tailored to fit each company's specific needs. In addition, a number of trade publications are available, some of which are complimentary to ES&H professionals, and others that vary in price that can address issues or concerns an environmental professional may have. Be aware though that each publication is different, based on its editorial content and subject matter focus, and can be written by in-house journalists, contributing technical writers or consultants, or a combination of all of these, depending on the editor's focus and intent. Those publications that allow contributing writers, usually environmental or health and safety practitioners, can provide an added dimension to the subject matter, since these contributors relate their articles to their actual experiences. A comprehensive list of these publications can be found in Appendix B.

The time it may take for staff members to research and pull these publications together may be well worth it to the EH&S group and others at the facility. From this effort, you can create a tailor-made research library that is accessible to all facility personnel for research that can be updated regularly. This library can be a source of pertinent information for the EH&S staff, other facility workers who may want to research a particular regulatory item or issue, and any new staffers that may need additional assistance in getting up to speed with regulatory requirements.

From our experience, we have seen sections of an office or room set aside as a research library to house these documents for use by facility personnel. Such libraries may house various trade publications, periodicals, *Federal Register* notices, the complete set of the Code of Federal Regulations, Parts 29 (OSHA) 33 and 40 (EPA), subscriber-generated state regulations, specialty training manuals, conference proceedings, and consultants' reports, if applicable. Any portion of this information can be very useful to an employee who may be in need of some assistance in addressing or clarifying an environmental issue. By conducting a diligent research of the library's periodicals, for example, an employee may find an article or news item that may have already addressed the issues and provides necessary responses that employee may seek. As another example, an EH&S staff member could find sufficient material in a facility's well-planned library to develop an in-house training program, draft a memo for internal use that can help explain a particular subject that may be of interest to other employees, or initiate the groundwork for a benchmark study. In those cases where an employee may have attended an off-site training session, he or she may supplement training materials with articles or other documents from the library to reinforce the subject matter, or other "train-the-trainer" in-house uses as appropriate. Such materials, used in tandem with supplied training materials, can enhance a training program, and help to dispel any Dilbert® training witticisms, such as: ". . . The training is already forgotten, but the binder will last forever."[5]

Also, with the added popularity of the Internet, many organizations have developed home pages that provide additional insight into their products and their standpoint on environmental matters, such as environmental policies, reports, and other related information. EPA, too, maintains an extensive home page that provides timely information on various EPA-sponsored activities, such as the Environmental Leadership Program or Project XL. (Additional information on EPA and other government agency web sites can be found in Appendix G.) With such information at hand, a resourceful EH&S staff member could pull down supplemental information that may address questions he or she may have related to some environmental concern or issue, or at least be able to point the staff member where to go for additional information that may provide the answers.

As another example, a growing interest by regulated industries in the ISO 14000 standards and certification to ISO 14001 has generated a business enterprise for the twelve (and counting) ANSI/RAB-accredited course providers[6] that offer a suite of ISO 14001 courses, including the more comprehensive EMS Lead Auditor course, at various locations across the country at different times. Attending and passing the lead auditor course is an important first step for an individual seeking to become a Registration Accreditation Board (RAB)-certified EMS Lead Auditor, whether internally for the facility, or as a third-party service provider, to help prepare a facility for ISO 14001 certification by one of the accredited registrars. As such, a company interested in pursuing some aspect of ISO 14001 may wish to have one or more EH&S staffers attend an EMS course as the first step in becoming an RAB-certified EMS Lead Auditor.[7]

Upon completion of such ISO 14001 courses, attendees could utilize what they learned to develop internal briefing sessions or mini-seminars to help brief other facility employees on the ISO 14000 standards from what attendees may have learned at these training sessions. Recent course attendees could develop such sessions as a means to help prepare other facility employees to take the EMS course at a later date. Depending upon the amount of supplemental ISO 14000 or general environmental auditing information available on-site at the facility, for example, they could develop a fairly comprehensive in-house program to help bring employees less experienced in EMS and EH&S methodologies up to speed fairly quickly on EMS matters. Such "drills" or "quizzes" could increase their chances of being more prepared to take ISO 1400 courses, and more importantly, coming away from these courses with a higher degree of familiarity and proficiency in EMSs, auditing to a facility's EMS, and continual improvement matters—some of the key tenets of the EMS Lead Auditor's course. In some of the courses in which this author assisted in teaching, we noticed that some attendees needed more help than others to fully grasp EMS and auditing concepts presented and could have benefited from such a pre-course briefing. This is not to say that developing materials for in-house use should take the place of the EMS auditor courses offered; rather, any effort expended on internal briefings or mini-sessions could be beneficial in preparing other facility employees who may be scheduled to attend these same ISO 14000 courses themselves. Any use beyond that may encroach upon ethical and copyright issues that should be evaluated closely for their appropriateness on a case-by-case basis. *Additional discussion related to training considerations and concerns is provided later in this chapter.*

FOLLOWING UP WITH ENVIRONMENTAL AND EMS AUDIT DEFICIENCIES

Another consideration environmental managers or their designated specialists should explore in some regular fashion is establishing some program to track internal and external third-party audits [in the context of this book, audits refer to environmental or environmental management system (EMS) audits, unless expressed otherwise] to ensure that audit recommendations are corrected and to ensure that any deficiencies noted in audit reports are not repeat deficiencies. Such a program could be instrumental as one of the tools for moving a company away from a "compliance defense" mindset. In certain situations, an internal tracking software system or audit software package could come in handy to assist in this process. Additional information on specialized EH&S and EMS-focused software is provided in Chapter 13.

In this fashion, environmental managers could have a greater sense of assurance that audit deficiencies are being addressed and any potential regulatory issues related to such deficiencies can be avoided. In addition, depending upon the number and regulatory severity of each of the findings, an astute manager could evaluate each of the findings and their corresponding audit recommendations and prioritize them for management review. With this information, they could discuss these priorities with facility managers, and other business unit managers, if necessary, to reach a consensus regarding which of the items would require immediate attention (serious regulatory concern—high priority) and which items could be executed later in the quarter, or the subsequent fiscal quarters, should conditions warrant. Any items requiring immediate attention because of their potential impact to human health or the environment, that are responded to casually, may precipitate a stern enforcement response should a regulator uncover the violations.

The benefit of a consensus-building approach is that it allows top management to add their input in prioritizing audit recommendations for corrective actions, especially those focusing upon regulatory findings. In this manner, a strong likelihood exists that high-priority action items would be acted upon as expeditiously as possible, and have also factored in any of management's concerns regarding the organization's financial situation. In addition, such priority ranking allows management to set completion milestones for subsequent priority action items, as well as setting in motion a process to track the progress of the action items.

With that, let us take a closer look at a hypothetical EMS audit that we can presume is conducted by the company's corporate EH&S audit group. The audit uncovers several regulatory and management system findings, and based on these findings, determines that: (1) shipping and receiving employees require proper training in chemical handling and off-loading of chemicals; (2) third-shift line employees require introductory training in the proper disposal procedures for chemicals used on-site; (3) employees using machine equipment require training in the use of personal protection equipment (PPE); (4) hazardous wastes are not properly stored on-site; and (5) no procedure exists outlining the proper protocol for hazardous waste manifests.

In this scenario, some of the findings as described require some type of immediate attention, given their potential for accidents or other mishaps to occur as well as their regulatory consequences, while others could be addressed over a longer period of time. Training, for example, is one area that needs a greater time span, since there is usually a long lead time to identify the individuals who require training, ensure trainer availability, and schedule the employees for training, depending on their workloads. Based on these findings and their subsequent audit recommendations, environmental managers can develop corrective action plans coupled with (prioritized) audit recommendations to allow them to accomplish any audit-recommended tasks more efficiently while being able to convey audit recommendations to their managers in an easy-to-follow manner. To ensure a more receptive listening audience, such briefings should be conducted without a heavy dependence on environmental terms, regulatory requirements, and acronyms; easy-to-understand terms directed to the target audience may be more effective. In addition, a set of audit recommendations and corresponding corrective action items should also be replete with tables and graphs indicating implementation costs and the projected savings realized in implementing such corrective actions over the course of the next week, month, or quarter which make a greater impression on business or financial managers within the management review team. All too often, many of us in the environmental sector have a tendency to speak in environmental, or "enviro-speak," terms, which does not translate well in the dollars-and-cents terms of business people, or "business-speak."

DELVING IN ROOT-CAUSE ISSUES TO IMPROVE EMSS

The next level for moving forward focuses on root-cause issues that are usually uncovered as part of a comprehensive audit. Referring back to the previous discussion about following up on audit deficiencies, when one steps back and looks at previous audit reports and recommendations to obtain a "big picture" focus, another facet of a facility's environmental management system or program can be discovered that may have otherwise gone unnoticed. The value such review can provide to environmental managers and plant managers can be measured in a number of ways, such as described in Chapter 5.

Let us refer to the examples of hypothetical findings given in the previous section and focus instead on root-cause issues as a means to improve upon existing situations and provide a framework for developing some feasible recommendations. For example:

• *Shipping and Receiving employees require proper training in chemical handling and off-loading of chemicals.* Instances could arise where properly trained employees could be able to identify defective chemical storage drums and not accept them, thus precluding a possible chemical spill event in the chemical storage area after hours. The key here is to look for situations where a little forethought and effort can eliminate unnecessary problem spots and create improvements. In other situations, unless company procedures require otherwise, water-reactive chemicals could be inadvertently stored near a sprinkler system, eye wash, or other water-generating area that could create havoc at some point in time.

• *Third-shift employees require introductory training in the proper disposal procedure for chemicals used on-site.* In such situations, root causes may be varied and may stem from items as diverse as a disconnect between human resources and manufacturing that allows such employees to fall through the cracks and not receive the proper training; to training that has been offered, but the disconnect may be attributable to: the trainer's teaching ability, a language barrier (English is no longer the predominant language spoken by a good number of plant workers in some geographical areas), or an attitude that some employees may display for any type of training provided to them.

• *Employees using machine equipment require training in the use of personal protection equipment.* This particular audit finding may be directly attributable to OSHA requirements, and the root cause may stem from a number of sources, similar to the previous example. However, the connection to environmental issues bears noting in this section because of the environmental consequences that may result from an employee not donning or improperly donning PPE as part of his or her work routine. For example, an employee who is required to wear protective gloves, face shield, and apron when siphoning chemicals out of a drum, decides to forgo this procedure in handling a hazardous chemical. The employee spills some of the chemical and, in doing so, is injured. The 5-gallon pail holding the chemical is dropped to the floor and a spill occurs. The end result in this hypothetical example is that the employee is sufficiently injured to miss two weeks of work, and the HAZMAT Team is summoned to manage this spill. Precautionary care could have avoided this event.[8]

• *Hazardous wastes are not properly stored on-site.* The root cause to this finding can be one of a number of items: improper or insufficient employee training for those employees responsible for managing and disposing of hazardous wastes on-site; employees unable to handle the workload; not enough employees to address all of the job responsibilities; an attitude issue exhibited by one or more employees; or, any one of a number of issues that may need to be examined in greater detail.

• *No procedure exists outlining the proper protocol for signing hazardous waste manifests.* In some instances, there may be situations at facilities where the first available person at the loading dock may sign a waste shipment manifest before being transported off-site by a licensed waste hauler. This may appear to be a nonissue at first, but would a facility or environmental manager really feel comfortable in allowing any employee who happens to be available sign such an important legal document with potential liability ramifications to the facility? Chances are, they would not, and it would be a prudent course for a facility to establish a protocol or procedure to identify the appropriate facility employee responsible for signing hazardous waste manifests to accompany waste shipments.

In summary, the ability of environmental managers or other key environmental professionals to delve into root-cause issues as a means to improve upon existing conditions, rather than providing a facility the so-called "Band-Aid"™ approach, is a very valuable trait that may develop intuitively over time, or as lessons to be learned in a trial-and-error fashion over several audit occurrences. Each situation encountered, just as in the hypothetical examples provided, needs to be evaluated on a case-by-case basis. While some situations may appear to be similar, the root cause behind some audit findings and recommendations may be completely different. What managers can glean from examples in previous audit reports, or from their first-hand accounts, include the methodology behind a particular finding, and how certain audit recommendations can substantiate a particular root cause, or a series of root causes. With this information, managers can be better "equipped" to improve upon a facility's existing environmental management system and overall EH&S program.

PROVIDING ADEQUATE AND APPROPRIATE TRAINING TO AFFECTED WORKERS

It is probably intuitive and an understatement that a regulated organization's EH&S and support staff is only as good as its least-trained employee responsible for some facet of the organization's EMS and EH&S program. While this may appear to be a very broad statement, and something that may be difficult to assess for validity purposes at any given organization, let us take a closer look at this statement using a hypothetical situation.

For instance, workers in a specific work setting in a manufacturing facility need to be aware of any hazards associated with the machinery and processes they use, as well as the chemicals they may use, based on available material safety data sheets (MSDS) for each chemical at their work stations. They also need to be aware of the proper disposal procedures for the chemicals they may use in these processes, and the limitations these procedures place on them. Some of these employees may also be new hires, and, as a result, be unfamiliar with some or all of the environmental requirements that come with their new job descriptions. Such scenarios may also generate situations that could lead to regulatory compliance issues for the facility.

Let us suppose, for this example, that the facility is a heavy machinery manufacturer, and the workers need to be aware of several critical work-related items, such as: the proper PPE to don in those work areas requiring PPEs; the proper procedure to follow in a spray-paint application, especially when paints need to be changed or the paint system purged; what wastes need to be disposed as hazardous and where they are supposed to be accumulated; how to properly handle and store drums of virgin chemicals and waste chemicals; how to address accidents or other problems that may occur in day-to-day operations; and, borrowing from a previous example, what to look for in inspecting drums of virgin chemicals being off-loaded at a loading dock, to name a few.

Any of the examples provided can precipitate any one of a number of accidents, injuries, spills, or other safety and/or environmental regulatory compliance issues that could be avoided with the proper training of all affected employees. Granted, this preventative measure takes considerable time and effort on the part of facility management, but ask yourself: isn't it worth it to you in the long run?[9]

The preceding paragraphs were provided as an introduction to explain in greater detail the importance of adequate and appropriate training to an organization's target audience—those workers who absolutely need specialized training—to ensure that potential liabilities are mitigated even before they may have the chance to develop.

Specialized environmental compliance and environmental awareness training is not just for front-line or plant and process line workers; rather, environmental managers and their EH&S staff can benefit as well from such training. They can use it to freshen up their EH&S skills, or learn to see a different perspective relating to the issues at hand. Whether they attend a professional conference, trade show, technical presentation or other professional session, EH&S professionals can learn much from attending any of these advanced training sessions. At any of these sessions, they can obtain tidbits of valuable information from conversations with their peers at coffee breaks between technical sessions, after-hour mixers, or other off-line networking opportunities. If such opportunities avail themselves, attendees should take full advantage of them. This type of "out-of-class" training can oftentimes provide additional and valuable information that should not be overlooked or dismissed.

The most important lesson in training, whether it is directed to line workers in the proper execution of their job duties, or to EH&S staff at an off-site training session or conference, is that trainees or attendees who are keenly attuned and attentive at all times, can usually get the most value for their training monies.

EXPANDING AWARENESS INTO OTHER BUSINESS AREAS OF THE ORGANIZATION

While the focus of this chapter thus far has been to look inwardly within the EH&S groups to champion innovative environmental management strategies, a paradigm shift in business may be able to provide environmental managers with opportunities to share their long-term EH&S strategies with other, interrelated business units as a means to achieve their EH&S goals over a certain time frame. Including other business unit managers, such as purchasing, research and development, distribution and human resources and the like, in the EH&S decision-making loop, can provide additional insight to all the parties involved and allow EH&S issues to be seen in a different light by non-EH&S managers. With this added insight, these managers can help environmental managers attain their EH&S goals that can also have a positive effect on their own business units. These goals may include, among others: accident and pollution prevention, regulatory compliance, cost savings, quality outputs, and high worker morale.

While some of these goals clearly may be within the purview of the environmental arena, there just could be some goals that are shared by other business units, and in sharing such information, some duplication of effort by each of the unit managers could be avoided. For example, accident prevention is one area of concern that probably encompasses every manufacturing, shipping, and storing unit within an organization. Brainstorming of ideas and issues to be resolved by each of the units could be discussed in an open forum or meeting to reach a standard accident prevention protocol that could be part of a general employee training manual, if appropriate.

In addition, managers from these same business units could also brainstorm ideas to help reduce costs, in any one of a number of areas that may have been previously overlooked. For example, manufacturing may have been purchasing a chemical for its specific properties that could be substituted with another, less toxic chemical, that may also be less costly to dispose; or various degreasing machines may have been using a certain volatile organic compound (VOC) degreaser that could be substituted with a nonvolatile cleaner that could even eliminate the need for fume hoods or other VOC fume-containment measures. By allowing purchasing managers to be privy to such information, purchasing specifications could be changed to eliminate such chemical purchases and realize across-the-board cost savings that could also be tracked.

In shipping and receiving, for example, incoming materials may be packaged in cardboard or other materials that may be routinely disposed as a solid waste, while outgoing products are being shipped in a similar fashion. It may benefit that department to initiate a recycling program for the incoming packaging, just as it may be beneficial for outgoing products to be shipped in reusable containers. This information was also provided in Chapter 1. In another example, let us revisit the inspection of incoming drums containing chemical products. Here, environmental managers and shipping and receiving managers may want to evaluate devising a protocol that shipping and receiving personnel should follow with every incoming shipment of chemical products to ensure that none of the containers are damaged in any way, and thus avoid the possibility of leaks occurring after hours or on weekends, which could result in adverse and possibly costly situations.

Another area that has considerable overlap in probably every business unit and cost center within an organization is attention to quality. As an indicator of how widely quality consciousness has been ingrained into manufacturing organizations, increasing numbers of manufacturing sites are very eager to display banners proclaiming certification to ISO 9001. This phenomena is not just being exhibited by Fortune 500 companies, but small and medium enterprises (SMEs), as well. In similar fashion, with the growing global acceptance of ISO 14001, how much longer will it be before large organizations and SMEs will display their banners proclaiming certification to ISO 14001 alongside those proclaiming certification to ISO 9001? Several of the pioneers in this endeavor, SGS-Thompson, Lucent Technologies, and EG&G, provide an in-depth view into their experiences with ISO 14001 later in the book.

In focusing on quality as job one, to borrow from an often-used phrase, business unit managers are in effect acknowledging some of the tenets of an environmental management program, that is mirrored in ISO 14001. Having a number of employees ingrained with quality concepts contained within ISO 9001, helps instill a positive worker attitude for the work they perform and the product they sell. This is also a good exercise for workers in helping them understand the concepts behind an effective EMS and how that relates to their work ethics and total quality. This, in turn, can begin to put them into a mindset to understand the relationship between quality and environmental quality, and how that can lead to an EMS in conformance to ISO 14001. That aspect of conformance to ISO 14001 could be a business decision that the organization may wish to champion. Hand-in-hand with quality considerations goes worker attitudes and a high worker morale, which really need to be in place for quality products and environmental quality to move forward. These soft qualities[10] make up part of the backbone of an effective and efficient workforce, which was previously subject to the whims of top management during the "re-engineering," "downsizing," "rightsizing" frenzy of the past few years—a result of corporate pursuit of increased profits. It would appear that we may have come full circle in this regard, and organizations are again looking at workers as the main drivers for moving their businesses forward into the twenty-first century.

SHIFTING GEARS INTO A PROGRESSIVE AWARENESS MODE

ENVIRONMENTAL PROGRESS IS MORE THAN JUST COMPLIANCE

In Chapter 1, we provided an introductory discussion regarding the focus of regulated organizations and how environmental compliance has evolved since 1970 into a mature business, complete with budgets and cost considerations to help the EH&S group manage its responsibilities. The focus of environmental management has also slowly shifted away from an "end-of-the-pipe" approach to a pollution prevention, or "beginning-of-the-pipe" approach as one of the means to get ahead of the regulatory curve and save unnecessary waste management costs as well.

With this understanding, savvy and progressive environmental professionals and top managers at many regulated organizations have realized that a catch-up mindset for maintaining compliance with environmental regulations is not always the most efficient and cost-effective strategy to follow for attaining environmental progress. For some, the focus has shifted further up the food chain, if you will, to concentrate on environmental issues from a more holistic viewpoint to find effective solutions.

Among the items to be considered in such an approach include an understanding that environmental considerations are not just within the domain of EH&S professionals; rather, each line worker and other employee has some role to play in order for an organization to achieve environmental progress. It should also be stressed that environmental considerations should be viewed by plant personnel as their responsibility in some capacity or another, whether as managing hazardous wastes generated on-site, working on developing a new product formula in the research and development department, or ordering a new batch of chemical products to be used. While each of these individuals may have a different job description and different responsibilities, each of their actions can have an impact, either favorable or unfavorable, upon environmental situations that can affect their own health and safety, those around them, and the local community. In addition, costs associated with any adverse situations as a result of their actions could be avoided if acted upon before these actions occur.

For example, with a compliance-only approach, issues of concern to environmental managers and their staff focus on addressing compliance-related requirements within their EH&S domain. Areas outside this small sphere are not their immediate concern, until something occurs that brings in EH&S staff members, such as an accident, spill, or other occurrence requiring the assistance of this group. Taking the more holistic viewpoint, savvy plant or corporate managers may wish to take the time to review each of the business units and determine where potential issues may lie and initiate meetings with other business area leaders to discuss these issues and map out an agreeable solution to the issues on the table. Solving potential issues before they happen are worth discussing, especially since they may affect the entire company if left unchecked.

For these reasons, environmental progress cannot function in a vacuum, nor is this something that can be accomplished sporadically by the different business units to suit their needs. To be effective, the entire workforce within the organization needs to be on the same page regarding environmental issues as a starting point to move forward and see tangible results in increased compliance, less waste generated, and less budgeted production dollars spent. To attain that level of awareness, environmental managers may also need to focus their efforts toward getting other business unit managers to understand their position as well as listen to their concerns and needs, especially if they are to be team players on the same team. The issue is not about compliance, or total quality, for that matter—it's about people, people who care to do the job right, who understand each other, and are committed to do the right thing. From that, all top-down directives can follow pursuit.

GETTING THE ORGANIZATION TO PULL TOGETHER

The process of getting an organization's business units to pull together toward the common goal of attaining total environmental commitment needs to begin with the highest levels of the organization, and includes the approval and commitment of the organization's chief executive officer (CEO). Without that level of commitment, middle management, and EH&S management efforts may not have enough momentum to champion this cause throughout the organization. One way to gauge whether there is a level of commitment at the top management level is through the organization's environmental policy. Of course, if no policy exists, then the task at hand becomes that much more difficult to overcome. An environmental policy is a specific top-management-level document that sets the tone regarding what an organization is committed to do for the environment and usually has the full support and signature of the CEO. Environmental policies from Hewlett-Packard and Chevron, for example, are provided in Appendix E.

Another document that can provide useful information about an organization is a company's environmental report. The environmental report may include among items of interest: the company's efforts over the past year to achieve improvements in its operations; exploring alternative considerations to reduce pollution levels; and its commitment to its neighbors and the environment, to name a few.

The practice of issuing voluntary environmental reports by regulated companies, specifically publicly-owned companies, has been steadily increasing[11] since the first of these reports were issued in the 1988 to 1990 time period from several chemical and petrochemical companies, including Monsanto, Dow, Chevron, and Texaco. These reports are usually a good indicator into how some companies commit to their environmental goals within the organizations. As a learning tool, the environmental manager, quality manager, manufacturing manager, or even the CEO of a smaller company can learn much from the examples contained in these reports depicting environmental compliance, reporting data, and other exemplary milestones achieved by these larger, publicly-held companies, either as part of a benchmarking study, or as a case study from which some lessons can be learned. With regard to benchmarking, we provide further discussion about this topic later in the chapter and in Chapter 7. We also provide excerpts of several publicly-held companies' environmental reports in Chapter 10.

The information contained in such reports can also be useful as data points for EH&S management staff in preparing briefings for top management review that can be disseminated to other business unit managers and serve a dual purpose in spreading the word to other unit managers regarding the organization's total commitment to environmental responsibilities, as each unit strives to maintain an increased product throughput while running more efficiently. These other business units can include: new product research and development, existing product manufacturing, marketing and customer service, shipping and receiving, customers and suppliers, management information system (MIS) centers, human resources, and administration and finance.

As an example, the internal reports or briefings these EH&S professionals develop from the information compiled that is added to their own observations of existing conditions at specific facilities specifically relate to the environmental policy and corporate directives established by the organization. These reports could also take into account specific observations or findings from related studies conducted by others within the EH&S group or outside parties, such as environmental audit findings, to gauge the extent of progress made in these endeavors with employees in the other affected business units. Sample findings could touch upon a number of areas or issues, such as: (1) Are scientists and chemists in R&D familiar with their environmental responsibilities as highlighted in the organization's environmental policy and EH&S corporate directives? (2) Are shop workers familiar with and do they have access to MSDSs in their areas? and (3) Do shop workers and scientists in R&D properly dispose of their waste streams?

From a different perspective, the internal reports could highlight whether these same scientists and chemists: (1) Look at innovative ways to re-design or re-think their product lines; (2) take into consideration product designs that maximize product take-back or recycling; (3) use less raw materials or generate less toxic wastes as by-products; and (4) include additional related life-cycle assessment costs and issues into design considerations as part of overall business decisions. Such concepts, not exactly new, fit the description of the latest flavor-of-the-month acronym, SEM, or strategic environmental management penned by some practitioners. In reality, though these ideas and concepts more appropriately fit the catch-all category of "common sense," we sometimes lose focus in pursuit of business.

In another example of looking for ways the organization can pull together, let us not overlook customer support and customer and supplier representatives, who can also log customer satisfaction with a company's products and how these are packaged. As part of the internal report, EH&S professionals may also want to target key customers regarding whether they may also be willing to share their thoughts or any concerns they may have on a company's product lines, or ideas for improving products or packaging. While this may be more of a concern for sales and marketing managers, such information could also be useful to the EH&S professional as additional items of interest in their reports.

In a similar vein, EH&S professionals may want to track information provided by their own company's product buyers as well as their supplier representative counterparts from the companies that provide them with various raw products or chemicals. This information could then be funneled directly to the individual within the EH&S group who tracks any new materials or chemicals their company purchases, or alerts their suppliers to certain chemical products that will no longer be purchased. This information loop should also extend to research and development engineers to keep them apprised of the latest information on products. In this fashion, should research engineers wish to revise a formula with a new chemical, or wish to substitute a hazardous product for a less or nonhazardous product, they could do so by coordinating with their internal buyer and be able to obtain the latest information on these products, their uses and limitations, and any corresponding MSDS data sheets. All of this information would then be synthesized and condensed into an internal report for top management.

Could such information really be useful to top management, and if so, what could it provide to the organization? Putting this in perspective, let us say that an organization may be concerned with consistently high levels of chemical releases, and other reportable volumes of waste that is disposed or discharged, and collectively reported to EPA as TRI figures each year. These values, in turn, may indicate some process or other shortfall that may need further investigation to determine what root causes, if any, may be contributing to these high values. One such investigation that could confirm root cause issues would be an internal EH&S audit of the facility's reporting sources, including key chemical and manufacturing processes that generate the largest volumes of wastes, discharges, and releases. Based on an audit's findings and subsequent recommendations, facility representatives could then develop a corrective action plan to reduce TRI releases that may have occurred as a result of process vent filters not being regularly checked and replaced; covers not being placed on tanks of VOCs used in degreasing operations; process waters not being recycled/reused to reduce industrial discharges to sewer lines; or various solvents used as cleaners (instead of being substituted for detergents and other aqueous washes), or other areas for improvement leading to decreases in TRI releases.

With such qualified information, a buyer within a company experiencing such high TRI figures could then ask the supplier representative whether other chemicals could be substituted for the chemicals the organization currently purchases to help lower these TRI reporting figures. Since TRI reporting is one database that EPA relies heavily upon as a barometer of an organization's efforts to curtail pollution releases, it may be beneficial to an organization to try to continually lower TRI releases at its facilities as a means to move further away from a regulatory loop that is beginning to encompass a larger geographical area. For example, TRI chemical release reporting has now expanded from the United States into Canada and Mexico as part of a new North American Free Trade Agreement (NAFTA) initiative for continent-wide pollutant release and transfer registry.[12] This is yet another example of emerging regulations and legislation that should send a wake-up call to those companies within EPA's regulated community that still see environmental responsibilities as a low priority.

While beneficial in and of themselves, such exercises as described could also be useful as dry-runs for preparing the organization for ISO 14001 certification, since the certification process relies on environmental management system (EMS) auditors, whether working for outside consultants or registrars, who look for signs of nonconformance to ISO 14001 within an operating facility. It is not uncommon for them to poke around different process or business areas and ask employees random questions about their organization's environmental responsibilities. An organization that is well-prepared and knows that its employees thoroughly understand their job responsibilities and the organization's environmental requirements stands a better chance during a registrar's audit to be found in conformance with ISO 14001.

DO THE BENEFITS OUTWEIGH THE COSTS?

The previous discussion focused on explaining several areas where environmental managers, their staff, facility managers, and other facility workers could direct their efforts in moving the organization forward to attain higher levels of environmental awareness and stewardship, and reap the benefits associated with such positive activities. In this section, we provide insight into analyses that can measure whether such benefits can outweigh the costs to an organization and how that information can be provided in a format that nonenvironmental personnel can understand as well. One tool to achieve this goal is activity-based costing (ABC). With ABC, it is important that several elements are evaluated by analysts performing these functions to ensure the accuracy of ABC results. These elements can include: (1) properly identifying costs associated with specific activities, such as, the costs associated with waste disposal practices for a particular waste stream generated by a specific chemical or manufacturing process; (2) performing a cost-benefit analysis of targeted activities, such as, determining the costs associated with continuing present waste disposal practices versus the effort required to review alternative waste disposal practices, provide training to employees affected by these practices, and evaluate the benefits of each; and (3) utilizing total cost accounting or total cost analysis methodologies, including software, to help quantify and track cost savings generated from pollution prevention projects, such as, software used in Design for the Environment (DfE) and lifecycle assessment (LCA) considerations that can allow designers to interact with existing EH&S databases.[13]

With ABC, the savings an organization realizes in process or manufacturing areas attributable to EH&S functions can be readily measured via various benefit-cost analyses. The results can then be tabulated into numerical values that can clearly depict any savings realized in raw materials, energy consumption, waste handling, or similar areas for which metrics exist. EHS managers can then readily review this information to substantiate their conclusions. Depending on the ease with which this information is tabulated and how readily it could be translated into a format for non-EH&S managers to understand can make a big difference in conveying the message appropriately to top management's financial and marketing executives. The key to the data provided is that it should be timely, concise, and easy to follow *without* too much emphasis being placed on "envirospeak" terms, that is, minimizing emphasis on regulations, environmental terms and acronyms, and other difficult-to-follow environmental "buzz words" that might dilute the importance of the data presented to an executive more accustomed to business terms.

If we, as EH&S managers, want to scale the "green wall" as Renee Bobal talks about later in this chapter, we may need to repackage the results of our environmental investigations into more business-friendly terms with which our MBA counterparts in other business units are familiar. So, for example, if we are looking at cost savings that can be realized in production lines, and we choose to closely monitor variables to ensure more efficient use of raw materials and energy consumption (less wastes generated downstream), this data should be tabulated and summarized to show downstream cost savings realized. This data should than be compared to the costs generated when these items are not efficiently used. This report would then be presented to management for review where the main focus would be on depicting savings realized. To ensure that all costs would be accounted, EH&S capital expenses should also be included. The ultimate goal in ABC is for management to fully understand the basis of the analysis and the benefits derived by the organization as a result, otherwise a valiant, timely, and accurate analysis may have been done in vain if it is lost on its target audience. A good source of case studies of a number of companies that took the initiative to measure pollution prevention opportunities and tabulate savings realized in implementing changes to their processes is the EPA document, *Pollution Prevention Success Stories.*[14] A similar EPA document, *Partnerships in Preventing Pollution,* is a compilation of the agency's various partnership programs with regulated industries, including the Environmental Leadership Program (ELP), 33/50 Program, Green Lights, DfE, Project XL, and the Voluntary Standards Network that focus on ISO 14000.[15]

In developing an ABC analysis, a good starting point may be to conduct a pilot benchmark study of some of the organizations depicted in either of these reference documents described. In this manner, one could get a feel for the types of savings that can be realized and how some companies are providing this information for top management review to verify that EH&S benefits can outweigh their capital expenses.

Additional sources of information for conducting benchmarking studied or providing supplemental information to substantiate an ABC analysis can also be found in local or national newspapers, such as *The Wall Street Journal, New York Times,* or other national or regional newspapers, and trade publications and newsletters. Some of these specialty publications are listed in Appendix B. The information obtained can sometimes be relevant to an analysis being performed on a particular environmental issue. Depending on the news story and the intended audience, some detail can be gleaned about a company that could be relevant to your particular ABC analysis, such as, a depiction or comparison of costs associated with enforcement actions taken by federal and state regulatory agencies. These actions usually involve EPA or OSHA enforcement actions against companies that performed various activities in violation of environmental and/or worker health and safety laws.[16] These are just some of the examples out there depicting the consequences that organizations can incur with the EPA and OSHA as a result of their inability, reluctance, or plain shortsightedness, to address their environmental, and health and safety responsibilities in some fashion.

Should the previous examples shown be insufficient to include the cost of noncompliance into an ABC analysis as one factor for inclusion in a benefit-cost analysis, we can also take into consideration the cumulative effect interested stakeholders may have on organizations, and their ability to alter or stall specific business ventures or processes. In this regard, such stakeholders may not necessarily just focus on pollution issues concerning the organization; rather, they could focus on other issues, such as, sinking an oil rig (Shell Oil), a chemical leak (Merck), or siting a new store (Wal-Mart).[17] In such undertakings that take into account a much bigger picture, a company's ABC analysis may need to expand its focus to encompass a much larger universe to consider more options and intangibles, their benefits and associated costs.

Where then, one may ask, can benefits be found in activity-based costing within an EPA-regulated organization? Remember that for ABC to be useful and effective to financial and marketing executives, value is realized when they readily understand the information and it is free of unnecessary and extraneous envirospeak terms that can confuse and confound them. For example, should the data presented in an ABC analysis of subject areas such as we described not be totally transparent to the recipient, whether a company's chief financial officer (CFO) or chief operating officer (COO), they could perceive the results as a negative cash flow. Should that message come across, it may become difficult to gain their support for any recommendations you may want to pursue to improve the item(s) in question, whether process, employee or other aspect-related.

On the other hand, to you and others within the EH&S group, these items can signify areas where an organization may need to make expenditures in order to improve.

This may be a valid reasoning for the EH&S manager, but in order to scale the "green wall" in some companies, managers may have to perform some form of "spin doctoring" to ensure the message is transparent enough to reach its target audience: the company's top financial executive.

Should such strategies fail in not adequately conveying the message, then perhaps there might be lesson learned after-the-fact. For example, the company could continue to receive fines from regulatory agencies, or stakeholders continue to perceive corporate environmental initiatives as empty promises and possibly take their business elsewhere, or other issues could arise that would have a negative effect on the company. Issues such as these can be indicative of shortfalls in the company's EH&S structure that may need correction.

One management tool useful in identifying, summarizing, and recommending corrective actions to noted deficiencies that can be linked to the issues described is the environmental, or environmental, health and safety (EH&S) audit program. With the advent of ISO 14000, an additional management tool is the environmental management system (EMS) audit. We provide additional discussion regarding audits as management tools in Chapter 9.

The discussion of EH&S audits in the context of ABC analyses focuses on the functionality of audits for tracking and measuring results of an ABC analysis. Take, for example, a benefit-cost analysis that tracks audit findings and translates subsequent audit recommendations into measurable benefits realized from these audits can be useful to an organization's financial executive, such as a CFO or COO. With such data, the extent of the benefits can be measured through findings of follow-on audits that, in turn, can be summarized into a report for management. With this information, EH&S managers can also track the progress of follow-on audits as compared to previous audits as a measure of the audit program's effectiveness. In such an ABC analysis then, subsequent audit findings can be compared to the findings of preceding audits as part of the on-going analysis to measure the benefits of each subsequent audit.

To clarify this point, let us look at a hypothetical audit program at a manufacturing company as the subject area for an ABC analysis. Audit results from Q2-1996 comprise the baseline data for the ABC. The findings from this "baseline" audit point to regulatory issues related to: wastewater discharges, air emissions, expired chemicals disposed of as hazardous wastes, improperly designated and co-mingled waste streams, on-site disposal practices, and training inadequacies. As part of the ABC analysis, dollar values are assigned to represent the total costs associated with these regulatory findings. These values take into account the additional waste handling and disposal costs, treatment costs, and quantities of chemicals purchased. These costs do not include any possible penalties, which is presumed will not occur, and are labeled as "ancillary" expenditures. In this situation, the total costs to the facility associated with the audit findings total **$25,000.**

In the follow-on audit, several audit findings were corrected and the company's "ancillary" expenditures are reduced to $15,000, or a net savings to the facility of $10,000 from the previous audit. By the next follow-on audit, all audit findings had been corrected, and some of this audit's recommendations refer back to previous root causes that had been eliminated and do not recur. Furthermore, the EH&S staff working closely with the audit team identified several processes and waste handling areas from a previous audit that were since modified or changed, and as a result, generated additional cost savings that previously had been overlooked. One work station area uses a different chemical in its degreasing operation that is less costly to use and to dispose. Another work station modified a particular process line to allow recirculated water to be reused rather than being discharged into the wastewater treatment plant. The net savings to the company as a result of these and other corrections and changes is **$60,000:** $35,000 in actual savings, and $25,000 saved from the elimination of the unnecessary expenditures.

In summary, an EH&S audit program can be enhanced to serve a dual purpose as an ABC analysis tool by taking into consideration such items as: (1) how thorough the audit team conducts each audit; (2) how prepared each audit team member is during each audit; and (3) the level of proficiency and audit skills of each of the team members. The focus of this in-depth evaluation is to look more closely at the audit teams: The more skilled and experienced the audit team, the greater the chances are that the team's findings will be more in-depth and more conclusive regarding the findings and possible root causes. The more relevant the findings, chances are that the facility being audited can obtain a wealth of information from the audit findings and recommendations contained in an audit report. Timely and cost-effective recommendations that an audited facility decides to adopt and implement can lead to measurable cost savings and benefits that can be tracked. How much a facility can save may also depend upon the relative usefulness of findings and how quickly corrective measures can be executed to take advantage of cost savings.

In another example of benefits outweighing costs, a company could implement programs where designated employees inspect all incoming shipments of chemical products, oversee off-loading of these products into the facility, and oversee all shipments leaving the facility, including chemical waste. Such programs could establish what they should look for with incoming shipments of chemical products to insure containers are sound, and there are no visible dents or other markings that could cause leaks to occur; they would also be present at on-loading or off-loading of bulk containers, and to check couplings, fill-level indicators, secondary containment structures, and other pertinent components of bulk chemical and waste handling. Tabulating the savings realized from product that is off-loaded in sound containers that do not leak, or from tankers that are not overfilled, versus the costs associated with the loss of product from ruptured containers and overfilled tanks and the related cleanup of such mishaps, can more than offset the time required to train employees into knowing what to look for, and who to call for help should a mishap occur. This is a fundamental underpinning to preventive maintenance, and it makes good business sense.

The cost of implementing preventive programs can be offset by benefits that such programs can provide by mitigating or eliminating incidents and their associated clean-up costs, emergency notification, and other regulatory-required contingencies. In these cases, the benefits and associated costs could be tabulated into dollar amounts showing the relative benefit-cost. Factoring in costs associated with developing and maintaining an internal inspection program for incoming and outgoing chemical shipments as compared to cleanup costs required if such a program were not in place would not be difficult to estimate and show that cleanup costs would far outweigh a maintenance program costs.

INNOVATIVE STRATEGIES . . . MOVING FORWARD

The time span between developing innovative strategies as discussed and implementing such strategies within the workplace of an organization is usually a function of how well-prepared affected employees are, their willingness to embrace these strategies, and whether there is sufficient momentum at the management level to carry the strategies through. Of course, the level at which each facility, company, or organization is at varies with time, and also takes into account each facility's individual, unique cultures. Any of these variables can play a role in the effectiveness of any innovative strategies implemented at any level of an organization.

As a starting point, environmental managers should have a full understanding of the responsibilities and nuances associated with designing and implementing such strategies that should encompass a diversity of work areas, the individual workers, and of course, each unit's work priorities and budgets. In planning a strategy, managers should take into account each of these diverse items. In so doing, they can increase their chances of achieving their EH&S goals while being sensitive to the needs of other business unit managers who may have a completely different focus, allocated budget, and priorities. If any of the considerations as described are not part of the overall plan, an otherwise valid innovative strategy may get nowhere and be of no use to anyone in the organization.

Managers should evaluate existing environmental and health and safety (EH&S) programs with other existing related business areas (research and development (R&D), purchasing, inventory and manufacturing, for example) as part of this strategic plan and brainstorm various strategies as options to move these programs forward from their existing condition, the so-called "baseline." One way to evaluate where an organization is in comparison to where it may want to be, is to conduct a benchmark survey of other organizations. A specific benchmark could be designed to review the EH&S or EMS programs of other companies that are considered noteworthy, or best-in-class, or designed to evaluate the relationship between a company's EH&S group and related business areas, such as R&D, purchasing and manufacturing.

Such a survey could be conducted to provide an organization insight into how target organizations manage their EH&S or EMS programs, and provide that organization's managers with noteworthy tips on improving their existing EH&S and/or EMS program frameworks. Depending on the needs of the organization, the benchmarking survey could provide information that can assist in melding EH&S business areas with other business areas of an organization, such as quality systems, and other business profit and cost centers as mentioned previously, through some form of sharing of common data. Benchmarking, as a management tool, is such a widely used management tool that the whole of Chapter 7 is devoted to this subject area. In this chapter, we provide detailed discussion about benchmarking techniques and data gathering and share the results of several noteworthy surveys as case study learning tools.

In addition to benchmarking techniques that environmental managers can utilize as part of their innovative strategies for improving upon existing conditions within their facility, company, or organization, the following items are two other management tools that can be utilized either on their own, with benchmarking, or with other innovative strategies that the managers may have within their "toolbox." A brief summary is provided of each tool or methodology.

A. PLAN-DO-CHECK-ACT

The Plan-Do-Check-Act (PDCA) process, or cycle, is a commonly used tool in business planning, especially in solving problems in quality considerations. The PDCA cycle is similar to the Shewhart cycle, developed by Walter Shewhart in the 1930s for industrial applications. The PDCA cycle subsequently gained greater prominence through its association with W. Edwards Deming and is also referred to as the Deming circle.[18] As a result of that association, the PDCA cycle is usually referred to in context with ISO 9000 (the quality management standards), and more recently, has been referred to in context with ISO 14000.

Looking more closely at the elements of ISO 14001 in comparison with the PDCA cycle, the *Plan* element is contained within ISO 14001's Section 4.1 and 4.2, Environmental Policy and Planning; the *Do* element is contained within Section 4.3, Implementation and Operation; the *Check* element is contained within Section 4.4, Checking and Corrective Action; and the *Act* element is contained within Section 4.5, Management Review.

A basic function of the PDCA cycle is to provide a set of systematic techniques that can improve business decision-making processes. PDCA allows managers to obtain valuable insight into and thoroughly understand the problems to be solved while helping them select appropriate team(s) to execute the recommended solutions. For example, a manufacturer of heavy equipment that is concerned with excessively high levels of air emissions and large wastewater discharge volumes at a particular plant may want to convene an ad hoc team with the goal of evaluating these issues from several areas of consideration, such as regulatory concerns and waste reduction options. In the *Plan* phase, the team would devise a plan to evaluate the cause of each of the existing conditions to provide solutions to the problems, and as part of this step, the team may also want to develop a baseline of existing conditions. A baseline could add additional insight to the problem-solving exercise by setting the starting point. The team could then set a target value of reductions that the plant wants to achieve, and as data is received, can track results as they occur to measure performance. As part of this planning process, the team may also consider reviewing any facility audit reports for additional insight, if available. This review step could provide the team added information, especially if such problems continue to occur. With this information, the team could develop corrective action recommendations and a time frame to implement these actions to solve the identified problems. To ensure initiation and completion of the project, the team could prioritize problem areas as action items for completion, and establish a time frame for correcting each problem.

In the *Do* phase, the team would implement the plan. In order to ensure that corrective actions are being executed, a team member would be given the responsibility to track corrections and report on their progress. In the *Check* phase, the results of each corrective action would be measured against existing baseline conditions to determine whether the corrective actions were successful. This phase can also be useful in testing any alternative recommendations to corrective actions in the event corrections are not successful. For example, in the equipment manufacturer example, a manager reviewing the PDCA exercise may ask the team whether closed-loop recycling was instrumental in decreasing both air emissions and large wastewater discharges, or whether new process equipment may need to be implemented to significantly reduce waste volumes. If the answer is "no" in both cases, this may require the team to rethink their strategy regarding the issues to be resolved, look at the problem from a larger scope, or perhaps consider bringing on new team members to solicit their input. One item that could provide additional value in this phase is the use of specific EH&S software that can help assimilate, collate, and track information. Such software can assist in tracking and managing hazardous waste manifests, in generating material safety data sheets, measuring and tracking continuous air emissions, and a host of other analytical, simulative, and document management uses, such as enterprise document management (EDM) or enterprise resource planning (ERP) software systems in the marketplace. The use of such software is gaining much attention and interest in the petrochemical area.[19] We provide further information on a variety of software currently available in Chapter 13.

The next step, should the plan be successful, is to *Act*. The team should be prepared to execute any proposed changes as recommended. The benefit to the organization in such PDCA exercises is that problem or log-jam areas can be identified and corrected, and as part of this exercise, additional areas in need of correction may also be identified for correction in follow-up exercises. For example, one problem a PDCA team may have identified points to improper employee training. That item would then become a follow-on PDCA exercise for the team to evaluate. The results of this exercise could also be tracked and made part of the more "global" corrective action recommendations at the facility in question. The benefits of such PDCA exercises could lead to improvements of a facility's environmental management system, or EMS, as part of the continual improvement aspect of ISO 14001's management review program element. It would seem reasonable that upon several iterations of a PDCA exercise program, performance improvements could be easily noted.

B. PROVIDE FOLLOW-UP TO TRACK AND MEASURE PROGRAM EFFECTIVENESS

As described, the Plan-Do-Check-Act cycle is one method that allows an organization to track and measure the effectiveness of an environmental management program or system. In addition to PDCA, there are other methods that environmental managers and their staff can also utilize to achieve favorable results to improve upon their environmental management systems (EMSs). The effectiveness of each of these methods, though, may sometimes depend upon a number of factors, and may include: existing plant or facility conditions, the personnel involved, the level of maturity and sophistication of the EMS or EH&S program at each facility, and other factors that may be specific to the facility itself.

Taking several of the innovative strategies for moving forward that we previously discussed to implement a program that provides regular follow-up for tracking and measuring the effectiveness of an environmental management program requires considerable time and energy to ensure its effectiveness. Keeping up to speed on regulatory requirements need not be a task that invokes visions of demons in the minds of environmental managers and staffers or is perceived as something almost impossible to attain.

At the basic level, keeping up to speed on regulatory requirements requires the diligent efforts of one or more EH&S staffers to pool their resources together and collect and catalogue all of the regulations and other requirements that affect their particular facility, or if service-providers, that affect each of their clients' specific facilities. Once this cataloging exercise has been completed, the material should be housed in a central repository or research library area for others to use. Efforts should also be made to ensure that periodic updates of applicable federal and state regulations, federal and state press releases, and noteworthy cases are added to the library as they become available. This "library" should also house pertinent periodicals, newsletters, books, and conference proceedings that add additional insight into emerging regulatory or environmental management issues or concerns.

At the next level, the task of keeping up to speed on regulatory requirements and keeping abreast of emerging EH&S issues of interest, may be obtained by EH&S staffers via their PCs. The amount of information that one can obtain from logging onto the Internet and accessing an organization's web site continues to grow and should remain a viable source of relevant information into the next century. In addition, some vendors of EPA and state regulations provide their subscribers access to the U.S. Code of Federal Regulations on a CD/ROM disk, or on-line via their web site, along with regular *Federal Register* updates and other ancillary information, providing their users with the most up-to-date information on regulations and rulings.

For other EH&S professionals with modest budgets, all federal agencies and some state agencies have their own web sites that are cross-linked and provide a wealth of valuable information that is also updated regularly. For example, if you wanted to obtain information on recent EPA press releases, the latest information on an EPA, SEC, or Department of Justice enforcement action, or a copy of an organization's environmental policy, mission statement, or environmental report, most or all of this information is accessible via the Internet. Readers can obtain similar information by contacting the organizational and agencies web sites listed in Appendix G.

In summary, keeping up-to-date and effective EH&S programs and environmental management systems requires diligence and a sustained effort from the EH&S workforce—both professionals and technicians combined. This combined effort can be seen in: (1) filing of required permits, forms, and other regulatory-mandated documents, and maintaining a filing system for current and archived documents that allows access to records; (2) establishing a regular and periodic environmental, EH&S, and/or EMS audit program, that incorporates regular follow-up as part of the audit program(s) and looks into root-cause issues; (3) tracking the effectiveness of in-house or third-party training and/or outside training and conferences provided to EH&S staff, as measured in their work; (4) tracking the effectiveness educating other business units within the organization to understand their EH&S responsibilities and how these relate to the EH&S group; and other noteworthy considerations.

In this next section, Renee Bobal provides insight into the role of the EH&S professional from the 1970s to the present from the perspective of someone within the regulated community.

PAST TO PRESENT: THE DEVELOPING EH&S PROFESSIONAL FROM THE 1970S TO THE 1990S

Some of the first professionals to enter the new environmental field were civil and chemical engineers and chemists who designed and operated treatment systems, for example, wastewater treatment plants. The environmental and human rights movements of the 1960s provided the impetus for the creation of the Environmental Protection Agency (EPA) and the Occupational Health and Safety Administration (OSHA). In the 1970s, Congress passed legislation to begin to address the clearly visible environmental problems associated with many years of unregulated industrialization and booming development. The engineer/chemist was thrust into this newly developing regulatory arena. As the programs of EPA and OSHA expanded to address public concerns, the engineer/chemist professional was hired by federal and state agencies to help write new regulations and by industry to interpret and implement compliance with the new requirements.

The regulations were developed as a reaction to past and present problems, such as polluted waterways, fish kills, and billowing black smokestacks. The focus was to drive polluters into compliance with the new requirements and punish those who did not obey. Professional expertise became highly specialized, mirroring the agencies' own programs, and was segregated by media, such as air and water, or by expertise, such as industrial hygiene. An entirely new consulting business sprang up to address the needs of industry and government—there were not enough technical individuals available to handle this quickly burgeoning field. Colleges and universities began to create new technical programs to service the demand for the new technologists. A new professional field and a sizable business sector was born.

This highly technical, specialized environmental, health, and safety (EH&S) field created its own, complex and rapidly changing regulatory/technical language. The flavor-of-the-month programs and respective acronyms became the standard. These professionals were often misunderstood, largely as a result of the spirit of their new charge and a zealous attitude to correct past wrongs—and feared, because of the enforcement repercussions of violations, fines, and criminal prosecution. They began, for the most part, disassociated from the businesses which employed them.[21] This became known as the era of command-and-control—the EH&S professional had much free reign in determining how EH&S issues would be addressed because the businesses could not understand the issues and were afraid of the legal consequences. Early on, corporate EH&S departments in businesses did not exist. The first professionals were generally plant technicians reporting to a line management function. Traditional hierarchical organizations were structured according to the regulations in which EH&S professionals specialized by area of technical expertise reporting to a technical manager. Often, the EH&S manager had other duties, such as plant security or maintenance. Individuals typically stayed within their area of technical expertise and added specific new areas of responsibility as regulations grew. Corporate groups developed as companies realized the importance and gravity of the enforcement consequences. Their role began in providing compliance and project support to their businesses and in conducting compliance audits. Over time, the corporate groups began to focus on more strategic roles, such as establishing management systems, developing performance measures, and reporting on company performance to the public.

In the 1990s, EH&S organizations began experiencing significant problems in advancing new strategic programs within their companies. In many instances, even in leading global corporations, these initiatives were misunderstood and considered excessive by the businesses. The "Green Wall" had been created and the growth of EH&S initiatives began to reverse itself. According to Robert D. Shelton of A. D. Little, "The Green Wall is a point at which the overall organization refuses to move forward with its strategic environmental management program, and the environmental initiative stops dead in its tracks as if it had hit a wall. Symptoms of hitting the Green Wall include negative or deferred decisions due to a lack of management support for the strategic environmental management concept and program, environmental health and safety (EH&S) programs that are lacking focus, and the inability to demonstrate to others in the organization the attractive returns on further investments in environmental programs."[21]

CONFRONTED BY A WORLD OF CHANGE

What happened to cause such a dramatic turnaround? Removal of trade barriers and improved political climates opened the door to global competition. Businesses were confronted by increasingly educated and demanding customers with more power to choose. Yesterday's cost reduction measures were not enough and companies were faced with making drastic changes in a hurry to survive. Much of heavy manufacturing began moving to developing countries where the cost to produce is significantly lower. Many companies took on re-engineering efforts, tossing out the old traditional organizational layers and activities and totally reconfiguring how work is done. This resulted in radical restructuring around core business processes.[22] Some companies have decentralized to the point where their businesses and operations are separate enterprises competing against one another for production and customers. James F. Moore, in his book *The Death of Competition*, goes well beyond this to state "the notion of *industry* is really an artifact of the slowly paced business evolution during the middle of this century. The presumption that there are distinct, immutable businesses within which players scramble for supremacy is a tired idea whose time is past. . . . The traditional industry boundaries that we've all taken for granted throughout our careers are blurring—and in many cases crumbling."[23] And the pace of change is accelerating. One of the greatest challenges today is managing change. EH&S management has run up against these business realities—financial constraints, fierce competition, downsizing, and flatter organizations. The budgets and organizational freedom necessary to run large centralized EH&S functions began to quickly disappear. As we are now well aware, separating EH&S from the business has become a painful experience for the EH&S professional. The EH&S organization was hurled into the midst of this corporate change, no longer untouchable or considered a "sacred cow."

At the same time these business changes were taking place, the EH&S field began to go through some changes of its own—reaching a level of maturity where most programs have become a commodity. Environmental services, such as compliance programs, permitting, reporting, audits, and even remediation, have become standardized. The field has become glutted with too many suppliers vying for customers, causing the cost of services to drop significantly. This change has also been forcing a departure away from the traditional command-and-control organization to a highly competitive, custom-focused service organization. More specifically, the tide has shifted from the EH&S professional telling the businesses exactly what to do and how to do it, to the businesses consulting with the EH&S professional on issues they feel are important and exercising significant freedom in what they do and how they do it. The role of the EH&S professional has become one of offering advice, guidance, and choices, *without the control of the past*. Professionals are struggling with the change, the loss of control, and how to demonstrate and sell their value to their new *customers*.

In order to better address the needs of their internal and external customers, the EH&S organizations, not unlike other service functions, have been restructuring from the typical inflexible, hierarchical structure of environmental affairs, safety, etc. where individuals have been segregated by regulatory programs and/or media—to various other models. There is no "one size fits all" and today's organizational model may be inadequate tomorrow depending on the changing needs of the businesses served by the EH&S organization. In the centralized, shared services organization, the concept is to pool expertise into a cost-competitive, flexible organization or "center of excellence." Individuals function as internal consultants where resources are shared across businesses to eliminate redundancies in expertise, to maximize efficiency of existing resources, and to foster communications throughout the company. Shared services organizations compete with external consultants in cost, schedule, and quality of services. Other organizations have decentralized and moved specific experts to the individual businesses or plant sites which require a particular area of technical expertise. These individuals may even provide part-time services to other businesses or plants located in the same business or geographical area. Yet fewer companies have taken a more extreme approach and have decided not to maintain specialized expertise for what they consider to be non-core value/nonstrategic programs/operations. Sun Microsystems is among the companies that decided to outsource its nonstrategic EH&S activities. This fits well within its competitive business sector, company culture, and business strategy. It allows the EH&S function to react as quickly as the rest of the company in the ever-changing business environment. Still other companies are establishing internal networks of limited resources to provide the appropriate EH&S services across their various businesses or facilities. Use of company-wide Intranets has made the concept of a "virtual" EH&S organization possible.

Companies also vary in how they integrate environment with health and safety. In some cases, they report to different management lines in the company and have little interaction. Others have integrated environment with health and safety, achieving benefits in both cost and knowledge in certain cross-functional processes, such as auditing, training, regulatory tracking, and policies and standards. Regulatory programs, such as OSHA's Process Safety Management (PSM) and EPA's Risk Management Program (RMP) rule overlap and it makes sense in these instances to have the same individuals managing the programs or at least working closely together.[24] When cross-functional responsibilities are combined, the career opportunities, professional development, and marketability of the individual are greatly expanded.[25]

MOVING FORWARD TO THE FUTURE

Although many competent and talented organizations have restructured to better meet the needs of their customers, they have still been confronted with the "Green Wall." Demonstrating the value of EH&S has become an even tougher challenge than expected. What is still needed to achieve successful EH&S leadership today and in the future?

In order for EH&S to truly deliver value to the business, EH&S organizations need to either find a way to serve as the basis for a unique competitive position for their company, or need to contribute significantly to core businesses activities which distinguish the performance of the company from its competitors. In both circumstances, EH&S objectives must be directly linked to the goals and objectives of the business. Unless EH&S initiatives and programs are carefully linked to business value, they are viewed as serving only EH&S and not the business. The peril of not linking EH&S initiatives with the business objectives is that futuristic EH&S investments are seen as luxuries in good times which can be cut when times get hard. It is crucial that EH&S professionals understand the basis for competition in their industry and align their EH&S goals and objectives with those of the company in terms of cost, quality, time to market, and customer service. Success may be reflected in the skill of the EH&S manager to communicate EH&S issues in business language—demonstrated with financial analysis, cost/benefit analysis, and how the initiative fits into the business strategy—the way in which the business community is accustomed to listening to and understanding all other business issues. These initiatives should be viewed as business initiatives with an EH&S component, rather than as strictly EH&S initiatives. These sentiments are further echoed by Frank Popoff, chairman of the Dow Chemical Company, co-author of the book "Eco-Efficiency, The Business Link to Sustainable Development," at the 1998 Conference of Corporate Environment, Health and Safety Excellence in New York, NY, in which he said "superior options and solutions are available to us when we're on the front end of the decision-making process, not the back end in a more narrow, prescriptive, regulated environment."[26]

One of the greatest hurdles standing in the way of integrating EH&S into the business is the lack of an EH&S cost-accounting system. Unfortunately, much of the costs are still buried within standard budget line items, such as labor, equipment, supplies, and maintenance. By defining and measuring the true costs, it will be possible to link EH&S and economic business decisions.[27] "Sustainable competitive advantage may be defined as a set of distinctive business assets or skills that enable a company to consistently outperform its competitors under a variety of market, economic, and regulatory conditions."[28] Businesses have begun to use economic value added (EVA) as a way to measure the effectiveness of a business to utilize its available capital. EH&S can contribute to improvement of EVA in a variety of ways, including programs and initiatives, such as eco-efficient production, pollution prevention, resource conservation, asset recovery, and product differentiation through Design for Environment (DfE). Pollution prevention through source reduction and designing products for the environment can improve productivity and reduce operating costs. Significantly producing more product with less raw materials, energy, and waste contributes to improved profit margins.[29] Integrating or embedding EH&S into the business is perhaps the most important strategy for improving EH&S performance—shifting responsibility and accountability from the EH&S staff to the people who implement and control the business processes, in other words, product development, manufacturing, and distribution, and marketing and sales. This change will require effective management systems, expanded training programs, improved communications, and efficient use of enabling technology to move responsibilities from the specialists to line operations. (The chemical industry, under the Responsible Care® "umbrella" considers across-the-board responsibilities relating to EH&S as part of its principles, verified by Management System Verifications (MSVs). Further discussion on this topic is found in Chapter 9.

Today, the EH&S role is not just about minimizing risks, it also deals with presenting various alternatives so that the business may make informed decisions on EH&S investments.[30] EH&S management has traditionally focused on minimizing or eliminating downside risk which includes safeguarding against liability and loss and maintaining regulatory compliance, and consequently management has sometimes missed competitive opportunities associated with upside risk. Although the negative aspects of risk management are an often expected primary responsibility of the EH&S function, not looking for positive opportunities which require risk-taking is a mistake. In the financial world, risk and return are closely linked—the great the risk of investment, the greater the potential return on investment, but also the greater the potential loss. According to Lee Puschaver and Robert Eccles of Price Waterhouse,[31] risk, viewed as opportunity, reflects the outlook and responsibility of senior management, who create the strategy of the company and commit the necessary resources to carry out that strategy. Managing the upside of risk focuses on attaining a desired return on capital. Heavy focus on compliance and negative risk management can stifle innovation, initiative, and entrepreneurship. Out-of-the box thinking, usually found in the more progressive companies, can also be stifled.

There are also certain widely-held misconceptions which substantially hinder the progress of EH&S and impede its efforts to integrate into business decisions.[32] Costs are not rising as greatly as in the past, costs are uncontrollable and nondiscretionary, regulations fall uniformly on all competitors in an industry, and if you do the right thing then everything else will follow. Environmental spending is still rising in relation to the gross domestic product. It went from 0.9% in 1972 to about 2.5% in the mid-1990s. By making the above assumptions, both EH&S managers and business managers are losing out on strategic and competitive opportunities and cost reduction measures.

THE NEXT WAVE OF EH&S

Environmental issues tend to hit like waves, one after another.[33] Right now, global climate change is cresting—but a host of issues is gathering momentum. For example, the presence of potential endocrine disruptors (as we discussed in Chapter 2) in the environment and the concern about the long-term impacts of low levels of chemicals on human and animal reproductive systems is now a very prominent issue as a result of the publication of the book, *Our Stolen Future*, by Theo Colburn, Dianne Dumanoski, and John Peterson Myers, and its coverage by the media. A variety of web pages already exist on this subject.

The use of the Internet and expanded use of information technology is making EH&S performance more readily available to the public than ever before. The public will be able to review company, facility, and product information to make informed decisions about product selection and financial investments. Section 313 of the Superfund Amendments and Reauthorization Act (SARA) [Toxic Release Inventory (TRI)] is already available to the worldwide public on the USEPA's web page. (A more detailed listing of EPA's web site and pertinent environmental information is found in Appendix G.) A wealth of detailed compliance information—permit data, Superfund sites, releases, incidents, and violations—is available to anyone in the world with the click of a mouse. Information is being integrated into geographic information systems (GISs) to display discrete information on maps linked to complex databases. This trend is expected to grow around the world as developing nations view the success of the SARA TRI program in the United States as a cost-effective regulatory tool favored over burdensome regulations. Today, federal and state governmental agencies are the largest producers, collectors, users, and disseminators of environmental, health, and safety information in the United States, and possibly the world.[34] This is rapidly becoming a core component of their activities. Public dissemination of information via the Internet is being used as a means to influence the behavior of the private sector.

Socially responsible investing (SRI) is the inclusion of social or ethical standards in investment decision making. Along with financial criteria, these standards become part of the process that investors use to evaluate potential investment opportunities. Socially responsible investing is growing dramatically—in the last two years, the number of SRI mutual funds has grown from 55 at a value of $12 billion to 144 at a value of $96 billion. Social investors may affect how companies act by communicating their interests and concerns to companies, by voting proxies, and in some cases, by filing proxy resolutions.[35] According to Sophia Collier[36] president of Citizens Trust, a mutual fund company, doing a little research on companies can help investors avoid problem companies. Lawsuits, fines, and government investigations are signs of troubles which may lay ahead. For example, Archer Daniels Midland shareholders lost almost $3 billion in value in only three months after reports of a hefty $100-million fine for worldwide price fixing was announced. Publicly available safety and environmental records may show patterns of compliance problems which could be an early signal to future financial difficulties.

SUSTAINABLE COMPETITIVE ADVANTAGE

Companies operate in a dynamic world with constant and changing pressures from competitors, customers, and regulators, which drive them to find innovative solutions to remain competitive.[37] Environmental regulations do not erode competitiveness, but can stifle innovation and productivity by being overly prescriptive and not allowing flexibility in developing and implementing creative solutions. The problem is not the strictness, but the overall system itself. The misguided process of regulators issuing cumbersome and counter-productive regulations and the business community fighting them has resulted in adversarial conflicts, costly litigation, and delays at the expense of finding beneficial solutions. Society, business, and the environment all suffer.[38] Likewise, current measurements and incentives may be improper for driving strategic programs. For example, generation of waste adds cost to a product—from waste management, inefficient use of resources, and disposal—which the customer must bear. These costs are not always measured and shown in this perspective. The EH&S manager must advocate changes in these approaches to foster rapid innovation and promote resource efficiency so that companies can be more competitive. Environmental progress and global competition both demand that companies innovate to increase resource productivity. Alan Hirsig, president and CEO of ARCO,[39] stated that "government's chief contribution must be to advance and ultimately institutionalize a performance-based system of regulation that sets high but reasonable goals and gives industry the greatest flexibility in meeting them. This will unlock industry's technical capability to innovate, rather than constraining us under the old command-and-control paradigm."

Individual companies need to factor in how current international management system standards, such as ISO 14000, can increase their competitive advantage, or at least not erode their competitiveness. In order to do this, Joseph Fiskel indicates that companies need to integrate their environmental management system (EMS) with their existing business processes. They need to select environmental performance metrics relevant to the company's economic competitiveness. They need to use activity-based life-cycle accounting measures to identify and reward those activities that add economic value. Lastly, they need to utilize enabling information technology to support efficient use of the EMS.[40]

Many feel that the answer to future EH&S problems lies with sustainable development, where the interests of business and EH&S work together on common goals. The International Chamber of Commerce's Business Charter for Sustainable Development defines "sustainable development" as "meeting the needs of the present without compromising the ability of future generations to meet their own needs." It has been well established that economic growth provides the best conditions for protection of the environment, public health, and worker safety—look at the changes that have taken place in the developed nations over the last one hundred years. Yesterday's environmental issues were more localized and were more easily dealt with by developing regulations and local solutions. However, the issues of today and tomorrow are becoming increasingly global as a result of global marketplaces and world trade. Increasing economic growth and development positively creates new levels of prosperity and in raising awareness, but puts significant negative demands on natural resources. The world's developing nations pose enormous environmental, health, and safety challenges as a result of their population growth, increasing consumption, and economic development.[41] The greatest need is for increased awareness to sustainable development. These nations need support with hands-on training in applying modern environmental management skills and techniques, including the use of new technologies.

Michael Porter argues for progressive environmental standards: the competitive advantages these high standards yield accrue both to the companies that adopt them in their business practices and to the national economies that enact them in their public policies. J. Kirk Sullivan, vice president of governmental and environmental affairs for Boise Cascade Corporation, adds that, "we must proceed on our course toward sustainability despite great uncertainty as to what strategies and approaches will be effective and what unanticipated impacts those approaches will generate." Sustainable business practices can offer substantial competitive advantage.[42] Take the case of BMW and Germany's takeback regulations.[43] Rather than challenge the proposed requirements, BMW developed a strategy around the new regulations, ahead of its competitors, and tied up a good part of Germany's best disassembly firms, locking out the competition. Or look at British Petroleum's controversial, breakthrough business strategy to join in on the emerging consensus on global climate change, a move that radically diverges from the position of the rest of the oil industry.[44] BP saw global warming as a reality and took an opportunistic risk by proactively advancing policy in establishing and widely publicizing their strategic plans. These plans included efforts to reduce greenhouse gases from its own operations, join international activities to reduce greenhouse emissions, increase investments in solar energy so that the BP total revenues in solar photovoltaic would become greater than those of the present world market, and invest in developing alternative low-carbon emission technologies. BP's approach in business strategy is considered a value-based model, where value creates money that is sustainable for the future. Most widely used business models today are money-based models, where money is considered to create value, which is short-term and is not sustainable, but changes are occurring.

In the President's Council on Sustainable Development's report of 1996, "Sustainable America, a New Consensus," the council recommended that the whole tax structure and subsidy structure in the United States be changed from income taxes to consumption taxes in order to support sustainable development.[45] Customers will switch to more environmentally-friendly products, provided the quality is there and the product and its price is competitive with others on the market. They expect more detailed product information and make informed decisions based on such things as ingredients, packaging, distribution, and disposal. Some companies use strategic initiatives to make breaks in the Green Wall, at least in the launching of some new products or customer programs. Two examples of new competitive and environmentally-friendly products are a termite pesticide from Dow and a TV from Sony. Dow has developed a new termite pesticide which no longer requires spraying the house and the yard. The product is a baited trap, which is buried in the yard, and the termites are attracted to the treated bait, which they carry back to their nest, where it kills the termites. Sony makes a "Green TV" which is 20 percent lighter and 60 percent more energy efficient than conventional televisions, and which was produced with water-based materials, rather than solvent-based paints.[46] A Baxter International waste team, composed of suppliers of environmental management products and services, provides evaluations of waste streams and develops strategies for waste reduction and safe disposal.[47] Activities like this could drive down the cost of a hospital bed by $100 to $750. Similarly, Carrier Corporation,[48] part of United Technologies, a leading manufacturer of heating, cooling, and ventilation systems, has found its customers' buying decisions expanding to include environmental needs and concerns for products which are energy-efficient, recyclable, non-polluting, and biodegradable, among other qualities. In balancing customers' needs with those of the environment,[49] Carrier has focused on six key areas: indoor air quality, sound, air distribution, refrigerant, product efficiency, and size/material reduction. Recent product developments have included making the use of HFC's more efficient, improving upon energy efficiency in a variety of ways, such as better insulation of products, and reducing materials use by making smaller and lighter products.

THE EMERGENT EH&S PROFESSIONAL: CHANGING SKILLS TO ADDRESS NEW DIRECTIONS

We started in the 1970s with highly technical EH&S professionals with expertise in very specific areas. What we are witnessing is an evolution to well-diversified and flexible business professionals with EH&S expertise. Today, the professional cannot afford to be highly specialized in a single technical area. According to John Champy in *Reengineering Management*,[50] "we now want to work with a 'deep generalist'" (a phrase attributable to Warren Bennis in Champy's book) . . . "a person who is broad enough to respond appropriately to shifting work demands, to changing market opportunities, to evolving products and services, and to the ever-pressing demands of customers, but who has gone deep enough into some area of expertise to bring a well-honed and valuable skill to the enterprise."

Business process re-engineering has taken hold in the EH&S profession. The future focus will be on globalization and sustainability, linking EH&S ever more closely with the business. As EH&S becomes imbedded into the business, the need for specialization will become highly limited. Factors of sustainability—economic growth, development, and environmental resources—will become increasingly important with emphasis on energy conservation, increasing population density, infrastructure, food supply, natural resource management, and technological development. Harmonization of global issues will become more prevalent—expansion of issues we are seeing unfold now—regulation of hazardous materials, international labeling, public right-to-know, information management, intellectual property, greenhouse gases, fate and transport of pollutants, and so forth. Life-cycle analysis coupled with cost accounting, EH&S performance metrics translated into business impacts, product stewardship, and design for the environment are tools for driving sustainability. What this calls for are individuals with strong business knowledge and a technical background who can apply these tools to increase their company's competitive edge through product differentiation, cost reduction, and innovation.

A LOOK FORWARD

Where these hybrid environment-and-business individuals may need to begin in this next cycle is at the undergraduate and graduate level of universities, before they even hit the street. Anticipating this need, some forward-thinking universities[51] have already developed such specialized environment and business programs that stress hard work and high motivational factors to prepare the next generation of twenty-first century business, environment, health and safety ("BEHS") professionals.

ENDNOTES

[1]One noteworthy reference that contains a compilation of pollution prevention success stories which, in turn, point to some innovative ways to achieve compliance and save money for companies, is the EPA document, *Pollution Prevention Success Stories*, EPA/742/96/002, April 1996.

[2]This phrase was penned by the late Piero Trotta, Environmental Audits Manager for EniChem, S.p.A, Milan, Italy to explain the phenomenon behind how environmental issues generate environmental regulations to address these issues, and how additional issues that emerge once these regulations are in place spawn additional and more complex regulations; hence the term, the perverse spiral.

[3]For example, there has been some discussion surrounding EPA's Toxic Release Inventory (TRI) reporting following EPA's decision to expand the TRI database to include additional chemical reporting by organizations, referred to as "TRI phase III," and according to a GAO report, only a handful of states considered proposals requiring the reporting of chemical-use data. There are pros and cons to this expanded reporting, yet studies on the progress to date in two states that require such reporting have mainly produced benefits. Reference: U.S. General Accounting Office report, *Toxic Substances: Few States Have Considered Reporting Requirements for Chemical Use Data*, GAO/RCED-97-154, June 1997. In another example, EPA took issue with a citizens' group regarding information about a Superfund site. See B. Hileman, "Local citizens' group taking heat from EPA," *Chemical & Engineering News,* September 1, 1997, p. 12.

[4]For additional information, see C. Bwen, "Chemical Industry Members Voice Concern over Information Listed on EPA's Internet Site," *Chemical Processing,* November 1997, page 15.

[5]From the popular Dilbert® comic strip by Scott Adams, 1997 United Features Syndicate, Inc., December 21, 1997.

[6]For additional insight into choosing an ISO course provider, see G. Crognale, "Auditor Training—Choosing the Right Course Provider," *Quality Manager, Bureau of Business Practice,* September 25, 1998, p. 2.

[7]The Registration Accreditation Board (RAB), associated with the American Society of Quality (ASQ), confers upon a course participant who has successfully passed the EMS Lead Auditor's course and submits an application to the RAB with the requisite experience, a certificate that recognizes the individual as a certified EMS Lead Auditor.

[8]For additional insight, see First Word by S. Paton, "What? No Safety Glasses?," *Quality Digest,* August 1998, p. 4.

[9]Among noteworthy news items relating to accidents and incidents that might have been preventable, include: "Vapor Ignition Cause of Refinery Accident," *Safety Compliance Letter,* April 25, 1998, p. 8; "Chemical Manufacturer Fined $455,000 for Confined-Space Fatality" and "Shell Settles with OSHA for $130,000," *Safety Compliance Letter,* February 10, 1998, p. 8; and "Employer Indicted for Incident that Leaves Worker Brain Damaged," *Safety Compliance Letter,* August 10, 1998, *Bureau of Business Practice,* p. 8.

[10]For additional insight into this aspect, the reader should refer to the Work & Family column in the *Wall Street Journal,* "Businesses Compete To Make the Grade as Good Workplaces," by S. Shellenbarger, August 27, 1997, p. B1.

[11]Among the findings in an Investor Responsibility Research Center (IRRC) 1996 survey of the S&P 500 companies was that voluntary environmental reporting by companies has grown rapidly, according to the results of a survey question that was received by IRRC of more than 50 percent of the 195 companies who were sent this survey. Survey respondents had indicated they published environmental reports, and at least half of this group expects to publish these reports regularly. *SOURCE:* The IRRC *Corporate Environmental Profiles Directory 1996.* Additional information about this report is provided in Chapter 7. In addition, a news item in *Business Ethics* follows up on environmental reporting and the CERES principles, *Business Ethics,* Vol. 11, No. 4, July/August 1997, p. 10.

[12]See D. Hanson, "NAFTA Spurs First Toxics Inventory Report," *Chemical & Engineering News,* September 1, 1997, p. 26.

[13]One such DFE software tool was created by Ecobalance, Inc., a company of the Ecobilan Group. It is called EIME, or environmental information and management explorer, and is intended to be used by designers to evaluate the environmental impacts of a product under design. The tool is currently designed for electronic and electrical equipment manufacturers. Reference to EIME is found in Chapter 13.

[14]See the EPA document, *Pollution Prevention Success Stories,* EPA/742/96/002, April 1996. One article that highlights similar success stories is G. Crognale's "P2 Financial Incentives," *Environmental Protection,* May 1997, p. 37.

[15]See the EPA document, *Partnerships in Preventing Pollution,* EPA 100-B-96-001, Spring 1996.

[16]Among the organizations that received fines in the recent past from federal regulatory agencies, or are undergoing some enforcement action, are: AgriGeneral Company, LP, "Ohio Egg Producer Fined More Than $1 Million for Alleged Violations" (*Safety Compliance Letter,* September 25, 1997); "Jury Awards Lockheed Workers $4 Million for Alleged Chemical Exposure" (BNA's *Chemical Regulation Reporter,* April 11, 1997, p. 35); Rockwell International, which agreed to plead guilty and pay $6.5 million in fines (BNA's *Environmental Compliance Bulletin,* Volume 3, No. 10, April 22, 1996, p. 2); and controversy continues to surround General Electric at one of its facilities in Pittsfield, MA, regarding past PCB disposal practices (*Engineering News Record,* August 11, 1997, p. 44, and *The Wall Street Journal,* December 4, 1997, p. A1).

[17]For additional information, see "Forget the Government. It's the Community that Can Shut You Down," E. Burke, *Business Ethics,* May/June 1997, p. 11.

[18]See I. Ritchie and W. Hayes, *A Guide to the Implementation of the ISO 14000 Series on Environmental Management,* Upper Saddle River, NJ: Prentice-Hall, Inc., 1998, p. 159.

[19]E. M. Kirschner, "Running on Information," *Chemical & Engineering News,* September 15, 1997, p. 15.

[20]F. Friedman, "Environmental Leadership in the Next Century," *Corporate Environmental Strategy,* Vol. 5, No. 1, Autumn 1997, pp. 78–80.

[21]R. Shelton, "Hitting the Green Wall: Why Corporate Programs Get Stalled," *Corporate Environmental Strategy,* Vol. 2, No. 2, Winter 1995, pp. 5–11.

[22]M. Hammer and J. Champy, *Reengineering the Corporation,* Harper Business, 1993, p. 223. With all due respect to the creators of "reengineering," the authors are not advocates of reengineering for the sake of reengineering, which can be viewed as a management flavor-of-the-month fad. Reengineering, in the final analysis, is really taking a common-sense approach to solve business problems, and users of the process can add value to

it by looking for a linkage or thread that relates noted problems between business units. For additional archival and unbiased information on re-engineering, the reader is referred to two well-written *Wall Street Journal* articles, A. Ehrbar, "Price of Progress: 'Re-Engineering' Gives Firms New Efficiency, Workers the Pink Slip," March 16, 1993, p. A1; and F. Bleakley, "The Best Laid Plans: Many Companies Try Management Fads, Only to See Them Flop," July 6, 1993, p. A1.

[23]J. F. Moore, *The Death of Competition,* New York: Harper Business, pp. 13–14.

[24]For additional insight see G. Crognale, "Take Advantage of EPA and OSHA's Regulatory Syngery to Enhance Training," Safety Compliance Letter, *Bureau of Business Practice,* November 25, 1997, p. 8.

[25]P. Harris, "Centers of Excellence Streamline EHS Role," *Environment Today,* October 1995, Vol. 6, No. 9, pp. 1, 9.

[26]For additional insight, see P. Knox, "Dow chairman speaks on integrating EH&S goals," *Chemical Processing,* July 1998, p. 21.

[27]D. Ditz, "Environmental Accounting and the Business Case," *Environment, Health, and Safety: A Platform for Progress,* The Conference Board, Report Number 1175-97-CH, pp. 37–39.

[28]J. Fiskel, "Competitive Advantage Through Environmental Excellence," *Corporate Environmental Strategy,* Summer 1997, pp. 55–61.

[29]J. Mitchell and H. Brown, "Total Environmental Quality Management, Methodology and Examples of Work in Progress," *The Total Quality Review,* July/August 1994, pp. 17–26.

[30]R. Kolluru, "Minimize EHS Risks and Improve the Bottom Line," *Chemical Engineering Progress,* June 1995, pp. 44–52.

[31]L. Puschaver and R. Eccles, "In Pursuit of the Upside: The New Opportunity in Risk Management," *PW Review: Insights for Decision Makers,* Price Waterhouse LLP, December 1996, pp. 1–16.

[32]S. Colby, T. Kingsley, and B. Whitehead, "The Real Green Issue," *The McKinsey Quarterly,* 1995 Number 2, pp. 132–143.

[33]C. Frankel, "The Next Wave," *Tomorrow,* November/December 1997, pp. 30–31.

[34]M. Greenwood and A. Sachdev, "Environmental Policy in the Information Age," EcoStates, Environmental Council of the States, September 1997, pp. 10–11.

[35]S. Lydenberg, Social Research on Corporations for Institutional Investors, 1997.

[36]S. Collier, "Legal Land Mines," *Bloomberg Personal,* July/August 1997, pp. 41–44.

[37]M. Porter and C. van der Linde, "Green and Competitive," *Harvard Business Review,* September–October 1995, pp. 120–134.

[38]As we noted in Chapter 4, these same sentiments were expressed by Alfred De Crane at the Marketing Conference of the Conference Board, October 26, 1993, New York, NY. His press release is provided in Appendix G.

[39]A. Hirsig, "Environmentalism Comes of Age: A More Rational Approach for the 21st Century," *Corporate Environment Strategy,* Volume 4, Number 4, Summer 1997, pp. 87–91.

[40]J. Fiskel, "Competitive Advantage Through Environmental Excellence," *Corporate Environmental Strategy* (n. 28).

[41]E. B. Harrison, "The Strategic Implications of Global Environmental Communication Needs," *Corporate Environmental Strategy,* Volume 3, Number 3, Spring 1996, pp. 77–83.

[42]J. Sullivan, *Perspectives: Attaining the Sustainable,* Conference on Corporate Environmental, Health, and Safety Excellence, pp. 1–13.

[43]We previously provided some insight into Germany's takeback regulations in Chapter 1.

[44]E. Lowe and R. Harris, "British Petroleum's Business Strategy," *Corporate Environmental Strategy,* Volume 5, Number 2, Winter 1998, pp. 23–31.

[45]D. Buzelli, "Eco-Efficiency: A Competitive Edge in the Chemical Industry?", Eco 1997 International Conference in Paris, France.

[46]In addition to Sony, a group of German TV manufacturers, including Grundig, Philips, Nokia, and Deutsche Thompson-Brandt, received support from the German Federal Ministry of Education, Science, Research and Technology to develop a "Green TV." This presentation was given by Dr. Volker Strubel of Oko-Institut e. V., "Development of a Green TV," at CARE Innovation '96 in Frankfurt, Germany, hosted by Sony Europe.

[47]V. Sandborg, "The Challenges of a Specialized Industry," *Environment, Health, and Safety: A Platform for Progress,* The Conference Board, Report Number 1175-97-CH, 1997, pp. 25–27.

[48]S. Fedrizzi, "The Greening of Customers' Needs," *Environment, Health, and Safety: A Platform for Progress,* The Conference Board, Report Number 1175-97-CH, pp. 34–36.

[49]See Carrier Corporation's web site at www.carrier.com. 1998.

[50]J. Champy, *Reeingineering Management,* © 1995 and 1996 by James Champy, Harper Business, 1995, pp. 156.

[51]The Gordon Institute is one such school that stresses environment-and-business learning. The author provides mentoring to graduate-level environmental management students.

REFERENCES

Adams, S. Dilbert®, December 21, 1997, © United Features Syndicate, Inc.

Bleakley, F. "The Best Laid Plans: Many Companies Try Management Facts Only to See Them Flop," *Wall Street Journal,* July 6, 1993, p. A1.

BNA's *Chemical Regulation Reporter,* "Jury Awards Lockheed Workers $4 Million for Alleged Chemical Exposure," April 11, 1997, p. 35.

BNA's *Environmental Compliance Bulletin,* News item on Rockwell International, Vol. 3, No. 10, April 22, 1996, p. 2.

Burke, E. "Forget the Government. It's the Community that Can Shoot You Down," *Business Ethics,* May/June 1997, p. 11.

Business Ethics, Vol. 11, No. 4, July/August 1997, p. 10.

Buzell, D. "Eco-Efficiency: A Competitive Edge in the Chemical Industry?" The ECO 1997 International Conference in Paris, France.

Bwen, C. "Chemical Industry Members Voice Concern over Information Listed in EPA's Internet Site," *Chemical Processing,* November 1997, p. 15.

Carrier Corporation's web site at www.carrier.com (1998).

Champey, J. *Reeingeering Management,* New York: Harper Business, 1995, p. 156.

Colby, S., T. Kingsley and B. Whitehead, "The Real Green Issue," *The McKinsey Quarterly,* 1995, No. 2, pp. 132–143.

Collier, S. "Legal Land Mines," *Bloomberg Personal,* July/August 1997, pp. 41–44.

Crognale, G. "Auditor Training—Choosing the Right Course Provider," Quality Manager, Bureau of Business Practice, p. 2.

Crognale, G. "Financial Incentives," *Environmental Protection,* May 1997, p. 37.

Crognale, G. "Take Advantage of EPA and OSHA's Regulatory Synergy to Enhance Training," *Safety Compliance Letter,* November 25, 1997, p. 8.

De Crane, A. Opening Remarks at the Marketing Conference of the Conference Board, October 26, 1993, New York, NY.

Ditz, D. "Environmental Accounting and the Business Case," *Environment, Health and Safety: A Platform for Progress,* The Conference Board, Report Number 1175-97-CH, pp. 37–39.

Ecobalance, Inc, EIME software press release, 1998.

EPA document, *Partnerships in Preventing Pollution,* EPA 100-B-96-001, Spring 1996.

EPA document, *Pollution Prevention Success Stories,* EPA/742/96/002, April 1996.

Ehrbar, A. "Price of Progress: 'Re-Engineering' Gives Firms New Efficiency, Workers the 'Pink Slip,'" *Wall Street Journal,* March 16, 1993, p. A1.

Engineering News Research, news item on General Electric, August 11, 1997, page 44.

Fedrizzi, S. "The Greening of Customers' Needs," *Environment, Health and Safety: A Platform for Progress,* The Conference Board, Report Number 1175-97-CH, pp. 34–36.

Fiskel, J. "Competitive Advantage Through Environmental Excellence," *Corporate Environmental Strategy,* Summer 1997, pp. 55–61.

Frankel, C. "The Next Wave," *Tomorrow,* November/December 1997, pp. 30–31.

Friedman, F. "Environmental Leadership in the Next Century," *Corporate Environmental Strategy,* Volume 5, Number 1, Autumn 1997, pp. 78–80.

GAO document, *Toxic Substance—Few States Have Considered Reporting Requirement for Chemical Use Data,* GAO/RCED-97-154, June 1997.

Hammer M. and J. Champy, *Reengineering the Corporation,* New York: Harper Business, 1993.

Hanson, D. "NAFTA Spurs First Toxics Inventory Report," *Chemical & Engineering News,* September 1, 1997, p. 26.

Hanson, E. B. "The Strategic Implications of Global Environmental Communication Needs," *Corporate Environmental Strategy,* Vol. 3, No. 3, Spring 1996, pp. 77–83.

Harris, P. "Centers of Excellence Streamline EHS Role," *Environment Today,* October 1995, Vol. 6, No. 9, pp. 1,9.

Harrison, E. B. "The Strategic Implications of Global Environmental Communication Needs," Corporate Environmental Strategy, Vol. 3, No. 3, Spring 1996, pp. 77–83.

Hileman, B. "Local citizens' groups taking heat from EPA," *Chemical & Engineering News,* September 1, 1997, p. 12.

Hirsig, A. "Environmentalism Comes of Age: A More Rational Approach for the 21st Century," *Corporate Environmental Strategy,* Vol. 4, No. 4, Summer 1997, pp. 87–91.

IRRC's *Corporate Environmental Profiles Directory 1996,* Washington, DC.

Kirschner, E.M. "Running on Information," *Chemical & Engineering News,* September 15, 1997, p. 15.

Knox, P. "Dow Chairman Speaks on Integrating EH&S Goals," *Chemical Processing,* July 1998, p. 21.

Kolluro, R. "Miminize EHS Rules and Improve the Bottom Line," *Chemical Engineering Progress*, June 1995, pp. 44–52.

Lowe E. and R. Harris, "British Petroleum's Business Strategy," *Corporate Environmental Strategy*, Vol. 5, No. 2, Winter 1998, pp. 23–31.

Lydenberg, S. *Social Research on Corporation for Institutional Investors*, 1997.

Mitchell J. and H. Brown, "Total Environmental Quality Management, Methodology and Examples of Works in Progress," *The Total Quality Review*, July/August 1994, pp. 17–26.

Moore, J. F. *The Death of Competition*, New York: Harper Business, 1996, pp. 13–14.

Paton, S. *First Word*, "What? No Safety Glasses?", Quality Digest, August 1998, p. 4.

Porter M. and C. vanderLinde, "Green and Competitive," *Harvard Business Review*, September/October 1995, pp. 120–134.

Poschaver L. and R. Eccles, "In Pursuit of the Upside: The New Opportunity in Risk Management." *PW Review: Insights for Decision Makers*, Price Waterhouse LLP, December 1996, pp. 1–16.

Ritchie I. and W. Hayes, *A Guide to the Implementation of the ISO14000 Series on Environmental Management*, Upper Saddle River, NJ: Prentice-Hall, Inc., 1998, p. 159.

Sandburg, V. "The Challenges of a Specialized Industry," *Environment, Health and Safety: A Platform for Progress*, The Conference Board, Report Number 1175-97-CH, 1997, pp. 25–27.

Shellenbarger, S. Work & Family, "Businesses Compete to Make the Grade as Good Workplaces," *The Wall Street Journal*, August 27, 1997, p. 31.

Sheldon, R. "Hitting the Green Wall: Why Corporate Programs Get Stalled," *Corporate Environmental Strategy*, Vol. 2, No. 2, Winter 1995, pp. 5–11.

The Wall Street Journal news item on General Electric, December 4, 1997, p. A1.

State Roundup Column, *Safety Compliance Letter*, Bureau of Business Practice.

———, "Vapor Ignition Cause of Refinery Accident," April 25, 1998, p. 8.

———, Chemical Manufacturer Fined $455,000 for Confined-Space Fatality," and "Shell Settles with OSHA for $130,000," February 10, 1998, p. 8;

———, "Employer Indicted for Incident that Leaves Worker Brain Damaged," August 10, 1998, p. 8; and

———, "Ohio Egg Producer Fined More Than $1 Million for Alleged Violations," September 25, 1997, p. 1.

Sullivan, J. "Perspectives: Attaining the Sustainable," 1997 Conference on Corporate Environmental, Health and Safety Excellence, pp. 1–13.

INNOVATING THE ENVIRONMENTAL LANDSCAPE

Part 2 continues in this arena by beginning with a look into benchmarking studies of a number of regulated organizations, Chapter 7. In Chapter 8, the authors describe several building blocks for improving upon a company's environmental management system. Chapter 9 takes an in-depth view of environmental audits and provides the viewpoints of several environmental professionals from industry.

CHAPTER 7

BENCHMARKING STUDIES: ADDING INSIGHT AND VALUE TO ORGANIZATIONS' EH&S PROGRAMS

Gabriele Crognale, P.E.

"Knowledge is of two kinds. We know a subject ourselves, or we know where we can find information upon it."

Samuel Johnson

INTRODUCTION

Benchmark: (1) A permanent reference mark fixed in the ground for use in surveys, tidal observations, etc.; and (2) A reference point serving as a standard for copying or judging other things. (*Webster's Dictionary*)

Among the questions a senior company manager may ask during an internal review of business reports or other data providing insight to the organization's "bottom line" figures may include an inquiry into what changes, if any, can be made to improve these figures. One vehicle that can provide at least partial answers to such questions is a benchmarking study. A question one may ask in this regard is whether information gained from benchmarking studies can provide measurable benefits. To answer this question, let us look more closely at benchmarking and what some of the benefits are that can be derived.

Webster's Dictionary defines "benchmark" as a reference or starting point— appropriate to benchmarks set by the U.S. Geological Service (USGS) to reference a fixed point from which subsequent surveying data points can be set, such as in civil engineering projects. Benchmark, or benchmarking, also denotes the process by which companies review the practices of other companies in comparison to their own practices as a means to gain some insight into improving its own activities as compared with those of the other companies being evaluated. As a well-followed business function, benchmarking has developed its own "cottage industry" and is a widely used and known business process. For example, David Drennan, author of *Transforming Company Culture*, defines benchmarking as " . . . going out to see what (one company's) competitors or the best companies do, or, collecting hard data from their customers . . . "[1] Ingrid Ritchie and William Hayes, co-authors of *A*

Guide to theImplementation of the ISO14000 Series on Environmental Management, define bench-marking as " . . . a technique in which one operation is compared to an identical or similar operation that has better performance."[2]

Hence, a benchmarking study can become the foundation for a company—the "sponsor"—to look at other companies selected—the "partners"—as examples from which to learn. Their best practices in specific target areas are examined closely with the intent to capture the essence of these practices to improve upon their own practices. Benchmarking also provides a screening process that helps identify those organizations that are considered best-in-class. Usually, these organizations become the nucleus of a pool of prospective benchmarking partners from which benchmarking partners are chosen, as was the case with the American Productivity & Quality Center (APQC) benchmarking study highlighted later in the chapter, courtesy of Vicki J. Powers of the APQC.

The first example of a benchmarking study is that of the Goodyear Tire and Rubber Company which commissioned a benchmarking study of its business operations.[3] As part of the study, Goodyear decided to analyze, deconstruct, and reassemble various business practices into different practices based on previously collected information. As this particular study progressed, the company gained considerable insight in the way the organization and its employees worked. The company had measured itself against other partner organizations and had solicited input at each step of the study from key employees. The ongoing study generated a level of enthusiasm the company did not know existed, and actually managed to empower its workforce as a result. A lesson learned from this benchmarking exercise was that the company had realized a bonanza when its workers were noted doing their jobs more efficiently. This was an unanticipated side benefit as a result of the study. As Mike Burns, VP of Human Resources and Total Quality Culture at Goodyear said, "benchmarking" is about learning and sharing ideas."[4]

While the main focus of the Goodyear study dealt primarily with human resources (HR) priorities and issues, some of these items could also relate to EH&S worker responsibilities as well, thus providing insight to EH&S personnel issues, such as new hires for required EH&S training, or finding the best candidate for the job from a pool of prospective applicants, or other HR issues. In addition, the article highlighting Goodyear's study also made references to a database and repository of benchmarking studies compiled by the International Benchmarking Clearinghouse of the APQC that could be a valuable source of information for future benchmarking studies. Additional information about APQC and the clearinghouse can be obtained by clicking onto APQC's web site at **www.apqc.org.**

USING BENCHMARKING TO PROVIDE SOLUTIONS TO EH&S INQUIRIES

Within the sphere of manufacturing or service-provider industries, any number of business practices or manufacturing processes can catch the attention of senior managers. These practices may prompt management to raise thought-provoking questions for discussion at an upcoming board, managers', or staff meeting.

These questions could focus on specific practices that may need to be evaluated for improvement and cost-cutting measures. Situations may arise, such as a specific environmental or health and safety (EH&S) issue or practice that an organization may feel needs to be reviewed to find cost-effective solutions. A few EH&S areas where management inquiry may focus upon can include: (1) EPA, DOT, SEC, or OSHA regulatory compliance concerns; (2) waste generation/minimization issues to be resolved; (3) exploring innovative ways for providing employee training; and, (4) the benefits of environmental audit programs, and whether auditing software can add value. These are just a few of the areas that could appear on a meeting agenda for review and possible resolution within a company.

In such situations, management may reach a consensus decision at a meeting and decide that a benchmarking study is an appropriate vehicle for the organization to undertake. Once management makes that decision, the logical next step would be to commission a group to execute the task. Upon being charged, that group could initiate the study. With some planning, the sponsoring organization can develop a benchmarking study that has identified its particular partners for the selection process. Subsequently, the information obtained can provide the answers to specific questions management may have for issues targeted for resolution. The data obtained from the study can also be useful as part of any follow-on study the sponsoring organization may want to undertake for additional insight, or for other relevant data-gathering uses.

Let's look at another hypothetical example and review it as a potential benchmarking study. An organization's top management is concerned about costs associated with EH&S operations at some of its manufacturing facilities, such as additional staffing, training, volumes of off-spec raw materials generated, increasing waste management and regulatory requirements, and other issues, and would like to look at how other organizations account for such costs in each operational area in which they are generated. By evaluating how these organizations account for such costs, the organization may gain some insight into more effectively tracking each cost to calculate the net return on capital investment. With that, EH&S management may be better equipped to account for such costs and be able to tabulate them in some fashion that highlights to senior management what each cost means and how these can actually generate cost savings to the company. For example, additional dollars spent in training front-line workers on the proper handling of chemicals could be more than offset by the costs associated with the cleanup of an accident or chemical spill event. Highlighting such cost expenditures in a format that is clear for non-EH&S managers to understand as a cost savings can be a valuable tool to a chief operational officer (COO), chief financial officer (CFO), or other top management figure who tracks expenditures for the company. The cost savings realized could then be entered into the company's ledger books as a positive cash flow item.

Another benchmarking study could focus upon the relationship between an organization's EH&S group and other business units within the organization. In this example, management may want to look more closely at how the different business units relate to each other by evaluating similar business units at other "best-in-class" organizations. In such a study, they could evaluate their partners' business units to gain some insight into their business practices and evaluate whether such practices could work for them. They could find that their partners have established effective cross-communication between the business units and the EH&S group, and as a result, generate tangible cost savings within each business unit. This could manifest itself via a greater awareness level of EH&S responsibilities that extends from the EH&S group to research & development, purchasing, human resources, production or manufacturing departments, and distribution departments among others. Such heightened EH&S awareness levels from employees in these other business units can help minimize adverse environmental and worker health and safety issues, and help maximize more efficient product throughput. Such heightened awareness levels within different employee groups such as these may have been the impetus that prompted some organizations to evaluate Design for the Environment (DfE), life-cycle assessment (LCA), product take-back, and other innovative considerations as a means to push the EH&S responsibility envelope to non-EH&S units within an organization.

This EH&S awareness level embodies a holistic overview of environmental and health and safety responsibilities within an organization's various business units. In those organizations where other non-EH&S business units interact with the EH&S group to coordinate EH&S responsibilities more effectively and systematically, a greater sense of shared responsibilities may be found that may also foster greater teamwork in achieving the organization's environmental goals. As such, an organization wishing to conduct a benchmarking study of this shared responsibility relationship between a company's business units may want to study those companies that practice a shared responsibility of EH&S responsibilities throughout its various business units in a holistic fashion. For example, a best-in-class company may extend EH&S awareness training to machine shop employees, custodians, office workers, and outside contractors, to name a few, that could have an impact on environment or health and safety considerations at a facility. One EH&S video[5] currently available shows a scenario where two clerical staff workers are walking by a tank car off-loading chemical product and notice a spill of product. If such a scenario had occurred at your facility, how would your clerical staff react?

In any of the following hypothetical scenarios, it would not be uncommon for a company's management to ask any number of questions from which benchmarking studies may be initiated. A sampling of such questions may include:

- Where are we now in our environment and health & safety (EH&S) programs, and where do we want to be in six months? A year? Two years?
- What do we currently lack in our EH&S program? What would we like to add? If so, can we justify the costs?
- What companies or CEOs do we admire most? Are these companies "best-in-class" and what can we learn about their respective practices that can help us achieve our long-term goals? What are the similarities/differences between our companies, and is there a match?
- Is ISO 14001 for us? What organizations should we follow to view their success in achieving ISO 14001 certification? Have these companies experienced any tangible benefits from certification, or is it too soon to tell? What were the costs

and how long did this process take? Had any of these companies been certified to ISO 9001 or 9002 as well? Did that make a difference in cost and effort?

- Are we ready to look at software applications to: enhance EH&S functions and responsibilities; link the organization's various business enterprises with EH&S to better account for savings and costs associated with EH&S activities; track documents, perform EH&S audits, perform ISO 14000 gap analyses, monitor real-time chemical and process flows, create simulations in a manufacturing process; or other considerations?

- Are our EH&S costs accurately measured and accounted for during each fiscal quarter? Year?

- What companies have initiated benchmarking studies dealing with specific EH&S issues? What did they look for specifically, and how useful were the results of their studies? Are these companies our competitors, or are they in other industries? Would/could they share their findings with us if we asked them?

- Are there any (EH&S) benchmarking studies that have been conducted that we could reference in our study, and are they relevant to our needs? (Some examples are highlighted later in this chapter.)

- What will the study cost in time, labor, and other expenditures, and will the costs generate a favorable return on our investment?

- What innovative training techniques have other organizations used in training their EH&S and non-EH&S personnel in specific areas of consideration?

- Should our company issue an environmental report to highlight our environmental accomplishments? Should our company consider evaluating the reports of other companies for ideas on form and content, or should we consider following a recognized standard in developing our report, such as the Public Environmental Reporting Initiative (PERI) Guidelines? (A copy of the PERI Guidelines is included in Appendix D.)

- How should we address environmental audit disclosures? Should we be concerned about a regulatory backlash if such information is not in accordance with EPA's audit disclosure policy? Are we confident about our EH&S auditing program, or are we still in an evolutionary stage regarding the maturity of our program?

- And probably the underlying question at the core of any benchmarking study an organization undertakes . . . Are we really ready for *change?*

Any of these questions could lead to a thorough benchmarking study. The actual benefit to an organization that may be derived from answers to any of these questions may depend upon several factors. This may include the specific areas to be studied, the level of detail in each of the questions asked, the partner organizations, and other considerations based on site-specific circumstances. In the following sections, we look at some benchmarking studies in greater detail to provide a "lessons learned" approach.

WHAT CAN BE GAINED BY AN ORGANIZATION IN A BENCHMARKING STUDY?

Depending upon the specific areas of interest an organization is looking to find answers to address specific management questions, some information is usually available that can be researched prior to initiating a study. Such data can provide a reference point to a well-organized benchmark study. For example, in preparing background material for this book, we compiled a number of surveys, reports, and related documents that are referenced and footnoted throughout the book, and especially in this chapter. These reference documents provide insight into a number of different areas that focus on a variety of EH&S considerations.

For example, referring to how benchmarking studies can benefit from information that has already been compiled, let us look at a winning formula prevalent at many conferences and seminars. In our research, in the years 1996 to 1997, an increasing number of conferences showcased one or more case studies depicting how a company solved a particular EH&S management or compliance issue. The organizations promoting these conferences and seminars have hit upon a "hot button" to attract prospective attendees. Such "lessons learned" case studies provide conference or seminar attendees varying levels of "sound bytes" of information highlighting a company's practices or experiences relating to a specific EH&S activity. Some of these case studies may have focused upon, for example, innovative approaches to achieve P2 as a cost-saving opportunity, innovative EH&S software to help streamline EH&S processes; they may have discussed innovative approaches to product take-back, or any other "magnet" topic that may have attracted a prospective attendee's interest. We provide a representative sample of some noteworthy conferences during this period later in this chapter.

Among the benefits in attending such conferences and listening to presentations is the prospect of providing attendees ample opportunities to ask presenters questions about their cases, while also providing them opportunities to network with the presenters. In addition, most conferences provide each attendee a published copy of the proceedings containing each speakers' presentation. This information can also be useful to attendees as references for a future benchmarking study. Attendees should also make use of any time that conferences set aside for social hours to introduce themselves to the speakers and fellow attendees. One never knows when such casual contacts may lead to an interview for obtaining supplemental benchmarking data, or establishing a personal link to potential benchmarking partners. We met several of our contributing authors in this fashion.

In addition to the information that can be gained from conferences, valuable supplemental information can also be obtained from researching various sources, such as public libraries, EH&S staff individual libraries, and a company's central repository of business reports, companies' annual reports, specific studies and surveys conducted, and any relevant additional reports. In addition, as we found in our research, the Internet also provides a valuable source of information. Such searches could find surveys of companies' environmental audit programs, benchmark studies of company EH&S practices, innovative business solutions practiced at some progressive companies, and other relevant data that could be valuable in a benchmarking study.

Given such information, you may wonder whether any of this could be of value to your company and whether it could provide any guidance to a particular concern or issue. So, let us step back a moment to reflect upon the concept of benchmarking. It is a reference point for copying other things. Let us refer again to the example of an organization in which someone in management has noted some aspect of a business function that has drawn the concern of senior management. In this instance, let us suppose this item of management's interest may be related to the organization's environmental audit program. During the audits, let us say that the audit team regularly uncovers regulatory or management practice findings that are of concern to top management, such as, production processes that continually waste large amounts of raw materials or resources that translate to slimmer profit margins, or other concerns that management would like corrected or improved. Solutions to these issues may sometimes come from studying the issue from a different perspective, which may require looking to other sources for these solutions, such as comparing procedures with another company—a benchmarking study. When this evaluation and comparing process with other organizations is completed, you should come away with a greater understanding of solving your particular problems by having studied how other organizations have solved theirs. In studying their activities, you may have gained sufficient knowledge from their experiences to provide your organization some measurable benefits. How much one can gain in a benchmarking study may be a function of how well the study may have been planned from the start and how carefully you may have chosen partners as models to follow.

INITIATING A BENCHMARKING STUDY: KEY CONSIDERATIONS

We began this chapter with an overview of what benchmarking studies can provide to organizations through their insight into specific areas under consideration. Let us now look closer at some of the steps you should consider in benchmarking. An important step in initiating a benchmarking study should be that the organization commissioning the study should first have a thorough understanding of existing conditions within its operating facilities, including all processes and business areas that may need to be evaluated. This may be a standard procedure at some organizations, and for those organizations where this is not, developing various contingencies to address this consideration can be worthwhile. Taking into account existing conditions and identifying areas that may require further study to evaluate make up the "baseline." Let us label establishing a baseline of existing conditions *Step One* in the benchmarking study process.

DEVELOPING A BASELINE

The first step in initiating a study is to get a better understanding of your own facility to gauge where your facility, or the company, may be with respect to its EH&S goals and objectives. The more this baseline accurately reflects existing conditions at your facility, the greater the insight it can provide as a foundation for a benchmarking study.

For example, establishing your company's baseline may take into consideration specific aspects of business unit practices, such as human resource department procedures in locating and obtaining new talent to add to the workforce in various positions, tracking any required training and medical monitoring procedures, or other such HR considerations that could impact other business areas.

The baseline could also focus on evaluating regulatory compliance issues generated from: worker attitudes or language barriers; lack of management oversight and follow-up; lack of worker understanding of procedural or regulatory EH&S requirements; union constraints; or any other of a number of considerations. Your facility could also establish its baseline following a comprehensive environmental, EH&S, or environmental management system (EMS) audit that may identify areas in need of some form of regulatory or management system improvement. In either case, issues may surface that depict actual situations in need of correction, for which a benchmarking study may provide some insight into providing solutions to issues identified.

While establishing your facility's baseline, you should also keep in focus those business functions and areas the baseline is supposed to encompass, how it will be measured, and what insight this can provide to top management in developing a benchmarking study for your company. Someone in your organization should also take it upon himself or herself to delineate measuring points that will define the baseline, so that relevant findings can accurately reflect where your baseline may be at a given period of time. Fine-tuning the baseline with accurate and relevant measuring points can be very useful in helping to scope out your benchmarking study.

Those areas or considerations that may help to further define a company's existing baseline can include:

- Worker training records and specialized job descriptions, and whether such records are part of personnel records or are maintained as separate records. Further insight could be obtained through spot interviews of plant or shop employees.
- Reviewing various EPA- or State-required documents and checking whether these documents are maintained and up-to-date. You may also want to look for systems that keep track of these documents, and mechanisms that flag any documents that fall through the cracks.
- Manifests, inspection records, and environmental audit report files.
- Results of contractors' or consultants' previous work, including third-party audit reports, site investigations, RCRA closure plans or groundwater monitoring reports, or other data for establishing a historical baseline.
- Regulatory enforcement files—Are Notices of Violation (NOVs) up-to-date, or can any unresolved violations be flagged in a follow-up inspection or subsequent environmental audit?

In developing this baseline you may also want to consider a review or check of interrelated business functions, personnel, and other sources of information that can provide an additional layer of insight into existing conditions.

This review can encompass, for example, any of the following scenarios we suggest: (1) interviewing internal EH&S specialists, process engineers, R&D engineers, plant managers, process managers, HR personnel; (2) comparing interview results with documents in the EH&S file, such as audit reports, emission records (annual TRI emission results), permits, and any enforcement files; (3) reviewing your corporate mission statements, environmental policies, annual reports, and environmental reports; (4) your corporate and risk communications policies and files, and any other nonprivileged corporate data relating to these areas; (5) a comparison of other companies' EH&S programs and how they relate to your company's; and (6) other sources of information that can provide you insight into how your company is perceived by others outside the company, to name a few. Other areas you can suggest to review can provide an added depth of information for defining your own company's baseline. How far you may decide to pursue this exercise may depend upon the time at your disposal, and the amount of funds available. You should also limit your review to that which is practical and let past experience be your guide.

Combining this information with any previously gathered data into a report that defines your company's baseline of existing conditions can provide a fairly accurate roadmap of your organization's current EH&S status. With this information readily available in a concise report format, you are well-prepared to launch into your benchmarking study.

We now describe *Step Two* of the benchmarking process. This step focuses on the team assigned to the task and its responsibilities relevant to scoping out the details of the study. This includes evaluating the areas in need of review based on the baseline, identifying the sources of additional information that should be researched, identifying the partner pool from which the final partners will be selected, and developing a list of questions that may be asked of the individuals from the partner companies chosen.

During this step, the team should also set aside sufficient time to research and scope out critical components of a study, such as where to go for research data and who to contact for additional information sources to help fill gaps. The team should also conduct a thorough research of various media sources, including books, business and specialty trade publications, newspapers, annual and environmental reports, relevant EPA studies, and any related studies compiled by private firms, such as conference proceedings, and online sources from the Internet. A less-than-prepared data search, or a casual attitude displayed by team members in gathering this material may have your team coming up short in this preliminary data-gathering stage. Depending upon the quality and extent of the data collected, experienced data researchers can sometimes find supplemental information that can be very useful in helping to fill data gaps. In either case, the information collected can be archived for later use or as research material for a related or follow-on study at a later date.

For example, some of the research material available in print that can provide a wealth of information for any EH&S-related benchmarking study is referenced in this chapter. We compiled this information from various public sources over the past several years and this made up a portion of the background material for this book. Our research library has thus grown into a valuable source of reference materials and network contacts. Noteworthy material in our reference library that can serve as possible references in a benchmarking study includes: two Price Waterhouse surveys of U.S. companies dealing with environmental issues,[6] a Duke University study,[7] a series of General Accounting Office (GAO) reports that deal with regulatory issues,[8] a United Nations Environment Program (UNEP) study,[9] Arie DeGeus's *The Living Company*, an Investor Responsibility Research Center (IRRC) study,[10] a 1996 environmental rating questionnaire developed by the Swiss Bank Corporation, a comprehensive benchmarking study compiled by the APQC focusing on environmental health and safety at six partner companies[11] and a survey of business by the U.S. Chamber of Commerce.[12]

In addition to utilizing the data, members of the benchmarking team should also consider referring to the information and making contact with study developers or participants as a source of potential partners for the study or future studies. The team could screen this list further, if necessary, to allow them to select a smaller pool of companies that could exemplify "best-in-class" organizations. The team should also collect and review any supplemental documents that could be used to collect benchmarking data, such as questionnaires, checklists, or other forms that could be used during interviews of partner company employees.

Depending upon the scope of a benchmarking study and what direction it may take, subsequent compiled data can provide insight into a number of different areas in which a company's management may be seeking answers to specific questions or concerns. Examples of such concerns that could generate viable benchmarking studies are:

1. How other companies conduct environmental audits and whether audits have a bearing on improving environmental performance.
2. Whether conformance to ISO 9001 and 14001 can improve environmental performance and provide a favorable return of investment (ROI) over a measurable time period.
3. What concerns regulated companies have about regulatory issues with regulators, and can any of these be resolved to both parties' satisfaction?
4. Whether companies can enhance their credibility with stakeholders through issuing annual environmental reports.
5. Evaluating companies' environmental management practices to identify best-practice techniques and establish the hierarchy of what currently exists in the regulated community.
6. Which of the EPA voluntary programs provided the most benefit to the volunteers? What could be learned from their experiences?
7. Has EPA's audit disclosure policy provided any tangible benefits to those companies that have disclosed their audit findings to date?[13]
8. Can regulated organizations put total faith in the press or other media outlets regarding news items of events at their facilities?

As additional insight in developing some key components of Step Two, the following bullets, along with our interpretation, are provided, excerpted from an article on benchmarking in *Quality Digest*[14]:

- **Start the selection process with a clean slate.**
  Make no pre-conceived notions regarding the ideal partner. A recognized name is no guarantee of a useful partner. Let their activities and processes be your guide. For example, not all the two hundred or so[15] companies in the U.S. that have certified to ISO 14001 are well-recognized names. Does that make them any less appropriate as benchmarking partners for an organization wishing to look to certify to ISO 14001?

- **Establish well-defined criteria upfront for benchmarking partners.**
  What is it that you seek to learn? Cull your partner "pool" to allow you to choose from a representative sample to make the study fit your needs. For example, your company may be considering to evaluate specialized EH&S software to support your environmental management system (EMS). Do you limit your evaluation to software that provides one-dimensional capabilities (manifest form generation, MSDS sheets, or other limited functions), or do you include those software packages that allow for integration with other types of software that are emerging in EH&S, EMS, and EMIS applications? (We discuss this topic further in Chapter 13.)
- **Define what "best practice" means at your company, then woo partners accordingly.**
  Who are the best-in-class? This example may be illustrated by looking at a company that seeks to implement an innovative P2 program or environmental audit program at each of its facilities. One measure of the effectiveness of a prospective partner's P2 program may be seen from TRI emission figures the partner reports each year; just as the effectiveness of the company's environmental audit program may be seen in the number and types of enforcement actions noted in the partner's environmental report, if one is issued. (We discuss this topic later in this chapter and in greater detail in Chapter 10.)
- **Use secondary research to identify potential benchmarking partners.**
  Fine-tune your benchmarking partner "pool." You can rely on networking with your professional peers for additional information to narrow your search, or take note of a particular speaker at a conference or seminar explaining the accomplishments of their company in reducing their waste streams, applying innovative product take-back solutions, or a number of other considerations to narrow down the field of potential partners.
- **Weed out the best from the rest.**
  Make your final selection from your "pool" of prospective candidates. One tool that can help pinpoint your selection is the Internet. This source can provide a wealth of information in selecting a partner, or in gaining up-front information about a potential partner before making first contact. For example, your company may want to evaluate a number of innovative companies' environmental management or compliance methods that you are interested in pursuing, but you are not quite sure which way you want to go. You can log on to EPA's web site (www.epa.gov) and review the status of companies that have provided information as part of one of the EPA voluntary programs, such as the Environmental Leadership Program, Project XL, or the Common Sense Initiative. From that preliminary data-gathering exercise, you may find some innovative solutions that fit your needs and you may decide to concentrate on those organizations providing these solutions as benchmark partners.

In addition, EPA's latest addition to its WebSite—Envirofacts—allows you to know more about a company's regulatory compliance status by accessing its Envirofacts database of enforcement data. (**www.epa.gov/enviro,** Pat Garvey, Envirofacts Director, EPA). Note: While still in its infancy, this particular site has generated concern from some chemical companies regarding the information listed. A study was commissioned by the Chemical Manufacturers Association (CMA) that found such data reveals information about a company's competitive status. That study is titled "Economic Espionage: The Looting of America's Economic Security in the Information Age."[16]

These tips and our interpretation in italics are offered to help define some of the salient points of benchmarking worth noting as part of planning a thorough and effective study.

Selecting partners for a benchmarking study does consider a number of qualifying points and supplemental material in the initial selection process. In narrowing the selection down to focus on the best partners, there are specific points an organization could evaluate. These points may focus on any one of a partner organization's procedures, processes, or other criteria that are the basis for the screening, and may include a closer look at their: (1) EMSs; (2) EH&S awareness training programs; (3) human resources staffing and tracking procedures for new hires; (4) developing and running management information systems (MIS) and environmental management information systems (EMIS) as efficient management capability tools; (5) environmental reports and other public outreach efforts;[17] (6) corporate communication groups' handling of newsworthy items focusing on risk communication;[18] or any other of a number of situations that could provide valuable insight into the host organization.

For example, taking a closer look at a hypothetical benchmarking study a company may have commissioned to evaluate other companies' MIS or EMIS systems, you may want to review and set the parameters for selecting best-in-class partners. These parameters could include comparing the strengths and limitations of the software each company uses; the ease of integration into existing software platforms; the level of employee expertise and time required with each system set-up; how the software is sold, in other words, as a complete package, or by specific modules that can be added as needed; whether the software allows user interface and tailor-made revisions to suit the client; its cost and licensing agreements; and whether tangible benefits can be derived by using the software to offset costs associated with start-up and implementation. As depicted in the case study of Millipore in Chapter 8, or in the examples provided in Chapter 13 of some of the specialized EH&S software that is currently available, a benchmarking study developer could obtain considerable insight into how such systems could more effectively identify and track business areas or processes. Of importance here would be whether such areas could be more effectively tracked by software to verify whether improvements could lead to increased productivity, a reduction in pollution, the generation of less waste or off-spec product, or other noteworthy improvements and allow management to view progress reports on a computer screen if necessary. Such improvements could also provide workers the impetus to look for further improvements in: quality control, internal and external communications, and improved business relations with suppliers and customers, to name a few. For example, one method companies use to improve business relations between customers and their suppliers is "strategic sourcing" or Just-in-Time II (JIT-II) inventory control in which manufacturers allow their suppliers to maintain offices at their operating facilities and become privy to previously guarded data, such as up-to-the-minute sales forecasts, to streamline processes and create harmony and efficiencies for both customer and supplier.[19]

Upon completion of Step Two, the team should have a fairly good database relevant to the study area(s) in question, and be able to make a final partner selection. Your team may also decide to cross-reference information obtained with other data to confirm a partner's selection, or possibly generate additional questions for partners during the study. As another example, the team may have compiled information from various sources including a list of industry environmental stewards, those companies with the lowest TRI emission figures several years running; or, in contrast, companies with poor compliance records. This additional information could provide added insight that may otherwise have gone unnoticed during the interview part of the study. With such information readily available and synthesized in a format to allow comparisons to be drawn with each partner and help facilitate final selection, the team is prepared to begin the study. This is *Step Three.*

Upon completion of the data-gathering and interviewing portion of the study, the benchmarking team will have obtained a considerable amount of information which can be summarized, analyzed, and condensed into a comprehensive benchmarking study report tailored to suit the needs of the sponsoring organization. This study report is *Step Four*—an interim final step. We purposely call this an interim final step in deference to the Plan-Do-Check-Act cycle attributed to Demming in Total Quality Management (TQM) applications. As such, the completion of a benchmarking study and issuance of a summary report or survey should not be seen as closure in this process; rather, as in quality management and environmental management system applications, benchmarking can be viewed as an interim step that, once begun, should open new doors that can help identify additional areas for improvement within the organization. This is the continual improvement aspect indigenous to the ISO 14001 "revolving door" process that looks at an environmental management system from planning to completion to management review and back to opportunities for continual improvement.

In summary, we define a benchmarking study process to encompass four main elements:

- Step One—Baseline
- Step Two—Scoping Out the Study
- Step Three—Study Development
- Step Four—Study Report

CHOOSING A PARTNER: WHAT TO LOOK FOR

As has been often said, we must first learn to crawl before we learn to walk before we can run. The concept of benchmarking follows the same basic philosophy. As we described, an organization interested in planning a benchmarking study needs to have certain key elements within its grasp before a benchmark study can be planned and executed. With these elements, one can begin the process. Let us now review the five tips for success in locating a potential benchmarking partner: (1) start the selection process with a clean slate; (2) establish well-defined criteria upfront for benchmarking partners; (3) define what best practice means at your company; (4) use secondary research to identify potential benchmark partners; and (5) weed out the best from the rest. We now apply these steps in developing a benchmarking study, looking in greater detail at two distinct study areas: (1) a company's environmental management system, or EMS; and (2) a company's certification to ISO 14001.

A. PREVIEW OF A BENCHMARK STUDY LOOKING AT A COMPANY'S EMS

In initiating the study, you need to know which companies you intend to use as partners, and in considering a company's EMS as the focal point, you may need to thoroughly evaluate a representative number of companies that have achieved certification to ISO 14001, since certification validates an organization's EMS to be in conformance with ISO 14001. As such, those companies that have certified can provide reliable reference points to an organization looking to benchmark its EMS against other organizations. This criteria is also helpful in selecting the best in class. Also as a consideration, companies that have not certified should not be summarily dismissed as potential partners, since their EMSs could be just as appropriate.

That topic brings up the next tip for locating a benchmarking partner. You should establish a clear set of criteria from which to select your partners. For example, in screening your potential partner pool, is certification to ISO 14001 a prerequisite for consideration, or would you relax that criteria to evaluate a company that is not certified, yet developed a noteworthy EMS that other companies may be evaluating for its own merit or as a "best-practice" model? In close concert to the last tip, defining what best practice means to your company is equally as important, and actually, it can be included in the criteria process to select potential partners. You need to establish how high the bar is set, and whether you have set it too low or too high for current standard industry practices in EH&S and EMS considerations. An improper setting of criteria can result in a potential partner selection that may not reflect current conditions, or may preclude legitimate partners from being considered in the selection pool of companies being evaluated.

One tip we suggest for providing information at a reasonable expenditure of cost and effort is to use secondary research to identify additional companies that could be considered in the partner selection process. Research can consist of: literature searches; networking with other environmental professionals at industry and trade group meetings; conferences and symposiums; or other sources of information that can provide additional names of companies and key contacts at these companies to discuss their EMSs or environmental auditing programs. Such research can also allow you to cross-reference information you may have previously compiled from potential partners, providing a second check on the data to verify whether it can be useful to the study. Having more than one source of information about a particular company can provide valuable cross-checks to aid in partner selection.

Having more than one source of information can also provide opportunities to address the final tip discussed, that of being able to weed out the best from the rest, since weeding out the best can be a time-consuming, but worthwhile exercise. All of the previous tips are grouped into a process that allows each potential partner's EMS to be evaluated against set criteria and against each other to make a preliminary selection. You can further reduce the selection based on additional evaluation of each criteria point, coupled with anecdotal information from your network contacts. You should now have a representative sample of organizations with noteworthy EMSs that fit the criteria of your benchmarking study, to allow you to begin your evaluation process.

B. PREVIEW OF A BENCHMARKING STUDY LOOKING AT A COMPANY'S ISO 14001 CERTIFICATION PROCESS

As we discussed in the example to initiate a benchmarking study of a company's EMS, a similar study could also be initiated to focus on a company's certification to ISO 14001 and take into account the following considerations:

- Company "A" plans to expand its market overseas and is considering certifying to ISO 14001 as part of its marketing strategy. As this organization is well-aware, certification to ISO 14001 is not cheap. There are costs associated with pre-audits, the time required to raise internal awareness levels, and using a registrar.
- Due to the time and costs involved, company "A" may want to take a long, careful look at its options, and weigh benefits against long- and short-term costs associated with certification.

As such, an organization that is seriously considering having one or a number of its facilities certify to ISO 14001 may want to conduct a benchmarking study and target as partners a representative sampling of the two hundred or so U.S. companies that have certified, or the roughly five thousand companies worldwide that have certified, to ISO 14001.[20]

A sponsoring organization could ask a number of questions to any one of the companies that have certified to ISO 14001 in the U.S. or worldwide to obtain some background information regarding their experiences, such as the questions listed on the list below, or to select them as potential partners in a benchmarking study. In this manner, a company not quite sure whether to strive for certification to ISO 14001 can compile relevant information through a benchmarking study to allow its top management to conclude whether certification may be right for their organization at this time. A benchmarking study can help put in proper perspective a company's certification benefits versus the costs associated with becoming ISO certified, and related services, such as: EMS training, auditing, gap analysis, pre-assessment work, and the ISO certification process. Such costs can vary based on a number of variables. If you are interested in this subject area, and desire additional information on certification costs, etc., you should contact one of the registrars providing this service that have the proper accreditation and practical experience in the environmental management arena in performing registration audits to the ISO 14001 standard.

As part of this exercise, some questions that an organization could ask potential benchmarking partners may include:

1. What was your company's driving force to propel you toward ISO 14001 certification? Were you previously certified to ISO 9001? Did that influence your decision?

2. How was your company's EMS, and in similar fashion, your health and safety program prior to attaining certification? Was preparing for your registrar difficult, or were you adequately prepared for the registrar?

3. How did you choose your registrar? Did you use the same registrar that certified you to ISO 9000? Did your registrar provide you any added value in their registration process?

4. How long did the audit and registration process take with your registrar?

5. How much did certification cost the company? How much time was involved in all the steps leading up to your certification? Was a gap analysis performed on your EMS? Did your existing EMS baseline exhibit any shortcomings that were corrected by the time you were recommended to be certified?

6. Have you noticed any positive changes in employee attitudes, work ethics, production quotas? Have you noticed any other changes that can be linked to continual improvement?

7. Has your management taken steps to make corrections as necessary as a result of any follow-on EMS audits?

8. What is your EMS or environmental audit policy or procedure? Do you have such a procedure? Is management involved in this procedure? Is there a management review component? If so, how are the reviews conducted?

9. Do you plan to certify any other of your facilities to ISO 14001? If yes, would this be in the United States or overseas?

10. What is your overall impression of the ISO 14000 standards? Could you suggest any areas for improvement, say in structure, form, or other consideration that can improve upon these standards? (*Note that in the ISO standards development process, standards are reviewed for revision every five years—since the standards were finalized in 1996, this event will not occur until 2001.*)

11. Did you provide specialized training to your EH&S staff, EH&S auditors or EMS auditors, such as EMS Lead Auditor training, or other ISO 14000-type training? If such training was provided to your staff, was this useful to prepare your facility for certification?

These hypothetical questions are some examples of questions that a sponsoring company could ask of a partner company to obtain relevant information regarding the partner company's ISO certification.

In summary, this section explained some of the considerations an organization may explore in choosing potential partners for a benchmarking study. Choosing a partner is a crucial step in the process and can be very instrumental in making your study fit the goals and objectives of the management team commissioning the study. The remainder of this chapter focuses on some benchmarking studies we consider noteworthy and how these can provide valuable lessons learned. We also take a sneak peek into emerging trends that can aid benchmarking.

BENCHMARKING EXAMPLES: NOTEWORTHY CASE STUDIES

A. The APQC's International Benchmarking Clearinghouse Report[21]

There are a number of benchmarking studies dealing with environmental issues that have been commissioned in the recent past by various organizations. Among one of the notable cases in our research database is data from a benchmarking study commissioned by the American Productivity & Quality Center's (APQC) International Benchmarking Clearinghouse, the *Environment Health & Safety— Final Report*, provided courtesy of Vicki Powers of the APQC. That report was the summation of an environmental, health and safety (EHS) benchmarking study sponsored by four organizations—AGT Ltd., Amoco, Atlantic Richfield Company (ARCO), and Conoco. The purpose of this study was to identify "best-in-class" or "best-practice" methods to establish, manage, and monitor EHS policies and procedures of six companies that had been selected to participate as best-practice companies as a means to provide additional insight to the four sponsor companies.

The six best practice companies were: Baxter International, Duke Power Company, IBM, Lockheed Martin Missiles and Space, Northumbrian Water, and Whirlpool Corporation. The report provides additional background on each company. As an added item of interest, IBM is among the companies highlighted in Chapters 10 and 15.

As a prerequisite for the study, the APQC and the sponsor companies developed a list of those processes most critical to developing and implementing effective EH&S programs. These included: (1) policy deployment; (2) communication; (3) organizational issues; (4) training; (5) performance measures; (6) cost management; (7) (*EH&S*) audit programs; (8) information systems and reporting; (9) waste minimization/pollution prevention programs; (10) product stewardship; and (11) remediation.

To begin the study, the participants convened a kickoff meeting in early 1994. The meeting allowed participants to gather in a casual setting that facilitated discussions regarding the benchmarking process, EH&S focus areas, and each of the sponsors' objectives. Interestingly, sponsors identified potential partners based on their own *research and experience*, which is similar to the approach we previously described. The sponsors and APQC worked together utilizing secondary research to identify other potential partners. These additional companies were screened and selected to participate based upon their commitment to their own EH&S programs and their success in achieving their EH&S goals.

As described in the report, APQC, with help from the sponsors, developed a detailed questionnaire for collecting relevant information regarding the EH&S programs of each of the participating companies. APQC sent the questionnaire to all the sponsors and partner companies, collected responses, analyzed data, and wrote the benchmarking study report. To provide added value, APQC held a "sharing day" to allow participants to discuss and share their observations on project findings and hear presentations by the best-practice companies. From our perspective, this elevates brainstorming to a higher degree and provides an open forum for sharing of ideas from which all participants can benefit.

The report identified eleven distinct findings that we group into: (1) developing plans; (2) policies and procedures; (3) the management of resources; the monitoring and documenting of performance; and (4) remediation.

Encapsulating the highlights of the report, the first finding dealt with policy deployment, and identifying what is required in a strong environmental program—a challenging environmental mission statement. Many of the survey respondents indicated that they all placed a strong emphasis on having an EH&S mission statement. IBM, in particular, updates its strategic plan and projects regularly by using a tool the company calls an Environmental Master Plan, or EMP. The EMP is key to IBM's environmental stewardship, and takes into account a number of factors, including: regulatory activity, costs, and relevant data points tracking emissions into the various media. In another example, Baxter International developed its environmental policy program by combining TQM with best environmental management practices to develop noteworthy environmental management standards. Baxter's standards are similar to the European Union's EMAS and ISO 14001. The survey also noted that companies with the most effective EH&S plans also participate in legislative and regulatory developments.

The next finding dealt with communication, and the report stressed the importance of communication to employees and the public—the "stakeholders"—in those environmental programs deemed to be cutting-edge. The survey found that most participants communicate their EH&S policies and goals to employees on a regular basis, and each participant uses various methods to effectively communicate EH&S practices, policies, and goals down through the employee ranks. Again, Baxter and IBM were highlighted: Baxter International, for maintaining a comprehensive environmental publications program that generates a number of manuals and reports for internal and external use; and IBM, which is one of ten companies—the other of this group is Amoco—listed as a signatory of the Public Environmental Reporting Initiative (PERI) Guidelines. (PERI was discussed in Chapter 1.) Notably, the report identified that . . . "cutting edge organizations not only communicate their policies to employees but also encourage employees to communicate hazards and suggestions."

Rounding out the first tier of findings was organizational issues. The report noted that companies with successful environmental programs have: (1) ongoing direction from senior management; (2) dedicated and well-trained EH&S personnel at the corporate level insuring compliance and encouraging the meeting of corporate standards[22], and (3) substantial contact between EH&S and engineering, production maintenance, and corporate legal counsel.

The second tier of findings highlighted training, performance measures, and cost management. The report noted that training is critical to a successful EH&S program.[23] In order to be successful, the companies execute different organizational methods to insure that both managers and shop personnel are trained. Of note, the most effective programs track and document all EH&S training and keep a file for each employee, including training logs and compliance notices. Realizing that some regulatory issues may occur as a result of contractors, many companies also promote training for them through joint training efforts that define the criteria to be met.

The finding dealing with performance measures noted that successful programs use a variety of EH&S measures within programs that receive senior management support. The survey companies use a number of specific measurements in their EH&S programs, including the tracking of accidents and incidents that occurred. Of the companies, four reported their performance measures to the public, which includes: press releases, community events, and environmental reports to publicize their environmental stewardship. We discuss environmental stewardship in greater deal in Chapter 10.

The last finding in this tier highlighted the cost management of EH&S programs. Those programs with the most advanced cost systems developed a strict process for tracking EH&S costs in some form of accounting system. The paradigm shift in EH&S costs, as noted in the study, focuses on cutting-edge cost management that involves charging the cost centers that incur environmental expenses and allowing other departments to utilize EH&S cost information. Several survey respondents also place particular emphasis on having accounting and environmental departments work together to determine reserves for environmental liabilities.[24] We discussed this topic in greater detail in Chapter 4.

The next tier of findings focused on audits, MIS, and reporting and P2 opportunities. As may be expected, the report noted that audit programs are deemed successful when the audit is used as a management tool in helping to identify compliance and root cause issues. Another consideration that is not uncommon at some facilities being audited, concerns audit programs that become less effective when facility personnel feel they are being unjustly criticized by corporate auditors unwilling to take part-ownership in the issues. In these cases, the auditees feel there is a disconnect between working together with the auditors to resolve the issues. In addition, as we have noted from our own audit experience—reinforced by the findings of the APQC report—audit programs should also include a structured process to effectively track corrective actions as well as identify root cause issues that can recur if not effectively addressed. Most of the companies surveyed have implemented programs to ensure timely follow-up of audit report recommendations.

Similarly, another of the tools management can use within the purview of its EMS, with a link to auditing, is a comprehensive program to manage data gathering and reporting systems, commonly referred to as management information systems, or MISs. As the study found, organizations with the most advanced EMSs have MISs in place for various media programs, chemical accountability, and integrated processes. Even those companies that lack comprehensive MISs have systems that manage basic EH&S information. Some of the more progressive programs also track permits, MSDSs, OSHA-related events, and TRI data. Two of the participants also incorporate more sophisticated real-time software systems. Of note, some advanced systems such as electronic document management and imaging are not widely used, and it appears that systems are not integrated. The subject of software in EH&S programs and document management is covered in greater detail in Chapter 13 where several different types of EMIS software is discussed. That chapter focuses in greater detail on various software packages, including real-time applications, for use in EH&S programs.

The remaining findings, pollution prevention (P2) and product steward-ship, round out the APQC's report. Both sets of findings need to have the support and leadership of senior management to continue. With respect to P2, the report found that senior leadership must convey its commitment to achieving waste re-duction at each facility, and must be implemented via effective processes. One key for success is that programs must know how much waste needs to be re-duced at each facility. These programs must provide evaluations of the impacts of releases, have plans with goals for reduction and be successful in implement-ing the plans where necessary in the process streams. In addition, the study found that most companies also seek to reduce waste by incorporating the effec-tive use of raw materials, energy consumption, recycling of waste streams, the minimizing product waste streams, such as off-spec materials or expired shelf life chemicals, and minimizing the use of banned chemicals, such as ozone-depleting chemicals in the chlorinated fluorocarbon (CFC) family, freon, and other similar chemicals. Most of the companies surveyed integrate these programs into their operating policies and procedures.

Product stewardship can be aptly summarized from a quote by Frank Popoff, the CEO of Dow Chemical. He said, "There is a need to assume responsi-bility for one's product even after it has left the plant."[25] In essence, companies most admired as environmental stewards, as the study found, are those with the most successful environmental programs that ensure, among other things, cus-tomer satisfaction, have diverse product development teams, attempt to gauge the environmental risks that might result from products in development, and in our estimation, are sincere in their efforts to total commitment. Cutting-edge pro-grams use an array of product stewardship teams that communicate regularly with environmental personnel. Such programs should also have procedures to foresee EH&S risks and possible exposures from new products and reformulated products. As noted in the survey, cutting-edge programs also consider the impact of new product designs and modified facility processes on waste and emission streams. Our research came to a similar conclusion with a discussion in Chapter 1 about product take-back that IBM, Toshiba, and other computer companies initi-ated as part of product stewardship, as well as product re-think and re-design strategies to maximize the recycling of various electronic products to help mini-mize or eliminate the generation of solid waste streams entering landfills.

In conclusion, this APQC benchmarking study acknowledged that even the most successful environmental programs face challenges that have proven diffi-cult to overcome. One of the main stumbling blocks noted in the study, which is not surprising, is that EH&S managers face continuous pressures to justify envi-ronmental programs and expenditures even in organizations that have top man-agement backing. The report states this "green ceiling" is unlikely to be shattered so long as environmental programs are not accounted for and not perceived as core business enterprises. In this regard, as our research found, ABC, or activity-based costing, may help to change this perception and help portray environmen-tal programs as part of a company's core business enterprises with profit and loss accountability, just like any other company business unit. While this may not be easy to surmount in the short term, there are organizations that have initiated programs that take into account all environmental expenditures and savings realized. Additional research may need to be conducted by organizations to view such data, and perhaps, organizations may decide to commission a benchmarking study to investigate further. The Millipore example in the next chapter provides some insight into this area. Also, the additional interest the environmental management standards of ISO bring to the table may make the "glass ceiling" a thing of the past as we head into the twenty-first century.

B. THE IRRC CORPORATE ENVIRONMENTAL PROFILES DIRECTORY 1996 EXECUTIVE SUMMARY[26]

Similar to the APQC study, the Investor Responsibility Research Center's (IRRC) 1996 edition of the *Corporate Environmental Profiles Directory* provides a benchmarking summary of key environmental information and indicators from various sources of companies that make up the Standard & Poor's 500 index, commonly referred to as the S&P 500. This executive summary report was provided courtesy of Kristin Haldeman, Director of IRRC's Environmental Information Service of the IRRC. As noted in the introduction to the report, IRRC plans to update this information annually and may expand the study to additional companies should conditions warrant.

Among the highlights of this directory/report, IRRC compiled the information to address a range of subscriber needs, which included: portfolio risk analysis, social screening, and the benchmarking of environmental performance at companies in the S&P 500. IRRC developed the Environmental Information Service in response to increasing requests from various corporate stakeholders, such as institutional investors, for reliable, detailed information about corporate environmental performance. This is not a new occurrence since financial institutions, such as the Swiss Bank Corporation, have been developing environmental rating questionnaires to gain additional information regarding environmental expenditures and issues within regulated organizations. Furthermore, as described in the directory, financial community interest in environmental risk management and environmental responsibility issues will continue to grow as more institutions recognize that environmental performance has a direct impact on financial performance. As companies continue to improve upon their EH&S programs in response to such issues, more and more companies will look to the best-in-class companies to compare their accomplishments with the other companies. The previous APQC benchmarking study example and the subsequent GAO audit study tend to emphasize this point. In addition, a number of companies utilize quantitative measures like those noted in the directory to benchmark their environmental performance.[27] A summary of some of the quantitative and qualitative measures used by the S&P 500 companies and noted in the IRRC report include: (1) average size of penalties assessed; (2) industries with the highest average RCRA Corrective Action Index values; (3) Form 10-K environmental disclosures; (4) environmental codes of conduct to which companies subscribe; (5) sources of environmental auditors used; and (6) number of companies participating in EPA voluntary programs.

Among the highlights of environmental practices the IRRC 1996 survey compiled of the S&P 500 companies, the following are noteworthy since they mirror mainstream conditions prevalent at a number of companies. These include:

1. Companies have responded to the environmental challenges posed by regulations and rising costs with a series of initiatives designed to improve regulatory compliance and reduce financial exposure. Survey respondents show that more than 80 percent of the companies have board-level committees with environmental responsibilities, up from 60 percent in 1995. At the other end of the scale, very few survey respondents make the results of environmental audits available to shareholders. The paradigm shift that has prompted companies to change their business focus to look at environmental expenditures as an integral part of business while recognizing that opportunities for improvement should be followed to improve compliance, reduce waste, and save money in the process. ABC studies are beginning to gain momentum within regulated entities, as are EPA- and NIST-sponsored software programs currently in beta versions that track P2 considerations of various criteria, such as technical performance, economic feasibility, and EH&S benefits.

2. Voluntary environmental reports are on the rise.[28] More than 50 percent of the 195 survey respondents indicated they had published an environmental report, and about 50 percent of that group expects to publish such reports annually. Several companies' environmental reports are highlighted in Chapter 10. In addition, a number of companies provide copies of their reports, environmental policies and ISO 14001 certification on the Internet. A comprehensive list of these companies is included in Appendix G.

3. As was also noted in the APQC study, corporations are constantly assessing ways to transform environmental costs into business opportunities, such as looking to find ways to recycle waste streams as raw materials for other uses, factoring in a product's life cycle, or other considerations to distinguish their companies in the marketplace. This is product stewardship in action—looking at all environmental issues during product R&D, design and final assembly, culminating with product take-back and recycling of base components.

In addition, the report focused on eight key findings based on the survey results of respondents from the S&P 500. A representative sample of the key findings included: (1) toxic chemical emissions, or TRI data; (2) compliance data; (3) Form 10-K environmental disclosure, and other SEC requirements, such as SAB 92; (4) environmental policies, management practice, and program; and (5) sustainable development indicators.

Reviewing these key findings in greater detail, the report noted:

1. The significance of TRI emission levels as indicators of whether companies with higher reportable levels may face greater environmental risk than companies with lower levels, due to the potential for negative publicity, tort actions using TRI data, and increasing costs for pollution control and waste management. Based on IRRC Emissions Efficiency Index[®29], companies with lower index values than their competitors could have an advantage in terms of environmental performance. As noted in the study, TRI emissions are concentrated in a handful of industries, notably miscellaneous metals and chemicals.

2. Environmental compliance data of the companies surveyed provides information on the success of each company's compliance program and identifies potential risk factors associated with companies in selected industries. For each media compliance program, such as RCRA, TSCA, CAA, and CWA, for example, IRRC presents graphical data from an IRRC Compliance Index® that also track trend indicators. These indicators, along with the graphs, should provide data that can be used in calculating trends in a company's environmental performance. Such information could be an analytical tool in benchmarking.

3. The study noted that the Securities and Exchange Commission (SEC) requires publicly held companies to make environmental disclosures to investors to appraise them of the value of their securities. Companies can report such activities through the SEC's Form 10-K and Staff Accounting Bulletin 92. The study's findings, however, indicate a decline in disclosures from the reporting period of this survey. It will be interesting what changes, if any, the IRRC will report in its next reporting period, since one of the key highlights of the IRRC study indicated that environmental reports are on the rise. Similarly, our own research notes an increase in the number of voluntary environmental reports being issued by publicly held companies.

4. The study compiled information from responses to IRRC's annual Corporate Environmental Practices survey. Based on this survey, which ran through 1996, 287 companies, in 82 out of the 90 S&P industry groups, responded to the survey—an increase of 10 percent over 1995. Broken down further, 94 percent of the manufacturing companies responding to the survey have an environmental policy in place, and 39 percent of that group also subscribe to one or more environmental codes of conduct, such as: CMA's Responsible Care®, the ICC Business Charter for Sustainable Development, API's Strategies for Today's Environmental Partnership (STEP), and GEMI, among others.

5. In addition, 96 percent of the respondents also indicated they maintain environment auditing programs, and depending upon the industry sector, audit their facilities with varying amounts of frequency, and may utilize corporate staff, company personnel, outside consultants, or a combination of all three.[30]

6. Taking a more holistic approach, the study also asked respondents whether they had participated in any voluntary programs, such as EPA's, and what they were achieving in sustainable development. Of the voluntary programs, the one EPA program with the highest industry participation was the 33/50 Program that ended in 1995, and from our account, the programs attracting the greatest interest are currently the ELP Pilot program and Project XL. As this manuscript was going to press, both programs were in progress. With respect to survey responses to sustainable development, this was only the second year IRRC included the question, and of those companies that responded, many stated that such information is not yet tracked. As our research found,[31] a growing number of companies are beginning to track costs associated with their hazardous waste generation, energy consumption, raw material and water consumption, the amount of material recycled, P2 opportunities, and how much can be saved by companies in each of these activities. A final note: concerned with this gap, IRRC intends to work with companies and other interested parties to further refine the sustainable development indicators.

In summary, a growing number of corporations and investors are hedging their bets that environmental performance is predictive of future financial performance in day-to-day business decisions. As noted in the IRRC report, IRRC is dedicated to providing and enhancing environmental, financial, and sustainable development information in ways to help investors, corporations, and analysts incorporate environmental performance into their daily business and investment decision making.

C. U.S. GENERAL ACCOUNTING OFFICE (GAO) REPORT ON ENVIRONMENTAL AUDITING[32]

The last of the reports that focus on benchmarking studies is the GAO report on environmental auditing. That report was submitted to the U.S. Senate's Committee on Governmental Affairs. It was completed in April 1995 and contains many relevant findings that complement and mirror many of the findings of the reports we previously highlighted. This report was initiated to provide justification to federal agencies for improving their environmental performance and reduce costs by conducting audits as part of a comprehensive environmental audit program. The "partners" selected for this GAO benchmarking study are a group of twenty-two companies that consisted of: eighteen industrial companies, two commercial TSDFs and two consulting firms; twelve affiliated organizations; and three trade associations: the CMA, GEMI and the ICC.

Among the salient points in the GAO report referring to auditing are the benefits and distinguishing characteristics of effective environmental audit programs. Specifically, the report noted that: (1) environmental auditing can increase environmental awareness and capability among employees; (2) environmental auditing can result in relaxed regulatory scrutiny; (3) environmental auditing establishes a record of a company's environmental performance; and (4) environmental auditing facilitates planning and budgeting for environmental projects. GAO compiled this information from interviews with employees of the twenty-two companies and organizations consulted for the study.

In addition, as noted in the GAO report, top management's commitment is one of the prerequisites for an environmental auditing program in order to have an organization maintain regulatory compliance. For example, the manager of Allied Signal's audit program echoed the opinion of many others interviewed by stating that the support of top management is key to the success of any program since the rest of the organization follows management's lead and supports the things it perceives management cares about.[33] DuPont EH&S audit managers stated that top management's commitment to environmental goals is assured since its CEO also serves as its chief environmental officer.[34] In a similar vein, the director of compliance audits for Union Carbide's health, safety and environment (HS&E) group stated that top management has always demonstrated support through its funding of environmental activities, including environmental auditing programs and corrective actions geared to maintaining compliance.[35]

The report's findings concluded that auditing as practiced within regulated organizations can serve as a role model for environmental audits to be conducted at federal facilities. Since that report was issued, various federal agencies have taken a proactive approach toward environmental audit and environmental management system programs, especially DOD and DOE. As this manuscript was being finalized, at least one DOE facility had gone as far as certifying to ISO 14001, a noteworthy accomplishment that goes beyond the expectations of the GAO report.

RECENT TRENDS: A SNEAK PEAK
AT ENVIRONMENTAL CONFERENCES

The earlier portions of this chapter focused primarily on the how-to's in developing, gathering data for, and conducting benchmarking studies for various EH&S activities. The last part of the chapter takes a more detailed look at several reports that conducted benchmarking studies to obtain additional information for their specific target audiences. Interestingly, each of the studies complement each other with sufficient overlap to confirm key conclusions within each study, especially the need for top management commitment as one of the drivers for many EH&S program activities. The bottom line is that benchmarking can provide a wealth of information to an information-starved organization, or to an organization looking for a novel approach to solve a particular problem. Its uses are unlimited, provided sufficient planning is undertaken from the get-go.

One of the best barometers for gauging cutting-edge thinking in the environmental field, which is no different from other professional fields, is to look closely at what some of the leading professionals in the field are investigating and doing. For example, using conferences or seminars as one guide, you may decide to attend one over the other, depending upon who the speakers are, and what the subject matter will be. How relevant the topics may be to you can also factor into your decision to attend. From our experience, cutting-edge topics presented at professional and industry trade association conferences are usually a good investment in time and money for an EH&S professional.

As another example, the author had an opportunity to participate in one such trade association symposium, CARE Innovation '96, in Frankfort, Germany, in which there were several noteworthy presentations on product take-back and rethink/redesign procedures and other innovative considerations from some of the best names in the electronics industry, among them: IBM, Hewlett-Packard, NEC and the host, Sony. Similarly, The National Association for Environmental Management's (NAEM's) Environmental Management Forum, provided noteworthy presentations in 1997 from various industry EH&S professionals. That forum included a number of technical sessions. Another offering is through the Conference Board, which hosts an annual environmental excellence conference. The Conference Board convened its most recent conference in early 1998 and included presentations from various industry EH&S professionals and consultants.

These particular conferences are just a small sampling of the many conferences and seminars dealing with EH&S issues and cutting-edge solutions that are offered each year in different parts of the United States and overseas. Each conference can provide you with a sneak peak into what a particular organization has developed with respect to the latest innovations in any number of environmental "hot" buttons, such as: environmental management techniques to enhance auditing, such as software applications; managing EH&S programs as part of a strategic business unit; exploring innovative processes in effectively managing EH&S issues, or a host of other topics that can be instrumental in providing its audience insight into any of these applications. Each conference is developed with its own specific theme, although many conferences cover different variations of mainstream EH&S topics, and each conference may have its own group of keynote speakers. For this reason, a perspective conference attendee should take the time to evaluate each conference brochure to see if a particular conference would be of value to attend. You should weigh all options carefully in making your particular selection. After all, you may need to justify these costs to your supervisor before you are given permission to attend, so you might as well be prepared to provide justification for your decision up front.

WHAT THE NUMBERS TELL US: EMERGING TRENDS

In researching the materials for this chapter, the author uncovered at least ten surveys and reports listed in the references section, that either directly relate to benchmarking studies or can be utilized as supplemental materials in preparing for a benchmarking study. Many of these reports focus on companies that make up the Fortune 100, 500 and the S&P 500, and they were conducted by various private and nonprofit groups since about 1992. Our database also includes a distillation of at least twenty-five seminars and conferences that occurred over the past two years.

Of particular interest to us is the fact that a recurring theme is at the heart of each of these conferences and seminars: looking at a novel approach to solve a particular environmental problem or issue. Such solutions can focus on a regulatory standpoint in streamlining some requirement, or from a practical, hands-on or technical approach in tackling a process, employee-related, or other business-type issue in which the actual experiences of an individual can provide valuable and relevant insight into others who may be struggling with this or similar issues. The so-called "lessons learned" approach is a very valuable commodity in the conference and seminars circuit.

By looking more closely at a representative sample of these conferences and piecing together some key "firsts" into a mosaic, a bigger picture emerges. Diligent research in this area can provide an astute EH&S manager or savvy senior company official a "heads up" of emerging trends in the environmental field that could give that manager a leg up on the competition, or help that manager be better prepared for what is approaching and prepare for it. Such information could be useful in helping identify areas in which your company could be weak, where innovative techniques or technologies could be useful to your company's business lines, or in evaluating specific areas where you may want to initiate a benchmarking study to improve any of these areas. You can view our suggestions as the top of the information "iceberg" that could provide added value. Undoubtedly, you could perceive additional areas where this information could be useful to your particular organization.

Looking more closely at a representative sample of these conferences from our database, here is what a few of them presented in 1996 to 1997:

- CARE Innovation '96—(1) Market-driven developments, such as reducing electronics waste; (2) technology-driven developments, such as DfE—current practices and tools; (3) evaluating and communicating progress, such as ecological benchmarking; and (4) training for the future, such as the next generation of environmental management.
- NAEM Environmental Management Forum—(1) management skills, such as structuring EHS for sustainability; (2) new horizons, such as strategies for business integration; (3) management dialogue, such as EH&S hiring and compensation trends; and (4) regulatory update, such as international regulatory trends.
- GEMI '97 Business Helping Business—(1) Creating an effective EH&S training program; (2) preparing a credible environmental report; (3) developing an EH&S information management system; (4) assessing the performance of your environmental management program against ISO 14001; and (5) putting EH&S into business terms.
- The Conference Board®—The 1997 Conference on Corporate Environmental, Health and Safety Excellence: *Taking Bold Steps to Create Business Value*[36]—(1) Corporate environmental, health and safety excellence; (2) environmental-driven business strategies within a global corporation; (3) can corporate EHS responsibility pay off the bottom line?; and (4) implementing EHS shared services.

Each of these conferences and any other follow-on conferences their organizers may provide in subsequent years, or other organizers may produce, can provide prospective attendees a certain measure of information that can be useful in their day-to-day operations. How useful such information can be to attendees may depend on their specific needs, where their specific organization may be in the hierarchy of EH&S programs, whether at the beginning of the food chain, or at the "best-in-class" category in reference to benchmarking, or whether work responsibilities and fiscal constraints take precedent, in which case, attending a particular conference may have to be postponed.

In closing, benchmarking studies can provide a wealth of information to an organization seeking answers. Also, you need to clearly focus upon how a study can be useful to your particular situation. That's where you have to do your homework as outlined in the examples in this chapter. In addition, an item you should not lose sight of in considering a benchmarking studies for your own organization is that background information for a study is all around us—we just need to know where to go to mine it. Some of the information we provided in this chapter is a start. Good luck.

ENDNOTES

[1]D. Dreenan, *Transforming Company Culture,* London, UK: McGraw-Hill, 1992, p. 66.

[2]I. Ritchie and W. Hayes, *A Guide to the Implementation of the ISO 14000 Series on Environmental Management,* Upper Saddle River, NJ: Prentice Hall, Inc., 1998.

[3]S. Greengard, "Discovering Best Practices Through Benchmarking," *Personnel Journal,* November 1995, pp. 62–73.

[4]Ibid, p. 62.

[5]The DuPont Company's *STOP® for the Environment* series of training videos.

[6]"The Voluntary Environmental Audit Survey of U.S. Business," Price Waterhouse, March 1995.

[7]D. Lober, D. Bynum et al., "The 100+ Corporate Environmental Report Study," Duke University, January 1996.

[8]*ENVIRONMENTAL AUDITING: A Useful Tool That Can Improve Environmental Performance and Reduce Costs,* GAO-RCED-95-37, April 1995; and *REGULATORY BURDEN: Measurement Challenges and Concerns Raised by Selected Companies,* GAO-GGD-97-2, November 1996.

[9]UNEP Technical Report Series No. 6, *Companies' Organization and Public Communication on Environmental Issues,* Paris: UNEP, 1991.

[10]IRRC, *Corporate Environmental Profiles Directory 1996,* Executive Summary, Washington, D.C.

[11]International Benchmarking Clearinghouse Environmental Health & Safety Study, Final Report, American Productivity & Quality Center, Houston, 1996.

[12]U.S. Chamber of Commerce, Voter Consumer Research, *Federal Regulation and Its Effect on Business,* June 25th 1996.

[13]Up-to-date information on EPA's audit disclosure policy can be obtained by logging on to EPA's web page at www.epa.gov and scrolling down to EPA's Office of Enforcement and Compliance Assurance.

[14]These tips were the basis of V. J. Powers's article, "Selecting a Benchmarking Partner: Five Tips for Success," *Quality Digest,* October 1997, pp. 37–41.

[15]Based on the author's interview with Jason Hart, Editor of the *International Environmental Systems Update,* published by CEEM, Inc., Fairfax, VA. The information is also listed in a Special Supplement of the September 1998 issue of that publication.

[16]For additional information, see C. Bowen, "Chemical Industry Members Voice Concern over Information Listed on EPA's Internet Site," *Chemical Processing,* November 1997, p. 15.

[17]Among noteworthy industry organizations, the Chemical Manufacturers Association regularly features in its monthly magazine *CMA News* at least one case story about a member company providing its local community with some form of a community outreach program.

[18]Two cases that are referred by some pundits include the Exxon *Valdez* accident in Alaska and the Ashland Oil storage tank rupture near Pittsburgh, Pennsylvania. One reference book that describes these cases in greater detail is B. Seldner and R. Cothrel's book, *Environmental Decision Making for Engineers and Business Managers,* New York: McGraw-Hill, 1994.

[19]F. Bleakley, "Strange Bedfellows: Some Companies Let Suppliers Work on Site and Even Place Orders," *The Wall Street Journal,* January 13, 1995, p. A1.

[20]*International Environmental Systems Update,* September 1998, Special Supplement. (No. 15.)

[21]Benchmarking Consortium Study, *Environmental Health & Safety Final Report,* APQC, Houston, 1996.

[22]See also G. Crognale, "Allocating Corporate Resources for Environmental Compliance," *The Greening of American Business,* Rockville, MD: Government Institutes, 1992.

[23]See also G. Crognale, "Training; Preparations for Maintaining Effective Environmental Management Systems," *ISO 14001 and Beyond,* Sheffield, UK: Greenleaf Publishing, 1997.

[24]See also G. Crognale, "Stay a Step Ahead of Environmental Disclosure Laws," *Safety Compliance Newsletter,* October 25, 1997, p. 2.

[25]Benchmarking Consortium Study, p. 43 (n. 21).

[26]*IRRC Corporate Environmental Profiles Directory 1996,* Executive Summary, Environmental Information Service, IRRC, Washington, D.C., 1996.

[27]See *Petroleum Industry,* one of several industry sector reports provided by Environmental Information Services, Inc., and *Environment, Health and Safety: A Platform for Progress,* an excerpt of presentations given at the 1996 Conference on Corporate Environment, Health and Safety Excellence, sponsored by the Conference Board.

[28]Two references regarding environmental reports are, D. Lober "The 100+ Corporate Environmental Report Study," Duke University, 1996, and S. Archer (Monsanto) "Reporting on the Environment," a two-part series in *CMA News,* February and March 1996.

[29]The IRRC Emissions Efficiency Index® compares the level of TRI emissions per $1000 of domestic revenue generated by the company. It provides a useful tool for comparing relative emissions between industry groups or for analyzing emissions intensities of companies in similar industries.

[30]See also G. Crognale, "Auditing Answers," *Environmental Protection,* July 1992, pp. 45–47.

[31]Two noteworthy references are the EPA documents, *Pollution Prevention Success Stories,* EPA/742/96/002, April 1996; and *Partnerships in Preventing Pollution,* EPA 100-B-96-001, Spring 1996.

[32]A General Accounting Office report, *Environmental Auditing: A Useful Tool That Can Improve Environmental Performance and Reduce Costs,* GAO/RCED-95-37, April 1995.

[33]Ibid, p. 25.

[34]Ibid, p. 25.

[35]Ibid, p. 27.

[36]The Conference Board® publishes its proceedings each year in an edited format. The most recent conference available to the author at the time this manuscript was going to press was the edited version of the 1996 proceedings, *Environmental, Health and Safety: A Platform for Progress* Report Number 1175-97-CH, 1997, The Conference Board, provided courtesy of The Conference Board.® The topics depicted came from a descriptive Conference Board brochure highlighting the 1997 conference proceedings.

REFERENCES

Drennan, D. *Transforming Company Culture,* London: McGraw-Hill, International, 1992.

Ritchie, I. and W. Hayes, *A Guide to the Implementation of the ISO 14000 Series on Environmental Management,* Upper Saddle River, NJ: Prentice-Hall, 1998.

Seldner, B. and R. Cothrel, *Environmental Decision Making for Engineers and Business Managers,* New York, NY: McGraw-Hill, 1994.

Thomas F. P. Sullivan, ed. *The Greening of American Business,* by Government Institutes, 1992.

Christopher Sheldon, ed. *ISO 14001 and Beyond,* Sheffield, UK: Greenleaf Publishing, 1997.

Greengard, S. "Discovering Best Practices Through Benchmarking," *Personnel Journal,* November 1995, pp. 62–73.

"The Voluntary Environmental Audit Survey of US Business," Price Waterhouse, March 1995.

Lober, D., D. Bynum et al. "The 100+ Corporate Environmental Report Study," Duke University Center for Business and the Environment, January 1996.

ENVIRONMENTAL AUDITING: A Useful Tool That Can Improve Environmental Performance and Reduce Costs, GAO/RCED-95-37, April 1995; *REGULATORY BURDEN: Measurement Challenges and Concerns Raised by Selected Companies,* GAO/GGD-97-2, November 1996.

United Nations Environment Programme Technical Report Series No. 6, *Companies' Organization and Public Communication on Environmental Issues,* Industry and Environment Program Activity Center (IC/PAC) Paris: UNEP, 1991.

IRRC Corporate Environmental Profiles Directory 1996, Executive Summary, Investor Responsibility Research Center, Washington, D.C., 1996.

International Benchmarking Clearinghouse Environmental Health & Safety Study, Final Report, American Productivity & Quality Center, Houston, 1996.

U.S. Chamber of Commerce, Voter Consumer Research, Federal Regulation and Its Effect on Business, June 25th, 1996.

Powers, V. J. "Selecting a Benchmarking Partner: Five Tips for Success," *Quality Digest,* October 1997, pp. 37–41.

Bowen, C. "Chemical industry members voice concern over information listed on EPA's Internet site," *Chemical Processing,* November 1997, p. 15.

Archer, S. "Reporting on the Environment," a two-part series in *CMA News,* February and March 1996.

Bleakley, F. "Strange Bedfellows: Some Companies Let Suppliers Work on Site and Even Place Orders," *The Wall Street Journal,* January 13, 1995, p. A1.

Crognale, G. "Stay a Step Ahead of Environmental Disclosure Laws," *Safety Compliance Newsletter,* October 25, 1997, p. 2.

Crognale, G. "Auditing Answers," *Environmental Protection,* July 1992, pp. 45–47.

Petroleum Industry, Environmental Information Services, Inc., New York, 1997.

Environment, Health and Safety: A Platform for Progress, a Conference Report, 1997, The Conference Board, Inc., New York.

EPA document, *Pollution Prevention Success Stories,* EPA/742/96/002, April 1996.

EPA document, *Partnerships in Preventing Pollution,* EPA 100-B-96-001, Spring 1996.

Webster's New Collegiate Dictionary, New York: Simon & Schuster, 1974.

DuPont's STOP® for the Environment series of training videos

EPA's website at www.epa.gov

International Environmental Systems Update, Special Supplement, September 1998, published by CEEM, Inc., Fairfax, VA.

The Conference Board® report, *Environmental, Health and Safety: A Platform for Progress,* Report Number 1175-97-CH, The Conference Board, Inc., New York, 1997.

CHAPTER 8

BUILDING BLOCKS FOR IMPROVING A COMPANY'S ENVIRONMENTAL MANAGEMENT SYSTEM

Edward Spaulding,[1] Chevron Corporation, and Gabriele Crognale, P.E. with contribution from Patricia Davies, Millipore Corporation

> *"In most board rooms, (environmental) standardization is usually considered a 'MEGO' ('my eyes glaze over') subject."*
>
> —*Christopher Sheldon*

INTRODUCTION

In the preceding chapters, we touched upon how the effectiveness of a facility's environmental management system, or EMS, and the complementary environmental and health and safety compliance programs can depend upon several factors. These factors include: knowledgeable environmental managers, EH&S support staff and front-line workers, and a corporate culture, in other words, top management, that really is the driving force behind environmental considerations being able to flourish within an organization. We touched upon this topic in some detail in Chapter 6, where we evaluated several innovative strategies that environmental managers could use for improving existing workplace conditions with environmental overtones.

In this chapter, we take a different look at some of these innovative strategies and recommend additional "building blocks" that can be viewed as essential components for improving a company's EMS or its related EH&S program. We also touch upon additional venues that can help break through the "green wall" syndrome described in Chapter 6 that could frustrate some environmental managers in their quest for obtaining funds for specific EH&S activities, including those activities identified as management system improvements.

As those of us in the environmental field have come to realize, the "command-and-control" way of doing business, the hallmark of regulators handed down to regulated organizations, as one way for sustaining environmental compliance, has slowly given way to a new order of business by companies for sustaining and improving upon environmental management and compliance. This new order of environmental management takes into consideration some concepts borrowed from business activities that environmental professionals slowly began to incorporate into their day-to-day activities. As a point of reference, this probably coincided with EPA's introduction of its first of many voluntary programs, such as the 33/50 Program launched in 1991,[2] or about the time that the first voluntary environmental reports were issued by publicly-held organizations, such as Chevron, Texaco, and 3M (1990). This new order, or sea change, in strategy exhibited by senior environmental managers at a number of regulated organizations may have been brought on by a number of varying factors. Among them may have been a growing resistance from non-EH&S managers, such as corporate financial or marketing executives, to continue to accept with blind faith EH&S expenditures identified as necessary bottom-line costs when compared to other business unit expenditures as venues for increasing the company's bottom line, such as sales and marketing expenditures. This resistance, if you will, was penned the "green wall" by one consulting firm as an explanation for what can occur at companies that are burdened with continued EH&S expenditures without adequate justification to financial executives that such expenditures can improve bottom-line considerations. As we see it, this resistance is not insurmountable, and we should not feel threatened or challenged by the "green wall," which is probably a misnomer. Overcoming the "green wall," however, does require some creative strategies[3] from a manager if he or she is to have successful discussions with financial and marketing executives.

What could be perceived as a matter of routine, and hence, acceptable business practice, environmental managers at some companies may have opted to present their findings to top management in a matter-of-fact technical and regulations-oriented fashion that may have been lost upon these top executives. Well-designed presentations that focused upon environmental compliance costs and expenditures, such as: employee training; costs associated with improving management of waste streams, raw materials, and products; adding staff or outsourcing to third parties to compensate for a downsized workforce; and other considerations, were probably presented to a company's top financial and operational executives in ways that may have caused their eyes to glaze over (the "MEGO" Syndrome that Chris Sheldon refers to in *ISO 14001 and Beyond*)[4] and miss the points of a funding request presentation. Furthermore, these reports and presentations may have also been accompanied by figures, graphs, and charts, emission limits in tons/year/chemical, and volumes of waste generated in drums/year/chemical, and so on, as a back-up to the compliance figures presented. As environmental professionals, we could easily follow such presentations since we understand the language being spoken, but we need to keep in mind that financial and marketing executives do not necessarily understand these terms as easily as we do. We, the environmental professionals, cannot expect them to take a leap of faith regarding the expected benefits without some tangible results in terms they can readily understand.

Consequently, at some of these meetings, environmental managers may not have had much luck convincing their companies' financial executives of the need for additional funding if the executives viewed such requests as unnecessary expenditures. Such roadblocks experienced by environmental managers in their attempts to obtain funding for sustaining a growing EH&S management and compliance regimen could have prompted the creation of the term "green wall." Overcoming this green wall and winning the interest of a company's financial executives may require that environmental professionals repackage their funding requests in terms these executives can readily understand and appreciate.

By taking this approach, environmental managers can provide financial executives further insight into EH&S expenditures and be better equipped to show how such expenditures can generate bottom-line cost savings for each line item over a designated window of time as a reference point. In this fashion, an environmental manager could make a pitch for a successful funding request that depicts EH&S expenditures in a positive light allowing financial executives the opportunity to see more clearly the return on investment (ROI) to the company in funds allocated for specific EH&S activities. The funding requests could also take the form of presentations to the financial managers that include comparisons between monies saved by the company in the long term versus short-term expenditures as justification for the expenditures. For example, a presentation that includes tables showing the dollars a company can save by investing in additional EH&S training for employees handling chemical shipments versus costs required to correct and clean up any chemical mishaps or spills that may have occurred before such training was implemented can make an impact with these business executives. Placing emphasis on the fact that such mishaps usually require some emergency response or clean-up activity that can outweigh the cost of training, and EH&S managers could present a convincing argument to executives depicting a positive ROI to the company resulting from such training.

As another (presentation) example, you could also compare the costs required for developing or sustaining a comprehensive environment, EMS, or EH&S auditing program for auditing your company's manufacturing facilities versus maintaining the status quo. An audit program could be developed in one of many different fashions as a tool to measure: (1) a company's compliance with the applicable EH&S regulations; (2) the underlying management system in environmental and health and safety; (3) the management system's ability to address overriding liability issues; and (4) other similar considerations. At each level, audit programs can be developed and scoped to be as comprehensive as necessary, and with each succeeding level of detail, operational and sustaining costs can increase dramatically. Regardless of the up-front costs required to maintain an effective audit program, audits should be viewed as an insurance policy for an organization. A well-trained and well-seasoned core EH&S audit team can uncover potential issues that, left uncorrected, may not only lead to costly regulatory fines, but to accidents that can have disastrous results. Taking all this information into consideration, the argument can again be made to show that EH&S expenditures related to maintaining an effective auditing program can be vastly overshadowed by expenditures related to correcting regulatory findings in product and waste management that may have led to spills or releases, in unnecessary waste products being generated due to worker ignorance or attitude, or in the payment of fines to federal or state regulatory agencies for continued noncompliance of EH&S rules and regulations,[5] among other considerations. (In the next chapter we provide further discussion about environmental audits, where our focus is on audits as management tools, seen from the perspective of several regulated industry audit practitioners.)

These are just two examples depicting ways in which presentations can be made to a company's financial executives to help dispel the green wall syndrome, and be in a better position to have these executives as an audience willing to negotiate. Using a presentation strategy that emphasizes regulatory items in EH&S terms, such as listing the number of notices of violation that were issued by a regulatory agency without qualifying information that relates it to bottom-line costs, the number of criteria pollutants a company released in prior years or an increase or decrease in TRI emission figures in a company's Form R, all without similar bottom-line comparisons, or other compliance-driven technical data without a tie-in to cost figures, can ultimately lose the interest of these executives. To get some EH&S funds approved, EH&S professionals may need to refocus their presentations for the benefit of their audience, or develop innovative and creative ways to get a top financial or operational executive to stand up and take notice. As Al Iannuzzi recommends,[6] EH&S managers should put in the effort to develop, as he puts it, an environmental quality business plan (EQBP) which lays out for its intended audience a focused plan with clear goals, objectives, and targets, and quite possibly, we might add, a timetable to achieve these goals with an itemized list of projected expenditures. We should note here that it is also terribly important to ensure that offsetting the expenditures should be a tabulated listing of projected savings, otherwise even a well-designed EQBP could find itself on the cutting room floor as they say in the motion picture industry.

With that plan as a segue, let's put this concept in a different perspective and look to Hollywood for some creative "guidance." With that, one venue that EH&S managers could conceptually follow in presenting funding requests to corporate financial executives or in planning an EQBP may be by following actor Tom Cruise's lead from his line in the movie *Jerry Maguire* rather than follow the lead from actor Steven Seagal who plays a hard-hitting EPA agent in the movie *Fire Down Below.*[7] EH&S managers may want to keep their funding requests to corporate executives focused on one of Mr. Cruise's often-repeated lines in *Jerry Maguire,* in other words ". . . Show me the money!", (which in reality Cuba Gooding, Jr first said in the movie) or follow the classic line spoken by Clara Peller in the hamburger commercial . . . "Where's the beef?" Both statements reflect a lesson plan that can be taken right out of Business Management 101—not Environmental Management 101—which could be what more environmental professionals need to keep in focus when making such funding requests to the CFOs, COOs, and other fund management and disbursement executives within a regulated organization. This impasse just may have been the causal effect behind the green wall that was vexing many an environmental manager speaking in "envirospeak" instead of "businessspeak" to business executives, and failing to show "the beef" or "the money" to these executives.

In essence, a key to breaching the green wall and getting participants to agree in principle on the projected expenditures and funding requests would be for the environmental representatives to state plainly to the financial representatives the total net of monies projected to be saved over a period of time for each EH&S expenditure versus no-action alternatives. With that, savings could be realized from a number of areas: the examples previously described, or in areas dealing with a company's use of new process equipment; or even from the EH&S

group's start-up of an innovative environmental compliance training or pollution prevention program; or other areas, where metrics can be readily defined and measured. The compilation of such quantifiable and verifiable data useful to define and assess environmental risk associated with EH&S activities and responsibilities has sometimes been an elusive target for EH&S managers to achieve in their planning and strategies to corporate executives. Taking that data and being able to translate it into dollars and cents, can help unlock the hidden value potential and investment opportunities that exist within each organization.[8]

Another factor that may come into play with respect to such perceived intangible factors, is the effect that ISO 14000 may have. In particular, the standards dealing with environmental performance evaluation (EPE), ISO 14031 and life-cycle assessment (LCA) (ISO 14040s) place a greater emphasis on metrics to assist EH&S professionals and analysts to more effectively measure the effects of products and services on human health and the environment and hence effectively measure the dollar savings that can be realized for each EH&S consideration.[9] As one example, EPA has been providing seed money for specialized software to measure LCA data points, while software that Ecobalance developed is able to measure DfE data points, which we highlight in Chapter 13.

Placing a dollar value on an environmental process and being able to show clearly how much that process or other innovative environmental management system enhancement can save an organization is a valuable tool that an organization's money managers can readily see. Once you have their attention, gaining corporate funding to implement such processes and enhancements can occur in a more orderly fashion than the typical knee-jerk reaction many environmental managers may have been accustomed to during funding requests. We view this crucial paradigm shift as the first building block in the new-order environmental stewardship movement in companies from which subsequent management strategies can follow, such as innovative EH&S auditing programs, innovative training, and the pursuit of a more orderly environmental management system.

We focus our insight in this chapter on these and other related innovative solutions that could be useful toward enhancing an organization's environmental management system. Some of these innovative solutions may also be linked to an organization's culture and whether the culture does provide for employee empowerment, which is also a crucial factor in progressing environmental responsibilities and stewardship forward. Any of these linkages may tell whether a particular innovative solution may be beneficial to an organization. In other instances, EH&S managers may need to incorporate radically innovative ways to "show the money" to financial managers for approving funding budgets by following the lead from other groups within an organization that can obtain funds, such as the marketing folks who obtain considerable corporate monies to advertise their products or services on national television[10] during key events being broadcast.

In other instances, EH&S managers looking for innovative ideas may have to look at and evaluate procedures that other organizations may have previously implemented to assess whether such procedures or solutions may work for their company in their specific situation. For these companies, then, a benchmarking study, such as described in the last chapter, may be the right choice to help make that distinction.

As a final consideration, EH&S managers may want to consider funding requests that take into consideration other business units within the organization that could benefit from a specific EH&S expenditure request, and how the whole company ultimately benefits. One example would be credit points a company could receive or lower insurance premiums a company could pay to their insurance carriers from fewer on-the-job accidents, spills or chemical releases, enforcement actions from regulatory agencies, etc. Clearly in situations such as these, the entire company would benefit from reduced insurance or similar premiums, not just the EH&S group(s). If such savings possibilities are not immediately recognized, then it would be the responsibility of EH&S managers to ensure these savings are properly credited and recognized. Summing it up, that brings to mind the advertising slogan that says, ". . . The man who doesn't advertise is like the man who winks in the dark . . . he knows what he's doing, but no one else does." This is one mantra EH&S managers should keep in focus when dealing with company financial managers.

Providing further insight into a company's methodology for maintaining an effective environmental management system, contributor Ed Spaulding provides an overview of Chevron's perspective in environmental management later in this chapter.

BUSINESS FORCES CHANGING
THE ENVIRONMENTAL LANDSCAPE

As a backdrop to introduce this new environmental order, several books are worth noting, two of which were previously discussed in some detail in Chapter 1—David Drennan's *Transforming Company Culture,* and Arie De Geus's *The Living Company*—and a third book, *Participative Management: Implementing Empowerment,* written by Lorne C. Plunkett & Robert Fournier (1991, John Wiley & Sons).

In this book, the authors describe their experiences in dealing with management issues that are similar to the other books. These authors see participative management's goal as a means to collectively tap into each individual worker's resources, mold that into a unique learning experience, and produce results that are greater than the sum of the whole. Depending upon how each company measures success, the final destination is usually worth the effort and the cost of the trip.[11]

These three books are referenced herein as must-reading books not only for the environmental professionals, but for non-EH&S professionals within a company as well, since the new order of thinking requires that EH&S and non-EH&S personnel need to be on the same page to effectively communicate with each other. In addition, personnel from different business areas that interact with EH&S staff can learn to better understand EH&S staff and also learn how each group plays a part in the effective function of a company's environmental management system. As environmental management matures and becomes more of an integral part of the other business areas of a company, EH&S managers will strive to fully assess and factor in the needs and concerns of other business units, as well as keep the company's culture in focus. In the final analysis, any or all of these factors can play an important role in helping to sustain or provide support to a company's EH&S programs. We can also learn a valuable lesson by following the chemical industry's lead where the Chemical Manufacturers Association, by way of its Responsible Care® program, takes into account all of a member company's business units as having an impact on and being impacted by EH&S considerations. (We provide a copy of the Responsible Care® Principles in Appendix D.)

DEFINING THE ENVIRONMENTAL MANAGER'S ROLE: SETTING THE BUILDING BLOCKS IN PLACE

There is usually a clear delineation of responsibilities within a regulated company for each individual in an organization's environmental and health and safety groups. In this section, we focus in greater detail on the roles of environmental managers within various corporate and facility settings, and how they interact with environmental managers at sister facilities, with other business unit managers, and with their support staff. An integral part of effective interaction in each of these settings between parties includes open lines of communication that should exist between each of the functional groups: facility managers and their staff; managers of sister facilities within an organization; and facility managers with corporate EH&S managers, legal counsel, and executive management. In addition to the open lines of communication that should exist between managers and their staff at the facility level, managers should promote a greater sense of ownership or empowerment among their staff that gives some credibility to top management policies and helps promote open communications among employees.

A breakdown in communication between any of these groups can negate any internal achievements to date as well as slow down or stop any other progressive milestones. As a result, having clear communications and internal policies to ensure its sustainability should be one goal a company strives to maintain. Effective communications, extended to external operations, such as a corporate or risk communications group, can effectively downplay or mitigate adverse situations that may occur from time to time. To maintain such effective communication controls, it might be prudent for organizations to develop a corporate and risk communication system to address any adverse situations in a timely fashion, and not follow the examples of organizations that had not implemented such circuit-breaker defenses. *Additional discussion dealing with risk management, communications, and corporate communications is discussed in greater detail in Chapter 11.*

Setting the environmental manager's position apart from other business manager roles is the amount of responsibility these individuals usually have to shoulder dealing with every facet of a manufacturing organization that is regulated by environmental regulations. In more than one instance, environmental managers usually have to wear many hats, which also depends on the complexity of the position in dealing with the requirements of federal, state and local regulations. If that were not enough for environmental managers to handle, some industries also have to address additional regulations, besides environmental, health and safety, and transportation regulations, such as, pharmaceuticals (FDA), arms manufacturing (ATF), and energy-generating companies and federal agencies (NRC). So, conceivably, environmental managers of facilities within industries such as these examples would not only need to concern themselves with a mountain of EPA, OSHA, DOT, and SEC regulations, but also regulations specific to their industry. Therefore, while environmental managers may be employed by different industries as defined by their industry's standard industrial classification (SIC) codes, one item remains constant about their positions: The pressures they experience in keeping abreast of increasingly stringent environmental regulatory requirements at their facilities.

In addition, the environmental manager's role is not static and, as a result, can become more complex with each new regulation that is passed by legislation. These regulations, in turn, create additional requirements for these environmental professionals that subsequently need to be addressed, as noted in the "perverse spiral" concept we introduced in Chapter 6.

There may also be situations that arise where managers in one facility may need to confer with their counterparts from a sister facility or with other environmental professionals from their network of contacts to provide insight into a vexing issue or help resolve a regulatory problem, such as convene an ad hoc committee meeting, or initiate an open-forum session at an environmental managers' group meeting. The information that can come from such caucuses may provide relevant anecdotal solutions that may find their way into a company's compliance manual or procedure. Such anecdotal information could also supplement relevant materials from textbooks, and become part of a manager's reference library to serve as a repository for solutions to specific problems, or house reference materials useful to conduct in-depth research if necessary.

To clarify the diverse roles and functions that environmental managers may hold within a regulated organization, one professional group decided to look at the roles of these professionals in greater detail to help further define the role of the environmental manager. That group, still in existence today, is the National Association for Environmental Management, or NAEM. In 1991,[12] NAEM commissioned a survey that helped define the term "environmental manager." The survey further defined the term into two distinct subroles: Corporate consultation and facility compliance. The corporate consultation role focuses on those responsibilities within an organization involving broad authority over the company's environmental programs. These corporate environmental managers, possibly with some direction from corporate counsel and senior company executives, may be responsible for a number of corporate-wide duties such as: (1) establishing environmental policies and mission statements for the company; (2) planning, managing, or overseeing environmental audits and/or selecting third-party auditors; (3) participating in community relations; (4) managing worker safety and health considerations; (5) establishing the hierarchy of environmental management within the ranks of the organization and setting in place lines of communication via internal mechanisms, and other considerations.

In contrast, the facility compliance role of an environmental manager revolves around a more specific facility-oriented capacity within a plant or facility. These environmental managers are likely the ones dealing with "front line" issues such as spills, worker injuries, site-specific permits, internal inspections, environmental and health and safety audits, waste handling, and the mountain of ancillary paperwork and documentation that accompanies this job description, and any and all other compliance-related matters that may arise on a day-to-day basis.

In addition to these traditional environmental management responsibilities, today's environmental managers can also be expected to perform other duties, such as: (1) funding requests to upper management for maintaining regulatory compliance and management programs, (2) collecting and analyzing EH&S operating metrics to support a myriad of corporate-driven activities,[13] (3) other ad hoc duties they may be requested to perform as part of their duties. Of course, any of these activities, may, at times, be adversely affected by a company's size, the availability of staff to perform such duties, and the company's own commitment to its environmental policies and mission statements, in addition to how the requests are "packaged" to corporate executives for disbursement as previously discussed, among other considerations.

Taking a closer look at funding requests to top managers, these requests could include obtaining company funds for: specialty or regulation-required employee training and awareness, such as under RCRA or OSHA; process equipment upgrades or revisions identified as part of compliance or pollution prevention audits; supplemental staff resources or outsourcing assistance, or other such considerations that managers may need to study further as improvements to their organization's environmental management system(s). In today's downsizing-minded corporate culture, coupled with the "green wall" syndrome, such funding requests could develop into a manager's nightmare, unless they are well-planned and thought out beforehand.

As an example, here is one scenario we envision where roles and responsibilities of environmental managers could be enhanced to allow more to be done with less, and thus, keep in step with continued downsizing pressures. At issue could be the possibility of funds not being totally available to allow managers to set in motion specific tasks identified in an environmental audit report, say, or as a result of a minor enforcement action, such as a notice of violation (NOV). Using a "Plan B" approach, managers or a delegated senior staff member should look more closely at the allocation of staff and other resources and whether there may need to be strategic shifts to accommodate any funding constraints or bottlenecks. One suggestion to managers could be to utilize existing staff in ways to max out their pooled talents, such as empowering some staff members to provide affected employees internal training, such as HazCom and HAZWOPER (required by OSHA), providing insight into conducting EH&S audits and writing concise audit findings, or train employees in using specialty EH&S software, or a whole cadre of other opportunities; or allowing employees in another business group to trade places with EH&S employees and vice versa in ways the available talent pool could be expanded without adding additional staff. Managers could also consider experimenting with staff and possibly shift or piggy-back staff responsibilities among those staff members who have been keeping pace with applicable regulations to act as champions for spearheading activities requiring their expertise, such as leading EH&S or EMS audit teams, managing required regulatory reports and other paperwork, maintaining an electronic database, or EMIS, of all environmental information maintained on-site or in a central local-area network (LAN) or wide-area network (WAN), or other items of strategic importance. One explanation for a renewed sense of a management push to reshuffle staff responsibilities or remold staff proficiencies to foster worker empowerment, team building, and product ownership and other team-building techniques prevalent in the 90s, could be from total quality management (TQM) and total environmental quality management (TEQM) principles[14] driving management at many companies to maximize the efficient use of staff resources at every level to offset any "red ink" in lean times where ISO 9000 and its environmental sibling, ISO 14000, can be seen by companies as value added.

INNOVATIVE SOLUTIONS TO ENHANCE ENVIRONMENTAL MANAGEMENT SYSTEMS

A company's environmental management system is not an isolated and static entity that runs on its own energy. The system consists of the environmental staff and managers that perform day-to-day routines and are linked to other employees in ancillary functions and to other business units within an operational facility with which the environmental group(s) interact on a regular basis. These employees, in turn, are linked to the organization's corporate group that provides periodic guidance, support, funding, or a combination of all three, to help push the facility along as a whole. Stripped away of all its covering, the system, or EMS, is the glue that keeps staff together in pursuit of its goals with respect to environmental, and in an ancillary way, health and safety considerations and responsibilities.

An environmental management system includes those components that deal with maintaining environmental compliance requirements, and how these are maintained to ensure an effective utilization of staff resources. Improving upon EMSs as a management goal, for example, can take on many forms, one of which could be to have environmental managers evaluate their EMS to identify areas for improvement. One approach for initiating innovative solutions to improve EMSs could be for environmental managers to evaluate one or more specific EH&S program areas and compare them with best-in-class EH&S facility operations at other companies (a benchmarking study) which could include:

• Evaluating whether documents maintained are adequate, such as inspection records, manifest files, air permits, wastewater discharge records, OSHA 200 logs, and similar regulatory documents, to address whether the facility's EMS is adequate or may need improvements.

• Evaluating whether environmental audit programs are effective in identifying and correcting regulatory findings, or whether follow-on audits note recurring systemic issues of a regulatory or management nature that may require more in-depth corrective actions to solve.

• Evaluating supplemental files that may contain insight into system improvements, such as: maintaining regulatory inspection and enforcement action files; maintaining a file on contractors and consultants and their work products for archival use and reference base.

• Evaluating whether EH&S staff maintain communication with sister facilities to brainstorm ideas or tackle a particular regulatory issue or tap into their network contacts at other facilities or through professional associations for insight, such as those organizations identified in Chapter 7.

• Evaluating employee training effectiveness and keeping tabs on training providers where managers regularly evaluate employee training records, trainer evaluations, and whether training offered provides value to affected employees, such as making the training relevant to the staff's needs, in essence, helping them to help you, which is a fundamental tenet of employee empowerment.[15]

• Evaluating whether attending a conference or seminar can provide insight to help address one or more vexing issues within your organization, and what additional benefits might be gained.

• Evaluating whether participating in one of EPA's voluntary programs can help solve or address a particular regulatory issue, and whether some regulatory relief may be offered for program participants.[16]

BUSINESS CHANGES FOSTERED BY REGULATORS AND LEGISLATORS

As noted in the previous bulleted item, EPA is in the forefront in its continuing efforts to bring organizations, and now municipalities,[17] to the table as participants in a growing number of EPA-sponsored voluntary programs. The agency is leading the charge in these efforts, partly as a result of the agency's belief that more effective environmental management can lead to increased regulatory compliance, and what better way to achieve such goals than through the shared and collective experiences of program participants. At this point, we can venture to say that as a result of the number of voluntary programs that EPA initiated since the flagship 33/50 program EPA launched in 1991, regulators, legislators, regulated entities, and the general public have all learned much about changes that can be made in how business is conducted, and that cooperation can lead to regulatory compliance if industries are given credit where it is due. However, attaining full closure in this particularly sensitive area from the standpoint of regulated companies may yet require that EPA's Office of Enforcement and Compliance Assurance (OECA) ease up on its enforcement throttle, or at least allow companies more flexibility in attaining EPA's compliance goals via voluntary and alternative methods as in EPA's voluntary programs.

As Jeff Van, a spokesperson for the Chemical Manufacturers Association, stated in response to a question posed to him regarding Project XL, "While Project XL is a good idea, it will never do more than operate at the margins because of what it is not allowed to do. The enforcement wing of EPA has flexed its muscle and prevented things from happening."[18]

In addition, DeWitt John, the Director of the National Academy of Public Administration's (NAPA) Center for Economy and Environment, co-authored a NAPA report *Resolving the Paradox of Environmental Protection*, published in September 1997, that recommends EPA should provide companies with flexibility to reduce the cost of compliance. As a result, a number of companies continue to hold back on creative approaches because of fear of regulatory enforcement. One way out of this dilemma, John believes, is to initiate legislation to give EPA and the states to allow the regulated community flexibility.[19]

To provide firsthand insight into EPA's voluntary industry programs, two participant organizations, Lucent Technologies and the Gillette Company, share some of their experiences as part of Project XL and the Environmental Leadership Program (ELP) pilot project in subsequent chapters. In addition, Appendix G contains links to Project XL, the CSI Program, and the ELP through EPA's web site.

Of note, EPA's Environmental Leadership Program completed its pilot phase at the end of 1997, and preliminary indications from various EPA sources in early 1998 indicated that EPA would be rolling out a national ELP program where participation would be limited to companies with mature environmental management systems. Given that qualifier, EPA is giving some credence to organizations that have put forth the effort to maintain EMSs that lend themselves to continual improvement opportunities. Such opportunities, in turn, can allow companies to attain greater compliance with applicable environmental regulations, allow for environmental, EMS, and safety and health auditing programs to become more effective in uncovering areas in need of improvement, and can allow companies to achieve a decrease in raw materials use, energy use, and unnecessary waste products to be disposed, among other noteworthy considerations.

In addition, other U.S. government agencies are moving forward in this new business order; among them, the U.S. Department of Energy (DOE), has been taking steps to develop and implement more effective environmental management systems at DOE sites, where the ultimate goal is to have these sites attain certification to ISO 14001. Among DOE sites that were in the pipeline to achieve ISO certification as of the time the manuscript was completed included Savannah River and Oak Ridge.[20]

What does all this say for business upheaval that is currently underway, and is there a light at the end of the (EPA) tunnel for companies? Do well-run environmental management systems really make a difference to the bottom line? Let us step back a moment and ponder these questions further.

The main thread throughout this text, echoed by our contributors and survey respondents referenced, is that we as the environmental and health and safety standard bearers have a mission to educate and empower those with whom we come in contact in our work. They need to be reassured that we can sustain our manufacturing processes and our use of natural renewable and nonrenewable resources, if we use our intellect wisely and fortuitously, and show some sensitivity to our future generations and the world in which we all live. Simply put, if we put our minds to it, we can do more with less, and still keep our surroundings clean. We also have to keep focused on EPA's main enforcement and compliance marching orders, and that is, the protection of human health and the environment. In a different perspective, we can look to the often-used adage in business, ". . . Pay me now or pay me later!", that encapsulates where we are with respect to our environmental responsibilities to ourselves, our coworkers, our neighbors, our community, and the environment in which we live. Taking the time to establish an environmental management system that encompasses these considerations can and will take a considerable effort and expenditure of company funds to achieve (the pay-me-now scenario) versus playing catch-up with regulatory requirements, paying for future cleanup costs and any related accidents and injuries, paying regulators costly fines in response to enforcement actions, paying for wasted raw materials and energy use, and a whole cadre of items that an organization would find itself being held responsible to pay (the pay-me-later scenario). In pondering your response to either of the questions we pose, and whether your answer will be "yes" or "no," ask yourself in which scenario you would rather be.

Finally, whether certification to ISO 14001 can make a difference in whether an organization's EMSs have been developed to be cost-effective and can realize positive ROI's is too early to call. ISO 14001 attained standardization status in the fall of 1996, and even now, some organizations could be tracking the accomplishments of any one of the twenty-three hundred company facilities to date[21] that have certified to ISO 14001 for measuring purposes.

In the following section, we provide an insight into some of the techniques two organizations, Chevron and Millipore, developed to meet the challenges before them. In the first example, Ed Spaulding takes a look at continuous quality improvement methods and compares this approach to the U.S. Sentencing Commission's draft sentencing guidelines in a company's development of a commonsense approach system to address environmental compliance requirements and manage a more effective and all-encompassing environmental management system. In the second example, we report on a presentation given by Pat Davies on Millipore's use of an environmental management information system (EMIS) to cut down on man-hours in day-to-day EH&S activities.

INSIDE ENVIRONMENTAL COMPLIANCE: A CORPORATE VIEW

As shown in Chapters 2, 3 and 5, federal, state, regional, and local laws and regulations create an almost impenetrable web of compliance requirements covering air emissions, water discharges, waste disposal, reporting of spills and releases, employee training, and record keeping. Chapter 10 addresses proactive strategies of several noteworthy companies to improve environmental, health, and safety performance, while this chapter focuses on compliance in areas where laws and regulations already exist.

In reality, *full* compliance with *all* environmental laws and regulations *100 percent* of the time is a goal. Management must adopt full compliance as a goal and actively, publicly, and repeatedly reinforce this goal. Management must also clearly assign responsibility and hold employees responsible for compliance performance. This is the foundation of any compliance program.

Establishing full compliance as a goal also allows a facility's compliance program to be viewed as a continuous process and placed under the umbrella of continuous quality improvement (CQI). CQI is familiar to many managers in

American business today. Viewing compliance as a process also helps management think of environmental, health, and safety issues as an integral part of their business, not something added. This is critical as companies downsize and individuals have less time to spend on noncore activities.

Thus the compliance program is a process that can be continuously improved. As shown in Figure 8–1, increasing investment in resources, people, and money, results in improved compliance performance.

However, two questions arise immediately:

1. When is improving compliance performance justified?
2. What commonsense elements can improve a compliance program?

Answering these questions will serve to lay out the elements of a commonsense environmental compliance program.

WHEN IS IMPROVING COMPLIANCE PERFORMANCE JUSTIFIED?

The simple answer is that improvements should be made when performance does not meet the full compliance goal established by management. However, this is not very helpful in determining what actions to take. A more direct answer, one that fits the vocabulary and mind-set of managers, is that improvements should be made when they are justified, when the benefits from the improvements exceed the costs of implementation.

The fact that both the benefit and cost aspects of this analysis contain subjective elements poses a challenge to the environmental manager who must justify any information used to support requests for management approval. Let's consider each side of the cost/benefit analysis.

There are many benefits to improving compliance performance, both internal and external. The external reasons center around avoiding the costs of noncompliance. The potential financial impact of noncompliance has increased over the past few years, and appears destined to increase further this decade. Fines by EPA and state authorities are getting larger. The EPA has recently reorganized its

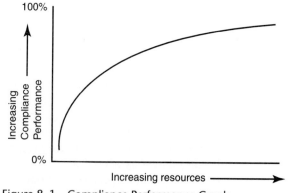

Figure 8–1   Compliance Performance Graph

enforcement division in an effort to become even more effective. The recently strengthened federal sentencing guidelines (see Figure 8–2) offer both a stick and a carrot: stipulated penalties for environmental violations, with significant reductions for an effective compliance program. In addition to fines, there is increasing application of criminal penalties with the potential for jail time for key management employees.

The internal reasons to improve compliance performance are less direct but just as compelling. First, compliance with the law is the stated goal of essentially all companies. The better the compliance program, the higher the chance the facility will achieve the goal. Excellent compliance performance can help improve employee morale, and help protect employee health and safety. Also, an excellent compliance record, or at least a lack of problems, can improve relationships with suppliers, attract new customers, and help retain existing customers. (Conversely, compliance problems can indicate more serious underlying issues.) Finally, as previously discussed, improving the compliance program is consistent with the quality improvement efforts of many businesses.

On the cost-side of the analysis is the recognition that improving the environmental compliance program may require the expenditure of resources—usually people's time and money. People must spend time for training, for investigating problems, and for auditing performance. Money is generally needed to improve equipment or change procedures to assure a higher level of compliance performance.

In summary, management should always be working to improve compliance performance. The key management decision is balancing the *potential* for improved compliance with the *real* costs to implement the improvement. In some situations, performance improvements can be achieved at no cost or even with

---

The U.S. Sentencing Commission substantially increased the potential for large corporate fines in its sentencing guidelines for organizational defendants, which became law on November 1, 1991. These guidelines give substantial credit to corporations that, through due diligence, police their activities with an "effective program." The following is a summary of the elements of an effective program:

1. The organization must establish compliance standards and procedures to be followed.
2. Specific high-level personnel within the organization must be assigned overall responsibility to ensure compliance.
3. The organization must use due care not to delegate authority to persons who have a propensity to engage in illegal activities.
4. The organization must effectively communicate its standards and procedures to employees.
5. The organization must take reasonable steps to achieve compliance with its standards through monitoring and auditing.
6. The organization must take reasonable steps to achieve compliance with its standards by having an internal reporting system where employees can report criminal conduct without fear of retribution.
7. The organization must consistently enforce its standards through appropriate disciplinary mechanisms.
8. After an offense has been detected, the organization must take all reasonable steps to respond appropriately and to prevent further similar offenses.

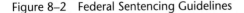

Figure 8–2   Federal Sentencing Guidelines

savings. Deciding to implement such win-win situations is easy, really no decision at all. However, this is not always the case. The answers to the second question are meant to be a guide for the environmental manager as these difficult decisions are approached.

WHAT COMMONSENSE ELEMENTS CAN IMPROVE A COMPLIANCE PROGRAM?

An environmental compliance program consists of a number of elements. To improve performance, existing elements are strengthened and new elements added. Referring to Figure 8–1, this represents a continuum, where the increased effort produces improved performance.

Every compliance program should have some basic elements. These are presented first, followed by additional elements that add depth and breadth to a compliance effort. The management of each facility must determine which elements are appropriate for their facility and situation.

A basic commonsense compliance effort should have the following elements: a clear statement from top management that full compliance is the goal, a list of legal requirements, clear assignment of responsibility, a training program, legal support, and an audit process.

MANAGEMENT STATEMENT

Top management must clearly state that full compliance with all environmental, health, and safety laws and regulations is the policy and the goal.[22] Such full compliance should be independent of the degree of enforcement or inspection by government agencies. Top management should reinforce this policy at every opportunity.

LIST OF REQUIREMENTS

Knowing what to comply with is the foundation of any compliance program. If you will, these compliance requirements set the environmental specifications for facility operations. Such a list is often difficult to develop, because requirements fall into several categories, each needing to be addressed differently. There is no single source for a facility-specific list. Developing a complete list takes research. Trade associations, similar facilities, and regulatory agencies are all potential sources of information. Legal support may be needed as applicability is sometimes unclear, and the various jurisdictions often overlap with slightly different requirements.

The list of requirements should include the following to address item 1 of the federal sentencing guidelines (see Figure 8–2).

Operating Requirements This is a list of all federal, state, regional, and local laws and regulations that establish environmental operating requirements for the facility. Examples include air emission sulfur limits, wastewater discharge BOD concentration limits, and requirements for hazardous waste pretreatment prior to disposal. The specific applicable laws and regulations will vary between facilities and depend on the location and age of the facility and on the equipment, processes, and chemicals used at the facility.

Reporting Requirements Many environmental laws and regulations require reporting to government agencies. This reporting can be periodic, generally monthly or annually, or when an event occurs, such as a spill or exceeding a permit or legal limit. All such reporting requirements should be known and a specific person assigned to make sure the report is done correctly and on time.

Employee Training Requirements Some environmental laws and regulations require that specific employees receive specialized training prior to doing an operation. In addition, refresher training is often required. [One example is the hazardous waste treatment, storage, and disposal facility operator-training requirements under RCRA (40 CFR 265.16)]. The compliance program must see to it that this training is completed and accurate records are kept to prove the training has been done. Allowances should be made so that trained replacements are available to cover for vacations, sickness, and schedule changes.

Record Keeping Requirements Record keeping is required by almost all environmental laws and regulations. Records are the way a facility can prove it is or has been in compliance. While complying with operating requirements is important as far as environmental protection, inspectors tend to focus on record keeping as it is more easily measured. A dedicated, organized filing system should be established so any desired record can be found very quickly.[23] Inspectors expect that legally required records be produced within minutes.

ASSIGNMENT OF RESPONSIBILITY

Responsibility for compliance must be clearly assigned to specific positions as outlined in items 2 and 3 of the federal sentencing guidelines. The logical choice is generally line management. These people have the ability to apply the resources, both people and money, to improve compliance performance. It is generally ineffective to assign compliance responsibilities to support groups such as environmental staff, engineering, technical, or maintenance. In most cases, it is beyond their control to change the procedures, processes, or equipment needed to come into compliance. Obviously, compliance responsibility should not be assigned to persons who have a propensity to engage in illegal activities.

TRAINING

Item 4 of the guidelines focuses on training. There are two categories of training. The first category is the legally required training presented earlier. The second category is training employees (operators, maintenance people, and management) in the applicable environmental, health, and safety requirements for which they are responsible. Engineers and technical people may also need to know the list of requirements for a facility if the projects they design are to comply with operating requirements.

LEGAL SUPPORT

Items 1, 5, and 8 of the guidelines touch upon legal support. The facility should have an established link with legal support, either from corporate staff or a local law firm. The environmental compliance process is in reality a legal structure. On the front end, legal support may be needed to determine if a specific law or regulation applies. For those that apply, sometimes legal analysis may be needed to understand a law's exact requirements. During operations, a noncompliance situation may be discovered by employees. Legal help may be useful in determining how to communicate this to the authorities to reduce legal and financial exposure.[24]

The settlement of violations often involves negotiating a legally binding agreement that may include civil or criminal fines and operating or equipment changes. Legal help is critical in this area. Legal support can also help employees prepare for hostile inspections by government agencies. While such agencies have broad powers to search records and take samples, their power is not unlimited. The ability to have twenty-four-hour legal support on short notice is very valuable.

AUDIT PROCESS

Every compliance program should have an audit system, as noted in item 5 of the guidelines. There are two categories of audits: self-audits and independent audits. Self-audits are done by facility personnel and are generally limited in scope and done frequently, while independent audits are less frequent, more expansive in scope, and involve people from outside the facility. Figure 8–3 defines the advantages and disadvantages of the two forms of audits. Many facilities use a combination of both. The proper level of compliance auditing is a facility and corporate management decision.

Note that the elements of a basic compliance program do not address all aspects of the federal sentencing guidelines. In spite of this fact, these elements, if fully implemented, create a strong, solid, defensible commonsense program.

But what if more is needed? There are two ways to improve the basic program. The facility can improve the effectiveness of the basic elements and can add additional elements. These options are discussed below.

ELEMENTS TO IMPROVE COMPLIANCE PERFORMANCE

The following elements help improve compliance performance by taking actions earlier to assure the facility stays in compliance.

USER-FRIENDLY LIST OF REQUIREMENTS

Generally the exact requirements of environmental, health, and safety laws and regulations are difficult to understand because of the legal nature of the text. A translation into "plain English" can make the requirements much more clear. The various requirements for each process, piece of equipment, or operator position can also be collected together. For example, all the regulations for a solvent clean-

Advantage	Disadvantage
Self-audit: • Not people intensive • No planning • Done by plant personnel • Not formal • Limited or no reporting • Problems fixed on the spot	Self-audit: • No challenge to accepted view of what compliance looks like • No changes in system so problems may reoccur • Concern about recording problems may lead to superficial audits
Independent audit: • May be able to uncover "blind spots" because of outside perspective • Formal written report may result in system or procedure changes, thereby preventing reoccurrence • More likely to involve facility management	Independent audit: • People intensive • Costly if consultants used • May not be able to discover subtle noncompliance unless highly knowledgeable • Fear by facility personnel if not explained properly

Figure 8–3   A Comparison of Self-Audits versus Independent Audits

ing operation could be collected, translated into plant terms, and given to the person who does the solvent cleaning. This is much more effective than giving that person the many pages of applicable regulations.

For small operations, these requirements could be posted next to the equipment. For larger facilities, a binder may be appropriate. With the expanding use of computers, putting the operating requirements into an operator-friendly database may be justified. These computer-based lists can be developed internally or purchased from software vendors.

This further addresses item 1 of the federal sentencing guidelines.

CHANGING REQUIREMENTS

Environmental, health, and safety compliance requirements are dynamic; they are revised periodically. Therefore, the list of requirements will become dated unless a process is in place to keep it up to date. There are several aspects to this (these follow).

Lobby and Advocacy   Efforts should be considered to influence proposed laws and regulations to allow for cost-effective compliance activities. Trade associations are helpful but need active support from companies and facilities to be effective.

New and Amended Laws and Regulations   As laws and regulations are revised, the changes must be communicated in a timely manner to those responsible for compliance. All government agencies are required to publish new or revised legal requirements. For a fee, contractors will also provide this information.

This information is also provided by vendors who supply their customers CD-ROMs with the applicable federal and state regulations and updates for one year of the applicable regulations as they are revised and updated. Some of these vendors are listed in Chapter 13.

Court Decisions   As new regulations, violations, and fines are appealed, legal compliance requirements may change, even without changes in the text of the law or regulation. Law firms can provide advice on these issues. This information, too, is also found on CD-ROMs. One such program is Lexus-Nexus®.

www

Enforcement Policy   EPA and other government agencies develop and revise enforcement policies. While not having the force of law, these are valuable guides when improving a compliance program. EPA's office of Enforcement and Compliance Assurance regularly issue enforcement policies and priorities. (For further updates, visit EPA's web page at www.epa.gov/oecaerth/index.html)

Open Communications   Lower-level employees, who may have the most knowledge of a noncompliance situation, must have the ability to communicate that information without fear of reprisal. This allows employees to freely express concerns or ask questions, daylighting problems prior to an agency inspection. Examples of communication paths are a hotline where information can be given anonymously, and a neutral ombudsman who can act on behalf of an employee. This addresses item 6 of the guidelines.

Incident Investigation   Environmental damage, health impacts, and injury, as well as noncompliance, can result from incidents. Investigating all incidents, even those which did not result in noncompliance, is valuable to identify and prevent more serious incidents from occurring. The investigation should be open, with a goal to discover the root cause and change work processes to prevent future incidents. The goal should not be to attach blame. However, if a person is at fault, appropriate disciplinary action should be taken. This covers items 7 and 8 of the guidelines.

Understanding the Potential for Noncompliance   This brings risk analysis and risk management into the decision-making process. Personnel must understand the process and equipment well enough to identify potential noncompliance risks if something breaks or if a procedure is not followed properly. Management should then consider changing equipment, procedures, or training to reduce the chance the noncompliance situation will occur.

Reduce Dependence on Audits   A facility should not rely on audits or inspections to determine compliance performance. Excellent compliance performance should be designed in from the start, not inspected in at the end. While some auditing will always be needed, the focus should be on prevention. (This is the goal of audits as management tools, as we discuss in the next chapter.)

CONCLUSION

This chapter proposes that a company or facility compliance program be viewed as a process under the umbrella of continuous quality improvement. This allows the well-known tools of quality improvement to be applied. The elements of a basic, solid compliance program are introduced, as well as more sophisticated elements. These elements were compared with the federal government's sentencing guidelines—the government's idea of an effective compliance program. As a final aid, Figure 8–4 collects this into a diagram for compliance program improvement. The following chapters provide additional insight based on this information as a rationale for sustaining and improving a company's compliance process.

In this example, we provide a brief summary of a presentation made by Patricia Davies,[25] corporate EH&S manager for Millipore, to an informal group of environmental professionals as part of a CMA outreach session in 1996 that showcased Millipore's experiences with an environmental management information system (EMIS) as a way to integrate ISO 14000 into a continual improvement process of their environmental management system. The focus of Davies's presentation was to provide an overview of actual dollar savings that Millipore realized by using this custom-made EMIS. Her presentation was titled, Integrating ISO 14000 Into Continuous Improvement Programs—Automating Measurement and Evaluation.

THE MILLIPORE EMS EXAMPLE

Millipore is a Massachusetts-based high-tech company that manufactures components used in purification techniques of specific industry applications, among them, pharmaceuticals, electronics, medical and health care, and food and beverage. All of its facilities worldwide are ISO 9000 certified. Its corporate mission and vision is to act responsibly to protect the environment, and be recognized by local communities as an excellent employer and corporate neighbor. Translated, this focus points to an environmental management system (EMS) that assures continuous improvement.

As part of the company's goals to achieve its corporate mission, the company looked to software applications to help manage its EH&S data gathering and assimilation, and decided upon a tailor-made EMS application. The company chose EG&G to develop its environmental management information system (EMIS) program. This system would provide Millipore many anticipated beneficial uses.

The beneficial uses the company realized from their EMIS included the following:

- Real-time materials tracking and mass balances
- Reduced man-hour requirements for report filing
- Provided management with vital information
- Integrated with existing software systems
- Generated updated emission factors for all manufacturing processes

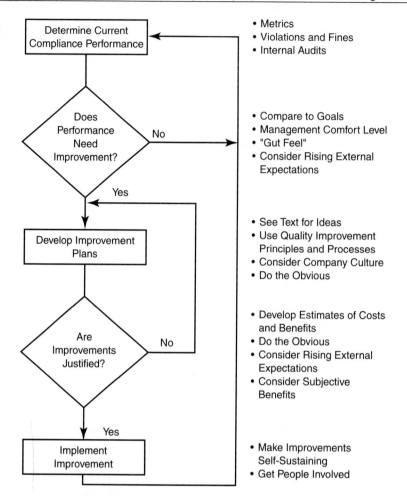

Figure 8–4   Compliance Improvement Process

For Millipore, assuring success of its EMIS depended upon twelve key items in its development process, which was straightforward, and similar to other methodologies that can be used in similar studies, such as benchmarking. Millipore's twelve points for developing an EMIS were:

1. Outline goals
2. Identify potential processes/procedures for automation
3. Identify (other) existing information systems
4. Investigate interaction opportunities
5. Outline initial scope of work
6. Identify resources for system development
7. Determine system cost

8. Cost justify/predict savings
9. Obtain management support and approval
10. Develop system
11. Implement and debug
12. Train users

The preliminary results of how well the company's EMIS performed were tabulated from the survey results of sixty-five employees who came from EH&S, manufacturing, warehouse, research and development, quality control, and engineering. The environmental activities devoted to these groups was approximately 23 percent of total man-hours, or $664,000. The estimated employee time savings with EMIS was 9 percent, or $64,000. Based on these numbers, Millipore expects a system payback within two years, basing this on a conservative estimate.

From their experiences, Millipore identifies these six key points for success:

1. Exploit the interest of all affected departments (manufacturing, MIS, purchasing)
2. Solicit user-input early and often
3. Remain focused on program goals
4. Integrate with existing systems wherever possible
5. Train, retrain, and train again
6. Provide (top) management with a follow-up of actual cost savings

In summary, each of the examples provided depicts different methods in which a company can improve upon its EMS. The Chevron example depicted a commonsense employee approach to improve upon an EMS while the Millipore example depicted the use of specialty software as a means to improve an EMS. Each approach is equally valid, since the main objective for each company in improving its EMS should be focused on a strategy that works for them. Improving a company's environmental management system is not an exact science, and as such, does not fall under a "one-size-fits-all" category. For this particular reason, a company interested in pursuing improvements to their EMS should also consider developing a benchmarking study to provide them valuable information into what other organizations are doing to improve their environmental management system.

WHAT THE FUTURE MAY BRING

The Millipore EMIS example we provided depicts a growing trend among regulated companies who are turning in greater numbers to more sophisticated software applications for providing some relief to their constantly growing EH&S responsibilities and related volumes of documentation that goes with the territory. As the EH&S group at Millipore found after implementing their custom-made EMIS, utilizing a software program that can allow the collecting and assimilating of relevant and timely data for use in environmental applications, can be a real time-saver. As many environmental professionals have seen since using the first EH&S software applications and off-the-shelf packages that were available in the late 1980s, these specialty software developers have come a long way in the software they provide. These developers have been able to learn much from their interaction with EH&S professionals in developing software that is both user friendly and readily integrates with other software platforms. As a result, many of the features attributable to the most recent generation of software applications can provide time and man-hour savings to an EH&S group.

In the larger context of improving a company's environmental management system, the next generation of environmental software, commonly referred to as EMIS, is evolving into software programs that meld EMS requirements and metrics with business processes, such as product throughput and waste generation with related information systems (IS) and management processes, such as data input requirements, regulatory requirements, and other product or project management requirements. In this fashion, EMIS applications can provide significant assistance to EH&S data input and analysis specialists. In essence, a well designed and defined EMIS helps integrate existing systems and helps pull the various responsibilities of an EMS together for greater efficiencies. In addition, as we alluded to earlier in the chapter, business demands have placed a greater pressure on EH&S managers to deliver relevant and timely information to top management. Such information encompasses regulatory requirements, management quota milestones, EH&S expenditures and projected cost savings, and the risk/liability areas inherent to EH&S considerations.

As an additional consideration for environmental managers, the advent and gradual acceptance of the ISO 14000 series of environmental management standards has provided opportunities for a number of enterprising software developers to design next-generation EMIS software that takes into account aspects of these new international standards. A number of these specialized software packages can provide assistance in conducting gap analyses, preparing for an EMS audit, or other function relevant to some other aspect of ISO 14001 certification. We provide additional information in this subject area in Chapter 13, including several case studies of end users of these specialized software products.

As we proceed into the twenty-first Century, our regulatory needs will still need to be addressed, if for nothing else than to continually measure the health and ecological risks associated with various activities onto our communities and surrounding areas. In addition, since federal and state regulatory agencies will still require the regulated community to collect, assimilate and analyze data to address regulatory, stakeholder and company concerns, there will continue to be a need for systems to facilitate that data input and analysis. For assistance in this area, regulated companies will continue to look to specialized software to perform these tasks. To borrow from the Common Sense Initiative, subsequent software releases will be designed to function cheaper, smarter and faster than previous releases to benefit the end user. In addition, there will be a new order of worker who takes an increased responsibility in his or her work by way of training, empowerment and teamworking to achieve individual goals and the goals of the organization. These are among the building blocks necessary to be in place to help a company's environmental management system move continually forward and improve.

We describe this particular subject area further in Chapter 15.

ENDNOTES

[1]With contribution from Janet Peargin of Chevron Corporation's Environment Policy Development Group.

[2]EPA's 33/50 Program, or the Industrial Toxics Project, was initially launched by former EPA Administrator William Reilly in July 1991. At that time, EPA had asked over six hundred companies to voluntarily cut their 1992 emissions by 33 percent of their 1988 levels, and to pledge to reduce their 1995 emission levels by 50 percent of their 1988 levels. Hence, the popular name of "33/50."

[3]Creativity knows no bounds and pushes the envelope. It may also help to have management support to foster creativity. Additional insight into creativity in the workplace can be found in H. Lancaster's article, "Getting Yourself In a Frame of Mind to Be Creative," *The Wall Street Journal,* September 16, 1997, p. B1.

[4]C. Sheldon, Introduction, *ISO 14001 and Beyond,* Sheffield, UK: Greenleaf Publishing, 1997.

[5]Among enforcement activities taken against companies in 1997 are: Darling International (Irving, TX) was sentenced to pay $4 million for illegal wastewater discharges by its Minnesota plant; and BFI Services Group, a subsidiary of Browning-Ferris Industries, was fined $3 million and ordered to pay over $600,000 to four publicly-owned treatment facilities in the greater Philadelphia area. (Source: *Safety Compliance Letter,* August 25, 1997, p. 8 and September 25, 1997, p. 8.

[6]See A. Iannuzzi, "The Environmental Quality Business Plan: A Step-by-Step Guide," *Environmental Quality Management,* Winter 1997.

[7]See C. Duff, "Steven Seagal as an EPA Agent? Get Real," *The Wall Street Journal,* September 18, 1997, p. B1.

[8]For additional insight, a good reference to review is B. Bentry and L. Fernandez, *VALUING THE ENVIRONMENT: How Fortune 500 CFO's and Analysts Measure Corporate Performance,* United Nations Development Program, Office of Development Studies, New York, Fall 1997.

[9]Crognale previously gave a presentation at the Rhode Island Pollution Prevention Conference in October 1996 in Providence, Rhode Island, that focused on savings that could be realized by organizations in applying various P2 techniques.

[10]Advertising costs during the largest sporting telecasts are running upwards of $1.2 million per thirty-second spot. How many environmental professionals' salaries would that cover in a year? To put this into perspective, see R. Balu, "Heinz Ketchup Readies Super Bowl Blitz," *The Wall Street Journal,* January 5, 1998, p. B6.

[11]See Plunkett & Fournier, pp. 14–15.

[12]The report compiled for NAEM as a result of this survey is *Environmental Management in the 90's: A Snapshot of the Profession,* July 1991. This survey was highlighted at a HazMat presentation in June 1991 in Atlantic City. More recently, NAEM commissioned a survey of its members in 1995, dealing with ISO 14000 awareness.

[13]For example, a senior Raytheon environmental staff gave a presentation of a draft report, *Raytheon Company EH&S Metrics Program* in September 1997 to a group of local-area environmental managers outlining their company's efforts to collect, review and compile such data for use at stockholder meetings, preparing the company's EH&S stewardship report and business unit proposal development.

[14]For a historical perspective on the subject of TQM, the reader may want to read "When Times Get Tough, What Happens to TQM" by D. Niven, *Harvard Business Review,* May–June 1993, pages 20–34.

[15]Additional information on training is provided in several sources, among them, J. Cornelison, "Who Should Be Trained and How Much" in the book written by B. Seldner, *Environmental Decision Making,* New York: McGraw-Hill, 1994, pp. 147–157. D. Drennan in *Transforming Company Culture,* London: McGraw-Hill, 1992, pp. 110–133; and G. Crognale, "How to Choose Environmental Management Courses: The 'Consumer Reports' View-point," *Environmental Management Report,* May 1997, pp. 17–20.

[16]There are some mixed feelings on the part of regulated organizations regarding the actual regulatory flexibility some EPA voluntary programs will provide participants. You can obtain additional information regarding Project XL in J. Pelley's "Companies Signing on to EPA's Project XL Program," *Environmental Science & Technology,* January 1, 1998, page 13A and non-specific references in the GAO Report, *Environmental Protection: EPA's and States' Efforts to "Reinvent" Environmental Regulation,* GAO/T-RCED-98-33, November 1997, and its complementary report, *Managing for Results: EPA's Efforts to Implement Needed Management Systems and Processes,* GAO/RCED-97-156; and on the CSI program in the GAO Report, *Regulatory Reinvention: EPA's Common Sense Initiative Needs an Improved Operating Framework and Progress Measures,* GAO/RCED-97-164, July 1997.

[17]EPA initiated a kick-off meeting of the Agency's ISO 14001 Environmental Management Systems Implementation Initiative for Municipalities and Counties as a Pilot Project in August 1997, in EPA's Region I offices in Boston, Massachusetts, with the intent to introduce the elements of an EMS to local governments.

[18]See J. Pelley, "Companies signing on to EPA's revamped Project XL program," *Environmental Science & Technology,* January 1, 1998, p. 13A.

[19]Ibid.

[20]Based on information obtained from several inquiries to DOE.

[21]Based on an item in European News, a German environment agency calculated that over twenty-three hundred ISO 14001 certifications were issued as of the end of 1997. For further information, see *Environmental Science & Technology,* January 1, 1998 p. 13A. More recently, the *International Environmental System Update* reports that about five thousand companies worldwide certified to ISO 14001, in the Special Supplement, September 1998.

[22]We include a copy of Hewlett-Packard's environmental policy in Appendix E as one example.

[23]Additional insight into various software programs that can aid in records management, such as EDM and EMIS, are described in greater detail in Chapter 13.

[24]This topic was previously described in Chapter 4.

[25]We would like to thank Patricia Davies for providing a copy of her presentation for use in this chapter, with the permission of the Millipore Corporation.

REFERENCES

Balu, R. "Heinz Ketchup Readies Super Bowl Blitz," *The Wall Street Journal,* January 5, 1998, p. B6.

Crognale, G. "Utilizing P2 to Highlight Financial Incentives," a presentation at the Rhode Island Pollution Prevention Conference in Providence, RI, October 1996.

Davies, P. "Integrating ISO 14000 into Continuous Improvement Programs—Automating Measurement and Evaluation," Chemical Manufacturers Association meeting, January 1996.

Dreenan, D. *Transforming Company Culture.* London: Mc-Graw Hill International (UK) Limited, 1992.

Duff, C. "Steven Seagal as an EPA Agent? Get Real," *The Wall Street Journal,* September 18, 1997, p. B1.

Gentry, B., and L. Fernandez, *VALUING THE ENVIRONMENT: How Fortunate 500 CFO's and Analysts Measure Corporate Performance,* United Nations Development Program, Office of Development Studies, New York, Fall 1997.

Iannuzzi, A. "The Environmental Quality Business Plan: A Step-by-Step Guide," *Environmental Quality Management,* Winter 1997, pp. 65–69.

Lancaster, H. "Getting Yourself in a Frame of Mind to Be Creative," *The Wall Street Journal,* September 16, 1997, p. B1.

Niven, D. "When Times Get Tough, What Happens to TQM?," *Harvard Business Review,* May–June 1993, pp. 20–34.

Pelley, J. "Companies Signing on to EPA's Project XL Program," *Environmental Science & Technology,* January 1998, p. 13A.

Plunkett, L. & R. Fournier, *Participative Management.* New York: John Wiley & Sons, Inc., 1991.

Seldner, B. *Environmental Decision Making for Engineering and Business Managers.* New York, NY: McGraw-Hill.

Sheldon, C. Ed. *ISO 14001 and Beyond.* Sheffield, UK: Greenleaf Publishing, 1997.

Sullivan, Thomas ed., *The Greening of American Business.* Rockville, MD: Government Institutes, Inc., 1992.

Miscellaneous Reports

European News, *Environmental Science & Technology,* January 1, 1998, p. 13A.

Government Accounting Office Report, *Environmental Protection: EPA's and States' Efforts to Reinvent Environmental Regulation,* GAO/T-RCED-98-33, November 1997.

Government Accounting Office Report, *Managing for Results: EPA's Efforts to Implement Needed Management Systems and Processes,* GAO/RCED-97-164, June 1997.

Government Accounting Office Report, *Regulatory Reinvention: EPA's Common Sense Initiative Needs an Improved Operating Framework and Process Measures,* GAO/RCED-97-164, July 1997.

Hewlett-Packard Company, Environmental Policy, downloaded from Hewlett-Packard's website at www.hp.com, October 1997.

International Environmental Systems Update, Special Supplement, September 1998, published by CEEM, Fairfax, VA.

National Association for Environmental Management (NAEM) report, *Environmental Management in the 90's: A Snapshot of the Profession,* July 1991.

Raytheon Company, draft report, *Raytheon Company EH&S Metrics,* September 1997.

Safety Compliance Letter, Issues and Trends, August 25, 1997, p. 8; and September 25, 1997, p. 8.

CHAPTER 9

THE ENVIRONMENTAL AUDIT AS AN EFFECTIVE MANAGEMENT TOOL

*Mitchell Gertz, the PQ Corporation, David B. Jefferies and
Thomas F. Harding, Lucent Technologies, and Paul Dadak,
Hewlett-Packard Company, with introduction by Gabriele Crognale, P.E*

> *"If you want to change a person's way of thinking, don't give him a lecture, give him a tool."*
>
> *Buckminster Fuller*

INTRODUCTION

Environmental audits, by their function, depict activities within an organization that are regulated by applicable federal, state and local laws, norms, and regulations. For multinational organizations, audits also address applicable laws, norms, and regulations of host countries, and any other applicable requirements against which the facilities can be audited. With the issuance of the first of the ISO 14000 environmental management series of standards in the fall of 1996, environmental audits take on additional significance in their expanded role as environmental management system (EMS) audits. EMS audits, in turn, are at the heart of certification procedures for a company or facility to conform to ISO 14001. Specifically, for those organizations that may desire to obtain third-party certification to ISO 14001, or may want to self-declare conformance to ISO 14001,[1] they may need to conduct an EMS audit as a prerequisite for certification.

Given these various driving forces, it should be transparent to the reader that environmental auditing has evolved within regulated organizations to become a useful management tool. Three such examples are provided later in this chapter. As such, audits can be utilized by organizations to help identify, address, and impart long-term corrections in regulatory and management system considerations. As a qualifier, audit surveys conducted of regulated companies depict environmental audits as useful EH&S management "tools."[2] This is also documented in the environmental reports of a number of regulated industries.[3]

In addition, a number of internal and external forces continue to drive auditing forward to generate greater usefulness within each industry group that utilizes EH&S auditing programs. The drivers include: opportunities for continuous improvement in EH&S programs; opportunities for preventing pollution, and saving raw materials and resources; eliminating regulatory enforcement actions; regulatory drivers, such as the Environmental Protection Agency's (EPA's) audit disclosure policy and the SEC's SAB 92 requirements, and market-driven incentives, such as ISO 14001 certification.

ENVIRONMENTAL AUDIT DRIVERS

To begin, what is an "environmental audit"? The EPA defines the term in the agency's 1986 audit policy on environmental auditing as ". . . a systematic, documented, periodic and objective review by regulated entities of facility operations and practices related to meeting environmental requirements."[4]

In our discussion about environmental audits, we also refer to environmental, health and safety (EH&S) audits, since most organizations conducting environmental audits usually perform them in conjunction with health and safety regulatory requirements. Hence, the term "environmental audit" in this chapter is interchangeable with health and safety. Audits, as an industry practice, have undergone an evolution of sorts since their inception in the early- to mid-1970s by companies such as AlliedSignal, then Allied Chemical, and Occidental Petroleum. These companies can trace the driver behind audits to several sources, among them, SEC requirements.[5] Since that time, environmental audit programs have matured as firmly established operational processes within a number of Fortune 500 and Standard and Poor 500 companies. This practice has also spun-off the creation of various professional groups that deal with EH&S auditing issues on a regular basis. Some of the groups with which the authors are familiar are listed in Appendix B.

For many of these companies, though, just meeting regulatory requirements is not their only goal. Corporate targets, goals, and objectives that focus on preventing pollution, product stewardship, conservation of raw materials and natural resources, reuse of materials, community awareness and outreach, *and* compliance assurance are among the considerations some companies evaluate in their EH&S audit program(s). Some companies conduct audits in a multi-tiered fashion as a means to facilitate continuous improvement, while others extol the merits of their EH&S auditing programs in their environmental reports.

In addition, the Chemical Manufacturers Association's (CMA) Responsible Care® Program requires that member companies adhere to the ten elements of Responsible Care®. One element consists of the six codes of management practices. These management practices are verified through a process called Management System Verification (MSV), another of the elements of Responsible Care®. While not exactly an EH&S audit, MSVs can provide additional insight to a chemical company's EH&S auditing program, because it delves into the heart of a chemical company's management system for procuring, producing, distributing, and disposing of its raw chemicals and waste products and looks for linkages between the various business units within a chemical company to see whether there is a common thread of Responsible Care® throughout the company. It is this aspect of Responsible Care® that bears some similarity to ISO 14001 that may provide additional linkages to the elements of the standard with respect to: environmental aspects and impacts, and training, awareness, and competence. We provide further insight into the CMA and ISO 14001 in Chapter 12.

With respect to EPA's audit disclosure policy, the structure and intent of that policy has also generated some controversy within the regulated community regarding EPA's position on audits, as described in the following EPA policies. These are: (1) "Voluntary Environmental Self-Policing and Self-Disclosure Interim Policy Statement"[6]; (2) its final version, "Incentives for Self-Policing; Discovery, Disclosure, Correction and Prevention of Violations,"[7] the final policy, which took effect on January 22, 1996; and (3) the "Audit Policy Interpretative Guidance,"[8] developed by EPA's Office of Regulatory Enforcement's "Quick Response Team" in January 1997 as a series of generic questions and answers to act as a further aid to both the government and the regulated community alike. These policies are EPA's way of saying to regulated companies that voluntary audit programs play a pivotal role in helping companies achieve their goals to comply with environmental rules and regulations.

In EPA's view, the final policy should greatly reduce and sometimes eliminate penalties for those companies that discover, disclose, and correct violations through their voluntary audits or other compliance management system, while keeping in focus EPA's overarching tenet to protect human health and the environment from the most serious violations.

Of note, the final policy also makes reference to the Price Waterhouse survey previously referenced that found over 90 percent of the respondents conduct environmental audits to find and correct violations before they are found by government inspectors.[9] With respect to the specific subject of audit disclosure, several viewpoints can be taken. For example, the survey references a number of companies that voluntarily disclosed their audit findings to see their findings used for enforcement purposes against them,[10] while EPA reports in its audit policy update newsletter[11] that a number of companies had voluntarily disclosed their violations to EPA under the audit policy and a certain portion of those companies had had their ensuing regulatory penalties waived as a result. Since there appear to be opposing camps with respect to audit disclosure, a company's decision to disclose or not disclose should be addressed on a case-by-case basis by counsel for each organization that may be weighing its options before making a decision.

Citing the Price Waterhouse survey, one can conclude that it can be in a company's best interest to follow through on audit recommendations, not just from a compliance perspective, but also from the standpoint of good business practices. The 90+ percent of the companies surveyed by Price Waterhouse realize the short- and long-term benefits of an environmental audit program, which includes improving a company's overall environmental management program.[12]

With respect to the other regulatory driver, the SEC, that agency's intent has been to ensure that companies more accurately report their environmental liabilities to stockholders in their annual reports. This was part of the push that the SEC imposed upon companies in the mid-1970s.[13] A follow-up by the SEC to ensure that companies more adequately address environmental liabilities was the SEC's issuance of SAB 92.[14]

WHAT WE CAN EXPECT FROM AUDIT PROGRAMS AS THEY CONTINUE TO EVOLVE?

Improving a company's overall environmental management program lies at the heart of a well-designed and managed environmental auditing program. Part of this improvement process includes "raising the bar" by a company's EH&S group with each subsequent audit, providing opportunities for continual improvement within an organization's business units. In this context, auditing takes on a larger focus than being just compliance-oriented and new areas to assess may emerge as opportunities to "raise the bar."

For example, we see "next generation" audits evolving to the point where supplemental management system techniques may be introduced within the audit process, such as Management System Verifications (MSVs), previously described. We also see auditing, through its continuous improvement, begin to assess operational and management system issues as described in ISO 14031, with environmental performance evaluation—evaluating a facility's environmental performance over a continual time period supplementing the audit's "snapshot in time" focus. This is not to say that ISO 14031 will become an auditable standard, rather, the intent here is for auditors to "look beyond the horizon" and utilize some of the "flavor" of ISO 14031 to consider in assessing areas to be audited—in essence, looking further at far-reaching considerations.

In a similar vein, we should also note that with a greater number of companies issuing voluntary environmental reports about their EH&S activities, compliance assurance, and liability considerations, going beyond adhering to the SEC's SAB 92 may already be a defacto way of doing business for these companies. As such, the "next generation" audits may even begin to assess liability disclosure among the areas to assess, or possibly assess a company's self-declaration in accordance with ISO 14021's environmental labeling, self-declaration environmental claim(s). Since ISO 14021 provides companies the opportunity to highlight their EH&S achievements in environmental reports as environmental claims,[15] these voluntary reports would be viewed as self-declaration vehicles under ISO 14021. A company's self-declaring its environmental performance in such reports, where a growing number are being posted on the Internet by the companies themselves, gives new meaning to the phrase, "Say what you do and do what you say."

With all these dynamic drivers in effect, environmental auditing programs may soon look at EH&S issues from a wider perspective than before and provide EH&S programs that exceed regulatory compliance, with the added benefit that their cost centers may self-sustain through monies saved by their activities. Activities focusing on: good business sense and continuous improvement, implementing P2 recommendations, introducing cost-saving programs for raw materials and resources, and appropriately training affected professional staff and workers in all of these activities, are among the ways "next generation" auditing programs can provide added value to an organization.

Equally as important, we should not overlook the added dimension that personal computers and network systems, such as LANs and WANs, can bring to this dynamic auditing process, as can specialty software programs designed to assist EH&S managers in their daily routines. The early software tools developed specifically for auditing have given way to software that takes into account wider-focused environmental management information systems, or EMISs, that can track, monitor, and assimilate data for different business units, not just EH&S. Many of these programs have become increasingly more sophisticated since their introduction in the late-1980s. We provide additional insight into this growing area in Chapter 13.

We should also point out that Chapter 9 is not intended to be a primer or introductory chapter about environmental auditing. Our focus, instead, is to take the essence of the environmental audit process to address current considerations and issues, and project where we may be headed with auditing as a management tool as it slowly matures. Our introduction serves to lay the foundation for auditing considerations and drivers, as well as provide a back-drop to our contributors who share their unique perspectives on auditing from an industry standpoint, and how each of their companies utilize audits as management tools. If readers are interested in pursuing audit techniques further, a number of books about auditing can be obtained that provide in-depth information in the subject area. Some of the books with which we are familiar are listed in the References at the end of this chapter. Additional reading sources are listed in Appendix F.

In this first section, Mitchell Gertz of the PQ Corporation provides his viewpoint on environmental auditing.

INTRODUCTION

This section of Chapter 9 provides a real-life description of an existing audit program at a medium-sized chemical manufacturing company. This company has operations in numerous states, Canada, Europe, South America, and the Pacific Rim. This section reviews the types of audits performed, legal considerations, auditing of suppliers, customers, toll manufacturers, TSD facilities, and due diligence and how the effectiveness of the audit is measured.

This section shows how and why decisions regarding audit protocol and methodology were made, essentially the philosophy of one company's audit program. While this model may not be appropriate for a particular company's culture it will provide a basis from which a company can develop or modify an audit program.

HISTORY OF THE ENVIRONMENTAL AUDIT

The history of the environmental audit goes back to the mid-1970s. Environmental regulations had been in place for a few years and a few leading-edge companies wanted to verify compliance. In the mid-1980s, a number of companies began to develop audit programs as a means of environmental management. There was and still is a concern over privilege and confidentiality. That concern was managed in numerous ways from no assertion of confidentiality to full attorney-client privilege.

Today hundreds of companies have some type of environmental audit program. The programs vary greatly depending upon the results for which the company is looking. Audits can be used for a multitude of purposes to satisfy the environmental management goals and can include:

- Compliance auditing
- Management systems
- Waste minimization
- Pollution prevention
- TSD audits
- Toll manufacturers
- Due diligence/real estate

Companies may do all of these or some combination of these. The audits can be done internally or through the assistance of third-parties, such as consultants.

There are pros and cons to using consultants.[16] Some of the advantages include completely nonbiased auditors, easier integration of attorney-client privilege, reduced need for internal resources, and audits performed by consultants may have more veracity with the regulatory agency. There are some disadvantages: The cost is usually higher, lack of specific process knowledge can lead to inaccurate conclusions, reports usually take longer to issue, cooperation at the plant level may not be as high as an internal audit. The decision to use external auditors is specific to your company's needs and culture.

TYPES OF AUDITS

There are two basic types of audits: compliance audits and management systems audits. Compliance audits account for most of the environmental audits done today. Management systems audits are relatively new and are just being initiated and include ISO 14000 and the Chemical Manufacturers Association (CMA) Responsible Care® program. In this section, the author reviews the purpose and the goals of each of these audit methodologies and how they are implemented at this mid-size chemical corporation.

COMPLIANCE AUDITS

The purpose of the compliance audit is to determine if a process or a facility is in compliance with applicable regulations and environmental permits. The primary goal for the regulated community is to be proactive. This means that situations that are out of compliance are fixed expeditiously and practices that could lead to a violation are corrected. This was the main driving force for implementing an audit program, particularly in the early years.

At our mid-sized chemical corporation, the formal audit program was instituted in 1989. Prior to that time, an informal program was in place. The older program was scrapped due to its inconsistency of application and lack of documentation.

This program was a significant change for this company. Documentation was going to be made concerning environmental compliance and this made people nervous. A formal protocol was developed covering scheduling to file retention. Confidentiality was a major concern. Plant managers and management were trained on what the audit was and a pilot audit was completed.

One of the concerns was the limits that resulted from just a compliance audit. The chemical company was concerned that potential liability issues were being overlooked as long as they were in compliance. So added to the audit were best management practices (BMP). These BMPs included items such as training, housekeeping, procedures, and awareness. In order to assure consistent application of BMPs, a baseline environmental document was prepared. This document essentially requires all operations to meet certain minimum environmental standards. With regulatory compliance and corporate environmental criteria defined, all facilities were on an equal basis. The audit team now had the tools to perform consistent audits.

For five years, the audit program as it was designed was effective. Areas of noncompliance were found and corrective actions were taken. Potential liability issues were identified and mitigated. Audit data was tabulated and the trend showed continuing improvement. As the program matured, improvement started to wane. Issues began to recur at subsequent audits. Change was needed.

The program was insular, and it was felt that an external review of the audit program was needed. The basic purpose of this review was to benchmark our program against other companies to determine if changes were necessary. A consultant was hired to review the program. Areas for improvement included:

- Having more than one person on the audit team. Even though the plants are small and highly automated, one auditor may miss some details.
- Develop a better system for tracking audit follow-up. This is important since the EPA considers follow-up documentation a critical part of the audit program.
- Reconsider the need for invoking attorney-client privilege.
- A goal of zero compliance issues was introduced.

As a result of this review, changes were made in the audit program. Except for the smallest plants, all audits are done by two people as a minimum. At least two audits include an outside consultant as a fresh pair of eyes on the operation. Follow-up documentation is now tracked by computer. Most important has been dropping of attorney-client privilege. In the previous five years, no one outside the company has asked to see an audit report. The findings have indicated very few significant issues of noncompliance. It was determined that invoking privilege for each audit was not worth the effort. In the event a significant environmental issue is observed, a separate document for that specific issue is prepared for counsel using attorney-client privilege.

One of the keys to the success of the audit program has been the development of the standards and protocols. These items provide a standard format for the audit. The process is predictable and does not vary from auditor to auditor. This is not to say that the process does not change. The process evolves but changes are consensual between operations, legal, and environmental affairs. All of this provides for a high degree of cooperation amongst all involved parties.

The process followed at the medium-sized chemical company is briefly described below:

- Audit schedule and audit team assignments are distributed in December for the upcoming year listing month selected for the audit.
- Plant managers are contacted at least three weeks prior to the month scheduled to agree to specific dates.
- A team meeting is held prior to the audit to review protocols.
- Upon arrival at the site, an entrance meeting is held. Audit protocols are reviewed. Introductions as necessary are made and the schedule is reviewed.
- An initial plant inspection is then taken with plant management.
- The process of documentation review begins.
- Interviews with plant personnel are conducted.
- A more detailed inspection of plant areas are made.
- Draft audit report is written.
- An exit meeting is held at which the draft audit is reviewed. A copy of the report is provided to the plant.
- A final report is written one to two weeks later. A copy is e-mailed to the plant manager. Other copies are distributed the following day.
- The plant manager submits a work plan to correct items identified in the audit.
- The plant manager submits quarterly updates until all items are complete.

The audit program is successful. The plants appreciate the attention and believe the process to be fair. Management benefits from reduced liability and exposure. Environmental improvements become continuous.

MANAGEMENT SYSTEM AUDITS

Management system audits (MSAs) differ from compliance audits. MSAs do not measure performance against criteria. MSAs determine if the infrastructure is there for effective environmental management. They typically do not measure how well you perform environmentally, but whether your systems are in place for environmental management.

The most widely known management system in environmental considerations is ISO 14001. ISO 14001 is the first of the ISO 14000 series of environmental management standards that were approved in the fall of 1996, and it includes supplementary standards that are still going through a review process. These standards are voluntary worldwide standards for environmental management systems. At this time, companies are beginning to look at the value of ISO 14001 certification and the cost to pursue it. Auditors are in the process of being trained. A very small number of companies have gone through the verification process.[17]

In the chemical industry, the program for improvement is known as Responsible Care®. Responsible Care® is a Chemical Manufacturers Association (CMA) initiative by its member companies to improve performance in the environmental health and safety area. This program was started in 1990 and implementation is a requirement of membership in the CMA.

The Responsible Care® program differs from ISO 14000 in some key areas:

- Responsible Care® includes health and safety; ISO 14000 does not.
- Each of the codes in Responsible Care® have metrics against which performance can be measured.
- Companies self-evaluate against each code annually.
- Companies must commit to a timetable for reaching full implementation of Responsible Care®.
- To be a member of CMA, a company must implement Responsible Care®.

The mid-size chemical company felt that supporting both systems was not justifiable. Therefore a decision had to be made. The company had all of their plants certified under ISO 9000 at great expense. That activity was driven by the fact that not having ISO 9000 could put a company at a competitive disadvantage with a certified manufacturer. Also, customers were requiring ISO 9000 certification in lieu of quality audits. However, since environmental management systems do not directly effect product quality, there was no perceived benefit of switching from Responsible Care® to ISO 14001, especially since five years of implementation were put into the Responsible Care® program. The decision was made that ISO 14001 certification would be sought only if requested by a major customer and on a plant-by-plant basis.

The company believed that a management system audit should be completed. To that end, they opted to undergo a Management Systems Verification (MSV) audit[18] as part of Responsible Care®. This audit will determine if the company has the management systems in place to implement Responsible Care®. The MSV audit is administered by CMA using outside contractors and peer reviewers. Six areas are reviewed:

- Pollution prevention
- Employee health and safety
- Process safety
- Community awareness and emergency response
- Distribution (transportation)
- Product stewardship

The MSV covers environmental, health, safety, transportation, community awareness, and life-cycle issues. This is a much broader scope than ISO 14000. A typical MSV audit takes five days and is applicable to the entire company, not just a specific site.

These audits result in a report that describes the environmental management systems. The report details whether the systems are adequate to meet environmental goals and points out areas where improvement is needed.

The management system audit is the next step beyond compliance auditing. If a facility is not in compliance with permits or regulations, it is very unlikely that any management systems are being utilized. Once a company reaches the stage where it is ready for a management systems audit, the company will be fairly sophisticated environmentally.

LEGAL ASPECTS OF AUDITING

In this section, the author discusses the legal aspects of auditing from a practical and operations perspective instead of the lawyers' perspective. The first question that companies ask is: Should I audit? To that, the answer is yes. Regardless of how you manage the legal aspects of auditing, knowing the issues and being proactive minimizes your legal exposure and makes you a corporation with a conscience. Also, auditing identifies risks and liabilities that, if not corrected, could lead to more serious problems. Unfortunately, the legal arena for auditing is more complex than before. With states passing audit privilege laws, EPA's audit policy, and attorney-client privilege requirements, companies can take numerous positions for protection of audit information.

The USEPA audit policy titled Incentives for "Self-Policing: Discovery, Disclosure, Correction, and Prevention of Violations" was issued in 1995. In January 1997 the EPA issued a document entitled "Audit Policy Interpretative Guidance." These documents set forth under what circumstances EPA will adjust penalties for violations downward and not seek criminal prosecution. A significant difference between EPA audit policy and state audit laws is that some states will not seek any enforcement action if a violation is reported and corrected in a timely fashion.

The EPA policy looks at the numerous criteria elements in determining the extent of any enforcement action. The enforcement activity is open to prosecutorial discretion. Also, it is likely to vary based upon the EPA region with which you are dealing.

The EPA has nine criteria it considers when looking at enforcement under the audit policy. They are briefly described below:

Audit/due diligence—The violation must have been discovered through a documented systematic procedure such as an environmental audit or other practice reflecting due diligence. The keys for identifying the violations are a systematic, objective, documented process.

Voluntary disclosure—The violation must have been identified voluntarily and not through a monitoring, sampling, or auditing procedure that is required by regulation, permit, or any variety of judicial decrees or orders of consent.

Prompt disclosure—The violation must be disclosed within ten days or less. The disclosures should be made to the USEPA.

Self-Discovery Disclosure—Discovery must not be the result of a regulatory agency inspection, notice of a citizen suit, legal third-party complaint, whistle blower, or an agency information request.

Prompt Correction—Corrective actions including any clean-ups must occur expeditiously. The company must certify to the agency that the action was completed. This certification must be submitted within sixty days. Written notice must be provided if the corrective action will take longer than sixty days.

Preventive measures—The company must agree to take steps to prevent recurrence. Those steps must be outlined.

No repeat violations—The same or similar violation cannot have occurred in the past three years or be part of a facility's pattern of noncompliance in the past five years.

Other violations excluded—Penalty reductions are not applicable to violations that resulted in serious harm or which presented a real and imminent danger to human health or the environment.

Cooperation—The company must provide information as requested by the EPA so that the applicability of the policy can be determined.

The EPA believes this policy to be successful. Over one hundred companies have disclosed violations at 350 facilities. EPA has settled matters with 40 percent of the companies and waived penalties in most cases. Industry still tends to be critical of the policy. Industry contends the policy is limited in its application, and individuals are not protected.

STATE AUDIT PRIVILEGE LAWS

Numerous states have passed or are developing audit privilege laws. These laws can vary significantly from EPA audit policy and that is a major issue with the EPA. In some cases, EPA has withdrawn or threatened to withdraw a state's authorization to administer a program or has refused to grant authority, particularly for permitting under the Clean Air Act.

The state laws have the same general requirements as the EPA policy such as prompt notification and prompt corrective action. However, they begin to vary in the penalty assessment area and applicability area. Some states allow for no penalties provided that the requirements of their specific law is met. EPA is concerned that violators may use audit privilege laws for protection from federal enforcement actions under the Clean Air Act (CAA). This area is very dynamic and unsettled. Many states have asked for congressional intervention to prevent EPA from usurping their authority. In Michigan, there is a bill to repeal the audit privilege law.

ATTORNEY-CLIENT PRIVILEGE

Attorney-client privilege is the other area of legal complication. Once they started performing audits, many companies were very concerned that EPA would request the information and use it against them. To protect themselves, companies used attorney-client privilege or attorney-client work product. Basically it worked one of two ways:

1. The company would have an audit performed, and the report would be sent to counsel for a legal opinion. The report would only make statements of fact and not draw any conclusions as to whether a finding was a violation. The report and the ensuing legal opinion would be marked as "Confidential—Attorney-Client Correspondence." Distribution of the report would be very limited.

2. The second method used would be a request from counsel to have an audit performed. The report would then be privileged as an attorney-client work product.

Both of these methods work, but they add to the administrative burden. Additionally, it gives the impression you have something to hide.

Given all of the above, our mid-sized chemical company was very concerned about how they would manage this issue. Initially they started out by addressing the audit report to counsel and labeling it as "Privileged." After five years of auditing and sixty audits, a report was never requested by the agency. Also, the violations typically noted were administrative in nature. Thus even if there was enforcement, it would be minimal. When the audit program was revised, the audit reports were no longer confidential. However, the company was still concerned about the potential for a significant compliance issue to occur. For that specific case, the company does not include the item in the audit. A separate memorandum is drafted to the legal department requesting their advice. That document is considered confidential.

With respect to using either federal audit policy or state privilege laws, the decision was made to avoid the complications that their use may entail. The company felt that the audit follow-up program of quarterly reports would adequately track the prompt corrective actions to any violations. Also since the severity of the violations was not great, it was felt that opening ourselves up to potential enforcement was inappropriate. It has been our experience that auditors typically find violations that agency inspectors may miss. It is the position of the company that all violations will be promptly fixed to preclude them from being uncovered as a result of an agency inspection.

AUDITING BEYOND THE FENCE LINE

A successful internal audit program is not the end. Environmental liability extends well beyond your fence line. Superfund liability is just one example of potential long-term liability with which most manufacturers are familiar. But there are others such as customers, suppliers, toll manufacturers, and real estate transaction. Any of these can result in significant exposure for a company. This section will describe what our medium-sized chemical company is doing to avoid liability in these areas.

SUPERFUND LIABILITY

Mention waste disposal to any large company and the concern of Superfund liability arises. Superfund is the way EPA gets companies to pay for cleaning up abandoned waste disposal sites. Costs for a typical site average $20 to $30 million. Even split between one hundred companies, that is a significant cost. Obviously the best way to avoid Superfund liability is to dispose of no hazardous chemical waste. That is not practical or possible. There will always be some disposal at landfills, incinerators, or treatment facilities. The only way to provide some level of protection is to have a waste disposal site audit program.

Our medium-sized chemical company requires that all disposal sites used for chemical residual waste or hazardous waste be approved. Approval requires a site visit, a review of their operations and technology, a review of the disposal company's financial status and a review of their regulatory compliance history. Companies are first visited, then a survey is completed, usually biannually, and is performed by the disposal site detailing any changes in ownership, operations, or regulatory status. This effort assures that only financially secure, regulatory compliant, and technically sound disposal facilities are used. While this will not protect us from liability at the sites we use, it limits the number of sites used. By picking the best disposal sites, long-term liability is reduced.

TOLL MANUFACTURERS

Toll manufacturers are typically small plants that make products for larger companies according to specifications and processes supplied by the customer. In many cases, significant engineering and processing advice is provided. In some extreme cases, an engineer may be resident at the toll manufacturer's site. Processing equipment may be provided by the customer. The reason tollers are used include a number of factors: There is no available manufacturing space at the plant; this is a new product/process and the company wants to see if the product/process is successful before significant capital expenditure; or the process can be too hazardous for the location.

Depending on the level of involvement in the toller's process, liability for environmental exposures can be significant. If the toller just makes product according to a specification then liability is low. However, if significant control of the process is maintained by the customer the environmental liability can be the same as if they manufacture the product themselves.

As part of our product stewardship program in place, our mid-sized chemical company felt that toll manufacturers should be, at a minimum, in compliance with environmental rules and permits. In order to determine whether or not the tollers were in compliance, an audit program was started.

The audit program is a three-step process. First the toller provides information on their process and environmental issues to the customer. A site visit and audit are then conducted. Based upon the results, a report is written and continued use of the toller is dependent upon their meeting environmental requirements. Aside from environmental liability, there is also the concern that business might be interrupted if a plant is forced to shut down due to violations of environmental regulations.

CUSTOMER/SUPPLIERS

Environmental qualification of suppliers is a relatively new area for environmental management. From the company perspective, we want to ensure an uninterrupted flow of raw materials from the supplier. Typically, a company will have no environmental liability at a supplier. Since most suppliers are not open to a physical environmental audit, a different method of compliance determination had to be developed. An environmental questionnaire is being considered and/or a certification statement that the operation is in compliance or has a plan in place to reach compliance. Our mid-sized chemical company has not yet reached a decision in this area. The environmental question will become part of the supplier audit program and will be integral in determining whether a supplier is utilized.

Environmental issues with customers have some basic drivers. First, customers want environmental information about your product so they can comply with applicable laws and regulations. Secondly, sales and marketing will want to use environmental issues as a tool to increase sales or gain a competitive edge. A third driver that is overlooked is environmental liability associated with the use or disposal of your product by the customer.

One likely scenario is that a customer buys your product, determines the customer does not need it and illegally dumps it on the side of the road. Your product with your label is found the next day, and you pay the bill. Or the end-use of your product is not compatible with the product and results in environmental damage. There is potential you could be liable for damages in a product liability suit. Good product stewardship requires that you understand how your product is going to be used. This information is critical to assessing the potential liability.

Environmental audits of customers are unlikely. A method to understand the end-use of your product and to assure that it is appropriate to the customer's needs. The sales force should be utilized when dealing with customers. The sales force should be familiar with the customer so that they understand the end-use. Also they should act as a conduit for communications on environmental issues.

Measuring Audit Program Success

Measuring the effectiveness of the audit program in some manner is important. Management's position is that if something can't be measured, you can't determine if there is improvement. Unfortunately environmental audit results are difficult to quantify. Environmental considerations, unlike safety, do not have a quantitative measure. In safety, you can measure injury frequency (such as with OSHA 200 logs). That parameter is widely used to demonstrate improvement. The problem with environmental measurements is that there are numerous measures and they vary based upon the industry.

Measuring compliance quantitatively is difficult because:

- Audits are not done on an annual basis and are not always multimedia.
- The regulatory requirements and the plants are very dynamic so year-to-year comparisons are not always valid.

- Regulatory agency inspections are infrequent and generally cover only a single media.
- Compliance is not measured on a unit basis (i.e., compliance/lb of product).

There are quantitative measures that can be used. You could develop a listing of all the environmental compliance items for a plant and determine a percentage of compliance. This would be an extraordinary effort even at a small plant. However, if this is done it does take into account any regulatory changes. This type of system requires frequent updating.

You can compare the number of findings on a year-to-year basis. This is acceptable only for developing overall trends. Its use is limited by the changes in regulations or plant operations. Other quantitative measures are more performance-related and include waste and emission reduction data, such as reviewing toxic release inventory (TRI) data from one year to another. Unfortunately these are not indicators of compliance performance. Measuring enforcement actions is a poor choice since they occur after the fact and a lot of times not at all.

In order to quantify compliance, our mid-sized chemical company reviews the history of each location to develop the trend line. The company uses a tiered criteria based upon the severity of the violations and the number of violations noted. Depending on the severity, violations are scored on a 1 to 3 scale using the definitions below:

Score	Definition
1	Poor recordkeeping, insufficient documentation, incorrectly completed reports, incomplete plans
2	Failure to train or inspect, not having required plans, occasional permit violations
3	Lack of permits, unreported spills, illegal disposal, falsification of data, recurring permit excursions

The number of violations is multiplied by the severity and a composite score is calculated. This data point can then be compared to previous years to track the trend.

Severity can be subjective. Even minor infractions become more serious when there are numerous violations. Numerous violations indicate a lack of any environmental program. Others may consider some infractions more or less serious. If the plant has more than five violations in categories 1 or 2, then the severity factor increases by 0.5. If a plant has more than two violations in category three the severity factor goes to 4. The company goal is a score of 0. The important factor in using this system is consistency. If the severity ranking changes, then the data must be recalculated or it becomes meaningless.

Using this system, the effectiveness of the audit can be measured along with determination of relative compliance. The historical trends can show whether a plant is improving in its compliance performance. Management has a tool to use so resources can be focused where needed.

BENCHMARKING THE AUDIT PROGRAM

Being able to measure the results of the audit is important. But it is also necessary to determine if the audit program is adequate. The best way to do this is to benchmark your program against other companies. This is best done by an outside consultant who should be completely impartial. The consultant should have experience performing these types of reviews.

The benefits of the benchmarking exercise are:

- Areas for improvement are identified
- A sense of how your program rates compared to other companies
- The audit program gets reenergized
- Plants respect that the audit process changes and improves

The audit program should be reviewed every five years at a minimum.

SUMMARY

Audit programs are a management tool. They benefit the company by reducing their legal exposure on a proactive versus a reactive level. They can be used to measure environmental performance. The use of audits is not limited to the operating plant but can extend beyond the fence line to a variety of entities in the business chain. There are many ways in which to implement an audit program. Legal and technical issues can be molded to fit your company culture and needs. If you are a manufacturing business and do not have an effective audit program, you are at risk of enforcement and liability.

In the following section, Dave Jeffries and Tom Harding of Lucent Technologies provide readers a glimpse into a step-by-step audit description as conducted at their company that provides much insight into the audit process.

LUCENT TECHNOLOGIES: CORPORATE GOALS

Lucent Technologies conducts environmental, health, and safety (EH&S) compliance audits of all company-owned facilities in order to identify potential areas of noncompliance with legal regulatory requirements (national, state, or local) and Lucent company requirements. Good management practices are also considered. When areas of noncompliance are identified, facilities are required to take corrective action in a timely manner. Periodic progress reports are requested by the Corporate Compliance Assessment Department, in order to follow the progress being made and to identify facilities that are falling behind their corrective action plan schedules and may require outside assistance. In addition, an assessment of the management system in place at each facility for EH&S programs is conducted. The overall objective of the program is to foster continual improvement of the EH&S programs as well as to minimize risks to employees, the communities in which we do business and to the environment, and by extension, to all stake holders. In many facilities, Total Quality Management (TQM) processes have been employed either to implement required programs or to evaluate necessary improvements and to direct fulfillment of commitments made in the corrective action plans.

INTRODUCTION

In the wake of the Bhopal, India, accident at a Union Carbide facility, most major U.S. companies became aware of the risks, both to their employees and communities in which they operate as a result of potential environmental incidents. In addition, the impact on the business from a financial, as well as public relations, standpoint became a focus of attention. The AT&T Compliance Assessment Program was initiated in 1985 at the request of the chairman of the board at AT&T, James Olson. The chief counsel of the corporation was requested to provide advice to Olson regarding the status of the EH&S compliance at the corporation's manufacturing facilities. The chief counsel enlisted the assistance of the corporate environmental engineering department to conduct these evaluations.

In the early years, the process was extremely cumbersome. Opening meetings had to be attended by legal counsel, the manufacturing vice-president (facility general manager) and the environment, health and safety engineering vice-president. Trying to coordinate schedules of these individuals was difficult at best and more than one audit had to be postponed at the last minute due to a scheduling conflict of one of these individuals. However, with time, the need to involve certain of these players was eliminated and the program began to function on a more regular basis. Because of the concern that something would be overlooked which could result in an incident, the audits were extremely intensive, requiring up to eight weeks to complete. During the course of any audit, from six to seven auditors would participate and would each spend up to two weeks on-site. The auditors were endeavoring to perform the "perfect audit." The reports generated often ran in excess of two hundred pages and action plans could contain in excess of two hundred action items. The average time required to complete a "draft" report and submit it to the facility was approximately 250 days. Clearly, this process was inefficient and, in most cases, the reports were never fully read or understood.

Over the first four or five years of the program, only four or five audits were performed each year. In 1992, it was recognized that this level of efficiency was unacceptable and that significant improvements were necessary. It was decided that the existing corporate staff was insufficient to accomplish the goals and a consultant was retained to assist in the performance of the auditing function. Figure 9–1 shows the audit history for the program both before and after the significant improvements were implemented.

In September 1995, AT&T announced a major restructuring of the corporation, and in 1996, Lucent Technologies was created. It should be noted that the number of audits has varied over the past few years due to the uncertainties related to the restructuring and other managerial decisions.

Lucent Technologies consists of fourteen business groups (BGs), focusing on specific areas of the telecommunications equipment market. BGs include Bell Laboratories (research and development), Network Products (copper and fiber-

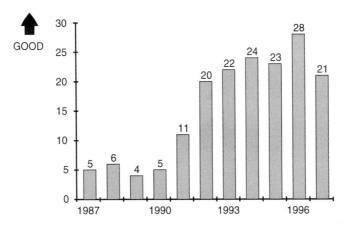

Figure 9–1   Number of Audits per Year

optic cable and apparatus), Switching and Access Systems (switching and network access products), Wireless Networks (wireless networks), Microelectronics (integrated circuits and optoelectronic interface systems), Global Service Provider (global marketing and sales), Data Networking, Business Communication Systems (voice-related products for business applications), and various service organizations. The BGs focus on customers, markets, and competitors. The company currently consists of approximately one hundred manufacturing, assembly, distribution, and research and development facilities located in the United States and in eighteen additional countries.

PROGRAM DESCRIPTION

Audits are conducted by the compliance assessment department, a part of Lucent's global environmental health and safety (GEH&S) organization. Audits are initiated at the request of the law department and the results of each audit are considered privileged and confidential. The law department supports the audit team and reviews all documents generated as a result of any audit, but does not participate in the day-to-day activities of an audit.

The audit program for Lucent Technologies consists of two similar but separate programs, one covering manufacturing-related facilities (including factories, large distribution facilities, repair facilities, and research and development centers) and one addressing concerns with administrative facilities (including office and administrative facilities and installation operations). Both programs integrate EH&S issues into a single audit.

In the early manufacturing facility audit program, all audits were conducted by a team consisting of environmental subject matter experts who were members of a permanent audit team. Health and safety issues were addressed by a representative of a separate department within the corporation working under the direction of the audit coordinator. Fire protection and life safety audits and Occupational Safety and Health Administration (OSHA) record keeping audits were performed on an independent schedule by other departments within the corporate organization. In 1992, when it became apparent that additional resources were necessary to increase productivity, an outside consultant was retained to perform some of the audits. Initially, the consultant team performed its audits under the direction of a coordinator representing the compliance assessment department. The coordinator acted as a liaison between the audit team and the facility and did not perform any technical audit function. The remaining members of the department functioned as a separate team and continued to perform audits, generally of facilities engaged in business-sensitive activities, including research and development facilities and facilities requiring government clearances. The program has evolved to the point where there is no differentiation between the corporate and consultant representatives, and audit teams are now completely integrated and staffed with the best available personnel, regardless of source. Health, safety, and fire protection subject matter experts from other departments within the GEH&S organization are used interchangeably with members of the consultant's staff. In addition, the "coordinator" now functions as a technical auditor as well, reducing the team size to a more manageable level.

The administrative audit program followed a somewhat similar track with some minor differences. The original program used members of regional environment and safety organizations, working independently of the compliance assessment department. These auditors had additional responsibilities beyond auditing. In many cases, the auditors were also responsible for developing and implementing corrective actions that they identified. With the restructuring of the corporation in 1996, all auditing functions were consolidated under the compliance assessment department. At that time, a separate consultant was retained to conduct the administrative facility audits under the direction of a coordinator from the department. The consultant provides all subject matter expertise for the audit team.

SCHEDULING THE AUDIT

Lucent Technologies currently uses a model of auditing all manufacturing and research and development facilities on a three-year cycle. However, a tiered scheduling tool, taking into consideration business importance and operational risks, is being developed and is being introduced into the scheduling of manufacturing facility audits. Service and administrative facilities are audited approximately every five years. In addition to considering risk, the audit schedule is structured to take advantage of geographical and regional proximity and climatological considerations. The proposed audit schedule is made available to business group and regional EH&S officers. These individuals generally notify the affected facilities of the proposed audits and in many cases act as a negotiator with the audit team for schedule revisions and postponements on behalf of the facilities.

PRE-AUDIT PHASE

During the pre-audit phase, formal audit correspondence (announcement letter, opening letter and pre-audit questionnaire, audit plan) is communicated to the facility. This correspondence is communicated to the facility manager and facility EH&S manager sufficiently in advance of the audit to make adequate preparations.

COMPLIANCE AUDIT COORDINATOR

The compliance audit coordinator manages all phases of the audit process and is responsible for preparing, implementing, and completing the audit. The audit coordinator, in addition to meeting the requirements of auditor, has a thorough understanding and the personal attributes and skills necessary to ensure effective and efficient management and leadership of the compliance audit process. The audit coordinator is responsible for ensuring efficient and effective conduct and completion of the audit within the audit scope. Audit coordinator activities and responsibilities include:

- Determining scope of the audit and preparing the announcement correspondence and audit plan;
- Obtaining relevant background information necessary to meet audit objectives;
- Forming the audit team;
- Directing activities in accordance with audit procedures;
- Distributing all appropriate working documents;
- Resolving problems that arise during the audit;
- Notifying the facility of significant nonconformances as they arise; and
- Preparing and distributing the audit report.

AUDIT ANNOUNCEMENT CORRESPONDENCE

Approximately four months prior to a scheduled audit an initial letter, referred to as the "announcement letter," is sent to the facility general manager announcing the scheduled audit. This letter is sent by the compliance assessment department manager. Earlier in the year facilities are advised they are scheduled for an audit and given an approximate audit date. The purpose of the announcement letter is to advise the facility that preliminary preparations are underway to conduct the audit. Prior to issuing the letter, the audit coordinator determines who the facility EH&S contact is and confirms the audit dates.

Three months prior to the audit, a second letter, referred to as the "opening letter," is sent to the facility general manager and the EH&S engineering manager from the corporate counsel of the Lucent Technologies law department, once again confirming the scheduled audit, including specific audit dates, and instructing the facility that all audit correspondence, including the report, is "privileged and confidential." Before the opening letter is sent, the audit coordinator contacts the facility and arranges for an opening conference time, which is referenced in the letter. This letter has the pre-audit questionnaire as an attachment.

PRE-AUDIT QUESTIONNAIRE

The pre-audit questionnaire is a twelve-page document that requests general information about the facility along with specific information about air and water pollution control, solid and hazardous waste management, bulk storage tanks, Toxic Substances Control Act (TSCA), Comprehensive Environmental Response, Compensation, and Liability Act (CERCLA or Superfund), and the Superfund Amendments and Reauthorization Act (SARA), radiation, regulated materials, industrial hygiene, safety, transportation issues, and drinking water. An alternate pre-audit questionnaire is used for international audits that has similar questions, without reference to United States regulatory requirements.

The purpose of the pre-audit questionnaire is to familiarize the audit team with the EH&S status of the facility and indicate to the facility which EH&S areas will be audited by the team once on-site. It is essential that the pre-audit questionnaire be completed and returned to the compliance assessment department sufficiently in advance of the audit to be used in developing the audit plan.

PRE-AUDIT QUESTIONNAIRE ATTACHMENT

Along with the pre-audit questionnaire is an attachment that requests that the facility make available during the audit all documents related to EH&S activities, including but not limited to:

- organizational charts;
- all applicable procedures and programs;
- local regulations;
- training records;
- permits and licenses;
- inspection reports;
- emergency, spill, and waste minimization plans;
- monitoring data;
- incident reports;
- facility drawings; and
- any other documents which could impact the EH&S program

EH&S COMPLIANCE AUDIT PROTOCOLS

Compliance audit protocols are detailed checklists that document regulatory compliance and Lucent Technologies requirements by EH&S compliance areas. Auditors responsible for specific compliance areas must familiarize themselves with the requirements documented in the protocols and be able to, depending upon audit experience, either fill out or reference the protocol during the course of the audit. The following protocols are to be used during the audit:

- general facility information;
- air quality;
- emergency planning and release reporting;
- Toxic Substance Control Act (TSCA);
- solid and hazardous waste;
- above-ground and underground storage tanks;
- PCBs;
- water quality;
- industrial hygiene;
- safety;
- hazardous materials transportation;
- motor carrier safety;
- fire protection;
- chemical handling;
- OSHA;
- radiation protection; and
- ergonomics

REGULATION REVIEW

After receiving the facility's completed pre-audit questionnaire, the audit coordinator obtains the applicable federal, state, and local regulations and distributes the appropriate regulations to the audit team for review. Current federal and state regulations are available on CD-ROM and made available to the audit team prior to, as well as during, the course of the audit. Local regulations are obtained either through local regulatory agencies or from the facility, upon request. Every effort is made to obtain comprehensive regulations before the audit to ensure compliance with the latest legal requirements. International audits require regulations applicable to the host country, although some countries have few regulatory requirements. International regulations are obtained through sources such as the World Environment Center (WEC), International Audit Protocol Consortium (IAPC), and networking with other companies and consultants who have experience or operations in foreign countries.

AUDIT PLAN

Approximately four weeks before the audit, the audit coordinator sends the audit plan to the facility. The audit plan is based on the information provided by the facility in the pre-audit questionnaire. The purpose of the audit plan is to describe the audit process and provide the facility with the following information:

- audit objectives and scope;
- audit criteria;
- time and duration for major audit activities;
- dates and places where the audit is to be conducted;
- audit team members;

- audit schedule;
- confidentiality requirements;
- identification of the organizational and functional units to be audited;
- identification of the functions and/or individuals within the audited organization having significant direct responsibilities regarding the compliance program;
- identification of compliance areas to be audited and the procedures that will be followed to complete the audit;
- reference documents;
- report content and format, and expected date of issue and distribution of the audit report;
- action item follow-up process;
- assignment of responsible person and expected due date at closing conference;
- audit debriefing requirements; and
- document retention requirements

AUDITOR QUALIFICATIONS

Auditors selected for the audit team are required to have the appropriate experience necessary to conduct a thorough audit. An auditor should have completed at least a college education or the equivalent and should have verbal and oral communication skills as well as interpersonal and analytical skills. Auditors should have appropriate work experience that contributes to the development of skills and understanding of environmental science or technology, environmental aspects of facility operations, relevant requirements of environmental laws and regulations, environmental management systems, and auditing procedures. An auditor is expected to maintain current training by enrolling in appropriate training classes annually, under the direction of the Lucent Technologies compliance assessment department.

An auditor has the responsibility to:

- follow the instructions and guidance of the audit coordinator;
- plan and carry out the assigned tasks objectively, effectively, and efficiently within the scope of the audit;
- collect and analyze relevant and sufficient audit evidence to determine audit findings;
- prepare working documents and document findings appropriate to their designated tasks; and
- assist with the preparation of the audit report and action plan.

TEAM DEVELOPMENT AND AUDIT ASSIGNMENTS

Audit team members are provided with the applicable legal requirements, Lucent Technologies Requirements, audit protocols and compliance program area rating sheets prior to conducting the audit. International audit protocols, which pertain to the country in which the facility is located, are obtained through various sources (WEC, International Audit Protocol Consortium) as well as by contracting for an in-country expert who can speak the appropriate language, act as interpreter, and is familiar with the local regulatory requirements. The process for selecting the audit team should ensure that the team possesses the experience and expertise necessary to conduct the audit. The compliance assessment manager, in conjunction with the consultant's project manager and the audit coordinator, determine the appropriate team size, experience, and knowledge applicable to the size and complexity of the facility. Team assignments and the audit schedule, documented in the audit plan, are determined prior to the audit. However, unforeseen circumstances at the facility may dictate schedule or team assignment changes. The audit team must have the flexibility to both meet the requirements of the audit and respond to the needs of the facility. Consideration should be given to:

- auditor qualifications;
- the facility being audited;
- language requirements; and
- the requirements of the state or country to be audited.

FACILITY COORDINATION

The audit coordinator obtains directions to the facility and makes arrangements for a conference room, phone, fax, copier, and computer, if necessary. Lodging, flights, rentals, and any other travel and living arrangements are made with the audit team. The facility should be contacted for suggestions or recommendations. The audit coordinator should also coordinate with the GEH&S liability assessment and management department to determine if there are remediation activities or any other EH&S activities currently underway at the facility to be audited.

Approximately four weeks before the audit, the audit coordinator obtains any previous reports and audit ratings (if applicable) and distributes these, local/domestic or international regulations, domestic or international audit protocols, the pre-audit questionnaire and audit plan, and any other pertinent auditing materials to the audit team. The coordinator prepares the opening and closing conference materials, the post-audit evaluation questionnaire and makes any final arrangements with the facility.

AUDIT PHASE

The audit phase, which consists of an opening conference, on-site audit activities, and a closing conference, generally lasts between three and five days, depending upon the size of the facility. The opening conference serves to communicate audit objectives, needs, and requirements to the facility participants. On-site audit activities involve collecting sufficient evidence by the audit team to evaluate facility compliance with applicable legal regulatory and Lucent Technologies requirements. Evidence is collected through interviews, examination of documents, and observation of activities and conditions. An overview of the facility's EH&S management systems is also conducted and includes interviews with facility top management and the performance of a root cause analysis of significant compliance findings. Communicating the findings occurs continually during the audit phase in order to ensure that facility personnel clearly understand and acknowledge the factual basis of the issues identified. A closing conference, at the conclusion of the audit, serves to formally communicate findings and convey the root cause analysis.

OPENING CONFERENCE

The opening conference serves to communicate audit objectives, needs and requirements to the facility participants, which includes the facility EH&S organization and top management. The opening conference is conducted in a rather informal manner and lasts approximately one hour. Each opening conference may vary in scope and topics generally presented. However, the following topics are covered during the opening conference:

- introductions and attendance;
- facility presentation;
- why do EH&S compliance auditing?;
- purpose of audit;
- audit topics and regulations;
- facility staff responsibility;
- auditor responsibilities;
- distribute schedule and discuss scheduling for the week;
- discuss briefing meetings; and
- estimated time of closing conference

WIDE-AISLE TOUR

At the conclusion of the opening conference, a wide-aisle tour is scheduled. The purpose of the tour is to take the audit team on a brief visit to the various manufacturing and nonmanufacturing areas of the facility and to understand where they are located and what processes occur at each location. The individual members of the audit team return to these locations during the course of the audit to make observations and/or conduct interviews. If the facility is small or has a limited EH&S staff, the wide-aisle tour can become the venue for the actual audit process. In such cases, the tour will take a longer period of time than normally scheduled.

INTERVIEW SCHEDULING

Upon return from the wide-aisle tour, the audit coordinator arranges appointments with the members of the audit team and the facility personnel. Scheduling is arranged in such a way as to allow the audit team members the appropriate amount of time to conduct their auditing and also not inconvenience the facility personnel any more than necessary. Typically interviews are scheduled with the facility manager, EH&S personnel, manufacturing managers, production personnel and any other individual who has the ability to influence the EH&S compliance program at the facility.

INTERVIEWS AND DATA GATHERING

The interview and data-gathering segment of the audit involves collecting evidence by the audit team to evaluate facility compliance with applicable EH&S regulatory requirements and Lucent Technologies requirements, which encompass Lucent Technologies Worldwide standards, goals, and policy. Evidence is collected through interviews, examination of documents, and observation of activities and conditions. Each audit team member selects the appropriate areas of the facility to audit and the appropriate candidates for interview. The auditors efficiently schedule time in order to complete the interviews and review the audit protocols pertaining to their subject matter expertise. Action items are prepared sufficiently in advance of the closing conference to allow the audit coordinator time to prepare the action plan.

The audit coordinator observes and accompanies the audit team members as much as possible in order to understand legal and Lucent requirement issues at the facility and to ensure that team members are thoroughly auditing their individual areas of responsibility.

MANAGEMENT SYSTEM AND CORPORATE GOALS REVIEW

An evaluation of the facility's EH&S management systems is conducted which evaluates the facility's ability to support compliance with legal and Lucent requirements on an ongoing basis. This evaluation includes interviewing facility top management and focuses on the EH&S structure at the facility. This, however, is not an evaluation of the facility's environmental management system against a recognized standard [British Standards Institute (BSI) or International Standards Organization (ISO)]. The audit coordinator usually conducts the management systems interview with the primary goal of determining if the EH&S programs at the facility are getting adequate management support. Management systems strengths or weaknesses in different areas or programs are evaluated and additional details or specific examples are provided to management to assist in identifying specific areas to improve the current EH&S management systems. Advance feedback on the progress of the audit can also be relayed to top management.

In addition to the management systems review, the audit coordinator also assesses the facility's progress toward achieving Lucent Technologies' corporate EH&S goals. The facility's progress toward achieving the Lucent corporate goals in the areas of environmental management systems, Design for Environment (DfE) criteria, energy efficiency, and wastepaper recycling are discussed.

FACILITY BRIEFINGS

With the exception of the first day of auditing, briefing sessions are held with facility personnel on all days of the audit. The purpose of the meetings is to communicate the daily findings to the facility management and personnel in order to ensure that they clearly understand and acknowledge the factual basis of the findings identified, and to provide the facility an opportunity to respond or ask questions about the findings. The meetings can also be used to notify the facility of any last-minute audit schedule changes. The briefing sessions are held early in the morning and provide a forum for appraising the facility of findings from the previous day and allow for final communication of the audit areas to be covered during that day.

ROOT CAUSE ANALYSIS

Prior to the closing conference, the audit team meets to discuss the significant compliance audit findings and determine if a root cause analysis is appropriate given the number and type of action items found. Root cause analysis is a technique that identifies underlying causes of compliance findings and may point to systemic issues in the overall management system.

The Lucent Technologies procedure for identifying root cause is to first categorize each significant finding into one of the following six broad categories:

- knowledge of the issue;
- develop or modify procedures;
- implement procedures;
- train personnel;
- document results; and
- oversee execution

Once findings have been assigned to the six categories, a determination can be made as to which category or categories contain the most findings. This approach systematizes the determination process somewhat; however, the category selection should be thoroughly discussed by the team to ensure an appropriate root cause is discovered. Although the category can be determined (train personnel), the true underlying root cause could be interrelated (i.e., lack of funding). The audit coordinator is responsible for directing the root cause analysis and ensuring an accurate root cause is determined. The root cause, or causes, is presented at the closing conference.

ACTION PLAN PREPARATION

Prior to the closing conference, a draft action plan is prepared by the audit coordinator. With input from the audit team members, the action plan lists all the findings segregated into the following six categories;

- exemplary findings (if appropriate);
- legal requirements;
- Lucent Technologies requirements;
- future requirements;

- good management practices; and
- corporate actions (if appropriate).

The legal requirements, Lucent Technologies requirements, and future requirements findings are followed by a regulatory citation, a comment section, responsible person notation, expected completion date notation, and actual completion date notation. For future requirements, the expected completion date is the effective date of the regulatory or company requirement. The good management practice findings are only followed by a comment section. Good management practices are not regulatory in nature and are only suggestions for improvement that the facility is not required to correct and are not part of the audit follow-up process. At the time of the closing conference, only the finding and the citation are completed in the action plan. The draft action plan is marked "Lucent Technologies Proprietary."

ACTION ITEM PRIORITIZATION

An audit action item prioritization is employed as a tool that assists the facility in prioritizing action items to be addressed. The priority in which action items must be addressed is directly related to the potential to negatively impact the environment and health and safety. The analysis is inherently subjective as it reflects the auditors' best judgment. Lucent Technologies deems all regulatory items to be significant and is not, through this tool, attempting to convey any judgment on the importance of each action item. Each action item in the legal requirements and Lucent Technologies requirements sections of the action plan have a priority level associated with the action item. These levels are:

- Level A—Critical—Immediate Action Required: This action item has the most risk and must be corrected at once.
- Level B—High—Priority Action Required: This action item is high risk and must take precedence over all other action items, except those which require immediate action.
- Level C—Moderate—Action Required: This action item has moderate risk and must be corrected as soon as possible.
- Level D—Low—Long-Term Action: This action item is low risk and can be addressed in the long term or after all other action items have been corrected.

CLOSING CONFERENCE

The closing conference serves to communicate findings formally and convey the root cause analysis. Each closing conference may vary in scope; however, the closing conference offers an opportunity for the facility to ask questions and clarify any issues that may have to be resolved. The closing conference should be limited to one hour and should include:

- facility thank you;
- discussion of the most significant findings;
- discussion of root cause analysis (if necessary);

- description of EH&S audit report format;
- distribution of report;
- report and action plan follow-up schedule;
- explanation of the audit follow-up process;
- distribution of the draft action plan;
- passing out of feedback questionnaire; and
- provision by facility of responsible individual and expected completion date for action items

ACTION PLAN INPUT

Directly upon conclusion of the closing conference, audit team members and facility personnel meet to discuss the action plan in detail, agree upon its content, and obtain individual responsibility and expected completion dates for resolving each legal regulatory and Lucent Technologies requirement action item. This input is subsequently added to the responsible person and expected completion date sections under the findings in the action plan. The draft action plan is reviewed internally and revised to include the above information, along with any possible corrections, and is issued in final format along with the audit report.

PROGRAM COMPLIANCE RATING

At the conclusion of the audit, the EH&S program areas are rated. Each compliance program area is rated to reflect the facility's compliance with legal and Lucent Technologies requirements. The purpose of the compliance rating is to baseline facility compliance performance at the conclusion of the first audit and track improvement in compliance performance after subsequent audits. Factors considered in the rating include inherent environment, health and safety risks and liabilities, and the degree of compliance program implementation on a facility-wide basis. The ratings are used to reflect the best judgment of the auditor based on the audit findings at that time and the result of interviews conducted with individuals during the audit. The following are the six rating possibilities and definitions for each:

- Numerical rating of 5: The facility *meets* all regulatory and Lucent compliance requirements, and full compliance is readily substantiated.
- Numerical rating of 4: The facility *substantially meets* regulatory and Lucent compliance requirements; all but a few less significant compliance items are satisfied and this high degree of compliance is readily substantiated.
- Numerical rating of 3: The facility *generally meets* regulatory and Lucent compliance requirements; multiple compliance items are not always satisfied and/or compliance cannot be readily substantiated.
- Numerical rating of 2: The facility *requires improvement* to meet regulatory and Lucent compliance requirements; multiple exceptions to applicable requirements were noted, some of which may represent significant issues.
- Numerical rating of 1: The facility *requires substantial improvement* to meet regulatory and Lucent compliance requirements; numerous applicable requirements are not satisfied and significant compliance issues exist.
- Not applicable (N/A): Program area does not apply.

The ratings, provided by each audit team member, are recorded on rating sheets and provided to the audit coordinator at the conclusion of the audit. The area ratings for each program are averaged and reported in a gap analysis report.

POST-AUDIT PHASE

The post-audit phase involves preparing and distributing an audit report and action plan prepared under the direction of the coordinator. The audit report discusses the status of the facility's regulatory compliance and summarizes audit findings with reference to supporting evidence, if needed. The audit report is marked "Privileged and Confidential, Prepared for the use of Counsel." The action plan outlines steps the facility needs to take to close compliance performance gaps. The action plan also serves as the input to the compliance assessment department's follow-up process. The action plan is marked "Lucent Technologies Proprietary."

REPORTING

When the audit team returns from an audit, a written report is prepared. Each auditor prepares his/her sections in accordance with a predetermined report format. The sections are submitted within one week following the closing conference of the audit to the coordinator who assembles the entire report. Once assembled, all reports undergo a series of peer reviews, at the appropriate consultant's offices, within the corporate EH&S organization, and from the law department, regardless of the audit-team makeup. This process helps ensure a level of consistency in all the reports. In the past, draft copies of the report were submitted to the EH&S manager of the audited facility for review. This process did not yield substantive comments and tended to extend the time needed to issue the final report. As a result, the process of soliciting comments from the audited facilities on the reports has been discontinued.

The report consists of an executive summary that provides an overview of the audit and identifies the most significant issues identified. The gap analysis report generated by the compliance rating system is included in the executive summary. The remainder of the report consists of an introduction, a facility profile, and individual sections addressing all of the technical elements of the audit. The report addresses observations made by the auditors and provides suggestions for corrective actions. The action plan, identifying all noncompliance issues, is included at the end of the report. The entire document is bound to discourage reproduction since the main body of the report is marked "privileged and confidential." The action plans are marked "proprietary." An unbound copy of the action plan is sent to the facility EH&S manager to permit reproduction and distribution on a need-to-know basis.

The report is prepared and distributed to the facility, business group (BG) management, and managers within the corporate EH&S organization within four weeks of the conclusion of the audit. Within two weeks of the issuance of the report, a teleconference debriefing is conducted with the facility and BG management. The debriefing is scheduled by a representative of the compliance assessment department and includes the following individuals:

- compliance assessment department manager;
- lead auditor/coordinator;
- law department representative;
- Operating Unit (OU) EH&S officer;
- OU attorney;
- facility EH&S manager; and
- regional EH&S manager (for non–U.S.-based facility)

The purpose of the debriefing is to communicate the results of the audit verbally to the BG management and to allow for discussion of the important issues identified in the audit. At the debriefing, any disputed findings contained in the written report are discussed and resolved. Resolution of significant disputed issues will be documented and a letter of resolution will be issued by the compliance assessment department within a week following the debriefing. Typically, the debriefing lasts less than one hour and does not yield any major disputed issues.

FOLLOW-UP

A formal audit follow-up process has been established, using a computer database system to track the progress of facilities in completing corrective actions in response to the issues identified during the audit and enumerated in the action plan. Only legal regulatory and company requirement issues identified in the final action plan are included in the follow-up system. The legal regulatory and company requirement action items are automatically loaded into the follow-up database when the final report is issued. Prior to the end of each calendar quarter, all facilities that have not completed all of their corrective actions are required to submit progress reports to the follow-up coordinator, providing the date on which each item was completed and any comments that they wish to provide relative to the corrective actions taken. This data is entered into the database manually. The data is then used to generate a series of reports that are distributed to various audiences.

One report is sent to the facility general manager and EH&S manager. This report summarizes the progress that the facility is making in addressing the action plan and includes the overall percentage of completion of all action items, the percentages of completion of each audit element for which required corrective action was identified, and a printout of all "open" action items still requiring action. This "open" item document can then be used by the facility for reporting its progress for the next quarter. If a facility falls too far behind its expected rate of completion, additional correspondence from the upper management of the corporate EH&S organization may be generated, reminding the facility management of their responsibility in addressing the corrective actions.

Another report is generated which provides a comparison of expected rate of completion versus actual rate of completion of corrective actions for each facility. This data is placed on a bar chart ranked by audit completion date with the "oldest" audit on top and the most recent audits at the bottom. This chart is sent to senior management of the corporation and to the heads of each OU and permits a quick synopsis of the status of all facilities that have not completed corrective actions.

ANNUAL REPORT

An annual compliance report is prepared at the conclusion of each year, providing an overview of the status of EH&S compliance of the corporation. The report provides an assessment of the progress of the corporation in meeting regulatory requirements and includes graphical and narrative overviews of the status of compliance for each regulated element that is audited as well as overviews of the status of each BG. The narrative provides comprehensive comparisons from year to year on improvements being made and attempts to explain the reasons for variations in performance. Data relating to manufacturing and administrative facilities is analyzed as part of the report. The report also includes sections addressing EH&S enforcement activities for the previous year and a summary of the corporation's involvement in Superfund activities. This report is distributed to senior management in the corporation and to BG leadership teams. Since the report is prepared under the direction of the law department for purposes of providing legal guidance to corporate officials, the report is considered privileged and confidential.

In this last section, Paul Dadak of Hewlett-Packard provides his insight into environmental audits from a historical perspective of the subject area.

THE HEWLETT-PACKARD COMPANY

The environmental audit, as it exists today, has definitely evolved over the past fifteen to twenty years to the point where regulated industries are either already using or prepared to use the EMS or ISO 14000 format.

To understand the evolution of the audit one must look at a historical perspective of "audits" and the process involved to understand this evolution. The comments and observations here are from the author's perspective of experience as a consultant, federal EPA employee (working as liaison with several state programs), and as an industry EHS coordinator or specialist. The author will not explain in detail all of the specifics of the environmental management system or EMS, but rather, will attempt to explain why environmental auditing and the environmental perspective has evolved to the present condition.

Although the author cannot tell the reader from personal experience, he would suspect with some certainty, for all intents and purposes, this activity did not exist prior to the 1970s in any organized way in the industrial business setting. The audit back then was the "financial audit." Today, as was predicted by some prophets in the late-1970s, the environmental audit is approaching the breadth and depth of a financial audit.

One last issue to mention. The audit, as an activity, began as a review of health and safety issues.

In fact, some years ago, a job advertisement might have read "Safety Professional Wanted" and meant someone who knew health and safety (or said they did!) and the ad might also have made a somewhat vague reference to "should be familiar with environmental issues." As the regulations and liabilities changed and multiplied, the environmental issues began to tower over health and safety issues.

This section arbitrarily divides the development of environmental auditing into four periods of time, based on the author's experience and knowledge. They are the early-1970s, late-1970s to the early-1980s, the mid-1980s to the early-1990s, and the early-1990s to the present. Some may differ with the author regarding the time periods chosen, however the author feels that these four periods really define a discernible progression of environmental auditing as a tool of the 1990s and into the next century.

THE EARLY-1970S: THE BEGINNING OF ENVIRONMENTAL AWARENESS

Concerns about the environment were emerging as a recognized public cause or "issue." In part as a response to this public issue, the United States Environmental Protection Agency was established, and it began to establish standards of measure. These were national uniform environmental standards, for example, the Clean Water Act and the Safe Drinking Water Act. The EPA also established related programs, for example, the NPDES discharge permit program. Note that some states had emerging programs earlier than this time period, but they were more general, less established in nature, and not consistent across the United States. In addition, during this time, there were few international standards.

In the early years, standards were media specific, so industry focused on meeting individual permits and reacted to each new standard as they were promulgated; in other words, regulations were written to explain the particulars of the laws. The visibility, if you will, of these regulations was in two areas. First, their passage as law made good political press on a slow news day, and if you were unlucky enough to violate the letter of the law your company may have been picked out for a little bad press. If an "audit" was carried out, generally internally by site staff (or by corporate staff to a lesser degree), it was generally environmental media/permit focused. A site review was not an overall review of conditions of "environmental programs." It was a "did you or did you not meet the individual standards for individual media or criteria?" There was little or no thought of how to set up a "program" to evaluate environmental issues; the emphasis was "keep the alligators from biting."

The compliance perspective was primarily "end of the pipe," industry evaluated what was being discharged or released to the environment and was told what standards to meet by the EPA and or the state agency. Noncompliance was established by case by case "exception" evaluation. As mentioned earlier, the term "OSHA audit" was even sometimes used to describe the evaluation to be performed. "OSHA" was a term that had become, even at that early a stage, a measure of compliance. At this point there was little "top down" emphasis on compliance. Actually to be fair there was mixed management involvement, ranging from "tell me if I am going to jail" to "I want compliance with every permit and permit conditions." However, even in the best organizations, the emphasis was environmental media specific. If a "program" was mandated it was a plan showing how a specific standard would not be violated in the future.

The environmental function (which then contained health and safety functions at times) was not an integral part of the manufacturing system. It was more like "you tell us what you must/will do after we create the process and wastes." More accurately, it was "here is the waste from the process in place, get rid of it or treat it for me." As a parallel to that perspective, the cost of hazardous waste disposal, now getting more expensive, was due in part to new regulations, and was generally paid for via a site-wide, facilities budget, not by the generating function.

At this point in the evolution of the process, the E(H&S) organization was often not well-defined. Environmental issues were not generally directly connected to health and safety, by organization or philosophically. Within any company or business, you could find a range of organizational and reporting arrangements. All three issues and organizations had a bad "rep."

EHS staff were those people who told you what you couldn't do! E(H&S) staff were not always educated or experienced professionals in environmental issues. They (or he or she) frequently were chosen for their availability rather for their abilities. The person responsible quite often was from the facility's maintenance or site engineering staff.

If "waste minimization" was even thought of, it was in relation to savings that could be connected to minimizing the cost of waste disposal from manufacturing processes and not normally driven by environment audits. In some ways, the function was like the early manufacturing "quality" perspective, in other words, inspect and OK at the end of the process not as part of the process. Fix it if it is broken!

During this period, companies did not want to be known for their environmental programs or compliance, such as they were, whether good or bad. Industry wanted to be known only by the products they produced. Industry didn't even admit they produced wastes, or in any way released anything to the environment.

LATE-1970S TO THE EARLY-1980S: THE ADOLESCENCE OF ENVIRONMENTAL AWARENESS

During this period, the USEPA was maturing and creating more and more complex standards for all media and wastes [RCRA, CERCLA (Superfund)] and compliance remained in the "command and control" mode even as the regulations were becoming more and more complex. As state and federal agencies gained experience, regulations were being amended and enlarged. Compliance remained a fire-fighting activity, reactive rather than proactive. The response to new regulations was not unlike the "end of the pipe mentality"; the regulation or issue was created, then industry had to come up with the solution.

Some companies at this stage were somewhat forward looking and began establishing the environmental specialist/manager/coordinator position, recognizing that an in-house specialist was needed to respond to the demands made by the regulators. Some even, I dare say the minority, expected their specialist to develop an overall environmental program. One reason for this shift may also have been that environmental issues were becoming very public. In some cases past perceived "sins" of industry were being emphasized in the media, for example, superfund sites, contaminated waterways, etc. Remember back to that period and think of what "Love Canal" means in the lexicon of the environmental movement.

During this time, some companies were beginning to recognize the need for environmental professionals, in other words, individuals with education and experience in the field, as part of the staff. However, just where these "staff persons" fit in the organization was not consistent or clear. As an example of lack of clarity, look no further than how the job descriptions were sometimes written. As indicated earlier, a want ad might say "Wanted—Health and Safety Professional, familiarity with CERCLA/Superfund and Clean Air Act a plus." Management was still at a level of understanding that can be described as "I don't understand what you do, but I will be glad if you keep us in 'compliance'!"

Even in this less than perfect or consistent world, company-wide audits begin to emerge in some larger more progressive companies. These were generally not program wide or program focused. Rather they were still more focused by environmental media; air, water, hazardous waste, and/or by exception; in other words, obvious issues that have created or could create a problem. Some examples of these were: wastewater treatment effluent standards, chemical releases, hazardous waste management. If audits were being performed they were site audits or reviews, but oftentimes no consistent overall company-wide audit program existed.

In our experience, this is the point where company-wide audit criteria development based on media-specific regulations began. The evolution continued through the period by including regulatory changes and modifications, after they were established. Bottom line, the "audit" criteria was still largely regulatory compliance-driven.

At this point, questions emerged from time to time about confidentiality of audit results. Legal concerns about whether the regulators must be privy to "internal audits" or more importantly, could "hang us with our own rope" were being discussed. There was concern and discussion but no definitive or consistent conclusion that manifested itself. And "audits" were still evolving and being performed.

There also seemed to be consistency in corporate (top management) or individual site management involvement. Top down management emphasis? What's that? Yes some companies were starting to recognize the need for it, but it was still developmental, a theory, at best. Environmental and health and safety "audits" were still perceived as an audit of the EHS function. They were not considered as an audit of the EHS program. Likewise, neither the E or HS programs were considered an integral part of manufacturing, or the "business." It was a service performed at the "end of the pipe"! But perspectives and conceptions were changing; "management" was beginning to realize the value of an audit program, to be able to "manage" environmental issues, if not programs. At this point "manage" was more of a compliance matrix management, versus true management as we might think of today. A good example of the application of this early management technique is "waste minimization." At this point, it was management consisting of reducing costs of off-site management, with some on-site waste reduction emphasis. The aspect still evolving focused on ways to reduce costs of after-the-fact environmental problems and issues, rather than environmental management by proactive measures. Again still primarily reactive as opposed to being proactive.

During this period, third-party audits were not used extensively, although sometimes consultants were brought in to develop auditing guidelines for industry use. These might take the form of checklists for overall applicability of local state and or federal requirements and media specific or regulation specific guidelines. Often times the documents were either actual copies of or reformatted regulatory checklists. Again media specific, responding to the "command and control" frame of mind prevalent during this time period.

Manufacturing, the folks that decide what to make and how to make it were beginning to have responsibility delegated to them for environmental issues. One example of this was the charging out of waste disposal cost or waste treatment costs to the generating entity. Management was now able to see the cost of environmental management on their balance sheet. This was a seemingly minor change in policy but was deemed a major "breakthrough" in how the environmental aspects of business were viewed. Now middle and upper management were seeing environmental concerns show up as a line item in their individual budget, have a "face," as it were. In a way this helped to make the issues related to environmental concerns visible to the internal "public."

Unfortunately old perceptions sometimes die hard. The perspective that environmental concerns and audits were still "end of pipe!" persisted in some arenas, but the world was changing and proactive thinking was on the rise.

Note here that the author does not want the reader to think that industry had no conscience at this point in time. The opposite is true, in the author's opinion. Large and small companies, on the whole, were (and are!) very environmentally conscious. But the whole "environmental thing" was still new in a way. "Industry" as a group was still overwhelmed by the breadth and width of what was expected of them. This wasn't a business issue as products and profitability were. We were all still learning. It wasn't as clear as one might think. Referring back to the mention of the quality function, certain techniques were used to assure quality in manufacturing. The "quality" function was, at that time, not woven into the fabric of manufacturing. In some ways, it was an "end of the pipe or line" quality review.

It wasn't a conscious part of the entire manufacturing process back to front. And so it was with environmental concerns. These were "end of pipe" segmented issues and not considered in an orderly pre-planned managed system.

Public scrutiny of environmental programs was still a thing of the future. Unless a company was perceived as obviously despoiling the environment the "public" was really the somewhat disjointed regulatory agencies.

MID-1980S TO EARLY-1990S: ENVIRONMENTAL AWARENESS ENTERS TEENAGE YEARS

During this time period, more emphasis was being placed on environment by the regulators resulting in yet more regulations. CERCLA, or Superfund was amended and reauthorized and now included specific community right-to-know provisions. The stormwater portion of NPDES was created. In addition, programs were being established by communities to review and understand what potentials industries might have to affect their well-being. This was called by some the "Bhopal Law." Industry was required, in a proscribed way, to report directly to the local emergency planning committee (LEPC) and the various states in terms of potentials as well as known releases or spills. This, in a way, forced the communities in a position of having more community involvement and scrutiny. Industry was likewise forced, if they hadn't already, to begin to acknowledge that they used chemicals and had to address that use and manage these chemicals responsibly.

The "teenage" environmental movement began to use this information to point the finger at industry, justifiably or not. What I mean here is that the "public" was made aware of millions of total pounds of potentially toxic materials that industry was emitting or using in some way! Unfortunately the "knowledge" was in some cases taken out of context. For example, during neutralization of acidic wastewaters, sodium hydroxide is used and sodium sulfate is produced as a result. If an industry was attempting in good faith to neutralize "wastewater" it had to report the resulting sodium sulfate as "being manufactured," with the resulting possibility of bad press that they were "emitting" thousands of pounds of "toxics."

How is this related to the evolution of the "audit"? An issue that was thought to be well understood and needed a straight forward response now became a very public issue in a not very favorable light. At this point industry was beginning to realize that the "big picture" must be looked at or "managed." One possible result, in a backdoor sort of a way, is that environmental issues now have center stage and management realizes that the "public" now involves more than the regulatory agencies.

Another positive is that the "audit" criteria evolved yet further, and in this author's experience became more program-focused across the EH&S spectrum. Furthermore, the site review/audit process began to include meetings with top-level site managers, including human resources, facilities, risk management (security; environmental, health, and safety; health services), and other functional managers. The audit was becoming a review of the site not only the EHS function. Top-level management was reading and acknowledging good and unacceptable performance. EH&S professionals began to work "smarter not harder" by doing regular EHS program reviews that evaluated individual issues that must be addressed and monitored, as a function of all inputs. These included local, state, and federal regulations and corporate requirements and the beginning of prioritizing work needed.

A measure of the evolutionary process was that "pollution prevention" was beginning to be used as a term rather than "waste minimization." This subtle semantic change was actually a major change in thinking. The "end of pipe" mentality was evolving to an "up the pipe" mentality.

As a further measure of the evolutionary process, delegation of applicable portions of the EHS program was beginning. For example, specific requirements integral to facilities management, such as CFC elimination in HVAC, R&D, and materials engineering, were being worked into EHS functions to actively look at alternatives to minimize by-products and waste produced. As indicated above, hazardous waste disposal costs were already being allocated to manufacturing source areas or processes, so management already had their eyes opened to what was coming.

Product stewardship, including product take-back, was just beginning to be discussed, although there was skepticism whether we would be taking back products in any significant number. Even though Europe was starting to regulate product take-back, product stewardship, "cradle to cradle," was not seen as an issue yet. American companies were not jumping on the bandwagon.

Once again, the developments mentioned above caused some industries to realize that a new era was upon us. Although "command and control" would stay with us, proactive management of EHS issues and concerns was beginning to infuse itself into all aspects of manufacturing.

Other developments at this point included expansion of the use of third party audits; liability and confidentiality issues were either accepted or resolved. Management involvement (top down) was more evident from a program perspective. ISO 9000 arrived with some EHS overlap. Industry was beginning to do more "tooting of their own horn" in the public arena about successful environmental programs and accomplishments.

EARLY-1990S TO THE PRESENT: ENVIRONMENTAL AWARENESS MATURES

Based in part on the experience of American industry with the ISO 9000, development of ISO 14000 (ISO 14001 Environmental Management System) is being watched more closely. It is, and continues to be, a time of "watching and waiting." The concerns include whether international business implications will limit or prohibit activity in certain geographical markets. Will these standards be imposed or suggested? What is everyone else doing? Are they doing *anything?* American industry continues to "watch and wait" as standards evolve, are accepted, and are implemented. Few are jumping ahead to embrace ISO 14000. Some are performing a "gap analysis" to evaluate, prepare and anticipate what level their EHS programs are in relation to accepted standards. Questions are being raised about the advisability or need for a third-party certification or the acceptance of self-certification.

What industry is trying to determine is what the compliance implications are, while USEPA and some state agencies are trying to determine where they fit in the picture. They are developing programs that shadow ISO 14001 in some ways, still trying to not let go of the "command and control" perspective. They are attempting to allay fears regarding liability and confidentiality via various programs. They talk about "self-certification" regarding compliance. Industry is still wondering what the advantages of self-certification and audit is for them.

The next step, after determining need and "gap" is determining whether third-party ISO audit or self-certification will be pursued. Remember, the ISO (or other international accepted) standard may replace a range of company-specific program audit criteria/formats in some cases and require a large amount of reorganization.

As the author said at the outset, the level of awareness and development of the environmental program and related audit review still has a considerable width and breadth across American industry. In other words, the gap analysis for some companies may uncover few or minor issues that have to be addressed to create an acceptable EMS. For others, the "analysis" may uncover extensive areas where programs are either entirely absent or in need of substantial overhaul to meet basic EMS standards.

Either way, in the near future, EH&S will evolve to become an integral part of the manufacturing process in large- and medium-sized companies. It will be a measurable and visible aspect of doing business.

IN CONCLUSION

As we have seen from the viewpoints of each of our contributors, environmental audit programs do play a pivotal role in the hierarchy of companies' environmental, health, and safety programs as a means to improve upon existing conditions. The effectiveness of an auditing program depends upon the pivotal roles of how audits are structured and scoped out, who have the main responsibility in EH&S audit functions and program management, the frequency of audits established, and how audit programs intermesh with other functional business units within a company. As we began this chapter, our objective was to concentrate on EH&S audits as management tools for an organization to follow as one of the means to obtain continuous improvement. How audits are conducted and followed through at each regulated company is an item each company considers in solving problems, looking for improvements, and maximizing the company's net potential as that company best sees fit to pursue.

In the next chapter, we carry this theme through a little further to highlight a number of companies, via our contributors and from excerpts of environmental reports, that implemented progressive EH&S programs as they sustain their positions as progressive environmental stewards.

ENDNOTES

[1] As designed by ISO's Technical Committee 207, ISO 14001 is the only ISO standard for which certification can be obtained. The standard does not stipulate that a facility must have a third party to certify its EMS. Facilities can self-certify if they so choose. Each organization makes its own decision whether to self-certify or to hire a third party to certify that their facility's EMS is in conformance with ISO 14001. For a recent listing, the reader should contact the Registration Accreditation Board directly.

[2]For additional information, refer to the GAO report, *Environmental Auditing: A Useful Tool That Can Improve Environmental Performance and Reduce Costs,* April 1995; and a Price Waterhouse survey, *The Voluntary Environmental Audit Survey of U.S. Business,* March 1995.

[3]Among the publicly-held companies that describe their EH&S auditing programs, Texaco provides an overview into its three-tiered auditing program in its *Environmental, Health and Safety Review 1996.* Additional information regarding Texaco's EH&S program is provided in Chapter 10.

[4]Published in the *Federal Register* on July 9, 1986 (51 FR 25003).

[5]Historical information can be found in several sources, among them: M. Crough, "SEC Reporting Requirements: Environmental Issues," *Environmental Claims Journal,* Winter 1994/95; pp. 41–52 and L. Cahill, *Environmental Audits, 7th Edition,* Government Institutes, Inc., Rockville, MD. Previous discussion regarding SEC requirements and Staff Accounting Bulletin (SAB) 92 was also provided in Chapter 4.

[6]Published in the *Federal Register* on April 3, 1995 (60 FR 16875).

[7]Published in the *Federal Register* on December 22, 1995 (60 FR 66706). For additional guidance, readers should consult one or more articles, legal briefs, and position papers written during this time that provide additional guidance on the final policy. This document is also available at http://www.epa.gov/reinvent/grpa.htm.

[8]For additional insight, see the EPA memorandum issued by Steven A. Herman, Assistant Administrator, dated January 15, 1997 that includes the "Audit Policy Interpretative Guidance" in its entirety.

[9]See n. 2. Of note, a total of 614 companies were selected as part of this survey.

[10]Ibid., section 1, p. 5.

[11]Refer to EPA's "Audit Policy Interpretative Guidance," found in that document's Attachment 2.

[12]Ibid., section 1, p. 5.

[13]See n. 5.

[14]Published in the *Federal Register* on June 14, 1993 (58 FR 32843).

[15]As ISO/DIS 14021 nears final approval by the ISO in 1998, there may be additional impetus for companies to prepare their environmental reports as self-declaration opportunities.

[16]For additional insight, see G. Crognale, "Auditing Answers," *Environmental Protection,* July 1991, pp. 37–41.

[17]As of September 1998, about two hundred companies had certified to ISO 14001 in the U.S. *The International Environmental Systems Update,* Special Supplement, September 1998, published by CEEM, Inc., Fairfax, VA.

[18]One of the first companies to volunteer for this third-party MSV verification process was Ashland Chemical Company. That fact is noted in the company's *1996 Annual (Environmental) Report,* © 1997 Ashland Inc.

REFERENCES

Cahill, L. *Environmental Audits*, 7th Ed., Rockville, MD: Government Institutes, Inc., 1996.

Crough, M. "SEC Reporting Requirements: Environmental Issues," *Environmental Claims Journal*, Winter 1994/95, pp. 41–52.

Crognale, G. "What's Up in EPA's Interim Policy Statement on Environmental Auditing," *Corporate Environmental Strategy*, Winter 1995, pp. 37–39.

Crognale, G. "Developing an EH&S Auditing Program," *Safety Compliance Letter*, Waterford, CT: Bureau of Business Practice, November 10, 1997, p. 2.

Crognale, G. "Auditing Answers," *Environmental Protection*, July 1991, pp. 37–41.

Harrison, L., ed. *Environmental, Health and Safety Auditing Handbook, 2nd Ed.*, New York: McGraw-Hill, Inc., 1995.

Herman, S. *EPA memorandum on EPA's audit policy guidance*, January 15, 1997.

Kuhre, W. *ISO 14010s Environmental Auditing*, New Jersey: Prentice Hall PTR, 1996.

Wallach, P. A position on EPA's Final Auditing Disclosure Policy Statement, December 26, 1996.

Ashland Chemical Company, Responsible Care® Initiative, 1996 Annual Report.

Draft International Standard, ISO/DIS 14021, Environmental labels and declarations—Self-declaration environmental claims—Guidelines and definition and usage of terms, International Organization for Standardization, 1996, Geneva, Switzerland.

Federal Register, July 9, 1986 (51 FR 25003).

Federal Register, June 14, 1993 (58 FR 32843).

Federal Register, April 3, 1995 (60 FR 16875).

Federal Register, December 22, 1995 (60 FR 66706).

GAO Report, *Environmental Auditing: A Useful Tool That can Improve Environmental Performance and Reduce Costs*, GAO/RCED-95-37, April 1995.

GAO Report, *Regulatory Burden: Measurement Challenges and Concerns Raised by Selected Companies*, GAO/GGD-97-2, November 1996.

International Environmental Systems Update, Special Supplement, published by CEEM, Inc., Fairfax, VA, September 1998.

The Voluntary Environmental Audit Survey of U.S. Business, Price Waterhouse LLP, March 1995.

Texaco, Inc., *Environmental Health and Safety Review 1996*.

PUTTING INNOVATION INTO PRACTICE: A CLOSER LOOK

Part 3 takes a more focused view of the topics previously high-lighted. In Chapter 10, environmental professionals from indus-try share their views on progressive strategies at work in their companies, while we also highlight examples culled from sev-eral high-profile companies' environmental reports. Chapter 11 takes a look at risk management and communications using some examples from several well-known companies. Chapter 12 takes a closer look at ISO 14001 implementation from the viewpoint of several certified companies.

CHAPTER 10

PROGRESSIVE ENVIRONMENTAL STEWARDS IN INDUSTRY

William Parker, P.E., EG & G, and Janet Peargin, Chevron Corporation with introduction by Gabriele Crognale, P.E.

> *"In everything we do at Chevron, we strive to be 'Better than the Best.' This is not just a slogan but the basis for a detailed and deeply held operating philosophy—a philosophy we call The Chevron Way."*
>
> —*Kenneth T. Derr, Chairman and Chief Executive Officer, Chevron*

INTRODUCTION

Our intent in this chapter is to provide a linkage between the preceding and proceeding chapters in the book by way of the examples provided by the contributors from the organizations that agreed to contribute their day-to-day experiences in environmental and health and safety operations. Their stories bring relevance to the general theme of maintaining and sustaining environmental compliance within organizations that is woven throughout each of the other chapters in some interrelated fashion. The ultimate goal in each case depicted is for the organizations to surpass a compliance-only stance and move ever-forward into progressive stewardship territory.

The contributing authors gives us a peek into their companies' operations, and where they currently are in their EH&S programs as they move forward into the twenty-first century. For example, Bill Parker, of EG&G, gives us an insight into his company's shift from "swords to plowshares" in moving away from being strictly a government contractor in the government's nuclear defense program to shifting its business focus into systems engineering, precision component manufacturing, and test site operating and management services to government agencies and laboratories. Bill also talks about EG&G's new paradigm and moving from compliance to excellence. His portion of the chapter provides considerable insight into his company's new direction.

Then, Janet Peargin of Chevron picks up where Ed Spaulding left off in his contribution to Chapter 8. Janet's focus touches upon Chevron's long-standing commitment and reputation for protecting people and the environment, encapsulated in Chevron's policy, appropriately titled, Protecting People and the Environment (PP&E). Additional information contained in Chevron's 1997 EHS Report can be downloaded from Chevron's WebSite at www.chevron.com. Interestingly, Chevron was found by the Council on Economic Priorities to be one of the top petroleum refiners in its Campaign for Cleaner Corporations.[1]

In addition, The Gillette Company's corporate environmental affairs department provides insight into where that company is headed, including its participation in EPA's Environmental Leadership Program.

In similar fashion, we provide additional insight to support the theme of this chapter by highlighting several case studies pulled from representative Fortune 500 and S&P 500 companies' environmental reports to illustrate how far these companies have come to maintain, sustain, and even exceed environmental compliance, and develop their reports to follow in line with various industry voluntary standards and practices, such as the PERI initiatives, CMA's Responsible Care®, and similar initiatives. A number of companies have made their environmental reports available via the Internet. Their web sites are listed in Appendix G.

In this first section, Bill Parker shares his viewpoint of company operations as corporate head of environmental, health, and safety for EG&G and provides our readers a look into EG&G's environmental focus and objective.

EG&G, INC.: BACKGROUND

EG&G, Inc. was founded in 1946 by three Massachusetts Institute of Technology professors to support the U.S. nuclear defense program. From that time, EG&G expanded its technological base. Today, the company is a Fortune 400 corporation with annual sales of $1.3 billion. It employs more than twelve-thousand people worldwide.

EG&G currently provides systems engineering, precision component manufacturing, and test site operating and management services to many government agencies and laboratories. It designs and manufactures laboratory equipment and field test instruments and electronic and mechanical components for commercial customers. EG&G has more than forty major manufacturing operations in the United States and is prominently located in twenty-two other countries.

Desiring to improve its worldwide business practices relating to environmental compliance and excellence, EG&G initiated policies, procedures, and programs to establish itself as an environmentally proactive company. In 1990, a director of environmental programs was created at the corporate level to lead the company's programs. The director reports directly to the chief executive officer.

Prior to 1990, trends in environmental legislation convinced senior management that increased emphasis on environmental issues was desirable. Excellence in environmental quality could be used as an effective competitive weapon, and the failure to improve EG&G programs would put the company at a competitive disadvantage. There was rapid expansion in the number of environmental laws, rules and regulations promulgated at all levels of government. These were complex and often conflicting with other environmental, health, and safety requirements. Most laws included stiff civil and criminal penalties for violations. In the United States, there were increasing enforcement actions and demand for higher civil and criminal penalties. At many federal facilities, emphasis was shifted from national defense to downsizing and environmental clean-up, concurrently, while reducing their operating budgets.

While enforcement actions increased in the United States, environmental awareness also increased substantially around the globe. For example, in Europe, the European Economic Community moved toward adopting uniform, stringent environmental requirements for its members using incentive systems in lieu of enforcement actions. In addition, with the advent of the ISO 14000 environmental management standards, and the increased movement toward environmental excellence as championed by many regulated industries in the United States and the European Community, the paradigm has shifted toward greater environmental excellence. Into this arena, the growing importance of environmental standards, such as ISO 14000, points to opportunities for industrial organizations worldwide to maintain standard, effective, and efficient environmental management standards that may someday make environmental enforcement less of a driving force for achieving environmental excellence.

In the 1990s, the United States, Germany, and Japan were viewed as the major environmental trendsetters with each country providing leadership in certain media (land, air, water). Historically, once a standard is established in one country, others rush to catch up. Leadership, therefore, provides both a technology edge and potential to create market entry barriers. EG&G recognized both the demand for technology and the sources where this could be obtained. This provides the company with valuable information to begin a marketing evaluation for new business niches for products and services.

At federal installations, such as military bases and energy families, government contractors have been put in the position of continuously being forced to accept more and more environmental liability. This has been driven by several things including poor public perception of government contractors, and Congress' frustration over the environmental conditions at these facilities. Public pressure resulted in the subsequent passage of the Federal Facilities Compliance Act of 1992, and enforcement agencies promulgated more stringent policies, rules and regulations. All of these changes presented challenges to EG&G's philosophy of operations of federal facilities. Interestingly, some federal agencies are beginning to recognize the values of adopting environmental management systems standards. However, few have, to date, provided adequate funding to implement the program. This has further compounded the complexities of working at federal facilities.

Historically, EG&G has operated as an extremely decentralized organization that empowers its division managers with the responsibility to operate their divisions in accordance with approved business plans and the authority to execute those plans. Each division is individually customer driven. While EG&G views the espousal of an environmental system as a progressive step and encourages its operating units to evaluate the various programs, the final decision on whether to adopt a system is left to the individual divisions.

Culturally, EG&G has always been committed to both high ethical and moral standards and to quality. Several years ago, EG&G initiated a corporate-wide program and provided ethics training to every employee. Seminars high-lighted issues relevant to government contractors, including ethical behavior in contract bidding and proper accounting of time and labor. However, the intent is not to provide exact answers for specific situations, but to sensitize employees generally to ethical implications of their work. This program remains in place, communicating management's expectations to the employees.

Also, EG&G has developed a corporate quality program and assists each division in establishing its own quality systems. Because of the rapid pace of global business change, EG&G has established an extensive internal training program. Top management is totally committed to this program.

Because of the emphasis on customer satisfaction by EG&G and many other companies, customers have come to expect on time, quality products as a norm. Total Environmental Quality programs have become the new discriminator for supplier selection. Customer emphasis is being placed on "green" packaging, waste minimization programs, and phasing out the use of ozone-depleting chemicals. EG&G believes that an effective environmental quality program can be used as a competitive advantage and, by 1994, had included environmental issues in the business-planning process. As such, environmental management is integrated into all programs. During 1996, operating units have been encouraged to open a dialogue with their customers and other stakeholders to more fully identify their expectations. And once completed, implement environmental initiatives to meet those expectations.

THE NEW EG&G PARADIGM

In 1990, EG&G acted on its desire to improve its environmental excellence business practices and initiated policies, procedures, and programs to establish itself as environmentally active. Among these activities included significant benchmarking to identify both what other companies' program goals were and determine what stakeholders were lobbying industry to do. For example, EG&G considered adopting the *CERES* Principles, a popular stakeholder model. However, EG&G's involvement with the clean-up of nuclear materials at U.S. Department of Energy facilities, a program that the company executed at the direction of the government, could affect the company's ability to comply with such broad, general principles. Because of the diversity of the company's businesses, no existing generic policy fit its needs. Ultimately, it was decided that the best alternative was to draft a strong policy statement for protection of the environment that took into account the uniqueness of the company's businesses. This policy adopted the best practices of industry and stakeholders.

In 1991, the updated policy was issued, consisting of the policy (eight statements) and implementation direction (see Appendix A of this section). Essentially, the policy requires:

> Compliance with environmental laws, rules, regulations and standards at all levels of government in countries where EG&G does business. Inclusion of environmental awareness and excellence into business practices. Dissemination of information on environmental issues to stakeholders.

The program implementation guidance includes a system for monitoring and reporting progress. The heart of the program is the requirement for self-assessment and the corporate audit program. Progress is reported through annual environmental management plans and periodic data calls. In 1996, EG&G adopted an environmental matrix developed by the Council of Great Lakes Industries. The matrix is an excellent tool that can be used to establish the current status of an operating unit as well as establish future goals. The goals are incorporated into the annual plans and monitored. As of 1997, operating goals include both quantitative (as defined in the WARP program described in the next section) and quantitative (as noted in the environmental matrix) goals.

In 1990, an environmental survey was conducted to ascertain rough quantities of waste generated, identify potential remediation projects, and identify other existing issues. The audit program was started using the environmental expertise from one EG&G division as a corporate auditor and putting a record keeping system in place. One early result of the audit program was to identify compliance issues that were common to many of the divisions within the company. Identifying these issues led to a search for effective ways for the corporation to help the divisions on these issues.

In 1991, the program's focus was on cultural change. Internal environmental committees were formed and chartered to assist in raising awareness of important environmental issues. The audit program was expanded. The corporate Leadership Institute, a week-long training program for future managers, incorporated a presentation by the corporate environmental director. The Procurement and Quality Institutes were similarly modified to raise environmental awareness in their audiences. Also, the revised policy was issued.

Major elements of the policy were implemented. For example, a formal review process was established for each operating division's environmental management plan. The plan elements included financial records for compliance and excellence programs, compliance status, and the status of program execution. Because of EG&G activities in acquisition and divestitures, a corporate due diligence procedure was developed and implemented. Concurrently, operating divisions began to explore environmentally related markets to identify potential niches for new products and services. This resulted in investments to launch new business elements in 1993.

WASTE REDUCTION PAYS (WARP) PROGRAM

In October, 1992, EG&G, Inc. held its first internal waste minimization conference. This conference focused on three topics: (1) direct feedback on customers' expectations; (2) other companies' excellence programs; and (3) the status of EG&G accomplishments. From these presentations, EG&G management assessed what was needed to achieve environmental excellence in waste minimization. At the conclusion of the meeting the CEO, John Kucharski, kicked off EG&G's Waste Reduction Pays (WARP) program and gave his strong personal commitment to its success.

WARP is an aggressive program designed to converse resources and increase operational efficiency through environmentally sound management of materials, energy, and waste. The goals of the program are to reduce liabilities and operational costs and to improve EG&G's competitive advantage in the world marketplace. As a first step, EG&G conducted an audit in early 1993 of its operating units to identify the volumes of hazardous materials purchased, volumes of all types of waste produced, and energy utilization using 1992 operating data. The 1992 data became the baseline for the WARP minimization goals. The managers of EG&G commercial divisions presented their goals in the 1993 environmental management plans. At the federal facilities, managers' goals were dictated by customers and the availability of authorized funds. On August 3, 1993, the President signed Executive Order 12856 making pollution prevention at federal facilities a goal of his administration. Federal facilities managers are working on plans to implement this order. *[Among other considerations with respect to the Executive Order are the attainment of ISO 14001 certification at various DOE facilities. We touch upon this subject in greater detail in Chapter 12.]*

In 1994, there was increased public and Congressional pressure to comply with existing laws. Furthermore, several laws are due to be amended with more stringent compliance requirements expected each year. The need for better environmental management is apparent and several risk reduction programs have been started to deal with this need. Also early in 1994, EG&G issued a data call to measure the first-year progress of the WARP program. The resulting 1993 data indicated a 17 percent reduction in hazardous or controlled waste generation compared to 1992, and a 22 percent reduction in reported use of EPA 33/50 chemicals. In subsequent years, there has been a downward trend in the generation of regulated hazardous waste. However, because of the program infancy and business fluctuations, progress somewhat lags behind stated goals.

In 1996, EPA recognized EG&G's efforts with an award for its success in EPA's Waste Wise Program.

FROM COMPLIANCE TO EXCELLENCE

Since the issuance of the new corporate policy, emphasis has been placed on compliance, self-assessments, and corporate audits. Each division is required to do at least one self-assessment prior to preparation of annual environmental management plans. Corporate audits are done on a priority selection basis with emphasis placed on (a) large quantity waste generators, (b) divisions using extremely large quantities of chemicals, and (c) the divisions' compliance records. All three factors effect the frequency of audits. As a result of these reviews, the managers are required to prepare a corrective action plan. When this action requires significant funding and/or will take a long time to correct, it is also required that the work be included in the annual plan.

QUALITY

EG&G's commitment to environmental excellence is driven by its recognition that environmental responsibility is a key to sustainable, worldwide growth. Environmental excellence is considered an extension of this program. For several years, the company has had an aggressive quality program. For example, several divisions have become ISO 9000 qualified. The European Economic Community has acknowledged the fact that environmental excellence is part of quality management and is actively moving to incorporate environmental performance into ISO 9000. Some efforts already in place include: the British Standard, BS 7750; ISO 14001; EMAS; and the Irish Standard, IS 310. EG&G has one of the first divisions of any company to become BS 7750 certified and has the first company to receive the IS 310 certification. More EG&G divisions are expected to seek certification beyond these certifications from 1996.

Another segment of an excellence program is the inculcation of environmental awareness in procurement. EG&G has emphasized this through a strong endorsement by the EG&G director of worldwide procurement. The director has assisted the environmental staff with joint data calls and through the training and education programs provided to procurement professionals.

TRAINING

EG&G also offers two additional major training programs: leadership and quality. Each is attended by selected, key professionals within the organization as part of the ongoing professional development program. Each features a segment devoted to the company's expectations for environmental performance. Overall, EG&G attempts to matrix environmental excellence into all aspects of its training programs. In 1994, the company developed a week-long Environmental Institute. The course was designed around a case study. The Institute is now offered at least twice annually to nonenvironmental, operations staff.

In addition to its in-house training programs, EG&G has attempted to initiate a series of other programs to promote environmental responsibility. One is the development of a global environmental trending study that analyzes not only current programs, but also considers emerging issues, such as biodiversity, that may effect business practices in the future. This analysis is provided to the EG&G managers at all levels of the corporation, not only to assist them in environmental planning, but to assist them in this strategic planning by the identification of opportunities and threats. It is expected that the trending data will be integrated into EG&G business plans.

MONITORING

Over the last several years, government has continuously tightened the requirements for corporate environmental responsibility. This has heightened EG&G's awareness and recognition of the need for increased internal monitoring. Nowhere is this need more obvious than the federal facilities segment of EG&G business. Through implementation of its corporate policy, EG&G is continuously improving its monitoring program. The internal audit program has been expanded from merely evaluating compliance performance to one that reviews plan performance and quality commitment. In short, EG&G is holding its managers more and more responsible to meet higher and higher performance standards.

Not only does EG&G monitor performance and require managers to self-report annually, EG&G is committed to reporting its progress as a matter of public record. In 1992 and 1993, the corporation benchmarked how other corporations reported their progress and concluded that this could best be achieved through the annual environmental report to its stakeholders. Because of the program's infancy, sufficient data (4+ years) is not yet available to show definite trends. However, it is anticipated that the data will be published sometime in 1998.

The evaluation of internal environmental training, discussed previously, is focused on the line managers and operations personnel. It is felt that while environmental staff can provide subject matter expertise, the responsibility for implementation must rest with the line personnel. A key to this approach is to assure that those responsible conduct frequent self assessments. EG&G is dedicated to providing self-assessment training to raise line personnel awareness and expertise in this area.

CULTURAL CHANGE

All of the corporation's efforts continue to be aimed at cultural change. In a large, multinational corporation this is a difficult, time-consuming process. Cultures are not changed by fiat. Like all quality programs, they are changed through continuous improvement. This is done at EG&G primarily through training, the business planning process, and recognition programs. For example, corporate staff continuously solicit feedback on training programs. Participants are questioned during and immediately upon completion of training to determine the effectiveness of the course. It is not uncommon to conduct a second follow-up six months to a year after a student has completed the program. Another way the effectiveness is evaluated is by the demand for the course to be continued.

At EG&G, a series of internal conferences was initiated to effectuate technology transfer. These focus on enhancing EG&G's environmental reputation. They are conducted annually to educate both managers and technical personnel on the advances that have been made in solving environmental problems. One of the key programs is the annual update of performance, against goals, of the WARP programs. Each year, a data call is initiated to monitor performance. The results are widely circulated among managers at all levels of the corporation. A WARP award program has been established to recognize program achievements at three levels: division, team, and individual. Annual recognition is given to those judged to be the most successful. For example, individuals have been recognized for suggestions to conserve resources. Quality teams have been recognized for waste minimization and pollution prevention initiatives. Finally, divisions have been recognized for their effort in conservation, waste minimization, and recycling. One division located in Ireland is so green conscious that their cafeteria recycles used teabags as garden mulch.

It has been found that a good communication system is necessary to support the program. At EG&G this has been achieved through (a) the establishment of environmental committees, (b) the use of electronic bulletin boards, and (c) the internal video newsletter, SPECTRUM. Each of these media reaches important segments of EG&G's corporate structure. For example, SPECTRUM is distributed worldwide in three languages and it is available to all EG&G staff for viewing. The summer 1993 SPECTRUM focused on pollution prevention within EG&G, and presented case studies at several divisions that had reduced waste and eliminated ozone-depleting chemicals from the manufacturing process. Subsequent corporate newsletters have featured articles on individual accomplishments such as the receipt of WARP awards and environmental system certification.

There have been many lessons since the initiation of the EG&G program, and it is expected that many more will be learned as the program matures. One of the most important elements is the need for the total commitment of the CEO and top management. At EG&G, this commitment has repeatedly been stated. Simply put, EG&G is committed to conducting its business in an environmentally responsible manner in all aspects of its operations. Implementing this policy is a primary management objective and the responsibility of every employee.

Another element is the commitment to continuous environmental improvement. It has to be recognized that regulatory requirements are continuously becoming more stringent and without continuous improvement, compliance cannot be sustained. Goals must be quantifiable and results must be monitored frequently to measure progress toward these goals.

Finally, the results must be communicated. Achievements must be recognized, and failures must be analyzed to determine how improvement can be made and what resources must be dedicated to achievement of goals.

RESULTS

At EG&G, program results, to date, have indicated steady improvement. Awareness has been elevated; programs have been initiated; and resources are committed. However, data continues to show that the corporation has not yet achieved its excellence goals. Thus, more emphasis is being placed on both training and inclusion of environmental performance criteria in all performance appraisals. When the latter is achieved, the corporation will have taken a major step toward continuous compliance.

While it is important, especially in the short term, to focus on continuous improvement, in the long term the emphasis must be placed on environmental excellence.

Once programs such as WARP are fully operational, standards above those of compliance can be met and at lower operating cost. The ultimate goal of EG&G is to achieve as close to zero discharge as can be economically justified. Data collected to date was adequate to allow internal distribution of the results and conclusions. The circulation of the data will provide company incentives to meet and exceed stated environmental goals.

Although its excellence programs have only been in existence for four years, significant progress is being made. For example, in EG&G's 1992 annual report, the chairman pledged to eliminate the use of ozone-depleting chemicals in all manufacturing processes by the end of 1994. This goal was met ahead of schedule without sacrificing product quality. By recognizing success, EG&G expects to make continuous progress toward its ultimate goal.

One of the positive results of EG&G's emphasis on environmental programs has been the development of new environmental products and services within existing business units. Another is the formation of a new EG&G division wholly dedicated to developing new environmental technologies. The first product is an in-situ pump-treat process that separates entrained organic from groundwater. The process is designed so that gasses are drawn off while treated groundwater is discharged back to the aquifer without having to pump the water to ground surface. The advantages of this system are ease of installation and reduced operating cost.

As EG&G focuses on the 21st century, environmental products and services provide new growth opportunities and the company is positioning itself to take advantage of these global markets.

In the following pages, Parker provides a copy of the corporate environmental policy, noted as Appendix A to his contribution.

Appendix A

CORPORATE POLICY

Subject: Environment

POLICY

EG&G is committed to conduct its business in an environmentally responsible manner in all aspects of its operations.

BACKGROUND

We believe that environmental responsibility is a fundamental requirement of all company operations. Implementation of this policy is a primary management objective and the responsibility of every employee.

With increasing worldwide awareness and sensitivity to environmental issues, there is an emerging need for corporate headquarters to coordinate certain aspects of the environmental programs of the company.

OBJECTIVES

The following are objectives of the company's environmental policy:

Compliance

1. Comply with all applicable environmental laws, regulations, and standards.
2. Evaluate compliance on a routine basis throughout all company operations through use of a self-assessment and audit program.
3. Train and educate employees in the identification and implementation of the requirements of environmental compliance.

Environmental Quality

4. Encourage the use of environmentally benign techniques and waste minimization methodologies in the (a) design of products and processes, and, (b) management and operation of facilities.

5. Develop programs for the conservation of resources and protection of the environment through recycling and reuse of material.

6. Develop programs to incorporate environmental considerations among internal management's criteria by which projects, products, processes, and purchases are evaluated.

7. Develop in our employees an awareness of their environmental responsibilities and encourage adherence to sound environmental practices.

Resources

8. Assure the integration of environmental consideration in business planning.

IMPLEMENTATION

Corporate headquarters, through the Director of Environmental Programs, is available to provide program guidance and will direct compliance audits. Corporate Legal is available to assist in the interpretation of applicable environmental laws, regulations and standards.

As appropriate, each Group Executive shall:

1. Develop, review and assure implementation of organizational environmental policies, plans, and guidelines on an on-going basis.

2. Communicate the appropriate environmental policies, plans, and guidelines to his or her employees.

3. Clarify specific individual responsibilities for his or her environmental program.

4. Identify, remediate, audit, keep records, implement, monitor, report, provide resources, and improve on the environmental program.

   a. <u>Identify:</u> Develop and specify an Environmental Program including objectives, schedules, milestones, cost, and implementation responsibilities; and, where applicable, provide information on the program in the strategic plans and six quarter plans.

   b. <u>Audit and Program Status Review:</u> At a minimum, self-auditing shall be done annually at each operating facility. Periodic audits and performance assessments shall be performed under the direction of the Director of Environmental Programs. Annually, the Director of Environmental Programs shall perform a program status review of each program. Environmental issues associated with the divisions strategic program shall also be reviewed as part of EG&G's business planning process.

   c. <u>Remediation:</u> If violations are identified, they either shall be corrected immediately, or a written compliance action plan shall be prepared that addresses scope, schedule, budget, and implementation responsibilities.

   d. <u>Record Keeping:</u> Each manager shall maintain environmental records in accordance with EG&G Policy 1-07, "Record Retention."

   e. <u>Monitor:</u> Monitor environmental program effectiveness against its plan during the course of the year and revise the plan as necessary to meet objectives, schedules, and milestones.

f. Reporting: Report (1) compliance action plans to corporate headquarters and (2) program progress as either part of the corporate business planning process or part of the program status review.

g. Resources: Provide resources adequate for implementation.

h. Continous Improvement: Continuously improve quality (effectiveness and efficiency) of his or her programs.

5. Provide education and training in environmental programs.

6. Obtain guidance from Corporate Legal, unless otherwise delegated, in addressing the public's right-to-know under environmental compliance requirements and the public's interest in the corporate environmental program.

This next section was written by The Gillette Company Corporate Environmental Affairs Department (M. Aleo, K. Christ, J. MacKenzie and G. Olson).[2] The information contained herein reflects the accomplishments of Gillette's environmental programs through the year 1995. It does not include the many environmental achievements made from 1996 to the present.

THE GILLETTE COMPANY AND THE ENVIRONMENT[3]

The Gillette Company has a long-standing, worldwide commitment to conserve natural resources, and to minimize the impact of manufacturing processes on the environment as a whole. The company has the backing of top management to promote its operations in an environmentally conscientious manner. Gillette has been referred to as an environmental steward as a result of innovative environmental and conservation programs the company developed and in which it participated, among them:

- Gillette policies on the environment;
- accomplishments;
- Gillette's pollution prevention program;
- EPA's 33/50 Program;
- Gillette's energy and water conservation programs;
- EPA's Green Lights Program;
- product stewardship (design for the environment);
- recycling and packaging; and
- EPA's Environmental Leadership Program.

GILLETTE POLICIES

Gillette is a globally-focused consumer product company that manufactures and markets quality, value-added personal care and personal use products. Gillette's major lines include: blades and razors, toiletries and cosmetics, stationery products (Papermate, Parker Pens, Waterman Pens, Liquid Paper), Braun electric shavers and small appliances, and Oral-B oral care products. The newest product line, Duracell batteries, was added in 1996.

Operations are conducted at more than fifty manufacturing facilities in twenty-three countries. Products are distributed in over two hundred countries and territories through wholesalers, retailers, and agents. The Gillette Company employs more than 35,000 people worldwide. Sales in 1995 totaled $6.8 billion.

As the company policy states, Gillette has a long-standing worldwide commitment to conserve natural resources and to manufacture products that are safe for consumers to use. The policy also requires the assurance of healthy and safe working conditions for employees, and the minimization of manufacturing's impact on the environment.

Prior to the 1970 inaugural Earth Day Celebration, Gillette had initiated programs to operate in an environmentally responsible manner. Gillette's Chairman Vincent C. Ziegler, in the 1970s, detailed the policies and programs Gillette had established, including the Gillette Environmental Quality Council, forerunner of the company's present Committee on Health, Environment & Safety (CHES).

Today CHES is a senior-management review board that develops environmental, health and safety policies, and reviews Gillette procedures. CHES also reviews worldwide regulation and policy compliance, and ensures that all of the company's business units are audited on a regular basis. Audits are performed by corporate auditors and reports are issued directly to the CHES committee.

Ziegler's successor, Chairman Coleman J. Mockler Jr., continued this commitment and created environmental task forces. Under Mockler and present chairman and chief executive officer Alfred M. Zeien, Gillette's business expanded geographically at a dramatic pace. New operations developed in China, Eastern Europe, Russia, India, and other countries. Gillette's environmental and safety practices and policies, which were more stringent than local requirements, expanded to these new countries. Amidst this rapid geographical expansion, Zeien has placed top priority on worldwide environmental standards. In a video shown to employees, Mr. Zeien discussed the company's stringent emissions targets and summed up the rationale for Gillette's global environmental policy by stating, "The air above us is the same whether it comes out of Bhiwadi, India or whether it comes out of South Boston."

The 1994 Gillette Environmental Report stated "Over the years, the Company has deepened its environmental commitment, in both manufacturing processes and in educational outreach to business, civic, government, and community groups."

ENVIRONMENTAL PROGRAMS AND ACCOMPLISHMENTS

Gillette has participated in various innovative, government-sponsored environmental programs and has a long history of developing and implementing environmental programs proactively. In April 1995, Gillette was selected to participate in the United States Environmental Protection Agency (USEPA) Environmental Leadership Program that is discussed in greater detail later in this section. Gillette was one of the first companies to volunteer for the EPA's 33/50 Program in 1991. This program was designed to reduce toxic waste, and was similar to Gillette's existing pollution prevention program.

In 1990, Gillette became a charter member of the USEPA's Green Lights Program, the first of EPA's market driven, nonregulatory "green programs." The Green Lights Program was designed to reduce pollution by encouraging voluntary reductions in energy use. The Green Lights Program had similar goals as Gillette's energy and water conservation programs that were established in the early 1970s.

Other environmental, health, and safety achievements from 1994 through 1996 are highlighted as follow:

- Gillette's Santa Monica Manufacturing Center (SMMC) won the 1996 Sustainable Quality Award presented by the City of Santa Monica and the Chamber of Commerce for efficient use of natural, human, and technical resources.
- The New England Council, a nonprofit regional business organization, selected Gillette for its "Excellence in Global Environmental Management" Award (12/95).
- For the eighth consecutive year, the Oral-B Laboratories in Australia won the "Five Star Award" from the National Safety Council (6/95).
- The USEPA honored Gillette among thirteen New England Environmental Champions for Energy Efficiency (4/95).
- Gillette's SMMC was awarded OSHA "STAR" status from the California Occupational Safety & Health Administration for superior safety and health performance (10/95).
- The US WasteWi$e, a voluntary program sponsored by the USEPA, recognized Gillette for its comprehensive "Waste Reduction Program" (9/96).
- The Oral-B manufacturing facility in Ireland was certified to the British Standard 7750 for Environmental Management Systems (3/96).
- The Commonwealth of Massachusetts recognized the Gillette Company for its outstanding contribution to environmental protection as a business and education partner in support of the National School's Recycle Center Network (9/94).
- Santa Monica Manufacturing Center (SMMC) was chosen by the California Integrated Waste Management Board as a 1995 Waste Reduction Award Winner (1/96).
- Gillette's Worldwide Conservation Program won a National Environmental Achievement Award from Renew America, a national nonprofit organization. Winners were chosen from the National Environmental Awards Council, representing twenty-nine of the nation's leading environmental organizations (1993).
- Gillette was honored for meeting the goals of the Massachusetts' Packaging Challenge, which calls upon companies to increase the recycling content of packaging material (1994).
- The "Governor's Environmental Achievement Award" was awarded to Gillette from the Commonwealth of Massachusetts for outstanding dedication and service to the Massachusetts environment (1990).

GILLETTE'S POLLUTION PREVENTION PROGRAM

Established in 1990, Gillette's pollution prevention program had an initial target of a 50 percent reduction in worldwide emissions by 1997. Reductions were established on an individual plant basis and were measured against a base year.

The company determined its own broad list of hazardous substances that in the company's judgment could pose an environmental risk.

Four years ahead of schedule, the company achieved its targeted 50 percent worldwide reduction. As of year-end 1995, the company's worldwide manufacturing facilities reported a 72 percent reduction, over five million pounds compared to the base year.

Since 1995, the program has strengthened with more aggressive objectives and lower reporting thresholds. Actual reductions were nearly six million pounds at year-end 1996, a 77 percent reduction, with continued reductions for subsequent years.

Another measure of pollution prevention's effectiveness is waste reduction. During the period from 1986 to 1995, when the company's business grew by more than 90 percent in the United States, the quantity of hazardous waste generated declined by 33 percent.

More than $30 million has been spent worldwide since 1990 to achieve reductions in waste and emissions. Two-thirds of this amount has been spent for projects that go beyond what is required by government regulations.

One pollution prevention innovative technology was replacing solvent washing systems with an aqueous wash system. After nine years of research and development by Gillette, the first aqueous blade wash system was installed in 1991 at the South Boston Manufacturing Center (SBMC). Systems are now installed in five major manufacturing centers (Brazil, Mexico, Germany, England, and the United States). Each aqueous wash system costs approximately $1 million. These five systems have reduced emissions of trichloroethylene by about 1.5 million pounds annually.

EPA'S 33/50 PROGRAM

Gillette was one of the first companies to volunteer for participation in EPA's 33/50 Program in 1991, when the program was announced. The program was named for its overall goals. The final goal was a 50 percent reduction of seventeen high-priority chemical wastes by 1995 when the program concluded. An interim goal was set for a 33 percent reduction by 1992. The baseline year was 1988.

Gillette U.S. facilities targeted three chemicals. In 1991, the company set higher goals than required for EPA 33/50. The goals were: a 73 percent reduction by year-end 1992, and a 99 percent reduction as a 1995 goal. Gillette achieved an actual reduction of 83 percent by year-end 1992, and a 99 percent reduction in 1994, a year ahead of schedule.

Gillette voluntarily expanded the 33/50 Program to all manufacturing facilities worldwide, in conformance with the company's commitment to a global environmental policy.

ENERGY AND WATER CONSERVATION

More than two decades ago, Gillette initiated energy and water conservation programs. The company's conservation programs since 1973 have produced reductions in energy use ranging from 30 percent to 50 percent per location. These programs have also resulted in water use reductions in the range from 30 percent to 96 percent per location.

In 1994, Gillette established new goals for water and energy conservation. The new goals call for an additional 10 percent reduction in energy consumption and an additional 35 percent reduction in water consumption over a five-year period on a worldwide basis. Water savings totaled seventeen million gallons in the first year of the new program, while energy savings exceeded eleven million kilowatt hours (KWH) annually.

EPA'S GREEN LIGHTS PROGRAM

Gillette became a charter member of the USEPA's Green Lights Program. This program was designed to reduce pollution by encouraging voluntary reductions in energy use by upgrading lighting systems and utilizing more efficient lighting techniques.

Gillette completed its initial Green Lights obligation in 1995, having upgraded nearly three million square feet of domestic office and manufacturing space. These lighting upgrades account for a reduction of over nine million KWH of energy use per year. The Green Lights achievements translate into nearly thirteen million pounds of carbon dioxide pollution avoidance at the source of power generation. Savings have also been significant as a result, often with paybacks in less than two years.

Gillette's Voluntary Green Lights partnership with EPA has served as a catalyst for U.S. and international conservation programs. Internationally, lighting upgrades have been completed in facilities such as Ireland, Mexico, Germany, and Brazil.

PRODUCT STEWARDSHIP

The Gillette Company is committed to "product stewardship," which means minimizing the environmental impact of a product during each phase of its life-cycle. This includes raw material production through product design, manufacturing, marketing, distribution, sales, customer use, and ultimate disposal.

DESIGN FOR THE ENVIRONMENT

An important step towards a life-cycle approach is incorporating environmental consideration into the design and development of new products. This is called Design for the Environment. In order for this to work, engineers and scientists must be provided with tools to help them select the most environmentally friendly design and development options.

Gillette is working to develop two computer-based tools: a design guidance tool (DGT) and a life-cycle impact assessment (LCIA) tool. DGT will also contain a computerized environmental education database to provide the most current environmental information.

A life-cycle impact assessment is used to assess the potential environmental impacts of well-defined product. LCIA studies the environmental impacts along a "cradle-to-grave" continuum of a product's life, through production, use, and disposal. Gillette developed a LCIA to evaluate the environmental performance of various razor concepts. Razor designs have been improved using LCIA.

PACKAGING

In 1991, Gillette formed a packaging and plastics task force. The task force identified specific goals that resulted in packaging design changes. Total packaging has increased due to sales growth, but as a result of these initiatives eleven million pounds of packaging has been eliminated each year.

Recycling content in Gillette packaging worldwide is now 29 percent of the total packaging content, or forty-nine million pounds. In 1995, the Gillette North Atlantic Group incorporated more than one million pounds of post-consumer recycled plastics in a variety of packaging components, from razors to antiperspirant packaging.

In 1996, the Gillette stationery products group became the first major writing-instrument manufacturer to introduce a recycled pen. The Paper Mate Eco-Pen has a housing made of 100 percent recycled plastic with 31 percent post-consumer content. The pen barrels are made from pelletized waste plastic containing clean post-industrial waste, and post-consumer waste from milk bottles and clothes hangers.

EPA'S ENVIRONMENTAL LEADERSHIP PROGRAM

In April 1995, The Gillette Company was one of ten companies and two federal facilities chosen to participate in the USEPA's voluntary Environmental Leadership Program (ELP).[4] The program was one of the President's "reinvention initiatives" to reduce regulations. ELP was designed to encourage facilities to form partnerships with state and federal environmental agencies in order to develop and demonstrate innovative programs. Gillette's program focused on environmental auditing.

By developing and implementing comprehensive facility-wide audits, EPA and facilities are striving for long-term environmental compliance and improved environmental performance. Rigorous auditing practices will provide facilities an opportunity to police themselves, and will allow EPA and the states to more efficiently use their resources.

Through ELP, several components of an environmental auditing program have been examined and tested by the EPA, the states, and facilities. These components include:

- third party audits;
- auditing criteria;
- scope of audits;

- reporting requirements; and
- self-certification.

The auditing programs developed and implemented by the ELP facilities encourage:

- Objective assessment of facility compliance with environmental requirements, and "beyond environmental compliance" commitments.
- Effective partnerships between states, EPA, and industry.
- Open communications between industries and regulators.

ELP projects that focused on auditing include: The Gillette Company; Simpson Tacoma Kraft Company; Duke Power Company's Riverbend Steam Station, John Roberts Company, McClennan Air Force Base; and the Salt River Project. Memorandum of agreements were developed and signed by all interested parties for each pilot project.

Environmental compliance self-certification standards were developed in the Salt River Project. The Simpson Tacoma Kraft's project explored information sharing of audit results between companies and communities.

The Gillette Environmental Leadership Program tested the concept of third-party verification of environmental management systems (EMS) including compliance auditing. The Gillette ELP team consists of representatives from USEPA headquarters and regional offices, the National Enforcement Investigation Center (NEIC), state and local environmental agencies and Gillette.

The Gillette ELP team developed guidance for conducting facility-wide audits to assess compliance with all applicable federal, state, and local regulations. The guidance included assessing commitments to "beyond compliance activities" and developing auditor qualifications.

Existing and evolving environmental management system (EMS) auditing guidance, such as those found in ISO 14000, BS7750, and EMAS were reviewed for possible use during the pilot study. The EMS auditing protocol used during the study was from the (then) draft International Standard, ISO 14001 and supportive guidance documents.

Gillette's consultant developed the compliance auditing protocols, which were evaluated and slightly modified by the Gillette team members. A general protocol was developed using all federal requirements. State-specific auditing protocols were also written using federal, state, and local regulations. State-specific auditing protocols were developed for Massachusetts, Illinois, and California in the Gillette ELP.

The Gillette Team consulted with EPA offices, private institutions, environmental groups, and industry associations during the pilot project. Guidance documents and ELP information were made available on EPA's electronic bulletin board system. The guidance documents were sent to interested parties for comments.

The following Gillette facilities participated in the Gillette ELP pilot project:

- South Boston Manufacturing Center (SBMC) in Boston, Massachusetts, which is the largest razor blade manufacturing facility in the world. SBMC employs approximately thirty-four hundred people in manufacturing, warehousing, and office areas.

- North Chicago Manufacturing Center that manufactures substances for Gillette products. Ingredients for shampoos, hairsprays, antiperspirants, and other products are made at this batch chemical manufacturing facility, which employees sixty-six people.
- Santa Monica Manufacturing Center (SMMC) which manufactures Paper Mate pens and stationery products. This facility employees approximately seven hundred people.

After auditing guidance and protocols were developed, facility-wide compliance audits were conducted by Gillette corporate auditors at SBMC, NCMC, and SMMC. An independent third party also conducted a facility-wide compliance audit at SMMC with environmental consultants from their local offices. Following the compliance audit at SMMC, the consultant performed an EMS audit and provided the facility with a gap analysis for ISO 14000. The consultant also performed an EMS audit at SBMC and provided an ISO 14000 gap analysis.

Gillette's Corporate audit and the consultant's compliance audit at SMMC were set up consecutively, so that the results could be compared. All audits at the three facilities were observed by EPA headquarters and regional staff, the National Enforcement Investigation Center (NEIC) and state and local environmental agency representatives. Audits were conducted between March and July 1996.

Audit reports were written and fully disclosed without holding any information as confidential. Corrective action measures were taken by the individual facilities after the audits and reported. Results were evaluated by the Gillette team, and final reports on the pilot projects were written by EPA. A closing meeting for ELP participants was held in Washington, D.C., in November 1996.[5]

The Environmental Leadership Program represented a public-private sector partnership that was an unusual opportunity to build trust where relationships are often adversarial. For federal, state, and local environmental agencies, ELP was an opportunity to assess the effectiveness of environmental auditing programs, and develop an innovative approach to environmental regulatory compliance.

DRAFT FRAMEWORK FOR NATIONAL ELP

The draft framework for a full-time National Environmental Leadership Program has been proposed and is currently under evaluation. The program proposes a six-year cycle after acceptance as an ELP facility. Compliance and EMS audits will be required at an ELP facility once every three years at a minimum (years two and five of a six-year cycle). There are mandatory elements for acceptance such as a mature EMS (i.e., an EMS in place for at least two years and demonstrated to have been fully implemented and tested at the facility or organization). Mandatory elements also include compliance and EMS auditing programs, community outreach, and employee involvement programs. The program is open to facilities or entities (public, private, or federal) regardless of size.

Past compliance histories will be reviewed on proposed applicants. The same disclosure and confidentiality policies that were developed for the ELP pilot phase is proposed to be used in the national ELP program. The policy includes: 40 CFR Part 2; a reference to any state-specific confidentiality regulations; and EPA's 12/22/95 policy (*Federal Register* Vol. 60, No. 246) titled "The Incentives for Self-Policing: Discovery, Disclosure, Correction, and Prevention of Violations Policy."

The draft national ELP framework proposes that auditors meet ISO 14012 auditor qualification for EMS auditing, plus additional ELP criteria. An approach for addressing issues of noncompliance has also been outlined for the full-scale ELP.

Benefits of participating are grouped in the following six categories:

- public relations;
- fostering partnerships;
- inspection discretion;
- reduced regulatory burden;
- economic incentives; and
- other federal agency initiatives

CONCLUSION

The Gillette Company is committed to the continuous improvement of its environmental management system and its environmental and conservation programs. The company fully supports innovative environmental programs such as the Environmental Leadership Program, EPA's Green Lights Programs and EPA's 33/50 Program. The Gillette Company encourages government and industry partnerships, and strongly believes that partnerships motivate participants to creatively solve environmental problems, and improve environmental, safety, and health conditions for all.

In this section, Janet Peargin of Chevron shares with readers additional insight into where her company is headed with respect to protecting people and the environment, and how it is constantly looking at ways to prevent pollution in all its business areas.

CHEVRON CORPORATION—THE CHEVRON WAY

The evolution of Chevron Corporation's experiences in implementing health, environment, and safety (HES) policies may serve to illustrate some of the points made in this chapter.

Prior to the 1980s, Chevron corporate policies for health, environment, and safety required "compliance with laws and regulations, without regard to the degree of enforcement." Some argued that this put Chevron at a competitive disadvantage—because some competitors did not invest in regulatory compliance to the same degree as Chevron. Still, Chevron management believed that this level of corporate commitment was a necessary foundation for the long-term viability of the company.

In 1986, then-Chevron Chairman George M. Keller told employees that the company needed to go "beyond compliance," integrating environmental and safety considerations into planning and business strategies. Carrying out this vision required a culture change. In 1989, separate environmental, safety, fire, and health policies were combined into one with the new policy better reflecting a long-standing commitment and reputation for protecting people and the environment. Soon after, over two thousand of Chevron's top managers received briefings on the new policy. Education was the first step towards shifting responsibility for E&S management away from the E&S specialists and towards managers and individual employees.

The policy and the program designed for its worldwide implementation are called *Protecting People and the Environment* (PP&E). Key to the policy and its implementation is the concept of risk management. It is expected the innovative integration of risk management principles through PP&E will also limit future liabilities and improve the long term competitive position of the company.

To ensure consistent and comprehensive implementation of the policy, ten categories of environmental, health, and safety performance and over one hundred management practices are specified that each business must integrate into its management systems. The practices describe actions that meet policy goals in the areas of compliance assurance, community awareness and outreach, emergency preparedness and response, energy and resource conservation, legislative and regulatory advocacy, pollution prevention, product stewardship, property transfer, safe operations, and transportation and distribution. The integration of these practices into our business is expected to deliver continued improvement in environmental, health, and safety performance. Regularly scheduled corporate reviews include assessments of environmental, health, and safety management systems to insure practices are in place and systems continually improve.

Operating companies are given the flexibility to decide how best to meet their obligations to PP&E, but the process also ensures consistency among the operating companies. This is important because the public views Chevron as one entity regardless of the uniqueness of the individual businesses. Operating companies follow a self-evaluation procedure to track implementation progress. The president of each operating company must report the company's progress to the corporate office. Any gaps in implementation become areas for improvement to be incorporated into future business plans.

Chevron's goal is for all operating companies worldwide to achieve full implementation of the PPE policy by year-end, 1997. A corporate audit conducted in 1996 showed facilities were well on the way to meeting that goal. As the book was going to press, more current information was not available. The corporation compiles an annual environmental report so employees can share "best practices" and monitor progress being made on PP&E.

In addition to the annual report, internal E&S web sites, newsletters, and databases are part of the "best practices sharing" and continuous improvement attributable to the PP&E process.[6] But what is the next step? What is beyond the 1997 goal? How can Chevron ensure that PP&E is a sustainable policy?

This is where health, environment, and safety management systems come into play. Rather than merely identifying "gaps" and monitoring "end results," the corporation and individual operating companies will have to develop and monitor the "systems" used to prevent, identify, and resolve HES issues.

To this end, Chevron is looking at ISO 9000 and 14000 concepts as a way to further define existing PPE policies and programs. Internal audits are no longer merely "prescriptive" compliance reviews—auditors no longer spend the entire review time looking at permits to see if they are up-to-date. Instead, auditors delve more deeply into the systems used for HES implementation. Auditors spend their time talking to employees about the process they use to identify which permits are required and how they are kept up-to-date. Compliance verification is still completed, but not at the expense of system review. Audits done by the local operating company will still emphasize a more prescriptive approach.

The difference in approach is subtle, but important. It symbolizes that HES has evolved from an "add-on" to the business, to become an integral part of the business. Compliance is no longer the HES specialist's job—it is the responsibility of every employee. Chevron will have successfully implemented PPE when HES issues are automatically included in business decisions. Some of this is already happening. Specialists can speak "business-ese" and business people can share perspectives on HES issues. Together, these groups can improve and sustain Chevron's HES performance.

The preceding company perspectives provided a wealth of information regarding the viewpoint within a regulated organization that only an insider can bring. As we previously saw through the eyes of Ed Spaulding, Paul Dadak, and Mitch Gertz, and will see with Joe Hess in Chapter 12, each of their contributions brings some added enrichment to the focus of their perspective chapters. Continuing in this vein, we felt that the book could be provided additional benefit if we were to share with our readers a small sampling of the two hundred or so environmental reports that have been developed by the Fortune 500 and Standard & Poor 500 companies since about 1988.

Our intent in this closing section of Chapter 10 is to highlight some of the more salient progressive EH&S points that each of these companies make in their reports to reinforce the theme and focus of our book, and to provide a bridge to the viewpoints of our contributors. To capture the essence of our research, we turn to the phrase that was displayed on the cover of Lotus Development Corporation's 1995 Annual Report: "Are We There Yet?" We will let our readers decide whether we are there yet in EH&S considerations, or whether we still have some way to go before we get there.

A SAMPLING OF ENVIRONMENTAL REPORTS: THE PRIMARY INDUSTRY DRIVERS

THE PERI GUIDELINES

In the context of preparing voluntary environmental reports, there are several schools of thought. One camp, the Coalition for Environmentally Responsible Economies, better known as CERES, views environmental disclosure as a business item that requires a corporation to adhere to the CERES Principles, a group of ten principles modeled after the Valdez Principles, which took its name from the Exxon Valdez. Some readers may have noticed, from time to time, publicly-held companies' shareholder proposals in favor of a company adhering to the CERES principles as a measure of their environmental stewardship and disclosure policies. Some companies, such as Sun Petroleum, General Motors, HB Fuller, and several others, have become signatories to the CERES principles, while other companies have voted down adhering to these principles, based on the principles being redundant to several company initiatives, or being too broad and general for particular companies, such as was the case with for EG&G previously in the chapter. These companies, instead, feel their environmental programs and disclosure policies actually provide additional information, and hence, do not feel that the CERES principles bring added value to the effort involved. For additional information, we include the CERES principles in Appendix D.

In a related occurrence, CERES is aiming to make the issuance of environmental reports more standard, and is coordinating a global reporting initiative to create a standardized environmental report by 1999.[7]

The other camp, PERI, an ad hoc group of diverse industries, is represented in North America by ten original signatory companies: Amoco, BP, Dow, DuPont, IBM, Northern Telecom, Philips Petroleum, Polaroid, Rockwell, and United Technologies. Among the companies that prepared their environmental reports using the PERI guidelines that are highlighted in this section are IBM and Texaco. The Gillette report referenced previously was also prepared using the PERI guidelines. The guidelines in their entirety are listed in Appendix D.

The PERI guidelines, to which each of the ten original companies agreed to adhere, were crafted to provide some consistency in the area of voluntary reporting that would be of use to the companies' employees and the general public. The guidelines themselves consist of ten components[8] that encompass:

1. Organizational Profile—Providing information about the organization allowing effective interpretation of environmental data.
2. Environmental Policy—Providing information about the organization's environmental policies.
3. Environmental Management—Summarizing the level of organizational accountability for a company's environmental policies and programs and the environmental management structure; and showing how policies are implemented within the organization.
4. Environmental Releases—Such releases are a quantitative indicator regarding an organization's impact on the environment. The compiled data lists the quantities of air emissions, discharges, and wastes released to the environment.
5. Resource Conservation—Focuses on conservation of materials, energy, water (the underpinning to pollution prevention), and forest and habitat conservation.
6. Environmental Risk Management—Focuses on several areas, such as: environmental audit programs and their efforts to improve conditions; remedia-

tion programs to mitigate past events; emergency response; and workplace hazards dealing with occupational health and safety issues.

7. Environmental Compliance—Focuses on the organization's compliance history,[9] including any enforcement actions and related penalties, the nature and extent of noncompliance, the related environmental impact resulting from noncompliance, and any corrective actions taken.

9. Product Stewardship—Where the "rubber meets the road" in EH&S management matters. This component looks at the end result of a company's products or services, and whether companies are sensitive to the environmental impact of their products, processes, and/or services. Within product stewardship also resides references to procedures and standards to sustain environmental stewardship, such as ISO 14001 ("the organization's commitment to reduce the environmental impacts of its products and services"), looking at opportunities for initiating pollution prevention, and introducing product take-back, among others as some representative examples.

9. Employee Recognition—Provide opportunities for employees and programs to be recognized instilling opportunities for sustaining environmental excellence. *This can be viewed as an "atta-boy" in work situations where it's needed to instill true employee empowerment to close the loop from the top-down (management's commitment) to the bottom-up (empowered employees taking the lead by introducing new ideas and suggestions) versus the methods some companies use by.* Methods that some companies use, such as the ubiquitous motivational posters that are usually found along walls and corridors in plants or shops—flavor-of-the-month posters.

10. Stakeholder Involvement—The last component takes into account the organization's efforts to bring to the table other stakeholders as well, including bringing in the local communities in which it operates. *This aspect is not unlike one of the guiding principles of the CMA's Responsible Care® Program, and is among the underpinnings of EPA's cornerstone voluntary programs, such as the ELP and Project XL described in previous chapters.*

The PERI group originally designed the guidelines as a guidance tool for companies' EH&S managers to refer in initiating or improving upon their environmental reporting. Since that time, a number of organizations have begun to view the guidelines as a baseline standard, if you will, to help provide consistency in developing these widely-followed environmental disclosure corporate vehicles. In time, since ISO 14021 views environmental reports as a company's self-declaration environmental claim, companies that desire making such claims may want to ensure their information is accurate and consistent with the guidelines, especially if they proclaim their adherence to the guidelines.

One final point regarding the PERI guidelines: These were created by the regulated community's EH&S professionals; fellow EH&S professionals in regulatory agencies, support services, and professional and trade groups can relate to them—we're speaking apples-to-apples. The CERES principles were developed by non-EH&S professionals who may lack intimate EH&S knowledge to be truly effective. To provide the perfect analogy, we had previously discussed that

EH&S professionals need to master "businesspeak" to become more effective with non-EH&S top management regarding funding; the same could possibly hold true for groups such as CERES and the Council on Economic Priorities—they may need to better understand the complex EH&S issues and be better prepared to speak apples-to-apples with EH&S professionals to address and help to solve these issues.[10] We are getting closer, though.

The Chemical Manufacturers Association's (CMA) Responsible Care® Program

As an introduction, the CMA is a nationwide group of chemical manufacturers that produce, distribute, and use chemicals as part of their business processes. From a historical perspective, the CMA first signed onto Responsible Care® in 1988. At that time, the United States and Canada were the only countries where the initiative was being implemented. Canada, by the way, was the originator of this initiative. Since that time, the number of countries that adhere to Responsible Care® has grown to forty-two, each a counterpart to the CMA in their host countries. For example, Federchimica is the Italian equivalent of the CMA and adheres to Responsible Care®.

The principles act as a foundation for the initiative ethic, and outline the commitment that each member and partner make to environmental, health, and safety responsibility in managing chemicals. CMA members and partners pledge to manage their businesses according to these ten principles, which include as a representative sample:

- To recognize and respond to community concerns about chemicals and company operations.
- To make health, safety, and environmental considerations a priority in company planning for all existing and new products and processes.
- To operate plants and facilities in a manner that protects the environment and the health and safety of company employees and the public.

In turn, the ten guiding principles are one of the ten elements of Responsible Care® that consist of:

1. Guiding principles
2. Codes of management practice
3. Public advisory panel
4. Member self-evaluations
5. Management system verification (MSVs) (We were provided entre into MSVs in Chapter 9.)
6. Measures of performance
7. Executive leadership groups
8. Mutual assistance
9. Partnership program
10. Obligation of membership

We provide additional information about Responsible Care® in Appendix D.

The CMA takes its EH&S responsibilities seriously and the organization has also been evaluating the overlap between Responsible Care® and the environmental management system elements of ISO 14001. The CMA also keeps its member companies up-to-date on various items of importance, such as employee surveys and general communications efforts. As an example, among the CMA member companies that have provided some insight into their plant activities[11] are:

- Merck & Co.—Over thirty legislators, state and local officials and business leaders participated in the company plant's open house in Albany, Georgia, to see first-hand how Responsible Care® is being implemented on site.
- Louisiana Chemical Association, BP Chemicals Inc., CIBA-GEIGY, and DuPont—These organizations, working with CMA, invited staff members of six states and local public policy groups to a series of plant tours to give them a better insight into the chemical industry.
- Exxon Chemical Company—This company keeps its employees informed about key issues facing the industry or its company through a program called "Community Ambassador Briefings." These briefings keep employees appraised on various subjects, such as public perception of the chemical industry, risk communication skills building, and Responsible Care®.

This comprehensive report provides considerable insight into the positive strides the CMA has taken with respect to health, safety, and environmental performance improvement initiatives. One of the fundamental features of Responsible Care® is the CMA's strong commitment to not only improve performance but to do so in ways that solicit input from a wide range of stakeholders. In this fashion, the tenets of Responsible Care® are not unlike the underpinnings of ISO 14001, from the perspective of providing opportunities to continual improvement.

THE AMERICAN PETROLEUM INSTITUTE (API)

The API represents the U.S. petroleum industry and was developed to function in a similar fashion to the CMA. The companies the API represents are committed to meeting America's energy needs while improving its environmental and safety performance. That goal is a priority to the industry, and as the API's report[12] shows, the industry is continuing to make progress toward that goal. Part of that progress is achieved by member companies through the adherence of the API's Strategies for Today's Environmental Partnership (the STEP program) a series of eleven environmental principles that became part of the API's bylaws in 1990.

These environmental principles[13] embody the general tenet of the petroleum industry with respect to their efforts to continuously improve the compatibility of their operations in developing energy resources with the environment. API member companies pledge to manage their businesses in accordance with these environmental principles by using sound decisions to prioritize risks and implement cost-effective management practices. Some of these sound decisions include:

- Recognize and respond to community concerns about raw materials, products, and operations;
- Make safety, health, and environmental considerations a priority in the planning and development of new products and processes;
- To commit to reduce overall emissions and waste generation; among others.

In line with these eleven principles, the U.S. petroleum industry compiled quantitative and qualitative data to substantiate the API's member companies' efforts to adhere to these principles over the reporting period within this study. As such, the report found from tracking Toxic Release Inventory (TRI) chemicals present in refinery wastes that TRI chemical releases have consistently declined from the base year 1988 to 1995, the most recent year referenced in the report. In perspective, a total of 1.3 billion pounds of TRI chemicals from refinery wastes were managed in 1995. Of that amount, 38 percent was burned for energy recovery, 36 percent was treated to reduce the volume of toxicity prior to disposal, and 20 percent was recycled. That left 6 percent, or about 60 million pounds, released into the environment as air emissions. The information was compiled from EPA sources and from oil companies who had responded to API's surveys.

Of note, reducing the percentage of chemicals being released into the environment (6 percent) even further can take continued diligence and follow-through on the part of EH&S managers who may be empowered to make decisions to achieve these improved emission reduction goals. This achievement, while feasible, is not something that can occur in a short period of time, but getting emissions reduced even further as a measurable achievement of the STEP program's reduction of overall emissions does show commitment and resolve, two important components for getting the job done.

A FEW PROGRESSIVE COMPANIES: HIGHLIGHTING VOLUNTARY ENVIRONMENTAL REPORTS

IBM[14]

www

As one of the forward-thinking companies with worldwide recognition, IBM stands out in its environmental management achievements. Most notably, IBM has received the first edition of a single worldwide ISO 14001 registration[15] for all of its manufacturing and hardware development operations across all of its business units. It is believed to be the first company to take this unique global approach to registration. IBM is also among the companies that make environmental reports available on the Internet. Specifically, reports from 1995 through 1997 are available on its web site @www.ibm.com/ibm/environment/annual.

Encapsulating a few key highlights from the electronic version of the 1997[16] report, some noteworthy items stand out, among them:

- IBM's environmental policy reflects a commitment to leadership, built upon respect that focuses on being an environmentally responsible neighbor; developing, manufacturing and marketing products that are safe, energy-efficient, and can be reused or disposed of safely; and strive continually to improve IBM's environmental management system and performance, among others.
- IBM's Environmental Excellence Award, that recognizes individuals and teams of employees for innovative accomplishments that contribute to IBM's environmental, safety, and energy objectives.
- As a measure of IBM's environmental performance, the company is measured against various internal and external requirements, such as its corporate instructions and regulatory requirements, through a comprehensive set of compliance audit programs.
- In the area of pollution prevention, IBM relies upon several tools to help in this regard, among them, the environmental impact assessment, and the company's research division, that plays a role in introducing P2 in developing new technologies and materials. Some of the research division's initiatives include: reducing process energy and solvents in the manufacture of printed wiring boards (PWBs); evaluating plant-based resins (Lignin) in PWB fabrication; developing lead-free solders; and creating new photoresists that do not use solvents for development.
- Participating in EPA voluntary programs, such as the ELP and Project XL, in which IBM Burlington, Vermont, was selected to participate in both projects; on a more global scale, IBM Brazil and IBM Mexico undertook leadership roles in voluntary programs in Latin America.
- Under Product Stewardship, IBM established five objectives as priorities for new products, including: developing products that can be upgraded to extend their useful product life; developing products that can be reused and recycled; developing products that can use recycled materials and are efficient in their use of energy; and other similar considerations.
- Putting some of this in perspective, IBM San Jose was able to prevent the generation of 3,053 tons of hazardous waste between the period of 1991 to 1996, for a total savings of $6.7 million.

Similarly, a few highlights of the key points of the hard copy version of the 1996[17] report include:

- IBM developed an environmental master plan (EMP) as the planning and reporting document for IBM's environmental programs. This document is submitted each year by various IBM sites, and measures performance in various environmental areas, such as emissions and waste management, among others.
- In the area of ISO 14001, IBM maintains a strong environmental management system, which it believes is a key driver for a company's commitment to environmental protection. *(As a follow-on, IBM is among the first companies to encourage its suppliers to allign their EMSs with the requirements of ISO 14001. See IBM's web site for additional information.)*
- In audits and compliance, IBM maintains a comprehensive set of audit programs, in which facilities are audited by their sister facilities and by corporate internal audits, where findings are presented to top management for subsequent review. As a measure of IBM's audit program and performance, as measured in regulatory penalties and fines, the company paid a total of $102,400 in fines over a five-year period beginning in 1990. In 1995, the last entry for this report, five fines totaled just over $2000.

- In pollution prevention (P2), similar achievements were attained as noted in the 1997 report.

- In hazardous waste management, of seventy-three thousand tons generated worldwide, 62 percent was recycled onsite, 22 percent was recycled off-site, and about 14 percent combined was incinerated, treated, and landfilled. These figures represent a 71 percent decrease in hazardous waste management since 1987, the earliest date for available data in the 1996 report.

- In community and global solutions, IBM is actively involved with the World Environment Center and the Environmental Law Institute; supported the 1995 Pacific Rim Conference on Occupation and Environmental Health in Australia; and IBM continues to support various environmental efforts, such as the company's Almaden Research Center in California, certified as a Corporate Wildlife Habitat by the Wildlife Habitat Council.

As further examples of its commitment to environmental considerations, IBM adheres to the PERI guidelines in preparing its environmental reports, as noted in its 1996 and 1997 reports, and in its on-going efforts to reduce the volume of solid waste entering landfills as computer components, as described by Dan McDonnell and Diana Bendz of IBM in Chapter 15.

HEWLETT-PACKARD COMPANY[18]

Another preeminent and forward-thinking company with worldwide recognition of its products is the Hewlett-Packard (HP) Company. The company's commitment to the environment stems from corporate objectives the company established in 1957: to be an asset in each community in which it operates. Today, that same commitment can be seen in the products HP makes and in the services the company provides its customers. In addition, HP initiated a human health and the environment that may occur at any point in a HP product's life cycle. The Design for Environment (DFE) guidelines created under this program helps HP to:

- Minimize energy consumption of their products, decrease raw materials feedstock, and increase the use of recycled materials;
- Decrease waste streams and emission sources from their manufacturing processes;
- Develop easier-to-reuse/recycle products, among others.

In specific applications, HP has taken into account customer needs in manufacturing innovative products that are environmentally responsible. HP is constantly working to reduce overall chemical emissions and to minimize waste from operations. Among some of the environmental highlight of HP's manufacturing processes are:

- HP recycles or reuses approximately 98 percent of the 1.25 million pounds of electronic hardware it collects monthly;
- Two of HP's series of PCs now carry Germany's Blue Angel eco-label;
- HP has reduced EPA 33/50 targeted chemical releases by 99 percent since 1988, the base year for calculating these reductions. This amounted to a cumulative reduction of 1.41 million pounds;

- In 1995, HP diverted 75 percent of its nonhazardous waste from landfills; and
- In 1993, HP eliminated all use of Class 1 ozone-depleting chemicals from their worldwide manufacturing processes.

- In *packaging,* HP has made considerable strides in the area of the creative packaging and distribution techniques that protect the product while not creating environmental impacts. Some examples include using: recycled polystyrene foam, foam pellets made from wheat, molded paper pulp from recycled newspaper, and recyclable paper packaging instead of plastic.

- And finally, in *recycling,* many of HP's products have been designed to facilitate being taken apart and recycled. A few notable highlights include: HP is a founding member of the Industry Council for Electronic Equipment Recycling (ICER), a group that explores environmentally sound and economically feasible ways to collect, recycle, reprocess, and reuse electronic equipment;[19] HP Germany is a proponent of "Cycle," a voluntary product take-back initiative; and HP offers customers a no-cost recycling program for LaserJet toner cartridges.

Moving forward, HP constantly seeks to offer its customers environmentally sound products and services, while constantly striving to minimize the environmental impact of its operations. In essence, HP constantly raises the bar for environmental excellence and for performance improvement. Additional information about HP can be found at HP's web site at www.hp.com.

Texaco[20]

Texaco is among the largest of the global energy companies, behind Exxon, Royal Dutch Shell, and several others, and as a global energy leader, has a major responsibility to protect the health and safety of its employees, the environment, and the communities in which it operates. That responsibility is part of Texaco's business, underpinning the company's vision for success in the twenty-first century. A key pillar of Texaco's vision is to "achieve an outstanding environmental and safety record."[21] Texaco's Vision encompasses:

- Texaco will be a unique, growing, financially successful energy company, driven by technological excellence.
- Texaco's success will be driven by the integrity, creativity and commitment of its people.
- Texaco will have vigorous leadership and a highly respected, diverse, world-class workforce.
- Texaco will achieve an outstanding environmental and safety record.
- Texaco will provide superior value to its investors, employees and partners, and the highest quality products and services to its customers.

Striving for an outstanding environmental and safety record takes a diligent effort and Texaco realizes what it takes: striving for EHS excellence involves a commitment to continuous improvement in their EHS performance, which includes integrating sound EHS practices into their business operations. Among some key highlights of their EHS program where measurable milestones were achieved are:

- Reducing chemical releases and emissions—Texaco cut emissions of EPA's 33/50 seventeen primary chemicals by at least 68 percent, exceeding the program's goal of 50 percent by 1995;
- Minimizing waste—Texaco designed processes to convert oily refinery wastes and used industrial oil into valuable products, eliminating the need to dispose of these materials;
- Advancing technology—Texaco expanded its proprietary clean gasification technology to convert a variety of feedstocks into synthesis gas for producing chemicals, while reducing emissions as compared to conventional technologies.

Paramount to making EHS considerations work for Texaco is making EHS excellence an integral part of the company's business. This involves different levels of Texaco employees, from the bottom up to the top down, which is then funneled to a group of senior EHS professionals from throughout the company working with senior EHS division staff, called the EHS council. This council provides a forum for brainstorming Texaco's EHS policies, standards, programs, and practices to help ensure their consistent application worldwide. In addition, the company's corporate EHS division is responsible for monitoring EH&S activities throughout Texaco's global operations, and provides necessary resources and expertise when required.

Texaco's three main EH&S programs consist of: (1) EHS policies and standards; (2) EHS auditing (facilitating continuous improvement); and (3) product stewardship (attention to detail). In greater detail, these encompass:

EHS POLICIES AND STANDARDS

Texaco's EHS policies set the EHS framework by which the company operates its business. The policies demonstrate the company's commitment to EHS training, research, the development of products that are environmentally sound, and a series of other commitments that encompass Texaco's commitment to EHS excellence. The mechanism to achieve the tenets of the policies are through EHS standards that lay out specific requirements for the company's worldwide operations touching upon a number of EHS issues. Each standard, in turn, is accompanied by a guideline, to facilitate achievement. These EHS standards and guidelines are, in turn, developed by the EHS council, that takes into account a consensus approach from each of these seasoned EHS employees.

EHS AUDITING

Texaco has maintained an EHS auditing program in place for a number of years, and the function of the auditing program is to evaluate both EHS laws and regulations and the company's internal policies and standards. As in any EH&S auditing program, Texaco's EHS auditing program helps the company identify areas of concern, and that may require some corrective action to ensure the safety and health of people and the environment.

To initiate additional benefits that could be derived from the company's auditing process, Texaco enhanced its EHS auditing programs in 1995 by implementing a three-tiered auditing program—compliance audits, systems assessments, and facility self-assessments. Each of these programs aims to increase the consistency of the EHS auditing process by taking into consideration the integration of sound EHS management practice with good business operating practices. Compliance audits integrate environmental with safety/industrial hygiene considerations to increase the consistency of reviews; EHS management systems assessments (MSAs) target the heart of an audit process by reviewing policies, processes, and procedures in place. This can reduce the likelihood of EH&S incidents. Facility self-assessments are then conducted to reinforce continued attention to EH&S issues and compliance considerations at each facility.

The three tiers of Texaco's auditing program are part of a process to enhance existing EH&S performance and, following the lead from ISO 14001-inspired continual improvement, identify areas for improvement and provide EH&S lessons learned opportunities that can be utilized by other Texaco operations.

PRODUCT STEWARDSHIP

Texaco views product stewardship as "cradle-to-grave" care for their products, and thus, makes EH&S concerns a priority throughout the product life-cycle. For example, within product stewardship is the impetus to design or reformulate petroleum products that can either minimize potential hazards or provide environmental benefits. These products include biodegradable hydraulic fluids, nonchlorinated gear compounds, long-life antifreeze, and ashless lubricants. In addition, as an aid to researchers formulating various additives for these lubricants, Texaco even developed a Formulators' Hazard Assessment Guide, providing a toxicity evaluation to each of the additive components. In this manner, research and development personnel can evaluate the toxicity of a new formulation at its very beginning and make intelligent decisions early on in the process.

The bottom line for Texaco is that the company can develop safer products much easier and less costly expending fewer development costs, and thus, ". . . proving that Product Stewardship is also good business."[22] Additional information about Texaco can be found at Texaco's web site at www.texaco.com, and in Appendix G, which contains a summary of a presentation by former CEO, Alfred DeCrane, at the Conference Board.

Boeing Company[23]

The Boeing Company is one of the largest aerospace companies in the world, and employes over 147,000 people at thirty-two major sites in fifteen states. The company is one of the leaders in providing customers around the globe with products that improve the quality of life. Getting there will also require an attitude of disciplined cooperation among many stakeholders. That attitude is also key in helping to achieve technical breakthroughs in aerospace products. In short, this attitude has become a way of life at Boeing, a simple principle the company calls *people working together*.

A key to Boeing's success in its people working together is getting Boeing employees involved in helping the environment. These can take on such activities as: ridesharing, recycling, transforming waste, meeting together, and staying well, to name a few.

In greater detail, here a few noteworthy highlights of Boeing's achievements:

- Boeing's Commuting Assistance Program, a rideshare program, has won a number of awards and features transit and vanpool subsidies and alternate work schedules. By Boeing's estimate, the program removes over 4,000 tons of pollutants from the air and saves over 600,000 gallons of gasoline each year.
- As part of the company's recycling program, over sixteen million pounds of paper, twenty-six million pounds of aluminum and twenty-four million pounds of steel were recycled in 1996.
- In transforming waste, Kent Space Center uses a process called Alchem to significantly reduce hazardous waste streams, thereby saving money by avoiding disposal costs and extensive chemical use. The unique process uses aluminum chips to treat water to achieve clean water standards by removing heavy metals from wastewater.
- As another innovative move, Boeing hosted its first environmental symposium in fall 1996 for its customers in which the company presented an overview of how it manages its environmental affairs. What was formerly a fragmented approach has been consolidated into a stable organization of environmental professionals, forming a single responsible network dedicated to managing complex EH&S issues. The end result has been a dramatic reduction of waste, emissions, and reduced costs.
- Boeing has a vision for 2016 in commiting the company to "promote wellness, not only in terms of safety and health, but also for the whole person on and off the job." The intent with such a program is to help provide a seamless network of health, safety, and wellness support for the entire company.

The company has taken a different approach to managing waste as a by-product of manufacturing. Rather than concentrate on waste alone, Boeing has opted to focus on managing the whole supply stream, from recycling to reuse, from reuse to conservation, and from conservation to process transformation and renewal. In this fashion, Boeing has shifted its focus from waste management to process management, and has provided opportunities for its R&D engineers to redesign products and processes allowing raw materials and other resources to be used more efficiently. Such opportunities can lead to less waste being generated that may need special handling, such as hazardous waste streams.

Another priority within Boeing focuses on managing hazardous materials. At each site, processes and procedures are in place to reduce, monitor, and dispose of hazardous materials and address EH&S considerations. In addition, each site utilizes an inventory management system to help track each chemical used, where it goes, and to repackage bulk chemicals as a means to help cut down on excessive inventories, and potential excessive waste streams.

In addition, in surfacing operations, Boeing is an industry leader in pioneering more environmentally-friendly paint technology. The company has worked with suppliers and customers to develop easier to apply paints, and has also introduced low-solvent paints. Along with spraypainting, cleaning is another critical step in this manufacturing process. Boeing has found aqueous cleaning to be very reliable, and in those areas where innovative solutions to vapor degreasers have not yet been found, emissions are controlled by various processes to control air emissions.

Finally, in recognition of Boeing's achievements to reduce waste streams, the company received a number of awards and kudos in 1996, including:

- Recognition from the EPA for Boeing's successful participation in EPA's 33/50 Program;
- Winner of the Waste Reduction Awards Program from California's Integrated Waste Management Board; and
- The Environmental Excellence Award from the Huntsville Land Trust.

Supplemental information about Boeing's 1996 report can be found by accessing Boeing's web site at www.boeing.com/environ/homepage.htm. In addition, we include a description of Boeing's cross function program in Appendix G.

CHEVRON COMPANY[24]

Like Texaco, Chevron is a leading international petroleum company, engaging primarily in oil and gas exploration, production, refining, and marketing. Chevron operates in about ninety countries and has more than forty thousand employees. In 1996, Chevron was one of the ten largest petroleum companies in the world, based on revenues.

Taking the lead from the company's chairman, Keneth T. Derr, striving to be "Better than the Best" embodies Chevron's philosophy, called The Chevron Way. This philosophy gave rise to a comprehensive program called "Protecting People and the Environment" (PPE) that was established in 1992, designed to achieve health and safety leadership and environmental excellence. To attain this objective, PPE emphasizes continuous improvement for significant long-term gains, as achieved through ten categories of performance and supported by 102 specific management practices. Janet Peargin had previously described these ten categories. As this report was being prepared, Chevron was on track to meet its objective of full implementation of PPE by the end of 1997. Verification would be through subsequent reviews or audits the company would conduct.

Chevron places emphasis on responsibility for health, safety, and environmental (HS&E) performance with line management, those individuals who operate the facilities. As part of their responsibilities, these managers are expected to include HS&E considerations in their business decisions, and are responsible for their actions. To provide the support they need to execute their HS&E responsibilities, Chevron gives them access to an extensive network of HS&E professionals. Putting all this in perspective, Chevron spent over $1.1 billion in 1996 for HS&E programs. Broken down, $322 million was invested and $803 million was allocated to operating expenses and environmental remediation projects.

Besides the ten principle categories of PPE, Chevron's various organizations pursue additional cutting-edge HS&E practices that complement the goals of PPE, among them:

- Chevron Chemical Company participates in the CMA's Responsible Care® program.

- Chevron Shipping Company is implementing the International Safety Management Code, covering marine facilities, whether afloat or ashore. Final code certification, in effect since July 1998, requires an independent audit of management systems.
- Chevron U.K. Ltd., and other European operations follow internationally accepted EMS standards and have adopted international safety and environmental rating systems.
- Chevron participates in the API's STEP Program.

Several areas in which Chevron has made considerable strides include: (1) pollution prevention (P2); (2) compliance assurance; and (3) community awareness and outreach. In further detail:

Pollution prevention—In the hierarchy of P2, each Chevron business area stresses source reduction at the top of the list, followed by reuse and recycling, and lastly, disposal. A representative sample of Chevron's P2 efforts include the substitution of a nontoxic solvent, improving process controls, switching to a new catalyst, and reuse of chemicals. Taking into account all Chevron waste streams in EPA's TRI reporting, these P2 efforts realized a decrease of 42 percent of emissions between 1988 and 1995. With respect to downstream production cycles, Chevron has attained a five-year downward trend in petroleum spills to land and water, and the company continues to lead the industry in spill prevention. As further mitigative measures to reduce spills, Chevron continues to emphasize a wide-ranging program of improvement in processes, process hazards analysis, and training and equipment (see accompanying graphs depicting Chevron's accomplishments to date in Appendix E).

Compliance Assurance—In this capacity, Chevron utilizes a system of HS&E reviews to assist each of the company's operating units in complying with applicable laws, regulations and company policies relating to health, safety and environmental issues. The reviews serve as checks to ensure the requirements are carried out at each of the operating units. These reviews, or audits, can be by internal Chevron employees, or by outside firms. To put this in perspective, Chevron conducted about forty thousand reviews in 1996, a substantial increase over 1995. Among the reviews conducted in 1996, included twelve corporate-level audits, which focus upon regulatory compliance and policy implementation, that provided top management with an independent check to verify extent of implementation of these programs. Among the benefits of Chevron's compliance assurance program include a decrease in the number of regulatory citations in health & safety, and greater attention to environmental citations aimed at correcting problems and avoiding future citations. Some steps to attain this goal include improved internal inspection, maintenance and operating procedures, and increased employee training.

Community Awareness and Outreach—Among the tenets of both the API's STEP Program and the CMA's Responsible Care® Program (as described previously in the chapter) is to provide the community with information about a company's activities.

Aware of its responsibility as a member of both of these organizations, Chevron's policy reflects these tenets and communicates openly with local residents, businesses and community organizations concerning the potential impact of its operations, and develop a mutual understanding of ES&H issues. To gauge the results of its on going efforts, Chevron includes a series of environmental questions in its public opinion survey, where respondents are asked to rate Chevron and other major oil companies in their area based on a series of questions that include which company:

- Is seriously concerned with protecting the environment;
- Cares about protecting wildlife and endangered species;
- Is a leader in developing fuels to reduce air pollutants;
- Encourages energy conservation; and
- Recycles used motor oil.

As a result of such surveys, Chevron has enjoyed significantly higher ratings than other major oil companies in each of these areas of consideration, spanning the last ten years.

Additional information about Chevron's 1997 Environmental Report can be found by accessing Chevron's web site at http://www.chevron.com

IN SUMMARY

The intent of providing a collection of contributor's works coupled with a compilation of several noteworthy publicly-held companies' environmental reports was to bring the bigger picture in focus regarding where some companies are with respect to their environmental, health, and safety responsibilities. Each of the companies highlighted herein are taking strides to improve upon their EH&S responsibilities to themselves, their customers, their community, other concerned stakeholders, and the environment, as they move forward in their respective industries and businesses. While each company may take a slightly different path or approach, each views EH&S responsibilities with a similar focus: being sensitive to these needs and looking to continuously improve. We don't need a leap of faith to believe them; we can see it in the numbers they provide, and if we still have an "I'm from Missouri" attitude, we can look to EPA's web page for additional compliance and enforcement information about any company that has an EPA ID number, or put in another context, any company that generates hazardous waste. There are a lot of them out there and if one really wants to take the time to verify or refute any company's claims, the information is there in plain sight.

ENDNOTES

[1]Business Bulletin—A Special Background Report On Trends in Industry and Finance, *The Wall Street Journal,* February 12, 1998, p. A1.

[2]This material is provided with the permission of The Gillette Company. This section also includes a discussion about Gillette's participation in EPA's Environmental Leader-

ship Program. Additional information on the ELP can be found at EPA's WebSite listed in Appendix G.

[3]This material can also be found in the Gillette environment report, the *1996 Report on the Environment, Health and Safety,* which followed the Public Environmental Reporting Initiative (PERI) Guidelines. We provide additional information regarding PERI later in this Chapter.

[4]Additional information about the ELP Program and Gillette's participation can be found at EPA's web site, www.epa.gov. From there, the user can click on to EnviroSense, es.inel.gov/elp. Additional background information about the ELP and other EPA voluntary programs was previously discussed in Chapter 1. More detailed EPA world wide web information is listed in Appendix G.

[5]As of this writing, the most recent information available on Gillette's participation in the ELP Program was the six-month status report at http://es.inel.gov/elp/gil6.html.

[6]Peargin's observations regarding the concept of "best practices sharing" is a recurrent theme throughout the other chapters of the book, and is especially relevant in Chapter 7 dealing with benchmarking. Sometimes, we need to look beyond our own four walls to get a different perspective, or come up with innovative ideas that charge our creative batteries.

[7]"Counting on the Environment," from *BusinessEthics,* January/February 1998, p. 10.

[8]© 1994 by the Public Environmental Reporting Initiative. There guidelines are also included in Appendix D.

[9]As noted previously in Chapter 7, EPA initiated an environmental audit disclosure policy that provides opportunities for companies to disclose their audit findings, as well as setting up Envirofacts on its web site that allows any user to see a company's regulatory compliance status by accessing the database of EPA enforcement data.

[10]"Business Bulletin," p. A1, n. 1.

[11]*CMA's Responsible Care® Progress Report: The Year in Review 1995–1996,* p. 17. The Chemical Manufacturers Association, Arlington, VA.

[12]API's *5th Annual Report; Petroleum Industry Environmental Performance,* 1997, p. 3. The American Petroleum Institute, Washington, D.C.

[13]Ibid., p. 54.

[14]*Environment, A Progress Report,* © International Business Machines Corporation, 1996, used with the permission of International Business Machines Corporation.

[15]Readers are referred to IBM's web site at www.ibm.com/IBM/Environment/release.html, dated December 10, 1997.

[16]*IBM & the Environment: A Progress Report,* © International Business Machines Corporation, 1997, at various locations in the report.

[17]*Environment,* A Progress Report, n. 14.

[18]Reference: *Hewlett-Packard's Commitment to the Environment,*© Hewlett-Packard Company, 1996.

[19]IBM is one of the first companies in the United States to be recycling electronic components, described further in Chapter 15.

[20]Texaco, Inc., *Environment, Health and Safety Review 1996* and the 1997 report. Used with permission of Texaco, Inc.

[21]Ibid., an excerpt from the first page of this document.

[22]Ibid., an excerpt from page 5 of this document.

[23]The Boeing Company, *Safety Health and Environment, 1996 Report.* Reprinted with permission from the Boeing Company.

[24]Chevron Corporation, *Protecting People and the Environment: A Report on Chevron's Practices and Performance, 1997.* Reprinted with permission from Chevron Corporation.

REFERENCES

The following publicly held companies' environmental reports were used as references in this chapter in some capacity:

Ashland Chemical. *Seeds for the Future, Responsible Care Initiative, 1996 Annual Report.* Columbus, OH, 1997.

Bayer. *1995 Environmental Report and Perspective on Ecology.* Leverkusen, Germany, 1995.

Boeing. *Safety, Health and Environment, 1996 Report.* Seattle, WA, 1996.

Chevron. *Protecting People and the Environment.* San Francisco, CA, 1997.

Fujitsu. *Fujitsu Environmental Protection Program.* Kawaski, Japan, July 1996.

The Gillette Company. *1996 Report on the Environment, Health and Safety.* Boston, MA, 1996.

Hewlett-Packard. *Hewlett-Packard's Commitment to the Environment.* Palo Alto, CA, 1996.

IBM. *Environment, A Progress Report.* Armonk, NY, 1996.

Monsanto. *Environmental Annual Review 1995.* St. Louis, MO, 1995.

Philips Electronics. *From Necessity to Opportunity, Corporate Environmental Review.* Eidenhoven, The Netherlands, 1997.

Philips Petroleum Company. *1996 Health, Environmental and Safety Report,* Bartlesville, OK.

Texaco. *Texaco, Inc., Environment, Health and Safety Review 1996.* White Plains, NY, 1996.

In addition, the following trade association and ad hoc groups' materials were referenced:

The American Petroleum Institute. *5th Annual Report, Petroleum Industry Environmental Performance, 1997.* Washington, D.C., 1997.

The Chemical Manufacturers Association. *The Year in Review 1995–1996, Responsible Care® Progress Report,* Arlington, VA.

The Public Environmental Reporting Initiative (PERI), *PERI guidelines © 1994* by the Public Environmental Reporting Initiative.

The following additional sources were also referenced:

Business Bulletin, *The Wall Street Journal,* February 12, 1998, p. A1

BusinessEthics™, Vol. 10, No. 4, July/August 1996, Trend Watch, p. 13.

Business Ethics™, Vol. 10, No. 5, September/October 1996, Editor's Note: Survey item, p. 7, Kelley, M.

Business Ethics™, Vol. 10, No. 6, November/December 1996, General Motors Award for Environment Excellence, p. 14. Gaines, S.

BusinessEthics™, Vol. 12, No. 1, January/February 1998, feature "Counting on the Environment," p. 10.

CHAPTER 11

EFFECTIVE RISK MANAGEMENT AND COMMUNICATION: TIPS ON WORKING WITH THE PUBLIC

Stuart A. Nicholson, Minnesota State Colleges and University System

INTRODUCTION

Assessment, management, and communication of environmental (including health and safety) risks have become vital concerns for corporations operating in the 1990s, because of the intense negative public and governmental scrutiny errant companies received from well-publicized disasters such as the Bhopal industrial accident (1985) and the Exxon Valdez oil spill (1989). Thus environmental and allied perils now receive serious attention in company planning, policies, and procedures; and now, virtually all major U.S. companies accept that investors, the public at large, and regulators and politicians alike are continually seeking to distinguish between environmental "bad actors" and "the good guys" in their investment, chastising, and penalizing decisions.

PRUDENT RISK MANAGEMENT AND RISK COMMUNICATION

Clearly, the name of the game for business regarding serious environmental risks is not to be singled out and put on the environmental "bad actors" list. And, of course, the all-pervading goal for mainstream companies is to avoid or minimize "tangible," (i.e., financial) losses from environmental risks. Achieving such goals may be considered as prudent environmental risk management and communication.

PROACTIVE RISK MANAGEMENT AND RISK COMMUNICATION

However, some companies desire to and do in fact go beyond this minimal level of environmental risk management and communication, by doing more. One departure from the narrow, traditional agenda is endeavoring to more thoroughly understand, evaluate, and voluntarily mitigate environmental impacts (environmental risks) emanating from company activities, e.g., products, services, and other functions. Thus some companies have embraced and attempted to implement newer concepts such as life-cycle analysis (LCA) and Design for the Environment (DfE). Simply stated, LCA involves assessing environmental impacts of say, products, throughout their existence or life-cycle. And, of course, there are minimal LCAs and more thorough LCAs—LCAs beyond mere budgeting of wastes and emissions, LCAs that have a goal of including true environmental impacts in nature (e.g., ecological effects). Similarly, DfE, attempting to design products for minimal environmental impact, may entail different degrees of sophistication, to the extent it encompasses true, broad, ecological effects. But the fact that LCA and DfE (and other progressive protocols) are done at all is what separates proactive from prudent companies with regard to environmental risks.

Thus proactive companies begin to address broader issues such as corporate image, reputation, respect, and responsibilities, and appeal to specialized consumers, investors, and publics. And, as the vast majority of firms adopt and implement minimally responsible environmental risk management and communication, utilizing prudent risk management, these broader issues are becoming increasingly important factors to stakeholders in differentiating firms; for it is getting to be that there are no more (or at least very few) "really bad" companies, environmentally, across-the-board, in the United States. Thus increasingly for firms, the challenge is to differentiate themselves from the pack by being more environmentally proactive, including the risk management and communication arenas.

PARADIGM-BREAKING RISK MANAGEMENT AND RISK COMMUNICATION

While familiar environmental risk concerns appear firmly entrenched in the vast majority of major U.S/Canadian and European Union companies, still newer, *broader* environmental risk issues germane to business, barely grasped by all but a handful of companies, are emerging in the late-1990s. First, and most obviously, these new issues entail increasingly serious appearing *global* environmental problems such as potential significant anthropogenic climate change, ozone depletion, mass extinction (biodiversity loss), tropical deforestation, habitat fragmentation, environmentally driven conflicts (e.g., "water wars" and oil conflicts in the Middle East), environmental refuges, and above all, runaway human population growth and overconsumption. Companies (and leaders within them) that can honestly and constructively acknowledge, manage, and communicate their connection to these new, emerging "super environmental risks" are paradigm-breakers.

In recent years, several noteworthy business spokespersons and management scholars alike have called on business to acknowledge the serious and escalating nature of broad environmental problems (e.g., global versus local effects, final versus partial destruction of ecosystems and related elements, ultra-persistent versus temporary chemicals) (see, e.g., Davies, 1991; Buckholz, 1993;

Hawken, 1993; Crosby and Knight, 1995; Stead and Stead, 1996)—as environmental scientists have noted for several decades (e.g., warnings by Eugene. P. Odum, Paul Erlich, Barry Commoner, G. Tyler Miller, Jr. and a host of others).

Thus notes Shrivastava (1995):

> In the classical industrial society, the logic of wealth production dominated the logic of risk production; thus risks were minor, and they could be treated as latent side effects, or "externalities" of production. In the postindustrial society, this relationship is reversed: The logic of risk production dominates processes of social change . . .
>
> [These] risks induce systematic and often irreversible harm in humans and the natural environment. They represent continued impoverishment of nature and often are indivisible, at least when they begin, because knowledge about them is riddled with uncertainty.
>
> Unlike the risks of earlier civilizations, modernization risks are rooted in ecologically destructive industrialization and are global, pervasive, long term, imperceptible, incalculable, and often unknown. . . . Modernization risks have proliferated through population explosion, industrial pollution, environmental degradation and the lack of institutional capacity for risk management. (pp. 120–121).

And, regarding business, the newly emerging, broad environmental problems and risks do not just sit in a vacuum; they call for company responsibilities, strategies, and actions for addressing them, including risk management and communication measures. Thus management visionaries increasingly call for management in business to do more than just acknowledge them, they call for companies to make appropriate shifts in views and actions, in order to have some reasonable expectation of successfully addressing these problems collectively, and where applicable at the individual firm level. These new challenges, fundamentally different from previous risk issues, call for much broader visioning, responsibilities, and actions than traditional approaches. And some management visionaries such as Shrivastava even call for a fundamental shift in environmental ethics among companies (and matching actions), to reasonably address the threats of modern industrial risks.

Thus Shrivastava identifies several limitations of the traditional environmental management paradigm, *viz.,* denatured view of the nature (ignoring and demeaning of complex nature), production consumption bias, financial risk bias (supremacy of monetary values over everything else), and anthropocentrism (myopic view of the world and nature placing mankind at the top, denying intrinsic value of everything else). Finally, Shrivastava offers two general prescriptions for companies interested in addressing broad new risks, industrial ecology and ecocentric management.

This chapter will demonstrate and distinguish good and bad examples of risk management and communication among firms at prudent, proactive (rare), and paradigm-breaking levels (rarer still), by presenting and analyzing a range of real and hypothetical case studies.

SOME EXAMPLES OF INEFFECTIVE AND EFFECTIVE ENVIRONMENTAL RISK MANAGEMENT AND COMMUNICATION

TWO EXAMPLES OF INEFFECTIVE RISK MANAGEMENT AND COMMUNICATION

EXAMPLE NO. 1: THE VALDEZ OIL SPILL

Everyone has heard of the Exxon Valdez oil spill in Alaska; the story is common knowledge. But until very recently few knew about the public relations nuances associated with the spill's aftermath, and how some have come back to haunt Exxon.

The basic essentials: An Exxon tanker carrying 1^+ million barrels of Alaskan crude oil ran into a reef in the early morning hours of March 24, 1989. Fourteen hours elapsed before a barge with cleanup equipment was deployed to the spill. More importantly, the amount of immediately available equipment and man-power was woefully inadequate to contain and clean up the spill (Newton and Dillingham, 1994).

Subsequently Exxon embarked on a high-profile, media-oriented campaign, prominent with images of workers "deoiling" seabirds (survival rate minimal), polishing individual rocks on oiled beaches (very costly, questionable effective-ness), and high-pressure blasting other oil deposits (claimed adverse ecological effects of treatment). As reported on ABC's prime time TV news show, *Turning Point*, some of Exxon's chosen strategies appear to have been motivated as much by public relations "show" value as opposed to true environmental effectiveness (American Broadcasting Company, 1994). Furthermore, in 1989 Exxon ceased cleanup operations when there was still 7 million gallons of oil remaining in Prince William Sound. (Kaufman and Franz, 1993). As widely disseminated by *Economist* magazine, Exxon faced potential liability of up to $16.5 billion for civil damages, including $15 billion in punitive damages (Anonymous, 1994). Ar-guably the size of any award for compensatory damages against Exxon will bear a relationship to the actual effectiveness of corrective action, that is, the extent natural resource damages were avoided. In contrast, potential punitive damages may be affected by what jurors perceive to be as Exxon's insincerity in caring about environmental protection.

The lesson learned from this case goes beyond inadequate prevention and preparedness, at the heart of the civil case is how truly effective were the cleanup and mitigative measures, not how well they looked on TV.

In summary, a huge tanker of a prominent U.S. oil company had a major oil spill of 1^+ million barrels in a biologically rich and commercially productive ma-rine bay, with predictable outpourings of shock and anger by locals, govern-ment officials and environmentalists, and consequent very extensive media cov-erage.

How did the company's image come across, and could things have turned out differently?

Let's recall some key things that happened soon after the spill.

First, we need to appreciate that the company, and the oil company consortium initially responsible for spill responses, were up against a catastrophe of great magnitude, danger, and relative remoteness. The spill itself occurred during the middle of the night on a weekend in marine waters. There was no effective containment response by the cleanup consortium until much later that day, partly because of delays in mobilizing equipment. Thus right off the public had questions about response quickness and thoroughness, which reflected on Exxon. Even though Exxon was not responsible for the consortium's initial response (because it was a different entity), in many people's minds the seemingly tardy response was Exxon's fault. Possibly the fact that two different entities were involved could have been emphasized more than it was—not to excuse the spill—but to indicate clearly that Exxon as much as anyone wanted to have immediate and effective containment of the spill.

After the initial opportunity to contain the spill was lost, Exxon, of course, was vitally interested in destroying or dispersing the spreading oil. Various tactics were considered, but an unfortunate change in the weather in the first few days distributed the oil too widely for significant additional surface containment, safe burn-off, or effective dispersant treatment.

How well were these initiatives covered by the press? To many, not much and not well. Obviously the company was mobilizing to join in fighting the spill, but who remembers *that* and *what* they were doing? Arguably the possibilities could have been more widely aired and debated. This would have been a good demonstration of the *two-way communication,* and perhaps even more importantly taken some of the heat off the company by converting the problem from a company-only problem to an "our problem."

In the ensuing months, the media reported on several kinds of company cleanup initiatives. Let's look at these, how the media reported them, and how they look now.

As mentioned, at the outset there were delays in getting equipment to the scene. That was bad enough, but five years later in June 1994, ABC's prime time TV news show, *Turning Point,* informed a nationwide audience (by way of recorded conversation) that Exxon's PR representative advised a person involved in managing the cleanup to just get out there and show activity, regardless of how much oil was recovered.

What message does this PR attitude convey to the public about Exxon's *really caring* about cleaning up the spill, and what does it matter that the message has come out *now?* Clearly it is not a flattering message, nor does it reflect a spirit of candor.

Coincidentally, not too long after the spill Congress developed the custom-made Oil Pollution Act, which tightens tanker requirements and boosts liability possibilities even more. Under a climate of suspicion *created in part* by the company's PR misguided efforts, even tougher oil industry requirements may loom ahead. And Exxon still has to sell product in a competitive, low-priced market.

Back to the clean-up. One of the most poignant and memorable images we all recall was the pathetic, blackened bodies of oiled seabirds being treated at cleaning stations. While such scenes generate instantaneous sympathy and concern, the public increasingly is aware of the relative futility of these efforts. The fact that survival is typically extremely low in such situations is becoming well publicized and known.

So the question then becomes: Why are they doing it, spending so much on it, *and making such a public relations deal out of it?* The point is that environmentally enlightened people now know that such *token measures* do not really accomplish much ecologically, so this could be viewed as just another PR gimmick to divert attention from the broader damage going on.

Another cleanup technique that also received much media coverage, also apparently facilitated by the company, was the one-by-one rock polishing by cleanup crews. While the technique appears simple enough, to many it came across as unrealistic and overwhelming—Was Exxon really going to polish every rock that got stained? Again the question occurred, and still occurs of why so much effort that seems to be only a drop in the bucket?

While there may well have been good reasons for decontaminating certain rocky reaches in this way, *why* particular areas were chosen for this special treatment was not so clear in media reports.

Finally, another technique that was used, and which has subsequently come under criticism was the high-pressure "steaming" to remove oil on hard surfaces. While this approach comes across as "high tech," it doesn't look so good right now. The ABC story alleged that concerns were raised to Exxon about side effects to minute biota yet such concerns were allegedly suppressed.

We ask, who cares this has come out now? Exxon should, because they were still facing trial on environmental damages issues. To plaintiff attorneys, such suspected cover-up can be a lethal weapon, namely that Exxon had reason to know that this aspect of the clean-up was actually harmful to the environment; that it in effect exacerbated damages from the spill.

What lessons can be learned from this incident? For one, the Exxon spill holds many relevant lessons for future incidents, which hopefully will be avoided. The lessons include sticking to tried and true PR principles such as candor, sincerity, and fairly considering broad input—two-way communication, as well as sharing the company's genuine dilemma of not having all the answers.

Finally, it appears Exxon could have benefited in this incident from advice on communicating to the public in a *socially, environmentally responsible* fashion, that is, recognizing and living up to a minimal environmental ethical standard.

EXAMPLE NO. 2: THE BHOPAL DISASTER

The fatal chemical leakage incident at Union Carbide's pesticide plant in Bhopal, India, resulting in thousands of deaths and injuries, shocked the world and received vast media and subsequent analytical coverage. Further, the resulting revelations of how the accident happened, and Union Carbide's responses, put the company in a bad light (to many observers), and apparently contributed to a massive slide in Carbide's stock valuation for many months after the accident.

What went wrong, and could some of the damage have been mitigated? For one, it is indisputable that fundamental risk management errors were made that contributed to the nature and severity of the accident. First, the plant was situated in a highly populated area—the major city of Bhopal. Other costly oversights directly bearing on the accident included: delay by employees in dealing with the leak, inadequately trained employees, inoperability of fail-safe, back-up safety devices, faulty monitoring instrumentation, primitive leak-detection process, and poor alarm system (Fink, 1986, pp. 172–174). Further, according to a crisis management expert, had any of the identified safety devices properly functioned, the disaster would have been averted (Fink, 1986, p. 174).

An interesting aspect of the Bhopal incident centers on the fact that early on it was reported that the incident was being blamed on employee misconduct, or even sabotage actions by a disgruntled employee, thus appearing as not the company's responsibility.

It is interesting to note the parallels to the Exxon Valdez incident, which also began with finger-pointing at employee errors carried by the press (in Exxon's case, the captain's blood alcohol content), giving the impression to the public that the responsible companies were taking the position they *were not responsible* for the accidents and subsequent damage, because of the untoward acts of their employees. Since the vast majority of citizens *are employees*, the perception conveyed by the media that giant, all-powerful companies are trying to elude responsibility "because our employee screwed up" does not sit well with the public, and ultimately is not accepted as a valid excuse for the companies bearing ultimate responsibility for the accidents.

However, Union Carbide has had some problems with recapturing the public's trust because of recurrence of accidents.

The disastrous India incident—and agonizing follow-on legal proceedings—have not been the end of Union Carbide's adverse publicity woes. As news of the India disaster sunk in, people in the United States pondered the safety of Union Carbide's U.S. plants. Unfortunately for the company, later in 1985, a pesticide plant in West Virginia leaked a toxic cloud over the town, causing suffering to 100+ residents (Starr et al, 1985). This and the recent Bhopal incident subsequently led to major US investigations, environmental fines, and litigation. Investigations of the plant showed there were 70 leaks from 1980–84, and Union Carbide was later fined $1.3 million dollars for 221 violations at the plant (Beaucamp and Bowie, 1988). While the company did invest several million dollars in upgrading safety and emergency preparedness at the Institute plant in the months after Bhopal, (Starr et al., 1985, p. 18) these corrective measures were not enough to insulate the company from government sanctions, civil suits, and certainly not adverse publicity.

Finally, several years later again, in 1991, there was a serious accident at a Texas facility (1 killed, 32 injured) (Anonymous, 1992).

TWO EFFECTIVE EXAMPLES OF RISK MANAGEMENT AND RISK COMMUNICATION

EXAMPLE NO. 1: ASHLAND OIL'S INCIDENT

In 1988—three years after Bhopal—an Ashland Oil refinery experienced a major oil spill into the Monongahela and Ohio Rivers. However, Ashland's handling of the crisis—and subsequent outcomes in the court of public opinion—differed significantly from Union Carbide's handling of Bhopal and its aftermath.

Reportedly, the company from top down immediately *took responsibility for the incident,* and focused on clean-up. Indeed, it is reported that, up front, the chairman *admitted* that the company had made a mistake with the tank that leaked. As a crisis management expert commented on Ashland's handling of this potentially scandalous incident: "Hall's [Ashland's CEO] handling of the situation received positive recognition and approval." Examples of Ashland's favorable press coverage included complimentary pieces in the home area Louisville *Courier-Journal* and the *New York Times,* which commented that:

> Ashland's CEO, John R. Hall was a little slow out of the blocks, but after a day and a half he began to move heaven and earth . . . He pledged to clean everything up, he visited news bureaus to explain what the company would do, he answered whatever questions were asked. Within twenty-four hours he had turned the perception from "rotten oil company" to "they are pretty good guys." (Dougherty, 1992, p. 93–94).

Clearly, the Ashland case demonstrates that it is possible, through sincere, straightforward, and seen-as-right actions by a key company spokesperson, here the CEO, to gain the public's trust, and even accolades for what was really a disaster (failed risk management: major environmental damage, subsequent tens of millions of dollars payouts). Thus, while sincere, forthright, risk communication cannot relieve a company from damages caused by a costly incident, it can, as demonstrated here, not only prevent further erosion of the company in the public's eye, but actually gain their support and admiration. Under the circumstances, one could scarcely expect more. As such, Ashland's handling of this incident, including risk communication, was clearly proactive if not paradigm-breaking.

EXAMPLE NO. 2: TWO EXAMPLES FROM VERY DIFFERENT PHASES

In the inherently negative field of risk management and post-incident management, it is difficult to find many examples of "unabashedly good" risk handling successes—these simply don't get much media attention and coverage. Consequently in this second "good" examples section, the author includes two examples to illustrate good risk management practices.

Example 2a: Ely Lilly's Disaster that (So Far) Never Happened In contemporary risk management, it is now virtually universally understood that preparation is not only the key, but also the necessity for success in dealing with risk disasters. (Head, 1995, p. 59).

Thus many major U.S. companies have been investing in and perfecting efficient responses to simulated emergency scenarios: practice exercises to get everything down and "get the kinks out" prior to occurrence of a real disaster.

Among major U.S. companies, Ely Lilly is no exception; the firm has potential risk exposures from chemical and biological materials routinely handled in its facilities. Thus to prepare for any untoward incidents that might arise, the company is one of many that have carried out emergency response exercises. And, in Lilly's case, the company managed to obtain favorable publicity coverage of its *practice* activities.

For Lilly, and many other companies, practice scenarios of setting up the simulated event and going through the motions of risk response teach valuable lessons. Thus Lilly's reported case of such a trial included a simulated rail crash involving chemical release, fire, and injuries, as well as media involvement. Reportedly, this case was valuable in sensitizing the company to the key role the media have in educating the public about such incidents (Feldstein and Cassidy 1993).

Though Lilly's exercise was laudable and the follow-on PR certainly favorable, this case is not all that revolutionary (practice exercises are becoming the norm), perhaps illustrating good prudential or a minimally proactive standard of risk management and risk communication.

Example No. 2b: The Tylenol Scare—A Serious, Real Disaster Successfully Contained and Managed Not too many years ago, the nation's attention was riveted on a most scary story: the tale of the fatally tainted Tylenol on store shelves that beset manufacturer Johnson and Johnson (J&J) and millions of American consumers. This poisoned product was discovered after some unfortunate buyers consumed Tylenol in the nation's midsection. This event immediately captured the attention of government authorities; death from consuming store-bought product is simply an intolerable act in the nation's marketplace.

But for our purposes, even more important than the hysteria and sensationalism surrounding the case was manufacturer Johnson and Johnson's *laudable and effective* risk reactions and responses. All right, said the company, howsoever the incidents occurred (immediately tampering was suspected), it *was* our product, so *we* will take the initiative to recall it, pull it off the shelves, clearly *for the good of the consumer.* Emerging from this potentially devastating crisis, the company's solid *precautionary* stance and prompt recall action ultimately won it many kudos. After all, here was one company apparently willing to put its bottom line interests secondary, in time of crisis, to the well-being of the consuming public, in spite of the fact that it firmly believed that the (untampered with) product lots were entirely safe and usable. Again, the keys to J&J's successful handling of this potentially very serious disaster were: firm belief in the integrity of its own quality control process and product, coupled with willingness to withdraw massive amounts of after-tainted product, pending successful resolution of the tampering investigations, and which was *unequivocally communicated to the public.*

All things considered, J&J's handling of this crisis was efficient, effective, sounding in the public interest, and successful. Clearly for a time, the company ditched considerations of company profit to the side and did the right thing: It executed on a decision to protect public health and safety despite major costs; this was proactive, arguably even paradigm-breaking behavior—which ultimately paid off for the company in untarnished reputation and continued financial well-being.

WHAT RISK MANAGEMENT AND RISK COMMUNICATION ARE AND HOW THEY'RE SUPPOSED TO WORK

VIEWS ON THE NATURE OF RISK MANAGEMENT, RISK COMMUNICATION, AND THEIR INTEGRATION

RISK MANAGEMENT

Although various characterizations exist on the primary role of risk management, its essential functionality converges to elements involved in the process of managing estimated risks, typically measures to control or otherwise mitigate risks. A classical, loss-control oriented characterization is: ... "the planning, organizing, leading, and controlling of an organization's assets and activities in ways which minimize the adverse operational and financial effects of accidental losses upon that organization" (Head, 1986, p. 50). Another traditional loss-control guru, Henri Fayol, characterized the goal of risk management as ensuring that "pure risk" losses do not preempt (overall company) management from pursuing their goals (Fayol, 1986).

Of course, risk management of environmental-related risks, or *environmental risk management* is characterized more narrowly than risk management generally. Risk management, and especially, risk communication have also been viewed in the context of risk crises. Thus in his landmark work on crisis management, Lerbinger (1997) identifies four elements: hazard identification, risk assessment, risk significance determination, and risk communication. He also acknowledges the heightened public fear of scientific and technological risks which has been prompted by recent "blockbuster" incidents, for example, the Chernobyl nuclear accident (a classic case of mishandled risk management).

Again, from a traditional loss-control standpoint, Head characterizes environmental risk management as: "the process of managing the materials, wastes, and releases of an organization to lessen their effects on environmental values and to ensure compliance with all applicable laws and regulations" (Head, 1995, p. 41). The view of environmental risk management among most contemporary scientists and engineers appears to be that based on the National Academy of Science's outlining of risk assessment components and processes (the National Academy of Science, 1983); this view attempts to keep some separation, or independence between risk assessment (which emphasizes hazard identification and evaluation) and risk management. Thus the leading text on environmental science characterizes risk management as: "the administrative, political, and economic actions taken to decide whether and how a particular societal risk is to be reduced to a certain level—and at what cost (Miller, 1997, p. 211).

Finally, under views characterizing society as being immersed in fears, judgments, decisions, and actions about allowing and distributing very severe or catastrophic technological risks (e.g., nuclear power, toxic chemicals, greenhouse gas emissions, ozone-depleting chemicals, etc.), much of the activities of business, from planning to ultimate disposition of products, processes, etc., are seen through "risk-lensed" glasses. That is, actual adverse impacts of activities and risks attending potential impacts, are viewed as only the end result of a long process which begins with the firm's environmental paradigm, or view of the environment, its responsibilities thereto, and how its environmental worldview and ethics translate into right actions. Attendant with the broad, social technological risk viewpoint is the need for the firm to appreciate its contribution to larger, collective, earth impacts of entire society. While the phrase "think globally, act locally" comes to mind, this emerging view of risk asks even more of companies, namely it calls for them to go beyond currently limited environmental management themes (e.g., end-of-the-pipe pollution control, command and control, pollution permits, omission of broad external environmental costs, and denial of accelerating worldwide environmental degradation).

Risk Communication In the context of business environmental risks, under traditional notions, or the "prudential standard," risk communication involves messages business (or others responsible for the risk) transmit about the nature of environmental risks. The orientation is sender-(company) centered; other than outright lying the risk communicator may provide whatever messages it deems appropriate (as constrained by applicable laws, policies, etc.). The bottom line is that the onus, priority, and power are on the sender; the receiver, typically the general public through the media, is perceived as "other," and not in the primary equation, definitely on a different (lower) plane than the sender. "Ideal" risk communication under this view sees the sender as completely determining and controlling message content, especially completeness, in the interests of the sender, definitely not the receiver. If anything, the receiver is viewed as a nuisance or obstacle to be dealt with, typically by providing carefully screened and to all extents possible, positive messages. This is reflected in common PR industry jargon and mores such as "spin doctoring," "damage control," "putting the best face on (tragedies,)" "image," and the like. In the environmental context corporate "greenwashing" and other dubious PR tactics are increasingly recognized as and perceived by the public as being underhanded, scarcely reflective of truly environmentally proactive companies. In summary, traditional, prudent risk communication is designed to protect company financial interests at virtually all costs and through manipulative communication techniques.

However, a recent National Academy of Sciences panel identified a need to recognize and manage risk communication as including needs and concerns of *both* originators (typically organizations responsible for risks) and receivers, including company supporters, antagonists, and the general public (Committee on Risk Perception, 1990). Needless to say, this is not the dominant view of PR houses or departments; for adopting this view means emphasizing risk messages are clear, unambiguous, as truthful as possible, and not merely intended 'to persuade', as is the case with traditional PR and marketing messages. Moving towards, even adopting this more balanced, even-handed, neutral approach in risk communications would be a proactive position.

Going further, what would paradigm-breaking risk communication look like, e.g., from ecocentric management (Shrivastava, 1993)? Although Shrivastava provides only general guidelines, this risk communication would presumably mirror decentralized, post patriarchal values, therefore entailing broad participation in decision-making. This could mean sharing intimate details of environmental risks with representatives of the public as well as seeking (and requiring) their consensus in decision making, that is, a truly collaborative, consensus approach. While this may appear radical or revolutionary, some out-in-front companies have already taken steps in this direction.

Integrated Risk Management and Communication Looking back, early approaches to handling of risks were typically ad hoc and unsystematic. Thus risk management and risk communication (and other aspects of risk) were frequently carried out as needed. But extensive experience has taught that more unified, systematic risk handling approaches are needed to achieve minimally acceptable effectiveness.

Now, for example, it is widely recognized that good risk communication (in-crisis situations) emanates from risk planning, an element of integrated risk management. Further, the more effective "front end" risk management, i.e., hazard identification and control, the less need for risk communication, by virtue of minimizing occasions in which hazards become out-of-control problems. Also, broad "up front" risk management includes, beyond hazard identification, thorough risk assessment and characterization, so consequences of manifested hazards are understood. These deliberations then form the building blocks for decisions about managing identified risks. Finally, risk communications has evolved to include expanded functions such as risk response planning, practice scenarios, training, as well as routine, specially targeted, and crisis communications.

In short, especially in recent years, we have come to understand that, to have truly broad and effective risk management all the above elements and concerns must be systematically included in and addressed by a company's overall risk management program. It is also realized that in practice risk assessment and risk management cannot and should not always be "artificially separated"; for it is unrealistic to expect "front end" risk processes to operate effectively without some knowledge of company resource limits and priorities.

For most companies of any size, adopting and implementing at least the above-basic, minimal functionalities, is required for attaining prudent risk management; proactive firms would do even more.

Going further, in paradigm-breaking companies, integrated risk management would include truly comprehensive risk assessments of company activities, as well as products and services. Thus risk management would encompass taking responsibility for and addressing its slice of broad biospheric environmental impacts (again, including ecological). Above all paradigm-breaking risk management is not focusing on maximizing adverse environmental externalities in order to save company costs, nor does it involve denying responsibility for environmental degradation, the firms contribution to total biospheric impacts, or the reality and severity of current accelerating worldwide environmental losses.

SOME BENEFITS EFFECTIVE RISK MANAGEMENT AND RISK COMMUNICATION CAN PROVIDE

Effective risk management and risk communication can provide many benefits to the subject company as well as to the public and the environment. From the traditional company point of view, the primary justification for effective risk management and risk communication is *problem avoidance*. Problems that may result from inadequate risk management and risk communication are major and numerous, including violations of environmental, occupational, and other laws resulting in criminal and/or civil liability. Liability, of course, opens up a huge range of "bads" from the company perspective, such as criminal indictment, monetary penalties, significant costs, and even imprisonment for selected actors, plus the accompanying stigma, and shame. Involvement in civil litigation, often an additional possibility in egregious cases, entails extensive "court time," document retrievals, and revelations, as well as lesser stigma and shame, but possibly greater legal fees, and very possibly a substantial payout (either through settlement or court judgment).

Of course, being a defendant in such cases has much additional negative fallout—bad publicity, leading to image diminution, which itself may lead to stock decline and investor desertion as well as loss of customers, employees, and face in the community. Experience has shown that such negative 'spillover' effects can last for years after the tragedy. Thus Union Carbide's stock stayed down for years after the Bhopal incident (*just one* risk management and risk communication shortcoming), and studies on Exxon's fate after the Valdez incident suggested likewise. Further, both companies found themselves being the butt of bad jokes for *many* years after their infamous incidents, as well as seemingly indefinitely sullied reputations [for example, in a round of "Password," if the prompt was "oil spill," how do you think many people would respond, even to this day?].

Having focused on the negative, what can be said about the "more direct" goods that sound risk management and risk communication can provide the company and other stakeholders? First, effective risk management, and especially, good risk communication can be extremely important as a confidence-builder and image remaker, after a company has suffered bad events. Thus, in recent years, Union Carbide has made determined efforts to reframe itself as a premiere health and safety conscious leader in the industry, and this strategy appears to be working—at least across the industry audience. Beyond that, DuPont has skillfully employed a wide range of "proenvironmental" type ads to enhance its image as a quality, environmentally caring firm (e.g., the happy dolphins applauding), apparently making a positive impression on many in the general public. And very recently (fall 1997), Exxon made a significant advance by supporting a new "save the tiger" campaign. Taking this step may well put Exxon on the road to re-making its image as an environmentally conscious company; for among the increasingly large block of environmentally informed consumers, biodiversity preservation is seen as critically important, far more so than merely reducing emissions by x-thousands tons (as can be measured by EPA-required annual TRI reporting.) Arguably, in a small way, Exxon's recent nature protection involvement is a paradigm-breaking risk communication move, and potentially a win-win situation all around (benefits to imperiled environment, nature, the public, and company itself).

In addition to rehabilitating what some came to view as "rogue" companies, for example, Union Carbide and Exxon, sound risk management and especially, forward-looking risk communication can go a long way in positioning a company as a progressive industry leader. Thus in late-1997, before the President's initiative to "sell" global warming as a real threat requiring action by industry, British Petroleum's executive broke ranks with the oil industry in acknowledging the threat of global warming from greenhouse gas emissions, again a paradigm-breaking pronouncement in what is increasingly being viewed as a "dinosaur" industry (fossil fuels). With the massive, mounting evidence and near complete consensus of the world's *qualified* scientists that the threat of global warming is real, plus endorsement of the threat by most countries *including the United States*, who will hold greater credibility among informed consumers and investors: BP or the diehards in the oil industry? Arguably BP's well-timed statement—broadly a sort of risk communication—will go a long way in positioning it as the informed and *credible* industry leader on environmental matters. And, when it comes to putting money up for investing or purchasing, individuals are most comfortable dealing with and investing in organizations in whom they trust. (Some of these organizations were highlighted in the previous chapter.)

Shortly after BP's publicized concern over global warming, Mobil came forward with an ad stating its position on global warming, essentially that carbon dioxide and other greenhouse gas emissions *may* be linked to global warming and climate change. And despite stopping short of its linking emissions releases with global warming, the company made several points on what it is doing for *reducing emissions* (why care about reducing emissions if they are no threat?), at least putting itself barely on what increasingly appears to be the "right" side of this issue, that carbon emissions are a threat to promoting global warming. (Mobil Advertisement, Global climate change, 1997). Mobil's PR risk communication move may be considered slightly proactive, but far short of paradigm-breaking, as was BP's.

Good risk management and risk communication are also becoming important today for attracting and keeping good employees (as well as sustaining sales and investments). Many employees do care about the moral and ethical positions and actions of their employers (one such example highlighted previously is the Hewlett-Packard Company). And communities certainly are very attuned to company risk management and risk communication; recent years have witnessed legions of examples in which communities have protested against and even prevented relocation of perceived hazardous and otherwise anti-environmental operations in their surroundings—the well-known NIMBY ("not in my backyard") movement. Thus citizens in a suburb of Minneapolis dug in and fought tooth and nail for years to successfully keep out a BFI facility, for which the struggle became the subject of a 'mini-classic' on how to organize and stop 'bad image' industry from moving into your community. This incident got off on the wrong foot, and got progressively uglier. It became obvious that company PR–risk communication was ill-advised in its decision that the company would build the facility regardless of community concerns.

SOME LESSONS LEARNED: WHY SYSTEMATIC RISK MANAGEMENT AND RISK COMMUNICATION ARE NEEDED

WHAT CAN GO WRONG WITHOUT APPROPRIATE RISK MANAGEMENT AND RISK COMMUNICATION: SOME EXAMPLES

Case Study #1: No Real Risk Management or Risk Communication Everyone knows nuclear power presents certain risks. And everyone now has heard of Chernobyl, which we know "couldn't happen here." Instead the United States did, in fact, experience an "almost Chernobyl"—that was the Three Mile Island (TMI) disaster, an incident we should remember not because of the apparent risk management oversight, but also for its botched risk communication which served to prolong and aggravate uncertainties—the seriousness of the crisis as perceived by the public.

So what happened at TMI, and what went wrong with the communications end?

As we are all aware, TMI captured the nation's attention not because of massive, immediate loss of life—or even serious injuries, but because of lingering uncertainties as to "how safe" the situation and local surroundings were. As many saw, authorities handling the TMI incident compounded the severe problems inherent in "explaining away" a near-nuclear-catastrophe, by a mix of delay, dilution, excess uncertainty, and ambiguity in their messages.

According to crisis communication expert Otto Lerbinger, communications about TMI were seriously hampered by a classic litany of oversights. First, there was a failure to speak out with one voice—one identified spokesperson to provide information. Further, apparently the operator, Metropolitan Edison, had no crisis plan. Next, the organization failed to notify the media itself; therefore the unfolding, mystery-laden story was "uncovered" by reporters. Then, TMI itself was unable to meaningfully field soon-to-follow press inquiries, deepening the mystery awe—and the sense of crisis. Finally, information about what had actually happened, as well as due safety advisories—*separately* from state, federal, and plant authorities, sometimes conflicting—dribbled out over several days after the incident, thereby creating still more confusion and wrenching apprehension among locals potentially affected (Lerbinger, 1997, p. 43–44).

Case Study #2: Inadequate Risk Management—Risk Communication Can Only Do So Much Such examples are becoming harder to find today in corporations of any size, but still exist in small and even medium-sized firms in the United States, and in developing countries.

Only a few years ago, there was just such a case in the Chicago area. A relatively small firm in the photo process industry had to work next to open vats of hazardous chemicals as part of operations. Employees in this area were said to have complained repeatedly of health-related problems, but apparently management did nothing substantive to protect them from the deadly exposures, nor warn them of the consequences of exposure. Then the bottom dropped out—a worker keeled over and died at the exposed chemical vat.[1]

The incident, when reported, made nationwide news, and attracted the wrath of criminal prosecutors, resulting in murder charges being filed against—and successfully maintained—against key company managers. With such a drastic occurrence, the company was put in a position for which there was no credible defense; any feeble attempts at explanation or justification fell on deaf ears. No amount of risk communication (what little the company attempted was ineffectual at best) could or can "overcome" a risk management oversight of this magnitude. This was simply a build-to-order all around lose-lose situation. The company was irresponsibly exposing its workers to lethal hazards (no risk management), failing to reasonably inform about such risks, and therefore directly caused and had to face the consequences of the unspeakable tragedy.

Case Study #3: Acceptable Risk Management, but Inadequate Risk Communication Mega-Borg was a supertanker that exploded, burned, and spilled millions of gallons of oil into the Gulf of Mexico. While it is unclear whether more appropriate risk management measures might have prevented this incident, this is a case in which risk communication was notably poor, resulting in more public misunderstanding and resentment than necessary. Problems that have been noted with handling of this incident by responsible company officials included: delay in getting out information, inflexibility in dealing with the media, and failing to provide steady information (Brown, 1990: 39–40).

WHAT CAN GO RIGHT WITH APPROPRIATE RISK MANAGEMENT

Case Study #1: Acceptable Risk Management and Risk Communication Over the past several years in Minnesota, and several other states, reports have accumulated about observed frog deformities, that is, ordinary citizens and experts observing and collecting frogs in the field with additional limbs, nonfunctioning eyes, etc. As news spread through word of mouth and occasional articles, the authorities got involved, and began conducting studies aimed at finding the cause for the deformities. While this work was proceeding—and deformities continuing to be found—in late-1997 the popular TV news show, *Nightline*, decided to air an expose on the "crisis."

Luckily, or through proactive communications, the state authorities investigating the case—the Minnesota Pollution Control Agency (MPCA)—learned of proposed airing a day or so before. In the interest of providing best possible available information on the status of the investigation and any threats, MPCA preempted *Nightline* by providing a press conference a day or so before the *Nightline* broadcast. Here MPCA was able to orchestrate and control the distribution of known facts, still quite incomplete, thus avoiding any sensationalism that might have been produced by the *Nightline* broadcast. The point is MPCA became proactive when it had to be, and did convey study results to date, as well as investigate strategy and participants—in short, a progress report. Consequently no one (in Minnesota) panicked from the *Nightline* report (which, itself, to remain credible, could not go beyond MPCA's findings), allowing the agency to return to its critical work on continuing its investigation of the problem. Finally, this instance is also significant for the MPCA, because all too frequently the agency has been criticized for not doing enough to protect the public interest, as in the case of its low key, minimalist response to discovery of a leak from an oil refinery threatening to contaminate the Mississippi River (Meersman, 1997, pp. B1, B7).

MPCA's handling of the recent frog deformities case was successful—up to a point—thus prudential, but arguably not proactive, because the agency will doubtlessly need to face additional inquiries as ongoing research results are made available.

Case Study # 2: Exemplary Risk Management or Communication and Acceptable Risk Management or Communication As stated previously, the "good-good" (risk management, risk communication) scenarios in real life are not widely covered nor reported upon; thus specific information is often hard to come by. Here we are talking about instances in which risk management—before the incident—is impeccable and risk communication acceptable.

This is the hardest case to find and describe, yet may be quite common in industry; the real goods (in real life) are simply underreported (after all, deaths and injuries make "good" news, not absence thereof). Thus the need to construct a composite hypothetical derived from existing good practices and principles.

Our company, "Good Company"—hereafter GC—is a global manufacturer, with some chemical-handling operations with potential for hazard. Despite very comprehensive measures and programs, a GC truck gets into an accident and has a spill, just outside company property, a few miles from a busy community. What did it do, and what should it have done?

First and foremost, GC has a well-planned and tested emergency response plan in place. As soon as word of the accident reached company authorities (via cell phone from personnel at the scene), the formal emergency response plan (ERP) was put into action. Appropriate public and company personnel were immediately put on notice through an automatic contact system, and the lead GC crisis manager was contacted. The crisis manager then communicated with and assembled key personnel—either physically or through real-time, virtual connections, to plan response actions. Meanwhile an emergency response team had already cased the site, along with public authorities, to address and neutralize any imminent hazards. Appropriate media were contacted and notified of essential facts, and that the situation is still fluid, but assuredly being responsibly addressed by GC and public personnel in coordination.

At the scene no one had been injured, but potentially exposed personnel are being checked for any impacts; the media is also advised of this. And the spill itself is being contained and cleaned up, under watchful eyes of public authorities. GC is able to, and takes the trouble to, demonstrate rapid, effective, state-of-the-art cleanup techniques; portraying (really and responsibly) to the media and the public a job well done. And because this—the successful management and cleanup efforts—became *the story,* rather than hidden facts, agendas, etc., the story for the media soon died a kind death, with GC looking good.

Finally, though Good Company's actions in this case appear "flawless," they still only demonstrate a prudential level of risk management and risk communication; not necessarily a proactive standard.

PUTTING IT ALL TOGETHER: DEVELOPING YOUR RISK MANAGEMENT AND RISK COMMUNICATION STRATEGIES

This section emphasizes practical, functional "how-to's" of risk program assessment and improvement; not assessment as to "level," that is, prudential, proactive, paradigm-breaking. This property of risk programs would need to be gauged by different, broader approaches that are beyond the scope of this discussion.

WHERE TO START

Regardless of the 'level' of progressiveness and proactivity a company desires in its risk management and risk communication programs, the obvious place to start, as with many business reform/reengineering processes is: What is the current state of affairs in the company's risk management and risk communication programs, and, what has been the history (e.g., *why* initiated) and the evolution of programs (e.g., disaster-driven, or prudence-driven)? Best available outsiders, along with other relevant company data and concerns, need this review, to understand how and why company programs are what they are. This information is important in deciding which aspects of the current system should be retained, which should be recommended to rebuilt, or merely patched, and which should be scrapped.

Besides the basic historic information, outside evaluators/designers need solid, complete information on program effectiveness. What has worked well in the past, what hasn't, and why? What are the objective facts and results about past programs? Armed with this information along with company history and goals (assumed to be part of initial charge), evaluators have a baseline from which to proceed.

These are the easy, primarily mechanical parts. The more important step is addressing fundamental goals of your risk management and risk communication program. In perspective, will your goals: 1) be merely minimally prudent, that is, just doing enough by playing it safe and being concerned only with your own little micro-corner of the world, as it if was disconnected from the rest of the world's environmental problems; 2) be as proactive as possible about addressing your company's environmental risks, perhaps even being an industry leader and having the best-looking program in your industry (judged by traditional standards), for example, having a minimum of "bad events;" or, 3) break free of the current failing environmental management paradigm, such as embracing closed loops; zero release goals and policies; broadest possible risk assessment, management, and communication, acknowledgement of company contributions to global problems, seamlessly integrated; risk function playing a major strategic company role.

DEVELOPING YOUR APPROACHES TO AN INTEGRATED RM PROGRAM: SOME RECOMMENDED STEPS

These steps build on and are based on, the initial information collected and analyzed above, and on the direction decided for the risk program (i.e., how good, how broad, true, and straighforward?):

1. Based on company history, goals, and business constraints, develop a vision for an effective and efficient integrated company risk management and risk communication program(s).

2. Determine which aspects of existing programs meet desired company risk management and risk communication goals.

3. Assess and compare costs and benefits of apparently satisfactory program elements (There will be wide range of possible approaches here).

4. Identify gaps, deficiencies, or missing links in program elements that meet company muster: Determine which additional elements are needed to provide a sound, integrated overall risk management and risk communication program.

5. Devise essentials of gaps, deficiencies, or missing links, along with cost-benefit analyses for proposed program additions.

6. Having a complete potential risk management and risk communication program mapped out, conduct further analyses to assure that:
   a. the program as a unit and critical parts will be acceptably functional;
   b. the program and all parts meet reasonable cost-benefit levels;
   c. the package is feasible—technically, psychologically, and financially.

7. Develop the program sufficiently to test it in its entirety as well as critical parts, preferably in realistic manner (e.g., staged scenarios with real actors), by simulations, expert grading, other techniques, and combinations of these.

8. To continually improve programs or address any still outstanding shortcomings, start back on the appropriate step, and work through subsequent steps.

HOW EFFECTIVE IS YOUR RISK PROGRAM?

This is the key question for all companies, all the time. Ultimately, program effectiveness will be borne out by results, hopefully positives, in terms of problems avoided. The next best approach is testing, under as realistic circumstances as feasible. And ideally, tests should be run by outsiders (there is too much temptation for company personnel—with a clear stake in quality of existing programs to fudge to the good side), with company personnel participating as "guinea pigs." The concept of testing as used here, is, however intended to be comprehensive. Thus a "test" in the context of assessing fitness of risk management and risk communication programs and elements may include multiple observations of qualities as hot line response times and reactions to full blown tests of the entire system using realistic scenarios and real company personnel. Historical data may also have bearing on today's programs, but will clearly be of secondary value to testing.

CONCLUDING THOUGHTS AND REMARKS

NEED FOR CONTINUOUS IMPROVEMENT

Today more than ever the world is witnessing an explosion in new technologies, social phenomena, and outright threats to organizational welfare. Thus the watchword for the wise has to be: eternal preparedness through keeping on top of trends, events, and anything that may present undue risk to company interests. And, as technology and threats keep developing in their sophistication, so too must company preparedness and responses. Also, from a legal perspective, companies must keep improving in tune with society's (i.e., jury) expectations of what are reasonable safety measures, to avoid being held liable for negligent behavior or worse. If we take parking garages as one example to emphasize this point, there was once a time in big cities when the only requirement for operating a parking garage was that is be safely engineered for cars. Then came the socially recognized need for adequate lighting. Then came some sort of personnel presence. Then (in some situations) came video surveillance. And now, sometimes, door-to-door escorts. Finally, this upward trend of social expectation of increased standards of safety—risk management—is not limited to parking garages; they are pervasive throughout industry, and well advocated for and advanced by the trial bar.

RISK MANAGEMENT AND RISK COMMUNICATION AND OTHER COMPANY FUNCTIONS

Clearly, risk management and risk communication bear close relationships with several critical company functions. Above all, risk management and risk communication are closely linked to a company's financial well being. Failed risk management or risk communication typically translate directly into financial payouts (through fines, penalties, litigation costs, settlements, and judgments). Other, less direct costs come later: increased insurance premiums (or cancellations), shunning by investors, customers, and others. Thus risk management and risk communication closely relate to the core business financial functions, for without positive finance no company can survive.

Depending on company size, legal advocates get involved when faulty risk management, even risk communication, lands us in court. There are considerable differences of opinion on how much legal counsel should be involved at the "front end" of risk management and risk communication. Proponents argue there is good justification for truly proactive, visionary sort of preventative law practice: at the other end of the spectrum 'legal conservatives' only want legal counsel involved as a last resort. Which philosophy a company adopts, including the vast gradient between these extremes is an individual company choice, which should be based on the relative risks from operations (perhaps they really are very minimal). Arguably, the more serious the risks, the greater the involvement of legal counsel should be across the board. For example, contrast risk issues faced by a chemical manufacturer vs. risk issues a writing service may face.

Engineering is intimately related to risk management and risk communication in companies that develop or sell goods. Arguably, engineering is the "first line of defense" in effective risk management; ideally ineffective designs and features are simply not allowed. Engineering and other scientific and technical experts are also extremely useful in providing accurate and effective risk communication; they should always have their input on important statements the company says about technical matters.

Finally, it goes without saying that the risk function itself—whether stand alone or submerged beneath some other function, HR, for example, is intimately involved with risk management and risk communication. The question is how much authority, and what are the reporting relationships. Arguably, in an optimally designed program, risk management (including risk communication) should be a stand-alone function reporting to the top, assuming the company faces significant environmental-related risks—or imposes significant environmental impacts.

Concluding Perspectives

The purpose of the discussion and examples presented in this chapter has been not only to illustrate some "good's" and "bad's" about risk management and risk communication in selected real and hypothetical case studies, but also to prompt the reader to look beyond where risk management and risk communication have so far gone, and ask: What sort of risk management and risk communication will be necessary and desirable in the next century, given the escalating and unabating pace of local through global environmental degradation? If it is accepted that, indeed, environmental degradation (resulting from the ever-increasing size, scale, and severity of human activities—themselves largely commerce-driven) shows no sign of decelerating, then we must assume that current environmental risk measures are inadequate.

The further conclusion is that merely doing more of the same, or the same, better by degree—(merely) reducing some emissions here, some wastes there, having fewer accidents, smaller spills, and better cleanups—will not adequately stem the accelerating tide of worldwide environmental destruction. Rather what will be needed will be much broader conceptions of risk assessment, and risk management, and of business itself. Business cannot continue indefinitely in its present form without destroying the earth. To help stem the current destructive tide, company risk functions and business in general must be brought up to speed: both need to adopt and practice environmental ethical guidelines that result in protection of the earth rather than its dismemberment. Already in our recognition of the serious threat expanding business-as-usual activities pose to the earth, suggestions have been advanced on alternative paths business, including environmental management functions, must follow. Among these are business greening, environmental stewardship, so-called sustainable development or sustainable business, ecocentric management (Shrivastava, 1993), and an inclusive viewpoint of environmental ethics (Nicholson, 1992, 199_).

Unfortunately, the majority of proposed environmental reform measures fail to offer needed fundamentally different ways of looking at business and the world (business greening, environmental stewardship sustainable development, sustainable business, etc.), because they all stem from and harbor an overriding anthropocentric worldview and environmental ethic (nonhuman entities have only instrumental value, therefore, they, including the biosphere are potentially expendable for the sake of human growth and commerce). Only truly different, nonanthropocentric approaches such as ecocentrism, inclusive environmental ethics approaches, and perhaps some cosmocentric systems offer values priorities compatible with earth survival. Therefore the paramount, overriding environmental challenge is visioning, designing and implementing risk management, risk communication, and business protocols in sync with truly earth-compatible ethics, not in fundamentally inadequate, incremental-type reforms.

ENDNOTE

[1]One trade publication that regularly features such examples and their accompanying fines and penalties is *Safety Compliance Letter*. Many of these cases contain valuable "lessons learned" nuggets for readers to refer to in their own workplaces.

REFERENCES

Beaucamp, Tom L., and Norman L. Bowie, eds. *Ethical Theory in Business*. 3rd ed. Englewood Cliffs, NJ: Prentice Hall, 1988, pp. 253–254.

Brown, Guy E. "Building Momentum after Delayed Response," *Public Relations Journal* vol. 46, December 1990, pp. 39–40.

Buckholz, Rogene A. *Principles of Environmental Management: The Greening of Business*. Englewood Cliffs, NJ: Prentice Hall, 1993.

Committee on Risk Perception and Communication, National Research Council. *Improving Risk Communication*. Washington, DC: National Academy Press, 1990.

Crosby, Liz, and Kenneth Knight. *Strategy for Sustainable Business*. New York: McGraw-Hill, 1995.

Davies, John. *Greening Business: Managing for Sustainable Development*. Cambridge, MA: Basil Blackwell, 1991.

Dougherty, Devon. *Crisis Communications: What Every Executive Needs to Know*. New York: Walker, 1992, pp. 93–94.

Fayol, H. Risk management. In Vaughan, Emmett. *Fundamentals of Risk Management and Insurance*, 4th ed, New York: John Wiley, 1986, p. 35.

Feldstein, Lee M., and James K. Cassidy. "Environmental Health Center Prepares for the Worst Emergency-Response Exercise at Eli Lilly's Tippecanoe Laboratories plant in Lafayette, Indiana," *Safety & Health*, 147, 1993, pp. 31–32.

Fink, Steven. *Crisis Management: Planning for the Inevitable*. New York: Amacom, 1986.

Global climate change. *Time*, November 3, 1997, p. 81 [Mobil advertisement].

Hawken, Paul. *The Ecology of Commerce*. New York: Harper Collins, 1993.

Head, George L., Ed. *Essentials of Risk Control, Volume 1.* Malvern, PA: Insurance Institute of America, 1995.

Head, George L. "Updating the ABCs of Risk Management," *Risk Management* October 1986, pp. 50–56.

Kaufman, Donald G. and Cecelia M. Franz. *Biosphere 2000: Protecting our Global Environment.* New York: HarperCollins, 1993, p. 214.

Lerbinger, Otto. *The Crisis Manager; Facing Risk and Responsibility.* Mahwah, NJ: Lawrence Erlbaum, 1997.

Meersman, Tom. "Private wells linked to frog deformities," *Star Tribune* (Minneapolis). October 1, 1997, pp. A-1, A-15.

Miller, G. Tyler, Jr. *Environmental Science,* 6th ed. Belmont, CA: Wadsworth, 1997.

National Research Council Committee on the Institutional Means for the Assessment of Risk to Public Health, *Risk Assessment in the Federal Government: Managing the Process.* Washington, DC: National Academy Press, 1983.

Newton, Lisa H., and Catherine K. Dillingham. *Watersheds: Classic Cases in Environmental Ethics.* Belmont, CA: Wadsworth, 1994, Chapter 4.

Nicholson, Stuart A. "Environmental Ethics," pp. 269–290. In Sullivan, Thomas F. P., ed., *The Greening of American Business.* Rockville, MD: Government Institutes.

Nicholson, Stuart A. *Business Environment Ethics: Defining and Implementing Business Environmental Responsibility.* Rockville, MD: Government Institutes, 199_ [under contract].

Shrivastava, Paul. "Ecocentric Management for a Risk Society," *Academy of Management Review* 20 (1), 1995, pp. 118–137.

Starr, Mark, Mary Hager, William J. Cook, and Carolyn Friday, "America's toxic tremors," *Newsweek,* August 26, 1985, pp. 18–19.

Stead, W. and J. Stead. *Management for a Small Planet,* 2nd ed. Newbury Park, CA: Sage, 1996.

Testing Responsible Care®: Union Carbide Finds It's Not That Easy Being Green. *Chemical and Engineering News,* vol. 70, October 5, 1992, p. 30.

The untold story of the Exxon Valdez. *Turning Point.* American Broadcasting Company (June 15, 1994), television broadcast.

Unexxonerated. *Economist* (June 18, 1994), p. 32.

CHAPTER 12

WHAT ISO 14000 BRINGS TO ENVIRONMENTAL MANAGEMENT AND COMPLIANCE

Joseph Hess, SGS-Thompson, Inc., Maria Kaouris, Lucent Technologies and John Williams, consultant, with an Introduction by Gabriele Crognale, P.E.

> *"ISO 14001 . . . It's not so bad."*
>
> *Jack Bailey,*
> *Environmental Manager,*
> *Acushnet Rubber Company*

INTRODUCTION

From the time in 1993 the International Organization for Standardization (ISO) first conceived the ISO 14000 standards with the formation of Technical Committee (TC) 207 and its supportive Technical Advisory Committees (TAGs), Work Groups (WGs) and Sub-committees (SCs), we have come a long way. These dedicated voluntary groups of TC 207, an ad hoc committee, if you will, of environmental professionals from various disciplines, developed the text of the standards that culminated with the ISO's approval of the first series of environmental International Standards between the periods of September and October of 1996. The first of the standards to be approved by the full body of the ISO was ISO 14001 (Environmental management systems—Specification, with guidance for use) and the accompanying support document, ISO 14004 (Environmental management systems—General guidelines on principles, systems, and supporting techniques), ushered in the dawn of a new era for environmental management as we know it today. The second set of standards approved by the ISO in October of 1996 were the environmental auditing series that consist of: ISO 14010 (Guidelines for environmental auditing—General principles); 14011 (Guidelines for environmental auditing—Audit procedures—Auditing of environmental management systems); and 14012 (Guidelines for environmental auditing—Qualification criteria for environmental auditors). Each of these standards was designed with the intent for organizations to utilize them for addressing, tracking and assessing environmental issues in each organization's daily business activities as they relate to the environmental management system.

317

On a parallel track, a number of environmental and quality professionals and instructors, many of them TAG members, collected their thoughts from participation in the standards development process into books about the standards, thus providing some insight into the standards from an insider's perspective, such as, (how to) implement an environmental management system (EMS), auditing a facility's EMS, or understanding each of the standards and how these may affect business as usual. Two books, in particular, are part of the Prentice-Hall PTR Environmental Management and Engineering Series: (1) Ritchie and Hayes: *A Guide to the Implementation of the ISO 14000 Series on Environmental Management,* and (2) a series of books on ISO 14000 by L. Kuhre, beginning with *ISO 14001 Certification—Environmental Management Systems.*

In addition, a number of organizations have emerged providing specialty training geared to the ISO 14000 standards, such as, an introduction into the standards, conducting EMS audits, and becoming EMS Lead Auditors, to name a few. In the electronic media arena with the explosive growth of the Internet, a number of nongovernmental organizations (NGOs) have established web sites for keeping their subscribers up-to-date on ISO-related matters, such as Chat Rooms with key USTAG members, and additional standards information, at the click of a computer mouse. Given this abundance of information related to the ISO 14000 standards, we made a decision not to duplicate the efforts of the other authors, and instead, chose to take a somewhat different approach in supplementing that which is already available. In this fashion, we felt we could provide our readers added value in dealing with the ISO 14000 standards by concentrating instead on several key areas in which readers may be interested: concentrating on day-to-day EH&S activities, and how these interrelate to the standards; providing a linkage between ISO 14001 and the auditing and environmental performance (ISO 14031) guidelines to look at areas for continual improvement(s); sharing some insight on certification and what can be expected; highlighting some cost savings; and closing with a bridge to the works of our contributors. We hope you approve.

We begin our focus with brief historical background about the ISO and how the standards came about, and segue into a brief description about ISO 14001, 14004, and how the 14010s, the 14020s, 14031, and the 14040s tie in with 14001. Should readers be interested in pursuing additional information about these standards, they should contact any of the ISO-related organizations listed in Appendix B, or log on to any of the ISO-related Web sites listed in Appendix G.

INTRODUCTION TO THE ISO 18000 STANDARDS

In order to properly introduce the ISO standards for environmental management, and to provide some insight as to the impact these standards can have upon industry and trade, we begin with the passages from two books written about ISO 14000 by several prominent authorities, and an interview with the environmental manager from one of the first organizations to certify to ISO 14001 in the United States. We chose these individuals as representative of the enviornmental professionals who make up the USTAG to the ISO's TC207, or otherwise affiliated with ISO 14000, that are banging the ISO 14000 drum and are among its champions.

We begin with Lee Kuhre, author of a series of books on ISO 14000 and other environmental management books, and cite a passage from the opening paragraph of his first book about ISO 14000.

> The addition of environmental management into ISO certification is monumental in significance. The impact will be felt in terms of the environment, operation of most organizations, customers, agencies and most components of society. There will be a positive impact on the environment since many organizations will have to start *doing additional environmental protection* to obtain certification.[1,2]

The action *of doing additional environmental protection* is at the heart of an environmental management system that organizations should strive to implement and improve as necessary, and for which, certification can be seen as one barometer to measure how successful these organizations can be in achieving such objectives.

In addition, we cite a passage from the opening paragraph of a book co-authored by Joe Cascio, Chairman of the U.S. Technical Advisory Group (TAG), and contributing author of several books on the subject as well. That passage reads

> ISO 14000 embodies a new approach to environmental protection . . . challenges each organization to take stock of its environmental aspects . . . (a) *new paradigm relies on positive motivation and the desire to do the right thing,* rather than a punishment of errors . . . a solid base for reliable, consistent management of environmental obligations.[3]

The *new paradigm resulting from positive motivation and desire to do the right thing* is one of the key drivers for organizations that want to move forward and take responsibilities for their products and services—the heart of being true product and environmental stewards.[4]

Interestingly, these books in particular[5] seem to focus on a theme similar to ours regarding the ISO 14000 environmental management standards, and what these standards can provide to organizations striving to attain their stated environmental goals and obligations. One key to understanding the intent of ISO 14000[6] is for the reader to understand the value that conformance to ISO 14001[7] can bring to a company's environmental management system. To clarify this concept, the reader should view the standards in the context of his or her company's environmental, and health, and safety (EH&S) programs, and honestly assess whether their EH&S or EMS programs could benefit from an EMS as described in ISO 14001 or not. Unless this comparison is made, it can sometimes be difficult for EH&S managers to make this leap of faith, unless they see it for themselves.

As another example extolling the benefits of ISO 14000, we refer to Jack Bailey, director of safety, health and environmental affairs at Achushnet Rubber Company, one of the first ISO 14001-certified companies in the United States. Based on several interviews the author had with Bailey, Bailey's thoughts on ISO 14001 can be summarized as follows:

> We see ISO 14001 as a sound business decision to implement, based on our belief that (ISO 14001) would help convince (our) customer base that Achushnet Rubber is a world-class manufacturer of rubber goods and allows our company to save money. (In this context) saving money takes on two forms: (1) as a cost avoidance in business liability through the prevention of costly fines and eliminating unnecessary waste streams; and (2) by helping (to identify areas) to reduce emissions and the use of toxics to make processes more efficient. Achushnet identified two particular projects with this format, (specifically), a toxics use reduction program and an energy saving program.

By Jack's estimate, Achushnet saved over *$2.5 million in four years* using these programs.

> With respect to ISO, Jack sees ISO 14001 "... as a process that leads you to process solutions. *In these solutions are where the money savings come in.* These could be done without 14001, (but you need to keep in mind that) *14001 provides the roadmap and consistency to get the job done.*" Or, to summarize Jack's philosophy regarding ISO 14001, as he says it, "ISO 14001 . . . it's not so bad."

To which we add, It's not so bad, indeed.

BACKGROUND ON ISO 14000[8]

From an historical perspective of the ISO 14000 standards, let us briefly recap the events leading to how the ISO 14000 standards came into being, and provide some insight into the International Organization for Standardization, or ISO. The ISO is a specialized organization that develops trade and manufacturing standards in a number of industries, including scientific applications, among others, and promotes the development and implementation of voluntary international standards. Among the more widely recognized of these standards are the ISO 9000 quality standards. The organization is based in Geneva, Switzerland, and was founded in 1947. It consists of a little over one hundred member bodies, or countries, and within each country, a designated group represents each member body. In the United States, for example, the American National Standards Institute (ANSI), a nongovernmental organization, holds this charge, and among ANSI's members, include governmental agencies, such as, the National Institute of Standards and Technology (NIST), the U.S. Environmental Protection Agency (EPA), and the Department of Energy (DOE).

ISO 14000 can trace its lineage to the Eco-audit scheme, a proposal to allow industrial companies within the member states of the European Community (EC) to devise a scheme with the objective of enhancing their environmental management programs,[9] and to the British Standards Institute's (BSI) BS 7750. These documents, in turn, can trace their lineage to ISO 9000 that was passed in the mid-1980s. At the very beginning, ISO began its work in the environmental arena in 1971 with the establishment of two new technical committees: TC 146, charged with developing air quality standards, and TC 147, charged with developing water quality standards. With the establishment of TC207, a new era for ISO emerged, focusing on environmental related standards.

During the 1980s, ISO looked to broaden its focus on management in organizations and decided to develop standards on one specific aspect of management, that of quality management. The committee convened to develop these quality management standards was Technical Committee (TC) 176, which went on to forge the now-familiar and popular ISO 9000 quality management standards to address generic quality management issues. ISO finalized these standards in 1987. Since then, ISO 9000 has gone on to win international recognition and credibility, and in some regions of the world, registration to ISO 9000 has become a *de facto* condition of trade and commerce.[10] From that recognition, ISO gained confidence in its standards development capacity in the quality management area. However, not until the emergence of several other factors would ISO embark into the more controversial environmental arena, on its way to developing the ISO 14000 standards that we know today.

Prior events such as the Basle and Lome Conventions in 1987, and the issuance of the Montreal Protocols regarding the use of CFCs, began to throw more light on environmental issues, and thus signaling a wake-up call for greater international action. The issues centered around ozone-depleting substances such as CFCs, the Bhopal incident [(the driver behind EPA's introduction of utilizing Toxic Release Inventory (TRI) release data under the Superfund Amendment Reauthorization Act, or SARA Title III)], and transboundary hazardous waste shipments, that transcend regional considerations, and are in fact, international in scope.[11] In addition, the beginning of the 1990s also saw the growing proliferation of national and regional environmental standards that had the potential to create unwelcome trade barriers that had many environmental strategists concerned.[12] As a result, when the United Nations (UN) announced in 1991 its intent to hold the United Nations Conference on Environment and Development (UNCED) in June 1992 in Rio de Janeiro, that event was the trigger that prompted many of these strategists into action.

Based on this request, ISO began an inquiry in the summer of 1991 for voluntary advisors from its members to assess the need for international environmental management standards and to develop an overall plan for developing such standards. With that, ISO formed an advisory group, the Strategic Advisory Group on the Environment (SAGE) in August 1991 to make that assessment. SAGE was asked to consider whether such standards could serve to:

- Promote a common approach to environmental management, similar to quality management;
- Enhance organizations' abilities to attain and measure improvements in environmental performance; and
- Facilitate trade and remove trade barriers.

By mid-1992, SAGE decided that ISO should develop environmental management standards, and made this decision public at UNCED, or the Earth Summit. At that conference, over one hundred countries agreed on the need[13] for further development of international environmental management programs that could support the UNCED goal of sustainable development and that could be compatible with diverse cultural, social, and organizational make-up. Then, in the fall of 1992, SAGE gave its recommendation to ISO's Technical Management

Board (TMB), which soon thereafter authorized the creation of Technical Committee (TC) 207. With that act, ISO commissioned its second effort in the broad field of management standards, with the intent of duplicating a successful quality formula within the environmental arena. By early 1993, TC 207 was empowered by TMB to craft the language to describe (and not prescribe) environmental management systems and the requisite ancillary tools in auditing, labeling, environmental performance, life cycle assessment, and environmental aspects in product standards.[14]

THE FAMILY OF ISO 14000: HOW THE STANDARDS COMPLEMENT EACH OTHER

In the general scheme of the architecture of ISO 14000, the environmental management standards are grouped into two subsets: Organization Evaluation and Product Evaluation, as described in Figure 12–1.

Our focus in describing the environmental management system standards of ISO 14001 and the supplemental standards, or guidance documents, is intended to introduce the linkages between each of these standards, and how, utilized collectively, can lead to effective environmental management systems within a company that can aspire to continual improvement, as depicted in Figure 12–3,[15] Environmental Management System Model for ISO 14001 and 14004.

In further detail, the following paragraphs provide an encapsulated view into the standards as conceptually described in Figures 12–1 and 12–2.[16] As the figures depict, the ISO 14000 series place their greatest emphasis on the organizational standards, specifically the management systems standards, ISO 14001 and 14004. Within these standards are contained the blueprints, the "DNA" if you

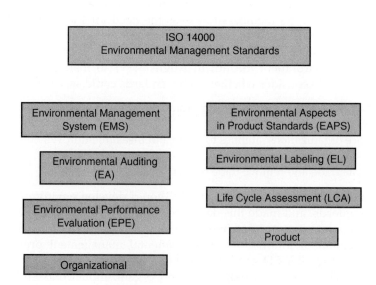

Figure 12–1   The sub-sets of ISO 14000: Organizational & Product Evaluation

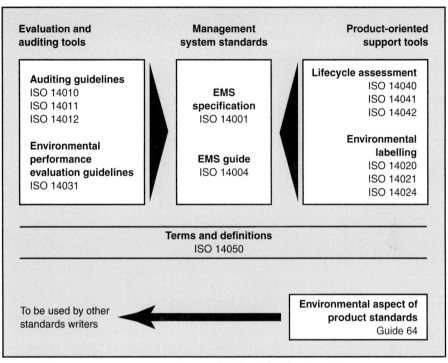

Figure 12–2   The Structure of the ISO 14000 Series of Standards

will, that organizations can follow to evaluate how its activities, products, and services interact with the environment, and more importantly, how to control those activities to ensure their stated environmental objectives and targets are met. These targets and objectives can be achieved following the basic steps as outlined in the Morbius Loop diagram that describes the steps contained within Clause 4 of ISO 14001, as shown in Figure 12–3.

ISO 14001

ISO 14001 specifies the requirements for an environmental management system(s) (EMS) of a facility or an entire organization,[17] that can be certified by a third party, such as a registrar, or by self-certification. The document provides specifications detailing those requirements an organization must meet in order to achieve certification. Certification is that process by which an organization demonstrates its successful implementation of an EMS in conformance with ISO 14001, as contained in Clause 4: Environmental Management System requirements, which include: (1) developing its environmental policy; (2) identifying its environmental aspects; (3) establishing legal and other requirements; (4) establishing environmental objectives and targets; (5) establishing an environmental management program to achieve its objectives and targets; (6) implementing an EMS that includes structure, training, communication, document and operational control, and emergency preparedness and response; (7) monitoring and measur-

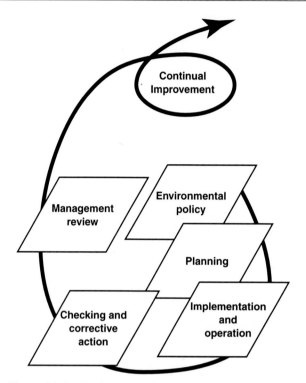

Figure 12–3   Environmental Management System
Model for ISO 14001 and 14004

ing of operations, including record keeping and EMS audit procedures; and (8) management review of EMS to ensure its continuing suitability, adequacy and effectiveness. Additional guidance on the use of the requirements of ISO 14001 is contained in Appendix A of that document.

With certification, an organization can use this demonstration to assure any interested parties that an appropriate environmental management system is in place in conformance with the elements of ISO 14001. With respect to demonstrating a successful implementation of an EMS, each of the elements of Clause 4 are auditable items for verifying conformance to ISO 14001.

ISO 14004

In ISO 14004, TC 207 developed this ancillary document to provide additional guidance to those organizations that may need assistance in designing, developing and maintaining an EMS. An organization cannot be certified against this document; rather, ISO 14004 provides the necessary "jump start" to those organizations that may need a little help in understanding the fundamental foundation to the principles, systems and supporting techniques necessary to develop an EMS that can be in conformance to 14001.

The intent of this guideline is to provide an organization with suggestions for helping to develop its environmental management system requirements, such as, defining its environmental policy. ISO 14004 states, in part, that ". . . organization should begin where there is obvious benefit, for example, by focusing on regulatory compliance, by limiting sources of liability or by making more efficient use of materials."[18] ISO 14004 also makes use of examples or prompts to help the EMS process along, that are listed as "Practical Help" in the guidance document. An analogy of ISO 14004 to regulatory compliance would be an inspection or audit checklist to ensure that all the bases are covered during the process. Since a good number of the U.S. companies represented by the U.S. TAG have mature EH&S and EMS systems in place, a practical exercise for organizations without such mature systems may want to evaluate (benchmark) the best-in-class companies to follow their lead in developing an environmental management system that follows the flavor of ISO 14001.

ISO 14010s

The next set of guidance standards, ISO 14010[19] through 14012, provide those support tools necessary to allow organizations to assess whether their EMSs are in conformance with the elements contained within Clause 4 of ISO 14001. These standards provide guidance specifying:

(1) ISO 14010—how an EMS audit is conducted, such as: providing guidelines on the general principles of environmental auditing, including the objectivity, competence and independence of the auditors, applying systematic audit procedures, and providing a format for preparing audit reports.

(2) ISO 14011—the objectives, roles and responsibilities of the audit team, developing the audit scope, collecting audit evidence and editing audit findings for report, and exit meeting.

(3) ISO 14012—the qualifications required for EMS auditors, including educational and professional qualifications, training, and personal attributes, competence and skills.

Looking at these standards more closely, EMS auditors, or EH&S auditors, basically cut their teeth from the same cloth at one point in time. While some pundits may argue that EH&S auditors and EMS auditors come from different approaches and EMS audits should not be conducted with an eye to regulatory considerations, we disagree in principle. Granted, auditing an organization's EMS against the elements of ISO 14001 is not quite the same as auditing for regulatory compliance, but real value can be gained by a facility in being audited to ISO 14001 by a seasoned environmental or EH&S professional that does keep one eye to environmental regulatory considerations, since they can also envision potential liabilities during the EMS audit that may otherwise be overlooked by an EMS-only focused auditor. We will probably be labeled as heretics for this observation, but let's play this hand out with these three hypothetical examples, then you decide in which camp you would rather be in conducting an EMS audit:

1. The facility under review is being audited against it objectives and targets (Clause 4.3.3), and among the considerations are the views of interested parties. *How far would you go to evaluate the views of interested parties? Would it be the same level of scrutiny in an industrial setting as in a residential setting? Would you also consider looking more closely in an area designated by EPA as an area requiring some form of environmental justice, such as in facility locations near poverty-level areas to develop a non-conformance finding?*

2. The facility under review is being audited against training, awareness, and competence (Clause 4.4.2), in establishing and maintaining procedures to make its employees aware of the potential consequences of departure from specified operating procedures. *How far would you go in this regard? Would you consider as relevant and auditable the lack of coordination with the local fire department in the event of an accident or spill release, and would you consider as relevant appropriate safety guidelines in chemicals handling, such as grounding drums of solvents during mixing, or would you consider these as minor or non-auditable events under 4.4.2 (d)? This Author viewed firsthand the catastrophic results of the second example, which resulted in the loss of life.*

3. In this last example, the facility under review is being audited against monitoring and measurement (Clause 4.5.1) regarding the recording of information to track performance, operational controls and conformance with the organization's environmental objectives and targets. *The facility regularly monitors process vents for air emissions, and notes the levels of chemical releases to be within regulatory limits, yet some employees complain of feeling sick in certain work areas. How far would you go to note this occurrence? Would you consider checking the calibration of the monitoring equipment to ensure that it is functioning in accordance with the manufacturer's specifications, and what call would you make in this situation? (For additional insight, refer back to Stu Nicholson's case study, (Bad) Example # 2 in Chapter 11.)*

What do we say in these hypothetical examples? In our opinion, we feel that EMS audits can provide added depth and value to a facility being audited if an audit team looks closer into auditing a facility's environmental management system elements by taking into consideration any underlying environmental, or related health and safety, regulatory issues or potential liability situations that may not be relevant at first blush to a less-seasoned EMS auditor. Having considerable mastery of EH&S experience and regulatory knowledge should not be viewed as a hindrance by the pundits who disagree. [The Author is currently compiling research in auditing practices for a subsequent book he plans to publish in the near future.]

ISO 14020s

The environmental labeling standards fall under the subcategory of product-oriented support tools to the EMS standards of ISO 14001 and 14004, along with life cycle assessment which we discuss within the next several paragraphs. This standard looks more closely at the focus of governmental environmental policy and the business community onto the environmental aspects of products and effectively communicating product information to all the stakeholders. Part of this push may have arisen from stakeholders' previous disappointing experiences with some manufacturers' claims of "environmentally-friendly" products. The standards are meant to provide a measure of acceptability of an organization's environmental declaration and claims regarding a product or service.

From an historical standpoint, one aspect of ISO 14021, *environmental labels and declarations—self-declaration environmental claims,* refers to voluntary environmental reports that are viewed as a self-declaration environmental label by the standard. While this may be a component of a standard that as is a work in progress as we speak, U.S. companies' voluntary environmental reports predate ISO 14000 in the US, and have been the subject matter of various groups to initiate a standard environmental report format. One of the first in this area has been the Public Environmental Reporting Initiative, or PERI, in having its original signatories develop guidelines for presenting the information in an environmental report. As you may recall, we discussed this aspect in Chapter 1 and include a copy of the PERI guidelines in Appendix D. As the ISO 14020s standards progress through final adoption, it will be interesting to see how these guidelines may affect companies' environmental reports, since these reports can develop into fairly good barometers for indicating how well (or how poorly) an organization's EMS may be functioning, and measure an organization's goals and objectives, P2 achievements, audit findings, and opportunities for continual improvement, in essence all of the elements within Clause 4 of ISO 14001.

ISO 14031[20]

Environmental Performance Evaluation (EPE), one of the evaluation and auditing tools depicted in Figure 12–2, is among the supplemental EMS standards that is still a work in progress.[21] EPE, in the context of an evaluation tool, picks up where environmental auditing leaves off. For example, if auditing evaluates an organization's EMS seen as a snapshot in time, then EPE is a real-time video of events unfolding that can supplement and add value to an EMS audit program as part of an organization's continual improvement.

EPE describes the process from measuring, analyzing and assessing a company's environmental performance against set criteria set by the organization's top management. These criteria could also be contained within the nucleus of that company's environmental policy, in which objectives and targets are set, that could also include a commitment to continual improvement and prevention of pollution. It can be reasonably safe to presume, then, that among the goals of EPE would be to provide reliable, objective and verifiable information relating to a company's success in achieving the objectives and targets of its environmental policy. In a nutshell, this is at the heart of how effective evaluation of an organization's environmental performance can lead to sustained and continual improvement of that organization's environmental management systems. To further promote EPE, some TAG numbers of other ISO-member countries proposed a new work item (NWI) for developing a Technical Report on case studies illustrating the application of EPE and of ISO 14031, that could provided supplemental examples to EPE, similar in concept to ISO 14004.[22]

In our improvement conscious business world, companies are constantly seeking ways to understand, improve, and demonstrate sound environmental performance as a complement to sound business decisions[23] to create better products with less waste generated by controlling the impact of their activities, products and services on the environment. Many companies have also found that improved environmental performance can contribute to operational efficiency and profitability, such as was depicted in the benchmarking surveys from Chapter 7. Taking this concept one step further, organizations that can orchestrate performance evaluations of their operational and management systems that are linked to their EMSs can obtain data points that can be useful for measuring and monitoring their own environmental performance. In this fashion, EPE and environmental audits can be seen as complementary management functions, or as described in ISO/DIS 14031, help an organization's management to determine the extent of its environmental performance status, and from that determination, help to identify areas for improvement. Environmental audits assess improvements one pass at a time, while EPE is a continual ongoing process, but the ultimate goal is the same: improvement (of the EMS). This is at the heart of an environmental performance evaluation.

ISO 14040s[24]

Complementing the EPE standards are the life-cycle assessment (LCA) standards, part of the product-oriented support tools as depicted in Figure 12–2. LCA, which actually consists of ISO 14040 (Environmental management—Life cycle assessment—Principles and framework); ISO/DIS 14041 (Environmental management—Life cycle assessment—Life cycle inventory assessment); ISO/CD 14042. (Environmental management—Life cycle assessment—Impact assessment); and ISO/CD 14043 (Environmental management—Life cycle assessment—Interpretation), describes a measurement technique for assessing and measuring the environmental impacts of a company's products and services.

The technique requires collecting an inventory of relevant inputs and outputs of a management system, evaluating the potential environmental impacts associated with those inputs and outputs, and providing an interpretation of the data results as they relate to the impacts, and how this compares with LCA study objectives. For example,[25] in a product manufacturing or chemical process system, the raw materials for the processes could be considered the "inputs," while product throughput, emissions, and waste streams generated could be considered "outputs." Measuring each input and output item in a controlled setting, such as measuring materials and energy consumption versus products and waste streams generated, could allow the accurate measurement of product and waste efficiencies, and whether there could be opportunities for reducing throughput inefficiencies and increase pollution prevention opportunities.

Also to be considered are the opportunities that LCA can provide organizations to systematically address the environmental aspects of products or services at every point of their life cycle. A key to the effectiveness of an LCA study depends upon the accuracy of the data collected and analyzed and how well the recipients of such data clearly understand the scope, data quality parameters, methodologies and focus, or output, of the study. Since the area in question that an LCA study may encompass can overlap into different, yet related study areas, it is very important that the goal and scope of the LCA be made explicitly clear from the onset.

As conceptualized, an LCA can provide supplemental assistance to an organization's environmental management system, EMS auditing procedure, and EPE, by providing an added dimension and link to each of these management system tools and procedures. For example, an organization may want to conduct an LCA study to determine the most effective way to streamline a particular widget-making process. The LCA study would incorporate and assess process emissions, raw material usage and energy consumption, wastestreams, and other resources and monetary expenditures that need to be factored into the study. Depending upon the parameters that are established to define the goal and scope of the study, and how data obtained may be interpreted, LCA study results could identify discreet opportunities to improve upon an organization's environmental management system and EPE.

In closing, as with the EPE standard, the LCA standards are still works in progress, and much new ground may still need to be reached. As this iterative process moves closer to interim completion, we who are the environmental community will have learned much about the entire suite of the ISO 14000 standards, and how these management tools can provide the insight some of us may need to conduct our environmental and health and safety work in an expeditious and safe manner, while keeping a close eye to the performance "bar," that unlike the popular limbo dance, keeps on getting raised with each pass of the "contestants" in the "dance contest." Let's not fall down while we participate, because unlike the limbo dance contest, environmental management is not a game.

CERTIFICATION TO ISO 14001: WHAT IT ENTAILS

What is the significance to certifying to ISO 14001? What are the benefits? For some answers in addition to those voiced by Jack Bailey, we again turn to Lee Kuhre, who states in his book about ISO 14001 certification[26] "... it is expected that customers will require their product suppliers to be ISO 14001 certified (and) some service providers who have an impact on the environment will also probably be expected to obtain certification. (This implies) to a customer that when *product or service was prepared,* the *environment was not significantly damaged in the process.*" In ISO 9000 quality circles, many customers already demand this type of certification under ISO 9001. Kuhre also goes on to state, "... The impact will be felt in terms of the environment, operation of most organizations, customers, agencies and most components of society. There will be a *positive impact on the environment since many organizations will have to start doing additional environmental protection to obtain certification.*"

The italicized phrases in the previous paragraph make a poignant statement about the certification process and the emphasis that certification places upon environmental protection. One item that bears noting during this discussion, is that our focus is not geared to help you choose an accredited registrar to perform third-party audits to certify that your facility conforms to ISO 14001. We neither recommend whether a facility should self-certify versus being certified by a registrar. Should you decide on third-party registration, our position is that just as you screen and determine your final choice for a consultant or other outside contractor to perform a specific task at your facility, you should consider a similar type of screening process to select the registrar or auditor who will certify your facility. That decision is entirely in your corner, just as the decision to use a registrar or vie for self-certification. The key is to know what you want, and how you want to get there. How you get there is up to you.

BENEFITS TO ISO 14001 CERTIFICATION

Here we focus on the tangible benefits that a facility can derive from being certified to all of the program elements contained within the clauses of ISO 14001, as outlined in the Morbius Loop diagram depicted in Figure 12–3—Environmental Policy, Planning, Implementation and Operation, Checking and Corrective Action, and Management Review—with the continual improvement loop to begin the process anew.

What are some of these benefits? As a start, having an EMS in place that conforms to the ideal scenario under ISO 14001 can lead to a reduction in unnecessary waste generation, and hence, initiate opportunities for pollution prevention, commonly referred to as "P2." This, in turn, decreases the amount of waste streams generated, whether hazardous or not, that need to be managed in some respect, and raw materials that could be more efficiently utilized. These translate to monies saved and less impact onto the environment.

This approach to environmental considerations can also send a strong signal to the federal and state regulators that certified organizations care about the environment and want to do the right thing. How that is manifested to the regulators is by increased or sustained compliance to the applicable environmental regulations. In addition, while EPA has not officially endorsed certification to ISO 14001, the Agency has alluded to on several occasions that organizations that certify to ISO 14001 may be inspected less frequently as a result of their diligent efforts. Most recently, EPA issued a position statement on environmental management systems and ISO 14001 and a request for comments to explore ISO 14001 pilots.[27] Among several projects that EPA spearheaded in 1997 in progressive environmental management, that is, focusing on ISO 14000 systems, are: Project XL, the ELP Program, and Star Track.

Finally, as EPA sees it, the benefits of ISO 14000 include the development of a comprehensive and systematic approach and integration of environmental management throughout an organization. Among EPA's concerns include: the actual performance of an organization to compliance and going beyond compliance; and extent of openness and public involvement.

Another organization that understands what ISO 14001 can bring to environmental management is the Chemical Manufacturers Association (CMA). Within CMA, a parallel of sorts exists between Responsible Care® and ISO 14000, specifically within Management Systems Verifications (MSVs) of the six codes of management practices under Responsible Care® (RC).[28] Among the key components of an MSV include the following: focusing on management systems, providing objective, third-party roles, appropriate public involvement, and the promotion of performance improvement and credibility. Additionally, the five elements of an RC management system are almost identical to the five main elements within Clause 4 of ISO 14001, as depicted in Figure 12–3. The relationship, then, between ISO 14001 and RC as a component of an MSV can be summarized into the following five points:

1) MSV provides guidance to CMA Members' interest in a "one-stop" audit.
2) MSV helps to protect the integrity of RC.
3) MSV helps to determine any overlap between RC and ISO 14001.
4) What additional criteria, if any, may need to be met by members to seek a recognition of RC as equivalent to ISO 14001 certification?
5) MSV may be able to provide guidance to members on this issue.

In addition, within the chemical industry a strong desire exists to see the integration of ISO 9000, ISO 14000, and RC as a means to both meet the tenets of RC and address ISO 14001 requirements as methods of fostering overall improvements within the chemical industry.

Finally, additional benefits that may be attained by organizations that certify to ISO 14001, include:

- Reduced on-the-job injuries and less frequent spills, releases;
- Improved community and corporate public relations (as described in Chapter 11);
- Improved customer and supplier relations (an off-shoot of positive environmental self-declaration);
- A reconfirmation of top management's commitment to continual improvement to all levels of employees; and
- And as in the Acushnet Rubber case, $$$ saved, with no upside limit.

Some companies and practitioners refer to this process under the umbrella of strategic environmental management (SEM). We like to refer to it as good business sense.

GETTING TO CERTIFICATION

For those organizations that may be considering certification to ISO 14001, here are a few areas that may need to be addressed as a preparation for attaining a registrar's certification which include: (1) a gap analysis; (2) an EMS audit program; and (3) choosing/bringing on a registrar to perform the surveillance audit required for third-party certification. Each of these items do not necessarily need to be evaluated in any specific order, nor does this imply that any of these areas need to be first evaluated by an organization seeking self-certification, or by an organization that is confident in its mature EH&S program(s), environmental management system(s) and in its effective EH&S auditing program(s). Each of the areas identified should be evaluated further by those orgnaizations that may need additional guidance and insight for deciding whether certification to ISO 14001 by an accredited registrar is right for them. The three areas described in greater detail consist of:

A GAP ANALYSIS—a measuring tool in which in-house EH&S staff or an outside consultant evaluates a facility's existing EMS and compares this to the elements contained within Clause 4 of ISO 14001. Those components of Clause 4 that are found to be missing from a facility's EMS are gaps that will need to be addressed prior to any outside registrar's surveillance audit. This area in particular, has caught the attention of several ISO 14000 software developers that developed an ISO 14001 Gap Tool that can facilitate this task for facility representatives or a consultant working with an organization in preparation for further work. Additional information about some of these software providers is found in Chapter 13.

EMS AUDITS—as described in ISO 14010 and 14011, EMS audits are: *"systematic and documented verification processes of objectively obtaining and evaluating evidence to determine whether an organization's environmental management system conforms to the environmental management system audit criteria set by the organization, and for communicating the results of this process to (top) management."*[29] Simply put, EMS audits are snapshots in time of a facility's EMS that can determine whether the facility being audited conforms to the elements of Clause 4 describing an environmental management system that adheres to environmental laws, addresses the tenets of the organization's environmental policy, and strives to continually improve every facet of its environmental performance.

CHOOSING/BRINGING IN A REGISTRAR—As of June 1998, there were twelve ISO 14001-accredited registrars in business in the United States, with an additional eleven applications on file, according to available figures from the Registrar Accreditation Board (RAB).[30]

That number may either be comforting to know or the beginning of a considerable research project for a facility's environmental manager in trying to narrow down the selection process for choosing a registrar, especially if their organization is unsure in which direction they should proceed. Once the organization arrives at a decision point regarding the certification of their EMS, the appropriate decision makers at the managerial level need to weigh in several factors in this selection process, not unlike choosing a consultant or a contractor.

Choosing the right registrar involves sifting through literature and qualifications packages (Quals Packages) of various auditors employed by the registrars to evaluate their qualifications against the work at hand at a facility, and whether their qualifications measure up to the requirements specific to the task at hand: practical knowledge about environmental issues and management systems and particular industries. Another key piece of information can come from one's network of contacts who have gone through certification for a candid opinion about a particular registrar's work, how the auditors conduct themselves, how thorough their surveillance audits may have been, or other considerations. Another factor in the registrar selection process could hinge upon whether their philosophies mesh with your organization's culture, or whether they may have a completely different agenda from your organization regarding how they interpret the elements of ISO 14001.

This particular topic may warrant additional scrutiny by the organization deciding which registrar to choose. While there are several guides in print[31] for how certifiers or registrars are to operate, such as their code of conduct, confidentiality and auditors' competence, each registrar operates differently from another. To help narrow down the selection process, a well-heeled environmental manager at a facility may want to focus upon these additional areas in narrowing down the registrar selection process:

- How are auditors chosen to work for a registrar, and what additional training are they provided? Is this training classroom only or does it include practical field experience?
- What criteria do the registrars use to certify/fail to certify a facility?
- Are the registrars adept at combining ISO 9000 and ISO 14000 audits if a facility so desires?
- How are disagreements/disputes handled with the client over interpretations of ISO 14001 Clause 4 elements, such as we previously provided in our hypothetical examples?
- How do the registrars ensure confidentiality with the client?
- Were there any complaints against the registrar?
- Did the registrar ever lose accreditation; if so, why?[32]
- How effective were the registrars' previous surveillance audits at other clients' facilities?
- How adept are the auditors at looking at potential "big-ticket" liabilities at client sites as they conduct surveillance audits, or do they just scratch the surface?
- And finally, as a rhetorical question, if you, the environmental manager, were to receive a surprise multimedia inspection the day after your facility were certified by the registrar of your choice, how comfortable would you feel with your choice?

As a final consideration in choosing a registrar, there may be some organizations that are somewhat concerned about the costs associated with third-party certification. In those cases, top management may be wondering whether third party certification can provide some tangible added value to the organization, or whether certification may be primarily providing added income streams for the registrars.

While we realize the preceding several pages only scratch the surface with respect to certification to ISO 14001, the subject matter is sufficiently detailed that various other environmental professionals have transformed their energies into complete books on the subject of certification. As such, this information could not have been condensed into the few pages of this chapter to do it justice, and instead, we provided our insight into several topics we feel are worthwhile to present in a different light. We respectfully defer our readers to procure one or more of these books, some of which we are aware, and are listed at the end of this chapter.

CONTINUAL IMPROVEMENT: WHAT THAT REALLY MEANS

Continual improvement is important enough within the context of an environmental management system in conformance with ISO 14001 that the phrase is contained within the requirements of the Environmental Policy in Clause 4.2, within the requirements of Management Review in Clause 4.6, and in the Morbius Loop diagram contained within Figure 12–3.

In and of itself, continual improvement is a phrase like TQM, TQEM, reengineering or Six Sigma, to name a few, that is meaningless unless there are people behind it who serve as the engine to move it forward. What it is not, is an exercise that facility personnel undergo just to prepare for third-party certification or self-certification, and once that goal is achieved, it is back to business as usual.

Continual improvement is a process that has a place at the grassroots level of an organization, at the level of front-line workers who mold, clean, degrease, assemble, paint, and reassemble parts or mix chemicals and raw materials to make other products that are used elsewhere. These are the people who can make a marked difference in improving conditions within the workplace, the local community and beyond. These are the workers who need to see management take a stand to push environmental goals and objectives through and who also need to be empowered to be able to do the right thing, a fundamental key to continual improvement.

In the meantime, it is incumbent upon each of us within the ranks of environmental and health and safety professionals to do our best to ensure that front-line workers understand their work responsibilities and how these responsibilities can adversely impact environmental situations, and how to mitigate such situations before they occur. Environment and EMS audits and environmental performance evaluations help to ferret out any such potential adverse situations, and based on accurate and well-worded findings and recommendations, additional areas can be uncovered where improvements can be achieved. These management tools work hand-in-hand with front-line employee situations, whether through training, employee empowerment or attitude adjustment, in the bigger picture that encompasses continual improvement as envisioned by the developers of the ISO 14000 standards. In summary, no man is an island, neither is continual improvement.

A CURRENT SNAPSHOT: WHERE WE ARE IN 1998?

Following final approval of the ISO 14000 standards by the ISO member countries as the first of the ISO International Standards in environmental management in the fall of 1996—ISO 14001, 14004 and 14010 thru 14012, respectively—the stage was set for organizations to begin to accept the standards as a method of conducting business. For example, as of January 1, 1998, more than twenty-three hundred ISO 14001 certificates[33] had been issued, broken down as follows: the United Kingdom (440), Japan (425), Germany (320) and the Netherlands (230), and a total of 83 sites had certified to ISO 14001 in the United States, as of December 1, 1997.[34] Interestingly, while the United States has been a key player in developing the standards from the inception of TC 207, it lags behind other major industrialized nations in achieving ISO 14001 certification. This may be due, in part, to the nature of the U.S. regulatory enforcement system, and the lack of a firm commitment from federal and state regulators that enforcement actions would become obsolete for ISO 14001 certified operating facilities.

While ISO 14001 certification may not be as rapid in the United States as overseas, the standard has generated considerable interest as an informational tool and business entity. For example, ISO 14001 created its own cottage industry in publishing and course offerings, such as the publication of a growing number of technical books and trade newsletters, and course offerings for EMS lead auditors in preparation for certification by the Registration Accreditation Board (RAB) to work as lead auditors for accredited registrars.[35] In addition, each of the ISO administrators generate some revenue from the sale of copies of the standards to interested parties.

If the growing interest from users of these products and services can be seen as an indication of where environmental management may be headed over the next few years, it may soon be difficult to challenge that interest exhibited by those organizations that view the ISO 14000 standards among the tools that can provide consistency and form to their environmental programs and related responsibilities. Judging by the number of certifications to date, this could hold true in Europe, the United States, the Asia Pacific Rim, South America, and in other parts of the globe over the next several years and beyond.

For further in-depth information on each of the standards and a description of the organizational make-up of ISO/TC 207 that lists the Secretariat [Canada—the Canadian Standards Association (CSA)], the TAG Administrator [United States—American Society for Testing and Materials (ASTM)] and respective subcommittees and subTAGS, readers should contact the ASQ directly at 1-800-248-1946. They could also inquire about how to obtain a copy of the TC 207 organizational chart, if they so desire.

Just what do these standards mean for environmental management as regulated organizations proceed into the twenty-first century, and what can and should these organizations do in preparation for this environmental management overhaul? At some juncture, readers may want to look further into ISO 14001 and explore the relevance it may have to their own organization. At that point, readers may want to follow the lead of our contributors and commentators, such as Jack Bailey of Acushnet Rubber Company,[36] or Joe Hess of SGS-Thompson and Maria Kaouris of Lucent Technologies. The latter share their experiences with ISO 14001 in the following sections of this chapter by providing insight into their respective organization's works in pursuit of ISO 14001.

ENVIRONMENTAL MANAGEMENT WITH ISO 14001 AT SGS-THOMSON MICROELECTRONICS

INTRODUCTION

With the issuance of ISO 14001 in the fall of 1996, more and more companies are beginning to look for examples of how to implement a management system that conforms to this standard. This section of Chapter 12 presents an overview of an ISO 14001-conforming environmental management system (EMS) at a SGS-THOMSON Microelectronics (ST) site in San Diego, California. It is based on actual experiences and the lessons learned during the efforts to develop and implement an EMS. It also provides suggestions for implementing an EMS at any regulated facility. This section is organized to correspond to the actual text of the ISO 14001 standard.

BACKGROUND ON ISO 14000

The ISO 14000 standards embody a variety of current environmental management approaches. These range from the 16 principles fostered by International Chamber of Commerce (ICC) to the approach of Responsible Care® practiced by the chemical industry. It also borrows heavily, in architecture and structure, from its close cousin the ISO 9000 Quality Standards. It is through ISO 14001 that a company's environmental management system is certified. The standard has five main principles as described in Clause 4 of ISO 14001 and depicted in Figure 12–3, shown previously.

BACKGROUND ON ST

ST is an international semiconductor manufacturer with facilities located in the United States, Asia, Europe, and Africa. With manufacturing sites around the world, ST's CEO Pasquale Pistorio made a commitment to "become a corporation that closely approaches environmental neutrality." With this commitment, the company created its Corporate Environmental Decalogue. The decalogue is the corporate ecological vision and presents corporate goals that address ten key areas. A summary of the decalogue is presented in Figure 12–4. Of note, one of the areas is validation. All the manufacturing sites have set out to be validated to the European Union's Eco-Management and Audit Scheme (EMAS) and many to ISO 14001.

EMAS is similar to ISO 14001 in terms of the requirements for an EMS, except for one addition. EMAS requires the preparation of an environmental statement that presents information to the public regarding a site's EMS and its environmental performance. The statement must be provided to the public after an external verifier validates the data.

It was with this background that the San Diego site obtained its EMAS validation in December of 1995. With EMAS validation accomplished, the site's management decided to proceed with the ISO 14001 certification. ST's site in San Diego, California was the first site in the United States to certify to a draft version of the standard, ISO/DIS 14001, in February of 1996. At present, this facility has achieved formal EMS certification to ISO 14001.

THE ELEMENTS OF ISO 14001: COMMITMENT AND POLICY

The first element of an ISO 14001 EMS is commitment and policy. Information related to management commitment and the policy statement is presented as follows:

MANAGEMENT COMMITMENT

Management commitment is the key to a successful environmental management system. It is up to management to dedicate the required resources for implement-

---

ST's Corporate
Environmental Decalogue

The following is a summary of SGS-THOMSON's Environmental Decalogue. It drives our environmental efforts worldwide.

In SGS-THOMSON we believe firmly that it is mandatory for a TQM-driven corporation to be at the forefront of ecological commitment, not only for ethical and social reasons, but also for financial return, and the ability to attract the most responsible and performant people.

Our "ecological vision" is to become a corporation that closely approaches environmental neutrality. To that end we will meet all local ecological/environmental requirements of those communities in which we operate, but in addition will strive to:

Regulations: meet the most stringent.

Conservation: conserve energy, water, and paper and paper products.

Recycle: use alternate energy sources and recycled paper products, and recycle water and the most widely used chemicals.

Pollution: phase out ODS, meet strict wastewater discharge limits, strive toward no landfilling, and keep our "noise to neighbor" <60 dB(A).

Contamination: properly handle, store, and dispose of all hazardous substances.

Waste: recycle manufacturing byproduct and use recyclable packing material.

Products & Technologies: accelerate our efforts to design energy efficient products.

Proactivity: support local initiatives, sponsor environmental days, encourage employee participation, provide environmental awareness training.

Measurement: measure our progress/achievement in meeting the decalogue, develop policy deployment to meet the goals, continue the environmental audit program.

Validation: obtain and maintain EMAS validation.

Figure 12–4　SGS-THOMSON Environmental Decalogue

ing an ISO 14001-conforming EMS. Committed managers who become involved are also necessary for strategic planning, setting the tone and scope of the system.

Once management is committed, it is important to keep them involved in the status of the program. This enables the environmental issues to be raised to a higher level. It also provides better management feedback and communication between the various levels of the organization. In this way, the site's objectives and performance can be reviewed and areas that need critical resources can be exposed and resource allocation assessed. By presenting information about the EMS to management both routinely and in a consistent fashion, the site's management has the necessary knowledge throughout the year to make decisions to assist with meeting site specific and/or corporate goals.

Management commitment can often be demonstrated by having an environmental status report as part of a monthly operations meeting. It enables not only the site manager, but all other key personnel to be involved in environmental issues. One of the key points to remember is that all activities and functions at the site that have environmental effects should have representation at these meetings. Third party auditors can easily review a site organization chart and a list of committee members to see proper representation. Since this type of meeting is also commonly used for strategic planning, it is easy to weave environmental considerations into the overall business plan. The environmental manager can focus top management on the important issues and, in turn, top management is made aware of business issues that may have environmental impacts.

This type of involvement will most likely move the environmental manager from a position of preparing necessary regulatory reports to taking a proactive role in developing objectives that will reduce costs, minimize regulatory hurdles, and improve the environment. The environmental manager may begin to act more as a facilitator, implementing environmental considerations into the various functional levels of the company.

At our site we use monthly operational reviews to keep management informed. The operational review includes representatives from the various functional levels of the organization. Typically, we communicate our progress towards meeting objectives, a review of environmental issues at the site, and the status of corrective actions related to internal EMS audits and inspections.

POLICY

A site policy must be prepared to define the site's commitment to compliance, pollution prevention, and continual improvement. The policy statement also defines the scope of the EMS beyond these three principles. Other principles which the site may be committed include: awareness and training, accident prevention, assessing and managing affects of new process, and raw materials management. It is up to management to determine the complexity of the EMS. In some cases, this may be driven from a corporate strategy. This is the case with ST. All ST sites are guided by the Corporate Environmental Decalogue, ST's vision for the future in terms of environmental performance. With the decalogue and our commitment to EMAS and ISO 14001 as guides, we developed an environmental policy statement for the site. This statement, see Figure 12–5, includes a brief description of our department's mission as well as identifying the principles to which we are committed. These principles define the scope of our EMS.

PLANNING

This section presents information regarding the second principle of an ISO 14001 EMS, planning. Planning includes assessing environmental effects, creating a legislative register, and establishing objectives.

Rancho Bernardo's HSS&E Policy statement is:

**To assure the health, safety, and
security of our employees
and the protection of the environment.**

We are committed to:
♦ Compliance with legislative, permit, and corporate requirements;
♦ Continuous improvement of environmental performance using TQM con-
  cepts; and
♦ Prevention of pollution.

In addition, we are committed to the following principles:

* Assessment and management of the impacts associated with the site's
  current, modified, and new (future) activities, products, and production
  processes;
* Energy management;
* Raw materials management (water, chemicals, and paper);
* Waste management;
* Noise management;
* Support Corporate product planning;
* Contractor, subcontractor, and supplier evaluation;
* Accident prevention and minimization;
* Contingency planning;
* Providing employee training and awareness, and defining responsibility;
* Monitoring and corrective action; and
* Communication with the public and customers.

Figure 12–5   Policy Statement

INITIAL REVIEW: REGISTER OF SIGNIFICANT EFFECTS

The initial review of environmental aspects or effects is completed in order to set
a baseline for the site. At this stage, the environmental consequences of the site's
functions, activities and processes are assessed. It is important to differentiate between the terms aspects, effects, and impacts. Under EMAS, the site assesses its
effects and decides which are significant. Under ISO 14001, the aspects of the site
are assessed and based on those aspects, certain impacts are identified. As an example, generating wastewater may be an aspect while degradation of a receiving
body of water caused by the discharge of wastewater may be the impact. In the
context of our discussion, impact can be interchanged with effect.

All activities at the site must be addressed when considering environmental
effects. This includes production processes, the effects of office personnel, pur-

chasing departments, and shipping and receiving activities. Some of these activities, specifically design functions, may have indirect effects on such issues as packaging, energy use, and raw material use and selection.

When concluding the review, the functions, activities, and processes should be reviewed under normal and abnormal conditions such as during routine and non-routine maintenance. It is best to use some type of checklist for this purpose to ensure a comprehensive review.

The initial assessment should contain an assessment of the site's programs related to the non-compliance related environmental issues such as recycling programs, training programs, contractor control, and vendor auditing. Another key aspect that should be included in the review is raw materials alternative consideration and use, including energy and chemicals. This is a requirement under EMAS.

The results of the initial review can then be divided into three categories. The first category contains the effects which the site considers significant. It is these significant effects that require the most immediate attention. However, it is up to the site to define what is significant. At our site we relied upon regulatory definitions to the extent practicable. For example, in California, the state Air Resources Board has defined significant risk related to toxic air contaminants as those that present a risk of cancer greater than one in a million or an acute health hazard index greater than one.

Other sites can develop a rating and ranking system that defines significance. Whatever the method it is important to remember that auditors will ultimately have an opinion with regard to this matter. Most likely it will be that they consider many more effects to be significant because the standard indicates that it is the significant effects that must be controlled. In reality this is just a matter of semantics and that is why the second category is used.

The second category is for those effects which require control, where controls are already in place. A control can include such things as a wastewater treatment plant and the subsequent operating procedures. For example, if a site discharges wastewater to a treatment plant it probably has a permit and permit conditions. If during the initial review it is determined that a permit is needed or that the site has not been operating with its permit limits, this could be a significant effect. On the other hand, if the site has documented compliance with its permit limits through proper operation of a wastewater treatment system and a procedure in place to ensure that the systems are maintained to replicate this compliance, then the wastewater discharge aspect may be placed in this second category of effects. Also, the initial review may not identify any aspect related to wastewater discharge, and that is the reason for the third category.

The third category is used to document that the site reviewed a potential environmental aspect and determine that there was no effect. This may be quite common, but it is important to document this fact because as the site's activities or processes change, this particular aspect may become relevant.

When complete, the initial review will have systematically addressed a wide range of programs and potential environmental aspects. In this way, each aspect is addressed as being significant, managed, or not applicable at this time.

LEGISLATIVE REGISTER AND COMPLIANCE

The register must include references to industry standards. For example, if the site is a chemical manufacturer, then Responsible Care® must be included.[37] From our standpoint, we had to address the ICC principles as a result of ST corporate commitment.

The method used by a site for ensuring compliance should be easily demonstrated to third party auditors. They will look for evidence that the management system is working, and that the site has a source of regulations. For example, if a site utilizes a subscription of federal and state regulations contained on CD-ROM for ensuring the site has access to regulations and legislative updates, this should be readily accessible during the audit. Also, site personnel should be able to demonstrate the procedure for identifying new regulatory requirements and communicating these requirements to affected personnel.

At the San Diego site, we rely upon various sources to provide regulatory updates as well as information regarding legislative changes. We have subscriptions to the regulations and are members of the San Diego Industrial Environmental Association, a local trade association. Through an active participation in the association we are able to stay abreast of regulatory changes and new legislation. We also conduct periodic compliance audits to ensure that we have identified the regulatory requirements applicable to our site. In our case we must also follow the various corporate standard operating procedures and the ST Decalogue. In many cases, these requirements are more restrictive than local regulations. However, all of these are contained in our legislative register.

OBJECTIVES

ISO 14000 moves strategic planning to performance. Policy statements that purport a company's commitment to various environmental issues must now be backed up by measurable objectives and programs aimed at reaching these objectives. For each policy principle to which the site shows commitment towards its policy, a subsequent objective must be established.

The standard encourages a company that may have historically practiced proactive thinking to move towards proactive action. It is the action, or program, that will take the policy statement and make it a reality. This is also where some of the economic benefits of environmental proactivity can be realized.

The objectives at our site are driven by the requirements of EMAS, the Decalogue, and our environmental effects. Each year these objectives are reassessed and new targets established. The objectives for 1997 are presented in Figure 12–6.

For each objective, a simple policy deployment or action planning form is used. This form contains information including the site objective, owner of the objective, what is the performance index, actual data (preferably represented graphically), and some the activities or programs that will be implemented to attain the objective. An example of such a form is provided in Figure 12–7.

1997 Objectives & Targets	
Objective	**Target**
Comply with regulations & corporate requirements	100%
Evaluate PFC reduction/recycling	
See all objectives	
Continue no use of ODS	
Reduce hazardous waste generation	5% / wafer
Assess HSS&E effects of new tools	100%
Reduce energy use	5% / wafer
Evaluate alternate energy	1 alternate
Reduce water use	10% / wafer
Reduce paper consumption	10% / wafer
Recycle water	60%
Maintain use of recycled paper	90%
Evaluate most used chemical	
Manage waste in level 1 through 4 of ladder concept	85% of waste
Recycle manufacturing & packing byproduct	73%
Maintain low noise levels	60 db(A)@ perimeter
Support corporate interface agreement	
Participate with contractor selection and on vendor audits	100%
Continue safe operation of the facility	Implement RMPP
Maintain an updated emergency action plan	ERT coverage at all shifts
Conduct environmental awareness and specific effect training	Annually
Use policy deployment/action planning forms	On-going
Conduct EMS audit	Annually
Support community activities	1 project
Sponsor an environmental day	1

Figure 12–6   Objectives & Targets

SGS-THOMSON: RANCHO BERNARDO

Subject: **Waste Landfilling**	1996 Objective: Increase the quality of solid waste sent for recycling by 10%
Owner: C. Lefler	WWS: none Decalogue: 6.1 Increase recycling 50% per year with 80% recycling long term

Index: Percent of Solid Waste Recycled of Total Solid Waste Generated

Data
Trend

Data

	1995	Jan	Feb	Mar	Apr	May	Jun	Jul	Aug	Sep	Oct	Nov	Dec
Actual %	49.89	32.89	74.90	76.72	63.63	77.52	62.99	65.34	52.40	39.96	53.39	24.81	51.51
Objective %	46.9	54.9	54.9	54.9	54.9	54.9	54.9	54.9	54.9	54.9	54.9	54.9	54.9
Landfill (tons)	94.7	8.06	7.05	15.67	9.63	9.60	8.93	11.42	9.14	9.69	7.02	8.70	7.05
Recycled (tons)	94.3	3.95	21.04	51.65	16.85	33.10	15.20	21.53	10.06	6.45	8.04	2.87	7.49

Comments	This data is for non-hazardous solid waste only. All hazardous waste is incinerated at a permitted offsite facility.	1996	**64 %** $8,000 Savings

Project	Prime	ECD	Status
Include construction debris (Metal) in recycling stream	C. Lefler	2/1/96	Complete
Identify new recycling vendor that can handle all streams	C. Lefler	6/1/96	Complete-Cactus
Identify additional streams for potential recycling	C. Lefler/ Vendor	7/1/96	Complete

Shoe covers
   Wood Pallets
   Hair Nets
   PVC Gloves			

Internal training to encourage source segregation	J. Hess/ C. Lefler	8/15/96	Complete 2/1/97
Identification of recycling barriers	J. Hess/ C. Lefler	10/1/96	No Action
Action Planning Form			Updated: 2/6/97

Figure 12–7   Policy Deployment

IMPLEMENTATION AND OPERATION

The implementation and operation section of the standard contains the substantive portion of what constitutes an EMS. It contains the following five subsections, each of which are major elements of an effective EMS.

- Training
- Communication
- Documentation
- Document control
- Operational control

TRAINING

The training required by ISO 14001 must address the needs of environmental, health, and safety (EHS) staff and internal auditors as well as those employees whose responsibilities and activities directly affect the environment. Therefore, it would be possible for training to be required for all employees.

As you can imagine, the type of training required by ISO 14001 lends itself very well to a few minutes during a monthly staff meeting, informal discussions, and employee orientation. More detailed courses may also be developed for key operational personnel and managers. The type and complexity of the training required can be accomplished through implementing a training needs assessment.

As with all ISO programs, linkages must be clearly present. The links help document how the system fits together. They should indicate which forms and records go with what procedures. An example is being able to move from the training needs assessment procedure, to a training matrix, a course description, and then documentation of course completion. These must all be clearly visible to an external auditor and therefore easily reproduced at the site.

Reoccurring training programs should be well documented. A course description should contain a course outline, the objectives, who should attend, and the type of documentation to be maintained. This documentation could include, for example: participants test scores (using 80 percent as a gauge); a signed page or other sign-up sheet documenting that employees attended a one-hour long course, or simply a notation in the computerized employee training database.

It is important to remember that the standard does not simply state training is required, but also refers to competence. This is different then what EHS managers typically find in legislation. When it comes down to the audit, auditors are going to randomly ask employees about environmental issues such as whether the site has an environmental policy statement, or whether a particular procedure is followed. At this point, should an auditor get a blank stare from the employee being interviewed to obtain answers to such questions, pertinent documentation in the employee's file related to environmental considerations may be relatively worthless to the auditor.

At our site, we use a one-hour training session to introduce all new employees to the environmental effects and programs at the site. They are also provided with an *Environmental, Health, and Safety Handbook* that contains the course material. An annual, one-hour EHS refresher training is also provided to all employees. Other employees, such as site services and chemical handlers, receive a more detailed training that includes a review of the operating procedures. Managers and supervisors receive a more detailed training program that was designed by the corporate group.

We use a site-wide database to document attendance. In this manner, environmental training can be integrated with the employee's training files and managed by training and human resources department personnel.

COMMUNICATION

Communication is best divided into communication with site personnel (internal) and communication with the community (external). For both, a procedure is required. For internal communication the procedure must indicate how the policy, objectives, and specific site environmental effects are communicated. Evidence of the communication should be maintained.

Likewise, external communication procedures must indicate who and how often communication occurs, who maintains the register of requests for EMS information, etc. The procedure should indicate which job function at the site has this responsibility and it should be reflected in the corresponding job description or other form of responsibility description such as personal objectives.

As indicated above, the standard requires communication not only with your employees, but with all your stakeholders in the community. Opening the lines of communication fosters an increase in dialogue with the community on such important matters as emergency response. One medium that promotes open communications is an open house. We invite members of the community into our facility once a year to review our objectives and foster/maintain our relationships.

DOCUMENTATION

There may be some confusion about the term "documents." I, the author, typically refer to documentation as all written proof of the EMS. This includes the manual, specification/procedures, and records. Different auditors may interpret these terms differently. Therefore, I suggest that the documentation system and its nomenclature be clearly described to the audit team during the initial review or at an opening meeting.

A manual provides the framework for the EMS. There should be sections that correspond with all the sections of ISO 14001. In this way, the manual can provide a roadmap for facility personnel and for auditors. This can be important when there is turnover in environmental personnel. If the manual is combined with quality or health and safety, it should somehow scope the environmental issues. This allows the auditors to focus only on the procedures related to the environment.

The manual, if it is to remain manageable, will most likely reference many documents. These include the operating procedures, plans, reports, and records. What it contains is dependent upon various factors at the site such as EMS and operational complexity and the site's documentation system.

Our current EMS manual contains some of the key elements of the EMS such as our policy statement, objectives, and responsibilities. It also contains our legislative and effects register and the audit program. A summary of the manual table of contents is presented as Figure 12–8.

DOCUMENT CONTROL

Controlled documentation is critical to successfully managing ISO conformance and regulatory requirements. For example, with a document control system, a procedure can be established for the control of hazardous waste. This procedure would reference the actual hazardous waste disposal manifest as the record to be retained. The procedure is required by the standard, but is also helpful for maintaining compliance. A controlled file can be established in the document control system to maintain the regulatory required manifest for a defined time period.

As discussed above, it is easy to see how establishing a formal document creation, modification, and retention program can improve documentation of records (e.g., reports to agencies, site assessments, manifests) and procedures. One drawback to documentation is overdocumentation. Knowing what to retain and what not to is critical. Having input from the legal department, document control specialists, environmental specialists, and management can help balance regulatory and business needs.

For sites already ISO 9000 certified, a document control system will most likely already be adequate for environmental needs. Our EMS manual, operating

---

Purpose
Reference Documents
Definitions
Site Information
Environmental Policy, Objectives, and Program
Environmental Management Review
Responsibility, Authority, and Management Representative
Training
Communication
Environmental Review
   Legislative Register
   Register of Environmental Effects
Operational Controls
Environmental Management Records
Environmental Audit

Figure 12–8   EMS Manual Table of Contents

procedures, and regulatory required reports are maintained in our existing documentation system. A benefit of this system is that report due dates and report updates can be programmed into the document control system so that a message is sent out as a reminder that an update or submittal is due. In this way, the environmental compliance activity is systematized.

OPERATIONAL CONTROL

Operational control refers to the actual activities at the sites designed to control the environmental aspects to minimize or eliminate the impact. These can be actual mechanical devices as well as operating procedures.

Operational and maintenance procedures for key aspects of the site's operation that may have environmental effects is critical. Many companies already have some type of written procedure for such activities as wastewater treatment or air emission control equipment maintenance. The procedures may, however, require minor modification. When using a system, all procedures should contain uniform sections.

Under ISO 9000 operating procedures, it is quite common to reference the type of training and/or qualifications required of the operator prior to operating a piece of manufacturing equipment or performing a process. These same types of details should be in the site services and facility specifications.

Detailed and structured procedures allow personnel external to the day-to-day operation of the equipment to understand the basic operational requirements and provides a standard to which internal auditing can be established. Having documented procedures for such things as regulation will also increase repeatability of the operation. One should note that this is not always valuable if a process continues to operate inefficiently. That is why the system must contain a process that encourages continual improvement.

The following are some of the operations that occur at our site and will most likely occur at other facilities. All would require some type of operational control. A review of some of the issues related to each are also presented.

CHEMICAL MANAGEMENT

One important question is how to actually control the use of approved chemicals and gasses. Again, a written procedure is required. Often, this will refer to a chemical supplier/handler contract or internal operations. At ST we use a material safety data sheet (MSDS) review sheet for identifying possible EHS issues associated with a new chemical. The purchasing department will not buy a chemical unless the MSDS is approved.

It is quite common for chemical storage areas to be labeled with signs indicating authorized personnel only. Who are these personnel, and is a list maintained? What type of training does the facility require before an authorized person can enter? As you can see from this brief example, a little more information may be required to eliminate the potential for releases or accidents that may occur should unauthorized personnel inadvertently enter any of these specially designated areas.

EMISSION ABATEMENT EQUIPMENT OPERATION

If emission control equipment is used to control environmental effects, then the procedure for the correct operation and maintenance of the equipment must be easily identified in the documentation system. By listing the activities that effect the environment and providing a link to the physical and managerial controls for this effect, the auditors and site personnel can understand the connection between the assessment of effects and operational control.

The details of these procedures must indicate what performance indicators are being used to ensure and document compliance. Typical indicators for a scrubber include the pH of the liquid, water recirculation flow, and media pressure drop. The procedure should also indicate the steps that should be followed for documenting corrective action when performance criteria is exceeded. If an inspection system is used, the checklist should indicate what may constitute a compliance issue and when the EHS group should be notified.

HAZARDOUS WASTE MANAGEMENT

As with chemical management, the hazardous waste management procedure will include information for internal activities as well as with contractors and vendors. A common approach to hazardous waste management procedure is to have it mirror the regulation. Therefore, in California for instance, it would begin with the steps to determine if a waste should be classified as hazardous.

One of the details that can easily be overlooked is identifying a vendor to perform analytical testing for waste characterization. This may include having documentation to support the laboratory's capabilities. This is probably best documented through an audit of the lab. This is typically completed for treatment storage and disposal facilities (TSDFs), but the practice could be expanded to provide a list of approved vendors for analytical services, transportation, and disposal companies.

OFF-SITE SERVICES/CONTRACTORS

More and more ISO-certified companies have active vendor and contractor evaluation programs. In the past, it was common to see hazardous waste vendors being audited. Today, a whole realm of vendors are being audited, from chemical and gas suppliers to rag washing and solid waste recycling facilities.

If the site has operations dictated or managed by off-site divisions or contractors, it is suggested that an interface agreement be created. This documents that an element of the EMS is managed by another party and both parties agree. For example, if corporate management provides information to customers regarding proper disposal of a product, this should be documented through an interface agreement.

In addition, many offsite contractors are integral to the production process at the site. Therefore, it is important to ensure that if the company is making claims of green products, this applies to the contractors. For example, if the company has a procedure for not using certain chemicals such as chlorofluorocarbons

(CFCs) for cleaning, this should be communicated to suppliers. Even contracted landscapers can be encouraged to use more environmentally conscious methods for vegetation, fertilization, pesticide eradication, and greens disposal.

CHECKING AND CORRECTIVE ACTION

MONITORING AND MEASUREMENTS

This section discusses some of the components of a monitoring and measurement program. These include data trending, statistical process control, and calibration.

TRENDING

The concept of trending, or measuring what you want to control, is an important element of monitoring and measurement. Typically, before much progress can be made in improving a portion of the system or environmental performance, it must first be measured. It is important therefore that the site's objectives have measurable targets. This is a common practice for such issues as waste reduction or water recycling, but it can be more difficult on the softer issues such as training and process review. A simple approach is depicted in the attached policy deployment form, Figure 12–7.

SPC

Statistical process control, or SPC, is one of the methods that industry may use to obtain regulatory waivers. By using SPC on critical systems such as air emission abatement equipment or wastewater treatment, the regulatory agency with oversight responsibility may be persuaded to decrease monitoring and reporting requirements. In many cases this may involve an upfront education of agency personnel, but may provide increased flexibility and even cost savings in the future. For instance, instead of conducting monthly wastewater sampling and analysis why not provide the internal control charts for the critical areas to the authorities.

CALIBRATION

Whenever possible, integrating environmental controls with existing facility resources is a good idea. One example of this is calibration. Many sites already have documented equipment calibration procedures for the production area, and these can easily be adapted by facilities personnel in order to ensure the proper operation of environmental control equipment. A good calibration program will require training of personnel prior to performing calibrations and the issuance of a personal calibration stamp. By using a computer database to store the information, calibration requirements can be automatically fed to the appropriate calibrator or department.

ENVIRONMENTAL MANAGEMENT SYSTEM AUDIT

The goal of auditing under ISO is to determine if a site conforms to the concepts described by the standards. The EMS audit is conducted by site personnel, however, external personnel may be used for objectivity and to supplement limited resources. It is important to distinguish that an EMS audit is unlike a regulatory compliance audit.

When performing audits the auditors must proceed according to a defined procedure. The procedure should be specific enough for auditors to see that key areas of the standard and the site's operational controls are being addressed.

When the audit is conducted, and nonconformances are identified, it is critical to identify root cause of nonconformities. Therefore it is essential for an audit program to have a well developed corrective action program. The program must identify the nonconformity, responsibilities, root-cause, and final corrective action. It must be managed in order to ensure that the corrective action is actually implemented. Subsequent follow-up audits may be necessary to ensure that corrective action was successful.

Regarding auditor objectivity and qualifications, the auditors must be independent of the areas they are auditing. These auditors should be formally approved or certified[38] for performing the audit. This does not imply that auditors should be third-party certified, but they should at least meet some type of internal criteria. A reference to the skills and experience level that demonstrates auditor qualifications should be maintained with audit documentation.

For our first EMS audit we relied upon an external consultant to audit the EMS and assist with the audit of the operational controls. ISO 9000's quality program for recording nonconformity/corrective action was used to document audit findings.

MANAGEMENT REVIEW

The management review portion of ISO 14001 refers to the time when management, either singularly or as a group, reviews the adequacy of the site's EMS. This includes evaluating the site's policy statement to ensure that it sufficiently scopes the extent of the EMS. For example, if a site is planning on a major expansion, the policy statement should indicate that the site is committed to environmental evaluations of new equipment and processes.

The review should also include the site's objectives and the results of a previous EMS audit. When reviewing audit results, management should be able to look at a summary of root causes. This may allow them to make a decision about the distribution of resources for the following year. For example, if many of the deficiencies were due to a lack of knowledge of a requirement then an emphasis could be placed on employee training and awareness. Or, a testing and certification component may need to be added to an existing training program to ensure proper and effective knowledge transfer.

One of the important aspects of the management review is that it must be documented. This is not simply maintaining copies of a presentation that includes the new policy, objectives, and a summary of the EMS audit. The documentation must indicate an agenda which includes who was present, when the review occurred, what topics were discussed, and what decisions were made. The key here is that the management review is intended to be a decision-making process and therefore the documentation must indicate that decisions were made.

CONCLUSIONS

In conclusion, an ISO 14001-styled EMS can help link the various programs of environmental and facility management. These programs, such as electricity reduction, pollution prevention, water recycling efforts, solid waste recycling, and many more, can now be tied together using the ISO framework. This can lead to improved efficiency in actual operations as well as recordkeeping, the establishment of measurable objectives, and increased reliability.

Overall, a site that embodies the principles of ISO 14001 will see its EMS move from a position of strategic planning to strategic action. In addition, the site's quality program will move beyond its manufacturing process to encompass all site activities. It is through the action and management of environmental aspects that the site may begin to realize the economic and subsequent business advantages of ISO 14001.

LUCENT TECHNOLOGIES' APPROACH TO PERFORMING AN INTERNAL EMS AUDIT

ENVIRONMENTAL MANAGEMENT SYSTEMS

An environmental management system (EMS) has come to be recognized as a valuable management tool. Politicians, industry leaders, and public interest groups from around the world have opened the dialogue on important topics such as sustainable development and overall improvement of environmental performance of both manufacturing processes and products. The development of environmental management systems standards in Europe, such as British Standard (BS) 7750 and the Eco-Management and Audit Scheme (EMAS), have become expected operating norms for many of the industries in the European Union. The finalizing of the ISO 14001 EMS standard in September of 1996 is the result of a multinational technical committee (TC-207) effort to develop an international EMS standard that would create a level playing field for companies around the world to effectively demonstrate their commitment to continual improvement in environmental performance.

PURPOSE OF THIS SECTION

The purpose of this section of Chapter 12 is to summarize the EMS auditing program that Lucent Technologies' (Lucent) Corporate Compliance Assessment Department (CAD) developed to support EMS initiatives of various Lucent Operating Units. To gain a full appreciation of the corporate EMS audit program, it is important to provide sufficient background information on the implementation requirements, program development, and ultimate goals of the EMS. With this information provided as the background, the remainder of the information included in this section will focus on the critical elements of the EMS audit program. Specifically, EMS audit program objectives will be described and the process for conducting EMS audits of Lucent facilities will be summarized, ranging from the initial request received by the corporate CAD to perform an EMS, through the preparation of the EMS audit report. Finally, other points to be discussed include: important items that should be considered when developing an EMS audit program, and ensuring audit objectives are met with the focus that overall continuous improvement to the EMS can be achieved.

LUCENT TECHNOLOGIES: A LEADER IN EMS IMPLEMENTATION

Various operating units within Lucent have made the decision to implement environmental, health, and safety management systems and/or pursue ISO 14001 certification at their manufacturing facilities located around the world. Several important considerations emerge as being the ultimate drivers leading to the business decisions by the Operating Units to pursue certification:

1. **Market Driver.** The first reason relates directly to market demand. As a global supplier of electronics equipment and services to the fiercely competitive telecommunications and computing industries, various Lucent Operating Units are receiving inquiries from their customers regarding the potential consideration of environmental performance in the evaluation and selection of their suppliers.

2. **Environmental Performance/Regulatory Relations.** A second major factor in Lucent's decision to pursue certification is to improve overall environmental performance and work in a cooperative and proactive manner with the regulatory agencies. One particular Lucent Operating Unit (Microelectronics) has voluntarily offered to participate in the U.S. Environmental Protection Agency's (USEPA) Project XL[39] program. Under Project XL, EPA has encouraged industry to submit proposals for projects that could result in a more objective, cooperative and goal-driven regulatory/industry relationship to achieve cleaner, cheaper smarter environmental protection. Microelectronics' proposal, accepted by the USEPA and the states in which it does business, uses the business-wide EMS as the foundation for demonstrating superior environmental performance and as the basis for regulatory flexibility in permitting, reporting, etc.[40]

3. **Design for Environment.** Lucent, and formerly AT&T, have maintained a long-standing commitment to pursuing the design and development of products that are "environmentally-friendly." Implementing formal EMSs at its manufacturing facilities supports Lucent's Design for the Environment initiatives.

4. **Corporate Goal.** Finally, Lucent established a corporate goal requiring at least 95 percent of its facilities to implement internationally recognized EH&S management systems by the year 2000. This goal is based on the importance that

Lucent Technologies places on responsible environmental management and the desire to be recognized as a global leader in that area.

The purpose of having the Lucent CAD conduct EMS audits within the company is to ensure a consistent, comprehensive, and objective EMS audit of each manufacturing facility. With trained and experienced EMS auditors, the Lucent CAD is able to perform a detailed review of an individual facility's EMS to assess its applicability and effectiveness and to share observations, lessons learned, and best management practices between facilities.[41]

EMS AUDIT PROGRAM OBJECTIVES

The first and most critical step in the development of Lucent CAD's EMS audit program was to develop a program that would be acceptable to the various operating units within Lucent. Since the Microelectronics Operating Unit was the first unit to pursue implementation of an EMS, the Lucent CAD worked closely with Microelectronics to develop a program that would meet the unit's immediate needs.

As a result, key strategic considerations were introduced, among them: developing a detailed understanding of Microelectronics' needs; considering other Operating Units' goals and objectives; and using the (now) finalized ISO standards 14011[42] and 14012[43] as guidance documents. From that, the EMS audit program established the following objectives:

- Evaluate the implementation, effectiveness, and maintenance of the EMS at Lucent Technologies manufacturing and research and development facilities and identify potential areas for improvement.
- Support the individual EMS initiatives at Lucent Technologies operating units and facilities.
- Determine conformance with the auditee's EMS compared to the applicable standard, Operating Unit and facility requirements, and
- Provide a written assessment of the facility's EMS to be considered as part of the management review process.

The Lucent CAD's responsibility as a corporate group providing services to the various operating units is to evaluate the effectiveness and applicability of each individual facility's EMS and to provide written documentation that is used during EMS management review process. Other important objectives of the Lucent CAD EMS audit programs are to provide support to the facilities via consultation (including management system program development), to communicate lessons-learned from other facilities programs and practices, and to promote continual improvement of environmental performance throughout the corporation.

EMS AUDIT PROCESS

Among the goals for developing the Lucent CAD EMS auditing program was to create an EMS audit procedure that would be flexible to accommodate the individual needs of the various Lucent operating units pursing ISO 14001 certification. Microelectronics was the first operating unit to pursue operating unit-wide certification; as a result, there were certain expectations and criteria that had to apply to the new EMS audit procedure. Since other Lucent operating units/divisions (e.g., Bell Labs, Business Communications Systems, and Network Systems) were going to be pursuing ISO 14001 as well, the decision was made to build a comprehensive, yet highly flexible corporate EMS audit procedure that could accommodate operating unit-specific requirements. In order to allow for this flexibility, the CAD EMS audit procedure was actually described in eight individual work instructions.

The EMS audit procedure and each of the individual work instructions were developed in the specified document control format and include all of the following information: title, description, documentation number, reason for issue, associated reference documents, process/procedure owner(s), approval signatures, notification requirements, record retention requirements, issue dates, and revision numbers. As shown in Figure 12–9, a work instruction is developed to document the requirements established for each element of the EMS auditing process. Similarly, a work instruction was developed to outline the roles, responsibilities, and qualification requirements for lead auditors, auditors, audit teams, clients, and auditees.

A brief summary of the purpose and major criteria of each work instruction is depicted in Figure 12–9 to emphasize the importance of planning, executing, and reporting of an EMS audit.

OPERATING UNITS REQUEST EMS AUDIT

The purpose of this work instruction is to describe the process by which the Lucent CAD receives a request to perform an EMS audit. To allow the Lucent CAD to effectively schedule and service all of its customers in the Operating Units, a work instruction was developed that specifies the request and scheduling process.

NOTIFICATION LETTER AND PRE-AUDIT QUESTIONNAIRE

Once an operating unit has requested an EMS audit to be performed by the Lucent CAD and the EMS audit has been scheduled, the Lucent CAD prepares a notification letter to officially notify the facility of the planned audit. The EMS audit notification letter includes a discussion of the:

1. Audit objective
2. Location to be audited
3. Time and date of the opening conference
4. Name of lead auditor
5. Required return date for pre-audit questionnaire and other supporting EMS documentation

Lucent EMS Audit Process

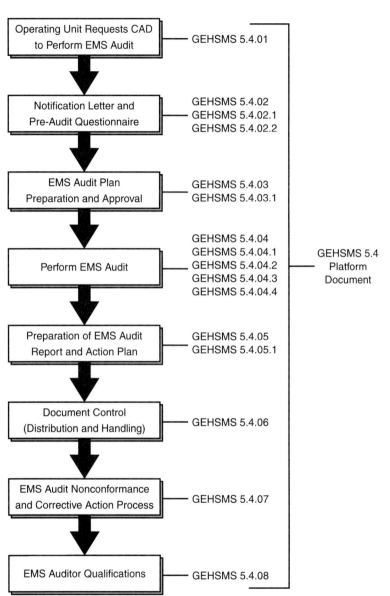

Figure 12–9   Lucent EMS Audit Process

6. Applicable distribution
7. Confidentiality requirements
8. Pre-audit questionnaire

Lucent CAD's pre-audit questionnaire is also included as an attachment to the EMS audit notification letter. The EMS pre-audit questionnaire becomes the basis for collecting sufficient information to prepare a facility-specific audit plan. The EMS pre-audit questionnaire requests specific information such as: facility name, address, phone numbers, organization chart, roles and responsibilities, information about the facility (e.g., size, number of employees, key operations and products), name of the EMS manager, scope of the EMS, etc. A preliminary review section of the EMS pre-audit questionnaire solicits specific information about the core elements of the EMS such as communications with external parties, site-specific environmental policy and development and review of objectives and targets, etc. Finally, an EMS pre-audit questionnaire requests that the facility forward additional site-specific information that will be useful in developing an understanding of the facility and producing a focused EMS audit plan. This additional information includes: an organizational chart, manufacturing process diagram(s), list of significant environmental aspects, and applicable EMS documentation, such as: EMS Tier I Manual and most recent internal EMS audit report and action plan, environmental policy, and notes or minutes from community involvement meetings, etc.

EMS Audit Plan Preparation and Approval

The purpose of this work instruction is to document the audit plan preparation requirements, audit team assignments, and working document requirements for conducting an EMS audit at a Lucent facility. The preparation of the EMS audit plan is an important step in the EMS auditing process. The EMS audit plan becomes the basis for the execution of the EMS audit. A properly planned EMS audit will result in a more thorough and efficient audit that will yield more detailed information and ultimately be a more useful tool for the facility being audited. The EMS audit plan is developed based on the information obtained from the pre-audit questionnaire and includes the following information:

1. Audit objectives and scope
2. Audit criteria
3. Time and duration for major audit activities
4. Dates and places where the audit is to be conducted
5. Audit team members
6. EMS audit schedule
7. Confidentiality requirements
8. Identification of the auditee's organizational and functional units to be audited
9. Identification of the functions and/or individuals within the auditee's organization having significant direct responsibilities regarding the auditee's EMS
10. Identification of standard elements to be audited and the procedures that shall be followed to complete the audit
11. Reference documents

12. Report content and format, expected date of issue, and distribution of the audit report

13. Document retention requirements

A copy of the EMS audit plan is sent to the facility EMS manager for final review and approval before the on-site audit activities are initiated. Again this planning process is a critical step in ensuring that the EMS audit is executed in an efficient and thorough manner.

REVIEW OF EMS DOCUMENTATION

A critical step in the EMS audit process is for the EMS auditors to perform a detailed review of the EMS documentation for each individual facility. In every case, the highest-level EMS documentation (e.g., Tier I EMS Systems Manual) and any other documentation that describes the procedures and systems that are integral to the scope, function, and control of the EMS should be reviewed and understood by the EMS audit team prior to starting the actual on-site phase of the audit. For example, the EMS audit team should have a good understanding of the scope of the EMS (what is included in EMS), the objectives of the EMS (e.g., integration and relationship of environmental management with other core business processes and management systems), and the status of the EMS (e.g., Is it certified?[44] What were the issues/findings of the last EMS audit? What are particular areas of concern? etc.).

In one Lucent Microelectronics facility's EMS, pre-audit documentation review consisted of comparing facility specific EMS procedure documents to the Microelectronics operating unit EMS Documents for consistency and completeness.

In addition to EMS documentation review, an EMS documentation-specific protocol was developed by the Lucent CAD EMS audit team. The operating unit-specific EMS audit protocol was based on developing specific audit questions to test each of the required components of the Operating Unit EMS that went above and beyond what was required in a specific element of Clause 4 of ISO 14001. For example, the environmental aspect identification and assessment of significance process is a required element in the Lucent Microelectronics EMS that must be conformed to by all operating units. The protocol that was developed for the Lucent Microelectronics operating unit includes a series of audit questions that are used by an EMS audit team to evaluate the conformance of individual sites to this overall requirement. The protocol developed for Microelectronics was developed to supplement an EMS protocol that the Lucent CAD has developed to evaluate conformance with the ISO 14001 standard.

EMS AUDIT

The purpose of the EMS audit work instruction is to document the procedures for conducting the on-site activities during an EMS audit at Lucent Technologies manufacturing facilities. The on-site audit phase, depending on the nature and scale of the operations and the results of the last EMS audit generally last from two to four days, consists of an opening and closing conference, facility tour(s), interviews with appropriate personnel, document review, and preparation and distribution of written findings (e.g., EMS nonconformances).

OPENING CONFERENCE

An opening conference is a scheduled and important part of each EMS audit. The purpose of the opening conference is to introduce both the audit team and the principal participants at the facility, and to communicate the EMS audit objectives, needs, and requirements to the facility participants. The EMS audit plan is reviewed and the schedule for the week is further refined. The opening conference is conducted by the lead EMS auditor. It is important, particularly at the facilities that have not been through the process, to reiterate that an EMS audit is *not* an environmental compliance audit. Although compliance is a basic requirement of the EMS, the EMS audit does not directly evaluate compliance but rather, focuses on the facility's conformance with the system and with procedures that have been included within the EMS audit scope.

ON-SITE AUDIT ACTIVITIES

The on-site EMS audit activities shall be based on the EMS audit scope, objectives, and criteria established in the audit plan. The EMS audit is focused on the audit team collecting sufficient evidence to determine whether the facility's EMS conforms with the EMS audit criteria. The audit team gathers evidence via personnel interviews, evaluation of employees' knowledge of the environmental policy and how their work is related to the facility's environmental aspects, documentation review, and observation of activities and site conditions, among other means.

Facility personnel are targeted for interviews depending on the nature and scale of the operations at the facility and the results of previous audits; both these variables are addressed in the audit plan. Using an EMS audit protocol developed by Lucent CAD, the EMS auditors systematically ask questions consistent with the requirements of the core elements of the ISO 14001 standard and the other pertinent requirements of the Operating Unit and/or facility EMS.

AUDIT FINDINGS

Based on the information collected during the site tour(s), interviews, and document review, EMS nonconformances are identified by the EMS audit team. To the extent possible, nonconformances are verified by following up with additional interview questions, confirming the intent of established systems and procedures, and reviewing EMS documentation. Verified EMS nonconformances are recorded onto EMS nonconformance forms developed as part of the Lucent Microelectronics EMS.

The lead EMS auditor prepares the nonconformance form and submits a copy of the record on nonconformances to the facility EMS manager during the closing conference at the completion of the on-site activities. Evidence supporting each one of the nonconformances should be duly noted in the EMS auditors' notes.

CLOSING CONFERENCE

At the completion of the on-site phase of the EMS audit, a closing conference is conducted to review the significant findings of the audit, the issuance of the reports and action plan, distribution of report and action plan, and audit follow-up activities, requirements, and responsibilities. The Nonconformance Forms summarizing the EMS nonconformances identified during the course of the audit are then provided to the facility EMS manager. The closing conference provides a forum to openly discuss audit findings and nonconformances, the potential root causes of the nonconformances that were identified, and opportunities for improvement to the EMS.

PREPARING EMS AUDIT REPORTS

The purpose of this work instruction is to outline the roles, responsibilities, and procedures associated with the preparation of the CAD EMS audit report. The EMS audit report is prepared under the direction of the lead EMS auditor. The EMS audit report includes the following:

- summary of the objectives and scope of the EMS audit,
- summary of the status of the facility's EMS conformance to EMS audit criteria (i.e., elements of ISO 14001);
- an assessment of whether the system has been properly implemented, maintained, and reviewed; and
- a listing of the EMS nonconformances identified during the audit.

The written EMS audit report is submitted to the facility's EMS manager, kept as an EMS record, and used during the management review process to evaluate the EMS implementation effectiveness and relevance.

DOCUMENT CONTROL

The document control work instruction under the EMS auditing procedure describes the Lucent CAD's document control procedure associated with the distribution and handling of written nonconformance records, plans, reports, and action plans generated as part of the CAD EMS audit program.

In the case of the Lucent Microelectronics Operating Units, all of the written documentation generated during the EMS audit process is marked "Proprietary—Use Pursuant to Company Instructions." The written documentation that is the output from the EMS audit process (e.g., audit plan, audit report, nonconformance forms, etc.) is provided to facility EMS managers who assign document control numbers and file the information within the individual facility document control systems, accordingly. The Lucent CAD has developed its own document control system to file and retain pertinent documentation.

Once the EMS audit reports and summary of nonconformances have been submitted to the facility EMS manager, the Lucent CAD's responsibilities under the Microelectronics EMS have been completed. The individual facilities are responsible for reviewing nonconformances and developing the corrective and preventive action strategies and implementation activities to promote and ensure continuos improvement of each of their own respective EMS program.

EMS AUDIT AS THE KEY TO CONTINUAL IMPROVEMENT

Since ISO 14001 registrars require that facilities demonstrate that they have been through one complete audit cycle before certification (e.g., management review completed), the results of the EMS audits conducted by the Lucent CAD were critical to evaluate the remaining gaps in the EMS and to assess performance of the EMS for areas in which the EMS had already been implemented. The "nonconformances" and the suggestions for improvement that were cited during the course of the EMS became important data to discuss at the management review meeting and became the basis for specific actions to work on and complete prior to the certification audit.

Based on experiences gained by applying the Lucent CAD EMS procedure and testing the effectiveness of the tools, data collection methodologies, and specified reporting criteria, several important lessons learned emerged. In some instances, the lessons learned were realized by the audit team working through the process and testing what worked and what did not work. Other lessons learned came from stated preferences of the facility personnel. Finally, during the course of the certification audits, it became very clear what the minimum criteria was that would be acceptable to the registrars (e.g., qualifications of the auditors and documentation of the audit process). The most important lessons learned in the development of an EMS audit procedure and the on-site EMS auditing activities that the Lucent CAD found to be relevant to the certification process itself, and to the overall development of an EMS audit program that is both functional and effective are summarized as follows:

EMS AUDIT PROCEDURE

Keep it Simple. A common mistake that is made by many organizations, particularly those that have been through the ISO 9000 process, is to design and implement an EMS audit program that is more complicated than it needs to be. Before the EMS audit procedure is drafted, it is critical to understand the various interests and requirements of the affected parties (e.g., corporate, facility personnel, registrar, etc.). Once this information is gathered and assimilated, the goal should be to put together a program and write a procedure that incorporates key issues and criteria, and is as simple as possible. The more comprehensive and complicated the procedure, the more work that will be involved to manage the program.

Keep it Flexible. When developing a corporate EMS auditing program for a large corporation with various operating units, and each unit approaches implementing an EMS in a different way, flexibility of the program must be emphasized. Even within a fairly centralized organization such as Lucent Technologies, there are distinct differences in the requirements and operating norms of the various operating units. Individual work instructions were deliberately developed for each of the major elements that combine to form an effective EMS, such as audit plan, on-site audit activities, audit report, etc. A program configured with enough flexibility can accommodate the variable requirements of individual operating units.

Take Advantage of Guidance Provided in ISO 14010–14012. ISO standards 14010, 14011, and 14012 provide useful guidance on conducting internal EMS audits and the pre-requisite qualifications of EMS auditors. These standards were developed as part of the TC-207 standards development process and were approved as international standards in the fall of 1996. A significant amount of time and energy went into generating these standards and they provide a lot of useful information and guidance. Although these standards are considered guidance, ISO 14001 registrars appear to be using the elements, criteria, and processes included in the guidance documents as benchmarks in EMS auditing programs. Therefore, it is good practice to use the guidance provided in these standards as the framework for developing an EMS audit program.[45]

Integrate EMS Audit Program with Other Management Systems. Companies that have already received, or are in the process of pursuing, ISO 9000 certification have likely already developed some kind of an internal auditing program. The requirements of an ISO 14001 EMS auditing program are very consistent with those required for a quality management system. Therefore, to the extent possible, it is useful to take advantage of existing procedures.

PERFORMING THE EMS AUDIT

Review Documentation Prior to On-Site Audit. EMS auditors must be very familiar with the overall scope of the EMS for a facility and the specific procedures and work instructions that are developed to address each of the elements of the ISO 14001 standard. To minimize the disruption to facility personnel and allow the EMS audit to proceed as efficiently as possible, the designated audit team should perform a thorough review of the EMS documentation prior to arriving on site. At a minimum, the audit team should have a thorough understanding of the scope of the EMS, the procedures and work instructions that are included within the EMS, the nature of the products, activities, and services that are included within the scope of EMS, roles and responsibilities for the EMS, and the results of previous EMS audits. It is also very helpful to gain an understanding of the type of environmental aspects at the facility, the most current objectives and targets that have been identified, and the environmental management program that has been developed to manage these objectives and targets.

Understand the Issues and Criteria Important to the Registrar. If a facility's goal is to achieve ISO 14001 certification, developing a thorough understanding of an individual registrar's specific expectations is critical to the development and execution of an EMS auditing program. The registrars will be keenly interested to know how EMS audits are performed, who is doing the auditing, and how the information obtained during the audit is managed within the context of the EMS. Engaging a registrar early in the EMS development process makes good sense for a couple of reasons: (1) it enables a facility to choose a registrar it feels it can work with; and (2) it creates the opportunity to develop an understanding of what a registrar will be looking for in the internal EMS auditing process. This is particularly important information and minimizes the chances of surprises and the need to have to make significant revisions to program design.

Document the Audit Process. Based on the experience working with the Lucent manufacturing facilities pursuing ISO 14001 certification, the importance of documenting the EMS auditing process cannot be overemphasized. The EMS audit plan for a facility should clearly define, to the extent possible, which activities and/or areas within a facility will be audited, who will be included in the audit (e.g., both auditors and auditees), and which elements of the standard the facility will be audited against. When executing the audit, EMS auditors must keep comprehensive and legible notes that document the process described in the audit plan. Since the goal is to always make the audit process as simple and traceable as possible, keep in mind the following: develop a standard template on which the EMS auditors take their notes with a summary table that includes an employee list, areas within the facility, and elements of the standard that were audited against. This standard template should include items such as: spaces to enter the date, auditor's name, auditee's name, the area of the facility being audited, the element of the EMS being audited against, and ample room to make notes and identify EMS nonconformances as may be identified. These are all good practices to facilitate the process for the registrars. Also, developing and using an audit protocol that is specific to both the requirements of the ISO 14001 standard and any other operating unit or site-specific requirements can be a valuable audit tool.

Follow the Audit Plan. As discussed earlier, preparing the EMS audit plan is an important step that allows the audit team and the facility to identify and agree on the audit objectives and the activities to be undertaken while on site. Therefore, the audit plan becomes an important reference document that should be used by the audit team to guide the program. During the initial certification audit and subsequent surveillance audits by the registrar, the registrar will likely request to review the EMS audit plans and then check the consistency between what was planned and how the audits were actually executed.

Establish and Adhere to a Document Control System. The EMS audit report is one of the pieces of documentation required by ISO 14001 to be kept as a record. An important step includes having the procedure(s), plans and reports, and other records that may be generated as part of the EMS audit in the document control system established for the EMS. A systematic process to compile and store the various EMS auditing documentation should be developed so that it is readily retrievable and simple for the registrar to review and understand. At a minimum, the EMS auditing procedure should have its own unique document control number and the audit plan, audit report, nonconformance files, and any other pertinent auditor notes should be kept in an EMS audit record file with an assigned document control number.

Another important consideration regards how a company wants to handle the confidentiality of any of the information associated with the EMS auditing process. As stated previously, it is not the objective of an EMS audit to identify regulatory noncompliances. However, some of the information collected and reported during the course of an EMS audit may be considered by some corporations to be sensitive. Lucent made the decision to handle all of the EMS audit related material as: "proprietary to be used pursuant to company instructions." Because the type of information that may be collected during the EMS auditing process may vary from corporation to corporation, all corporations should consider the legal implications associated with information that may be collected during the EMS auditing process and treat it in a way that meets the objectives and requirements of corporate policy.

Involve Facility Personnel in the EMS Auditing Process. For the EMS auditing function to be effective and enduring, every facility should build some internal EMS auditing capability within the facility. An EMS auditing program should not be structured in such a way that EMS audits are performed exclusively by an external auditing team at a prescribed interval. Rather, in most cases, there should be at least one qualified internal EMS auditor at a facility that is capable of performing an objective and thorough review of the EMS. The EMS auditing process should be continuous. The standard indicates that the frequency of EMS audits should be based on the results of the previous audit and the nature of the activities, products and services undertaken at the facility. At a minimum, some form of comprehensive EMS audit should be performed on an annual basis. For larger, more chemically and energy intensive facilities, the ideal scenario may be for an ongoing process in which various parts of the EMS are audited continuously.

The Lucent CAD audit program utilized highly trained and experienced auditors who were invaluable in performing a comprehensive audit, as well as, preparing facility personnel for the pending certification audit to be performed by the registrar. As the Lucent program moves forward, the role of the Lucent CAD will be to perform periodic audits of the overall EMS programs to evaluate consistency, best practice, and continual improvement within the corporation and to provide training to Lucent internal auditors for Lucent facilities.

IN CONCLUSION

As the previous case studies depicted, utilizing an environmental management system (EMS) that follows the architecture of ISO 14001, can provide genuine opportunities for continual improvement within a facility's or company's EMS. As each of these cases show, developing an EMS is not a half-hearted exercise. If the EMS is to function as designed, it requires hard work, diligence, and true management commitment from the very top of the organization.

ENDNOTES

[1]W. Lee Kuhre, *ISO 14001 Certification, Environmental Management Systems,* Upper Saddle River NJ: Prentice Hall PTR, 1995, p. 3.

[2] Of added interest, a number of Fortune 500 companies have begun to use the vast resources of the Internet to post their environmental goals and achievements on their web sites, including their environmental policies and mission statements. The environmental policy, a key requirement of ISO 14001, is described in Clause 4.2 of ISO 14001. We include copies of some environmental policies in Appendix E. Additional information can also be found in Appendix G.

[3] J. Cascio, G. Woodside, and P. Mitchell, *ISO 14000 Guide: The New Environmental Management Standard,* New York, NY: McGraw-Hill Companies, Inc., 1996, p. ix.

[4] Both the APQC study and IRRC survey described previously in Chapter 7 highlighted this topic in greater detail and add more credence to Mr. Cascio's passage.

[5] Disclaimer. These are just two of the many books published since 1995 that focus on the ISO 14000 standards. By referencing these specific books, the Author does not imply any preference for these books over others published in the field, other than the specific passages lend themselves well to the focus of this chapter. Other ISO 14000 books that we are aware of are listed in the References section.

[6] ISO 14000 encompasses a series of standards, the first of which is International Standard, ISO 14001, Environmental management systems—Specification with guidance for use, First edition 1996-09-01, © ISO 1996. The complementary standards to ISO 14001 tie into this standard as depicted in Figure 12–1 on p. 320.

[7] For additional insight, see K. Freeling "Implementing an Environmental Management System in Accordance with the ISO's Draft Standards is Not Necessarily Costly and Could Yield Benefits as Well," *The National Law Journal,* July 24, 1995, p. B5.

[8] Unless otherwise specifically stated, the reference for this section came from two sources: (1) J. Cascio, G. Woodside, and P. Mitchell, *ISO 14000 Guide,* New York, NY: McGraw-Hill Companies, 1996; and (2) D. Hortensius and M. Barthel, "Beyond 14001," one of the chapters in *ISO 14001 and Beyond,* edited by C. Sheldon, Sheffield, UK: Greenleaf Publishing, 1997.

[9] In accordance with the Commission of the European Communities, COM(91) 459 final, Brussels, 5 March 1992, proposal for a COUNCIL REGULATION (EEC) allowing voluntary participation by companies in the industrial sector in a Community eco-audit scheme, and the actual regulation itself, COUNCIL REGULATION (EEC) No 1836/93 of 29 June 1993, as published in the Official Journal of the European Communities, 10.7.93, NO L 168/1.

[10] As an indicator of ISO's growing acceptance and popularity, we have seen a greater number of companies hoisting banners onto their buildings proclaiming their certification to ISO 9001 as a way of telling the world their achievements. Many are visible along roadways and highways.

[11] For additional insight, see G. Crognale, "Allocating Corporate Resources for Environmental Compliance," in *The Greening of American Business,* Rockville, MD: Government Institutes, Inc., 1992, pp. 169–195.

[12] For historical reference, see B. Davis "Growth of Trade Binds Nations, but It Also Can Spur Separatism," *The Wall Street Journal,* June 20, 1994, p. A10.

[13] In the first Committee Draft of ISO 14004, the Executive Summary of that document refers to the Earth Summit and how ISO's 14000 standards are one response to that expressed need, ISO 14000: 199X, TC 207 SC1/WG2 N71, Guide to Environmental Management Principles, Systems and Supporting Techniques, Committee Draft, September 23, 1994.

[14] Additional reference for this passage came from G. Crognale, "Environmental Management: What ISO 14000 Brings to the Table," *Total Quality Environmental Management,* Summer 1995, pp. 5–15.

[15]Reprinted from ISO 14001: 1996 (E) *Environmental management systems—Specification with guidance for use.* Permission granted from Dick Hortensius, Netherlands Standardisation Institute (NNI). *Source:* D. Hortensius, "Beyond ISO 14001: An Introduction to the ISO 14000 Series," from *ISO 14001 and Beyond*, Sheffield, UK: Greenleaf Publishing, 1997, pp. 22, 24.

[16]Ibid.

[17]The first company to receive the first edition of a single worldwide ISO 14001 registration that covers all of the company's global manufacturing and hardware development operations across all of its various business units is IBM. It is believed to be the first company to take this unique, global approach to registration. We also discussed this previously in Chapter 10.

[18]International Standard ISO 14004 Environmental management systems—General guidelines on principles, systems and supporting techniques, 1st ed. 1996-09-01, © ISO 1996, Geneva, Switzerland.

[19]The first of these standards is International Standard ISO 14010 Guidelines for environmental auditing—General principles, 1st ed. 1996-10-01, © ISO 1996, Geneva, Switzerland.

[20]Material for ISO 14031 and ISO 14040s was partially obtained from a presentation the author developed for an ISO 14001 Conference in Toronto, Canada in April 1996.

[21]At the time this manuscript was being drafted, this document was voted as a DIS (Report of SubTAG 4(EPE) N 118, June 25, 1998).

[22]Ibid, p. 2.

[23]Among improvement techniques that some businesses utilize is the Six Sigma quality program. Six Sigma is a quality circle term to signify a measurement that describes a standard deviation on a bell curve as it applies to the number of defects in product manufacturing. A two or three sigma, for comparison, is about sixty-five thousand defects per million "widgets" manufactured. The higher the sigma number, the fewer defects created.

[24]As this manuscript was being drafted, ISO 14040 is finalized, 14041 was elevated to FDIS, and ISO 14042 and 14043 are in CD status. This information came from the Association Francais de Normalisation (AFNOR), December 2, 1997 and the International Environmental Systems Update, January 1998, CEEM Information Services, Fairfax, VA, p. 1.

[25]Additional information that can provide additional insight into LCAs can be found in these LCA documents: (1) B. Steen, R. Carlson, and G. Lofgren, *SPINE: A Relation Database Structure for Life Cycle Assessments*, Swedish Environmental Research Institute, Goteborg, September 1995; and (2) ISO/TC 207/SC 5/WG 3 N 18, Technical Report, ISO TR XXXX, First Edition 1997-10-12, *Illustrative examples on how to apply ISO 14041—Life cycle assessment-Goal and scope definition and inventory analysis.*

[26]See Kuhre, *ISO 14001 Certification*, n. 1, pp. xvii and 3.

[27]For additional information, refer to EPA's notice in the *Federal Register*, Vol. 63, No. 48, March 12, 1998, 12095.

[28]Additional background information about Responsible Care® and MSV can be found in CMA's *The Year in Review 1995–1996 Responsible Care® Progress Report: Members and Partners in Action*, Arlington, VA.

[29]Definition of environmental management system audit in ISO 14001: 1996(E).

[30]The RAB provided this information to the Author in a telephone interview.

[31]Among the documents providing these guidelines are the *International Accreditation Forum (IAF) Guide 62*, and the European Accreditation of Certification document, EAC/65, *Guidelines for the Accreditation of Certification Bodies for Environmental Management Systems,*

Rev. O, 06/1996. Of note, these guidelines were adopted for use by the United Kingdom Accreditation Service (UKAS) and the National Accreditation of Certification Bodies (NACB).

[32]News digest item by E. Larson, Cox, B., and Linstrom, Z. "UKAS Revokes Accreditation of UK Registrar," *Quality Digest,* November 1997, p. 9.

[33]*European News* highlights, *Environmental Science & Technology,* January 1, 1998, p. 13A.

[34]Information obtained in a telephone interview from CEEM, publisher of the *International Environmental Systems Update,* an ISO 14000 information service, Fairfax, VA.

[35]Registrars are those bodies accredited by registration groups, such as the Registration Accreditation Board (RAB) in the United States, and the United Kingdom Accreditation Service (UKAS) in the United Kingdom.

[36]For more recent information about Acushnet Rubber, refer to *International Environmental Systems Update,* January 1998, CEEM Information Services, Fairfax, VA, pp. 11–12. Additional information can also be obtained at Acushnet's Web site at www. acushnet.com, and EPA's Design for the Environment publication, *Environmental Management System Bulletin 1,* EPA 744-F-98-004, July 1998.

[37]The ten elements of Responsible Care® include its guiding principles, codes of management practices, self-evaluations, measures of performance, and others; the core of the guiding principles is found within the codes of management practices, which includes: community awareness and emergency response, pollution prevention, process safety, and product stewardship, to name a few.

[38]At present, the RAB is the main accreditation body in the United States that administers the certification of EMS Lead Auditors. The first step in achieving certification is for the prospective candidate to attend an accredited lead auditors course and pass the accompanying test.

[39]U.S. Environmental Protection Agency—Regulatory Reinvention (XL) Pilot Projects, *Federal Register,* May 23, 1995, Vol. 60, No. 99, pp. 27282–27291.

[40]Part of Lucent's Microelectronics group participation in EPA's Project XL included the company's testing of crucial innovations to reflect a project that not only achieves Superior Environmental Performance (SEP) as it proceeds, but dramatically improves upon XL projects that have preceded it while directly addressing the goal that XL projects test new concepts for improving EMSs. Reference: Lucent letter to EPA dated June 30, 1997.

[41]The concept of sharing information between (sister) facilities does have considerable merit, as we previously described in Chapter 8.

[42]Guidelines for environmental auditing—Audit procedures—Auditing of environmental management systems, ISO 14011: 1996(E). ISO, Geneva, Switzerland.

[43]Guidelines for environmental auditing—Qualification criteria for environmental auditors, ISO 14012: 1996 (E), ISO, Geneva, Switzerland.

[44]With respect to certification, ISO 14001 does not require a facility to seek third-party certification from an outside registrar; the facility can also self-certify. That decision rests solely with each facility considering certification.

[45]Since the ISO 14000 family of standards are less than three years old as of this writing, there have probably been a number of anecdotal revisions and network discussions regarding what may and may not work for a particular facility. As such, embarking on a benchmarking study to evaluate effective EMS auditing techniques, deciding upon which registrar to choose, or which EMS software to provide assistance in this process, among other considerations, can provide considerable insight to a facility's environmental manager considering EMS audits.

REFERENCES

Articles and Books

Cascio, J., G. Woodside, and P. Mitchell. *ISO 14000 Guide: The New Environmental Management Standard.* New York: McGraw-Hill, 1996.

Crognale, G. "Environmental Management: What ISO 14000 Brings to the Table," *Total Quality Environmental Management,* 1995, pp. 5–15.

Davis, B. "Growth of Trade Binds Nations, but it Also Can Spur Separatism," *The Wall Street Journal,* 1994, p. A1.

Freeling, K. "Implementing an environmental management system in accordance with the ISO draft standards is not necessarily costly and could yield benefits as well," *The National Law Journal,* 1995, p. B5.

ISO 14001 and Beyond, C. Sheldon, Ed. Greenleaf Publishing, Sheffield, United Kingdom, 1997.

Kuhre W., Lee. *ISO 14001 Certification: Environmental Management Systems.* Upper Saddle River, NJ: Prentice Hall, 1995.

————. *ISO 14010s Environmental Auditing,* Upper Saddle River, NJ: Prentice Hall, 1996.

Larson E., Cox B., and Linstrom, Z. "UKAS Revokes Accreditation of UK Registrar," *Quality Digest,* 1997, p. 9.

Ritchie I. and W. Hayes. *A Guide to the Implementation of the ISO 14000 Series on Environmental Management,* Upper Saddle River, NJ: Prentice Hall, 1998.

Steen, B. R. Carlson, and G. Lofgren, (1995). *SPINE: A Relation Database Structure for Life Cycle Assessments,* Sweedish Environmental Research Institute, Goteborg.

The Greening of American Business; Sullivan T., Ed. Rockville, MD: Government Institutes, Inc., 1998.

Weaver, G. *Strategic Environmental Management.* New York: John Wiley & Sons, Inc., 1996.

The Chemical Manufacturers Association, *The Year in Review 1995–1996 Responsible Care® Progress Report; Members and Partners in Action,* Arlington, VA.

Commission of the European Communities, (1992). COM (91) 459 final, Brussels, COUNCIL REGULATION (EEC) No. 1836/93, Official Journal of the European Community 10-7-93, No. L 168/1.

EPA position statement on environmental management systems and ISO 14001. *Federal Register,* Vol. 63, No. 48, March 12, 1998, p. 12095.

EPA Regulatory Reinvention (XL) Pilot Projects, *Federal Register,* May 23, 1995, Vol. 60, No. 99, pp. 27282–27291.

EPA's Design for the Environment publication, (1998). *Environmental Management System Bulletin 1,* EPA 744-F-98-004.

European News. *Environmental Science & Technology,* 1998, p. 13A.

European Accreditation of Certification, EAC/65, Guidelines for the Accreditation of Certification Bodies for Environmental Management Systems, Rev. 0, 6/1996.

Ask an Expert column. *International Environmental Systems Update,* Fairfax, VA: CEEM Information Services, 1998, pp. 11–12.

ISO 14001, Environmental management systems—Specification with guidance for use, First edition 1996-09-01, © ISO 1996, Geneva, Switzerland.

ISO 14004, Environmental management systems—general guidelines on principles, systems and supporting techniques, First edition 1996-09-01, © ISO 1996, Geneva, Switzerland.

ISO 14010, Guidelines for environmental auditing—general principles, First edition 1996-01-01, © ISO 1996, Geneva, Switzerland.

ISO 14000: 199x, TC 207 SC1/WG2 N71, Guide to Environmental Management Principles, Systems and Supporting Techniques, September 23, 1994.

ISO/TC 207/SC5/WG 3 N18, Technical Report, ISOTR XXXX, First Edition, 1997-10-12.

Report of (USTAG) SubTag 4 (EPE) N 118, June 25, 1998.

PART 4

WHERE ENVIRONMENTAL MANAGEMENT MAY BE HEADED

In the closing part of the book we look at some emerging tools and emerging trends. In Chapter 13, the focus is on software, and how this tool is helping environmental management become more enmeshed into other business units within companies. Complementing this chapter, Chapter 14 looks at additional sources of information for environmental professionals. Closing out the book, Chapter 15 provides a glimpse into emerging trends environmental professionals and other environmental stakeholders should be aware of to stay sharp in this ever-changing field. Like the Morbius Loop made common to environmental practitioners by ISO 14001, we start anew from here. Until next time . . .

CHAPTER 13

UTILIZING SOFTWARE TO ENHANCE ENVIRONMENTAL MANAGEMENT RESPONSIBILITIES

Gabriele Crognale, P.E.

> *"If you want to change a person's way of thinking, don't give him a lecture, give him a tool."*
>
> *Buckminster Fuller*

INTRODUCTION

In today's rapidly evolving world of information management, computers, peripherals, and software play a crucial role. Without them, it would be safe to say that the day-to-day management of environmental and occupational health and safety data would be a much more difficult task to accomplish, let alone manage.

In this chapter, we take a closer look at these widely used management tools and how rapid advances in computer hardware and software design have fueled a growth in automated information management to the point where much of the information that a data specialist or environmental manager needs at any particular time can be obtained by clicking a computer mouse or stroking the keys of a keyboard. The development of more powerful PCs, whether of the IBM-compatible or Apple Macintosh variety, network servers, and the software necessary to operate these computers, have collectively paved the way for industrious specialty software designers to develop highly specialized and sophisticated software that is necessary with today's more complex environmental management databases and information systems. Such software powers environmental management information systems, or EMISs, that assimilate relevant data from various business units within an organization and link them to EH&S databases. Since most of the software that is used in these applications is run on PC-based software platforms, such as Microsoft Windows™ , IBM's OS/2® or Apple-based software, we limit our discussion to software developed primarily for these platforms. The more powerful software used in workstations that run on UNIX®-based or other similar platforms, such as Windows NT®, are not discussed in this chapter. Readers, though, are welcome to explore these other software applications at their own leisure, if they so desire.

With respect to PC-based software systems, the following is an example of a company developed EMIS. It was provided to us by Audrey Bamberger of Anheuser-Busch (A-B) Companies.[1] A-B is developing a system designed to support the defined environmental management requirements of their EMS. The system is based on: understanding the environmental business requirements and their relationship to other parts of the business; emerging technology and the IS strategy of the corporation; and implementing systems projects based on a quality project management methodology. The initial efforts of the project brought tools that supported the implementation of the EMS requirements to the facility environmental management teams. As part of the on-going effort, A-B is implementing the computer-based systems and, more importantly, the business requirements necessary to integrate environmental processes into the organization. This is a much more difficult task.

As we saw in the Millipore EMIS example in Chapter 8, having such information at one's fingertips can be a powerful tool for an environmental manager interested in "mining" that data for any number of management-related uses. For example, management might be concerned about a particular product's process waste streams and may want to look closer at their hazardous waste volumes generated in the process, and also track the resultant manifest forms accompanying each hazardous waste shipment. Additionally, management may want to look at and evaluate TRI release data from year-to-year as a means to track pollution prevention (P2) efforts companywide, or evaluate raw materials/process throughput/waste volumes to gauge the efficacy of a particular chemical or manufacturing process, or any other number of "hot button" considerations that could be important to senior management and/or the facility's environmental manager.

We should also keep in perspective that managing such information, which can be voluminous, is crucial to various EH&S responsibilities and regulatory requirements. As such, irrespective of the particular regulated industry sector, it is key that these managers have a firm handle on the amounts, types, and concentrations of raw materials, water and energy consumption, product throughput, and the resultant waste streams. Whether these waste streams consist of mine tailings from ore extraction, or other natural resources waste products, such as drilling fluids and crude oil spills, or manufacturing processes, such as process air vents, industrial wastewater discharges, or hazardous waste shipments, to name a few, the end outcome is the same: ensuring the facility's accountability of the generating source through proper regulation of each processes resultant waste stream(s)—in other words, being responsible for the environmental outcome of the wastes generated. How far a company has progressed in this regulatory compliance scheme can be gauged by how effective a facility's environmental manager has been to ensure that the facility is doing the best that it can in properly managing its EH&S issues and responsibilities.

Of note, this specialized software market is not the sole domain of EH&S considerations. Among other business segments that also rely heavily on the computing power of these sophisticated systems are national defense, weather tracking, high-level scientific research, business process applications, and industrial processes requiring continuous automated access to data points. To complement EH&S considerations, our discussion also takes into account business process applications, such as enterprise resource planning (ERP) software, and industrial process control applications dealing with real-time intelligent systems and control software, such as man-machine interface (MMI) and supervisory control and data acquisition (SCADA). These other software applications can link business with EH&S considerations to provide an additional layer of total systems integration in addressing EH&S regulatory responsibilities and, in some cases, helping to predict or eliminate a possible adverse environmental scenario. Some names in software within this realm include: Allen-Bradley, Fisher-Rosemont, IBM, Wonderware, Gensym, Pavilion Technologies, MIT GmbH, Rockwell Automation, and Foxboro, to name a few. A listing of additional companies, by their respective software products, is provided later in the chapter.

To understand the relevance of such powerful and specialized software for use in PCs, we also need to step back for a moment and look more closely at the driving forces propelling this business segment forward, at least from the perspective of EH&S regulatory considerations.

THE EVOLUTION OF COMPUTERS AND PERIPHERALS

As much a product of rapid advances in computer hardware, such as powerful microprocessors—the "brains" of PCs—and the development of larger and larger hard drives and CD-ROM drives, the PC also owes its huge success to software development that has both kept pace with hardware development and has been the catalyst for computer and microprocessor manufacturers to develop more powerful hardware to keep pace with the rapid advances in software. Also, since our focus is geared primarily to software used in PCs, we do not provide a discussion into more powerful computer workstations that may have their own specific software that may also be used for EH&S applications. Again, readers who may want to pursue this further can do so at their own leisure.

For many computer manufacturing companies and their component suppliers, such as the companies manufacturing microprocessors, computer chips, hard drives, modems, and so on, this industry has made many of these companies common household names, and has generated multibillion-dollar revenues as a result. This rivalry, of sorts, has been among the drivers for computer companies to push the envelope farther with respect to creating more powerful computers containing ever-larger hard drives to accommodate increases in data and more sophisticated software.

From an historical perspective, for example, the 1980s ushered in 286 and 386 (referring to the microprocessor chip) personal computers that were basically of IBM design and IBM-compatible, and the Apple-design personal computers, with 200 megabyte ("meg") hard drives. These PCs gave way in the early 1990s to 486-based PCs with 500+ meg hard drives with CD-ROMs, replaced in the mid-1990s by 586-based or Pentium® PCs with gigabyte hard drives, complete with Internet access, speedy 56K+ modems and emerging voice recognition features, and other advanced applications.

Similarly, advances in software led to increases in user applications, evolving from the earlier MS-DOS® based applications from the mid-1980s that ran on the earliest IBM-compatible 286s, the XT and AT versions, and the alternative, Apple computers with their own distinct application and feel. The popularity of Apple-based software that powered those first "Macs" or Macintosh computers, soon gave rise to Microsoft launching Windows 3.1 in the early 1990s, gaining instant popularity with many MS-DOS® users. Since then, a whole new suite of Windows applications, beginning with Windows 95®, have propelled Microsoft into an applications software juggernaut, as it continues to roll out revised versions of Windows 95® . . . Windows 97® and by the time this manuscript went to press, Windows 98®. Meanwhile, in another camp, Lotus Notes® software continues to attract some developers of specialty software, especially in the ISO 9000 arena. In addition, rapid advances in PC design have placed PCs within reach of the more powerful workstations, and as such, some high-end software applications have even been accessible to high-range PCs and some workstations. This newest frontier will provide the fertile ground for another succession of next-generation software applications that will then make the applications in use today obsolete, and on it goes.

What has this advancement done, in turn, for EH&S professionals who have a need to have their pulse on various information sources as a means to address and manage the many environmental and health and safety considerations and requirements on a daily basis? We shall now take a closer look.

THE EVOLUTION OF EH&S SOFTWARE

The first of the EH&S software systems to roll out were developed about 1986 out of a need by regulated industries to obtain a better handle on the growing number of data points being generated due to more stringent and complex environmental regulatory requirements. As a starting point, we look at this period as our baseline to begin our review. This time period coincides with the first of the environmental auditing software tools developed in response to the increased use of environmental audits by a growing number of regulated industries. As we previously described in Chapter 9 environmental auditing as a service was first introduced at a handful of companies in the mid-1970s and slowly gained popularity with the regulated community. By the time that auditing software began gaining prominence, environmental auditing was well on its way to becoming a common practice at many EPA-regulated organizations. Another driver could be as a result of a greater number of regulated companies paying closer attention to environmental auditing programs as an indirect result of EPA's audit policy issued in July of 1986.

The increased use of environmental audits by regulated industries, and their understanding of auditing as a management tool, was gaining popularity with a growing number of regulated industries. This increased understanding of audit programs may have been the catalyst that sparked the interest for service providers to begin spinning-off separate auditing and audit training practices on how to conduct an audit and become an auditor. Here, a parallel can be drawn to

ISO 14001 training and consulting services that are prolific today as a result of the growing momentum surrounding ISO 14001 and related business opportunities for third-party service providers and assessors/auditors, following the same direction of ISO 9000.

Environmental auditing, in reality a practice with fundamentals borrowed from general accounting principles, was conceived as a business process to aid in the assessment of a company's compliance with environmental regulations. This practice took a little different slant from routine environmental responsibilities, such as, developing a company's required permits, contingency plans, or hazard communication programs, but focused instead on assessing or inspecting a company's total regulatory compliance program, and whether the company was in or out of compliance with the regulations. This could encompass evaluating whether routine environmental responsibilities were being carried out, such as the permits alluded to previously, or whether certain documents were being maintained as required, and the proper forms were filed, among other regulatory considerations.

The environmental auditing practice, as a service function, also developed its own tools in order to perform this task. Terms such as "working papers," "audit protocols," "checklists" became commonplace to describe audit "tools" and how to conduct audits;[2] while a new profile was beginning to emerge for the person conducting the audits—the "auditor." This profile included specific traits that eventually found their way into one of the ISO 14000 standards, ISO 14012.[3] Obviously, this practice, by its very nature, generates its own paper trail replete with a certain number of data points, findings, recommendations and conclusions, that make up the final audit document, the audit report.

As a result of this increasing paperwork, due in part to an ever-increasing set of environmental regulations, rules and policies to audit against, a number of software companies with some fundamental environmental knowledge or expertise, began experimenting with software programs to incorporate audit checklists, protocols, regulatory citations in the various Federal regulations dealing with air, water hazardous and solid waste, and developed specialized software into the first audit tools for general use. In this fashion, EH&S auditors could load this software into their computers and be able to run through an electronic version of their tools while conducting the audit. The earliest versions of such audit software tools, released sometime in the late-1980s, were MS-DOS® based and were usually limited in their applications to Federal EPA and OSHA regulations. At that time, the audit software packages commercially available did not contain modules for the individual state regulations, which were a necessity in those states with primacy for regulatory compliance and enforcement.

In addition, since these software packages were not developed by end-users, they were not totally user-friendly and readily adaptable to some individual site situations. Subsequent revised releases of these audit packages (1995–1996) incorporated various enhanced features that were much more user-friendly and adaptable to specific site conditions. For example, some of the more recent versions (1997) of commercially available packages allow the end user to bypass a specific set of regulations if they do not apply at a specific site, or allow the import of a specific set of regulations that was not part of the original regulations set if they do apply at a site. One particular available software also provides regular updates of regulations and flags key regulatory milestone dates to come to give auditors a "heads up" on pending regulatory dates to eliminate findings that may have otherwise gone unnoticed. This is one of the many examples of innovations some auditing software developers introduced to assist their customers after listening to their concerns.

While auditing software continues to gain popularity with users, other EH&S software products continue to evolve as well. A listing of some of these other software products is regularly found in one or more of the specialty trade publications in print, a sampling of which includes:

- *Pollution Engineering,* January 1998—highlights environmental software as a guide;
- *Chemical Processing,* March 1998—highlights the implementation of advanced control methods in chemical plants and refineries, including the use of third-party software;
- *Chemical Engineering**, April 1998—provides a regular feature, Chemputers, describing some of the recent features of the more advanced, "intelligent" systems;
- *Environmental Protection,* April 1998—lists various hazardous waste software providers; and June 1998—lists various air management software providers;
- *Control**, April 1998—highlights control software for use in electronic filing of information to federal and State regulatory agencies; and
- *Today's Chemist At Work,* April 1998—provides insight into the use of various software applications for your PC.
- *Software Strategy,* May 1998—provides a review of scheduling software for use on the shop floor.
- *Occupational Health & Safety,* June 1998—provides case studies of four companies solving information management crises with OH&S software.

Of the magazines sampled in our limited survey,[4] the two that consistently provide the most useful and relevant information regarding all types of EH&S, control and intelligent software, by our estimation, are the two listed with an asterisk(*)—*Control* and *Chemical Engineering.* Readers are encouraged to subscribe to these magazines and compare information. In this fashion, you can do what we did. Compile a reference library of each of these magazines, contact any of the vendors listed, and begin to develop your own database of demo disks from each of the providers you contact to get a better feel for a particular software you may be considering at some providers you contact to get a better feel for a particular software you may be considering at some point at your own facility. It is one alternative to spending hundreds of dollars per year to subscribe to specialty newsletters or reports that review this software for you and provide their readers an analysis of what the software contains. Instead, if you have the inclination, you can do your own homework and kick the tires yourself. You might gain added insight, and all it will cost you is the time to review the vendor's materials or diskettes.

SOFTWARE ADVANCES: SPURRED BY INCREASED USER DEMANDS

Some of the articles depicted in the previous section point to the capabilities of the more powerful intelligent and control systems that are used primarily in high-end systems and in applications where process flow control is crucial, such as in chemical operations, batch plant operations, petroleum refineries, and in automated manufacturing processes, among others. This is the domain of MMI and SCADA control operation software, and its very powerful sibling, intelligent systems. This software arena consists of about twenty to thirty different vendor applications, some that operate on Windows™-based software, while others operate on the more powerful UNIX®-based software, the realm of more powerful workstations designed and built by Hewlett-Packard, Sun Microsystems, Digital, Silicon Graphics, and a handful of other computer manufacturers.

While the universe of computer workstations is currently out of reach for software that is used primarily for EH&S applications, the increase in popularity with software that manages more voluminous enterprise-wide systems, such as the ERP systems, may someday bring PCs to the level of workstations in their function and form. For the time being, however, the architecture of PCs is not quite there yet, although Windows NT® does provide some promise in this regard to bridge this gap even sooner between PCs and workstations.

In the meantime, a number of industrious software developers have been able to provide the marketplace with suitable Windows™-based software to be able to capture and assimilate data points to the satisfaction of the EH&S end user. For some, the use of the higher-end software, while primarily for process control applications, has also been useful in helping to manage EH&S regulatory considerations and tap into supplemental data from other business units for assimilation into EH&S concerns. This particular aspect is at the heart of an EMIS, taking into account data from related business units within an organization to help pull together relevant information that can be useful to the environmental manager in a number of different ways. (Readers may recall our discussion regarding how environmental managers could tap into sources to provide relevant information to senior management, especially during times of funding appropriation requests as we previously described in Chapter 8 and Chapter 6.)

As a brief introduction into some of the practical applications such specialized software brings to environmental, health & safety considerations, the highlights[5] of some specialized EH&S software packages in end-user applications is provided in the next section describing some of the software capabilities in actual situations. Among the key drivers for developing many of these applications is as we have previously described: EH&S responsibilities have become more complex, and to stay ahead, a resourceful EH&S manager needs to evaluate innovative tools in the marketplace—enter EH&S and MIS software among the innovative tools emerging today. From another perspective, as developers of one specialty software application describe in their article,[6] . . . information technology makes sustainable development attainable." Herein lies another driver for pushing regulated companies further into the information data management arena, where the proper software tools are *crucial.*

Data generated as a result of more frequent environmental audits, the growing interest in ISO 14001 and its supportive EMS standards, interest in pursuing Design for the Environment (DfE) software tools to measure environmental and design indicators[7] the push to more sophisticated EMISs and a growing trend to move to electronic document management, or EDM, and what is shaping to become the central repository of all MIS databases, ERP systems, has begun to gain critical mass with software providers such as SAP AG, a German company, Baan and PeopleSoft. Also emerging are systems for managing enterprise-wide applications and EH&S applications in Java®, a programming language that is platform independent, developed by the computer workstation maker, Sun Microsystems.

Finally, we should not overlook the tremendous capabilities that the world wide web (www) offers to EH&S managers and other business unit managers that can interface with these systems, providing additional linkage opportunities, such as through the Internet and internal "Intranets." The Internet has opened up whole new doors to seasoned and novice computer users alike and opens up the whole world to an individual at a PC with an Internet link through one or more search engines, such as Alta Vista or Yahoo, and Web Browsers, such as Netscape Navigator® or Microsoft Explorer®. Access onto the Internet allows a user to look into Federal and State regulatory agencies, such as the EPA, OSHA, the Department of Justice, the General Accounting Office, stakeholder groups, such as Greenpeace, the Environmental Defense Fund, the Sierra Club, and regulated corporations, such as members of the Fortune 500 and other companies around the world, that can list relevant data on their home pages. This data can take the form of recent regulatory actions, voluntary programs, or activities within organizations, such as environmental reports, policies and other progressive happenings. (Web site linkages to many of these groups are found in Appendix G.) The term "data mining" has never been more appropriate than in such applications.

HIGHLIGHTS OF A FEW EH&S SOFTWARE PACKAGE CAPABILITIES

Among the tools that currently exist, and how they are being utilized by some end-users include specialized *EH&S auditing tools,* such as those provided by the various audit software vendors listed later in the chapter, that have the capability for: developing specific audit protocols, checklists, tailored audit findings, and drafting audit reports that can be left with the auditee upon completion of an audit. In addition to these capabilities, EH&S audit packages can also provide: (1) linkages to allow users to look into applicable federal and State regulations in greater detail should that be of additional benefit; (2) design processes that allow specific regulatory requirements to be grouped by facility and supplemented with supportive investigative questions that can aid the auditor(s); and (3) prompts to allow prospective auditors and facility representatives (the auditees) to be trained in understanding how to manage the processes being audited. For those end-users who prefer to utilize home-grown audit systems, Lotus Notes®

also provides that capability. As an example, one company that performs in-house audits in this fashion is the William Wrigley Jr. Company.[8]

Among the benefits a facility being audited can obtain by using such tools include: the development of consistent audit reports; the findings to be video recorded (also a component of some ISO 14001 software)—which may require counsel's approval, depending upon company policy—and the training component inherent in the software can provide opportunities for employee improvement in their work duties, should that be a concern of the audited facility.

A Case Study—Ocean Spray Cranberries, Inc.[9]

One company that utilizes specialized auditing software is Ocean Spray Cranberries. Ocean Spray finds one vendor to be particularly useful for their EH&S auditing needs in several ways. These include providing the company: (1) a basic corporate audit tool (used at 17 facilities in eight states); (2) a tool for plant self-audits that teaches the regulatory requirements and involves employees; (3) a catalyst for continuous improvement of the audit process; (4) an opportunity to help standardize the audit process; (5) an opportunity to introduce increased employee involvement and ownership (a catalyst for employee "empowerment"); and (6) an opportunity to improve upon risk reduction.

One factor that any end-user of a specialty software package such as described needs to keep in perspective is that a tremendous effort in time, employee commitment, and resources is necessary to allow such programs to be of any beneficial use to the end-user. If a company is not ready to make the commitment that Ocean Spray did, then perhaps specialty EH&S, EMIS, or even ERP software may not be the right tool for that company. Before you buy into a program or package, common sense dictates that you first "kick the tires" of the packages available, ask questions of the developers, evaluate demo disks or hands-on packages, or perhaps have an end user of the software share their observations with you about the particular software they utilize, and whether the all-important customer support services representatives are there when you really need them. That is key.

IBM's Specialty Software—Enterprise Document Management Solutions[10]

A company that is a household name in PCs and mainframe computers, IBM, also provides a number of different types of product and services offerings in specialty niche markets. The company's Enterprise Document Management Solutions (EDMS) Group, for one, provides software assistance in electronic data management (EDM) software, which includes NovaManage™, an object-oriented enterprise document management and workflow management solution. Among the solutions NovaManage™ provides in the environmental, health & safety area is the effective management of EH&S management and compliance issues dealing with: worker safety, the reduction of pollution and industrial wastes, the proper management of hazardous materials, the identification of environmental performance measurements, and opportunities to improve other areas within the EH&S and EMS management areas that can lead to good environmental practices.

With respect to good environmental practices, business process drivers that have emerged with linkages to environmental performance, include the ISO 14000 family of standards, the EMAS regulations of the European Union, and the CMA's Responsible Care® Program. Each of these standards and good industry practices have in common a linkage to and a dependence upon documentation of policies, standard operating procedures and practices, corrective actions, management review and improvement plans and the tracking of environmental performance results, such as through EMS auditing (ISO 14010s) and Environmental Performance Evaluation (ISO 14031). IBM's NovaManage™ can provide some assistance in these areas through its specific tools.

Among IBM's customers utilizing NovaManage™ are: Carolina Power and Light, Kimberly-Clark, Maraven (a unit of PDVSA, s.a.), The Norton Company, and Public Service Electric and Gas.

IBM's service in this area first began as an offshoot of the company's internal management of their EH&S business processes and extensive use of information technology as one way to better protect their workers, customers, community and the environment. As a result of the achievements the company attained in these areas, IBM launched a specialized consulting unit that includes a comprehensive consulting service portfolio to create a competitive advantage for a company in EH&S matters. A key element of this service offering was IBM's Chemical—Environmental, Health and Safety—Management System (CHEMS) model, designed to integrate EH&S management across business operations and incorporate EH&S "best practices" to address and exceed regulatory compliance requirements, and maximize the power of information technology.

In summary, the IBM's EH&S consulting services provides capabilities to support three distinct project phases: (1) understand client's needs—EH&S management strategies and drivers, priorities, business processes and information systems strategies; (2) build on EH&S business process model based on the CHEMS Model; and (3) make recommendations from the model. By implementing EH&S business process improvement information itself, IBM had some significant EH&S accomplishments. These include:

- The elimination of Class I ozone-depleting substances in manufacturing since 1993;
- The company's energy conservation efforts saved $22.3 million and 450 million KwHrs;
- The company's work-related injury rate was 67 percent below industry average; and
- IBM was instrumental in identifying and developing new materials and technologies for pollution prevention, such as, lead-free solder in PC-board manufacturing.

In addition, IBM's accomplishments to its clients included the following:

- The successful tailoring and implementation of the CHEMS model for multinational and international companies;

- Developing information technology solution frameworks for firms implementing CHEMS;
- Developing and implementing innovative information systems for various uses, among them: MSDS management, volatile organic compound (VOC) tracking, environmental management, EH&S audits, and wastewater management reporting.

CARIBOU SYSTEMS

Another company that provides specialty software that can be utilized in EMIS and ISO 14001 applications is Caribou Systems. Caribou's primary end-products include Caribou98 and the ISO 14001 Gap Analysis Tool, commonly referred to by Caribou as the "Gap Too." Caribou98 is defined as a relational database for management of environmental information at both the plant and corporate level. The system connects with a company's existing systems and exports to common Office Suite software. The system is modular by design and includes the following system modules, and can be used as part of a company's EMIS: (1) the core; (2) hazardous materials; (3) chemical/waste; (4) monitoring; (5) air/water emissions; (6) environmental management; (7) occupational health and safety; and (8) system tools.

Caribou developed the Gap Tool in conjunction with the Canadian Standards Association (CSA) as a tool for implementing ISO 14001 within a company. The tool utilizes a scoring system to identify strengths and growth opportunities within a company's EMS and as it is operated, tracks action items. It is equipped with the capability to graphically display scoring results upon completion.

Among the companies that use Caribou Systems' software include: General Mills, Nortel and Peregrine, Inc. The following case studies, courtesy of each of the companies highlighted, depicts their experiences with EH&S software before and after utilizing Caribou Systems.

NORTEL[11]

Northern Telecom (Nortel) is a global provider of communications network solutions. The company provides network and telecommunications equipment and related services in North America, Latin America, Europe, the Middle East, and Asia.

Nortel wanted to create a database to collect information gathered at its European, Asian, Central and Latin American and North American facilities. Under the project management of Nortel's David Conrad and Tony Basson the Nortel team worked hand-in-hand with Caribou to meet the following criteria. They wanted to:

- Provide a residence for the business Environment and Safety Management System (ESMS);
- Provide guidance in the development of the ESMS;
- Allow for the consistent management of the issues across Nortel;

- Contain user friendly features for operations use as a management tool by the business lines;
- Provide the kind of information that can help support business lines, and corporate and legal in decision-making;
- Provide real business value to end-users;
- Assist in demonstration of responsible care for environment and employees;
- Be easily adaptable to our changing business needs;
- Allow the collection of data necessary to adhere to the PERI guidelines;[12]
- Have the ability to adapt itself to display information from more than one country and in more than one language;
- Link to other databases, to avoid repetitive data-entry;
- Include the capacity to store Health and Safety information;
- Have the capacity to be rolled out in an incremental fashion, i.e., from less complex to more complex-seamlessly;
- Have the ability to run over a wide variety of networks;
- Be able to reconcile information changes between different locations;
- Have the ability to apply user-chosen normalization rules in order to provide common-sense adjustments for changes to business conditions.

The final system used replication technology to ensure the application remained fast over Nortel's network. The base program underwent extensive modification to conform to Nortel's needs. The final product was renamed 'Evolution' representing a changing 'plastic' system that would be updated and adjusted at a minimum on a yearly basis. As of this writing, Evolution is in use at Nortel's UK, Central and Latin American, Asian and North American facilities, and has over 400 system users. Additionally, the final product earned the Nortel Award of Merit for Innovation.

PEREGRINE, INC.[13]

Peregrine supplies a number of interior and exterior automotive products to the automotive industry. It's primary business operations are located in Michigan and Ontario, Canada and has approximately 5,700 employees.

Among the company's primary tenets includes a commitment to enhancing employee working conditions by way of establishing and conforming to performance standards at each of its manufacturing locations. To achieve this goal, Peregrine identified the need for a computerized environmental database system to track environmental performance at each location. Among the objectives this system would need to achieve included a stable, user-friendly platform as a basis for a customized system. As both companies assessed, Peregrine and Caribou would modify the Caribou System to fit Peregrine's need.

From this joint effort, Peregrine used in-house expertise and Caribou technical staff to modify Caribou's base system to fit their specific needs. This system is currently being used by about twenty Peregrine staffers to manage and report on the company's environmental performance.

THE NEXT WAVE IN EH&S AND OTHER SOFTWARE SYSTEMS

The companies highlighted in the previous section provide a brief insight into some of the specialty EH&S software that currently exists in the marketplace, and how some of these programs were used by a representative sample of end-users. Since there are not too many companies that can provide such information (a warm "Thank You" to the representatives of the companies previously highlighted), case studies are the only way some people can be informed about these systems. For the most part, case studies are generally found only at specialty conferences[14] dealing with EH&S software, in which a number of EMIS case studies may be highlighted, or as presentations, such as the ones we gave at the trade conferences referenced.

As one alternative to going to conferences or other time-consuming and costly endeavors, especially if one's training and travel budget may be somewhat limited, one viable source that allows prospective end-users to catch up with the latest in software systems, is to peruse one or more of the trade publications that focus on EH&S software and contact the vendors listed to obtain additional information, as we described. These trade publications are also listed in Appendix B.

Most of these publications provide complimentary subscriptions to qualified readers. It is worth the effort to explore and obtain a subscription. From there, once your magazine subscriptions arrive, you can start your search to peruse each month's issue to look for information that lists various software providers in the marketplace, as well as insight into what some of them have recently developed, and how these packages can be utilized. Without that, you cannot get a realistic feel for what each vendor's product is about, and whether that particular software suits your specific needs. In other words, prospective users must really "kick the tires" with software applications and do their own due diligence before considering the purchase of (a) specific package(s). Contacting an end user is another way to gather sound information, since a prospective user can gain tremendous insight from current users by asking them specific "apples-to-apples" questions that may be relevant to their own specific needs. Taking this aspect one step further, a benchmarking study of several current end-users may be a worthwhile endeavor to pursue as well.

To provide the reader some assistance in this area, whether to piece together data for a possible benchmark study, or just to begin a reference base, we culled some of the magazines highlighted previously and scanned through some of the software literature in our research file. We came up with this brief overview of what some software vendors can provide for environmental, health and safety end-users and some of their business colleagues, such as process control and document management specialists.

Among some of the capabilities specialty software provides EH&S users include:

- An updated accounting of TRI data for use in EPA Form Rs, or supplemental identical State forms;
- The generation/annotation of MSDS forms to assist in hazardous chemical data inventory to address regulatory requirements, such as Right-to-Know reporting and other considerations;

- Chemical Abstract Service (CAS) information on a number of chemicals that could also supplement MSDS information;
- An updated database of applicable EPA and state regulations, OSHA and DOT regulations, and other appurtenant information to ensure compliance;
- Chemical inventory and waste-tracking and manifesting capabilities;
- P2[15] tracking and measuring opportunities
- Data acquisition and report generation, useful in many applications, such as audits, ISO 14001 gap analyses and EMS audits, EMIS applications, and ERP/ EDM x-business functions.
- Tracking employee training: Who took what, where, and was it useful to the participants?
- Safety data management: HazMat, HazCom and other OSHA requirements, such as filing and tracking OSHA 200 logs for reporting on-the-job injuries;
- Scanning and tracking resumes for Human Resources databases—zeroing in on ideal candidate(s);
- EH&S audit management: generating annotated protocols, checklists and draft reports, and the ability to track corrective actions via audit reports and follow-on reports;
- Predictive Emissions Monitoring Systems (PEMS) and Continuous Emissions Monitoring Systems (CEMS) to monitor, track, and predict air emissions from stationary sources, such as smoke stacks;
- Life Cycle considerations of products manufactured that track and quantify P2 cost-saving benefits over a wide business enterprise spectrum; and
- Intelligent knowledge-based "real-time" expert systems[16] and process control systems that can be developed by a facility to minimize leaks, emissions, consumption of energy and excess raw material, and waste or off-spec product, while increasing product quality and throughput—which equates to product efficiency and monies saved.

The following section provides a partial listing of the two hundred-plus companies that developed specialty software in each of their respective categories.

A PARTIAL LISTING OF SOFTWARE PROVIDERS

EH&S COMPLIANCE, INCLUDING AUDITS

Aldrich Chemical	Automated Workplace Safety	Benefit Software, Inc.
Bureau of National Affairs	Business & Legal Reports	Computational Mechanics, Inc.
Corbus	Dakota	Environmental Profiles
Electric Software Products	Earth Soft	Environmental Data Services
Environmental Software	EnviroData Solutions, Inc.	LFR Technologies
& Systems	Essential Technologies, Inc.	Molecular Arts Corp.
Logical Technology	MICROMEDIX, Inc.	Pacific Environmental
Open Range Software	Oracle	Services

Per Datum, Inc.

Quantum Compliance
Systems

Safety Software

Petroleum Information

RegScan, Inc.

Summit Training Source

Primatech

SAP

Wixel, Inc.

ISO 14001 Considerations—EMS and Gap Analyses

Caribou Systems, Inc.   Greenware, Inc.   Modern Technologies   Reality Interactive

Intelligent Real-Time Expert Systems[17]

Aspen Tech, Inc.   Comdale Technologies, Inc.   Eimco Process Equipment Co.

Elsag Bailey   Fisher-Rosemount   Foxboro   Gensym

GSE Systems   Honeywell, Inc.   MIT GmbH   Moore Products Co.

Oxko Corp.   OSI Software, Inc.   Pavilion, Inc.   Siemens

Simulation Sciences, Inc.   Wonderware Corp.

Enterprise Resource Planning (ERP) and Electronic Document Management (EDM) Systems[18]

Allen Bradley   Baan   Cadis   IBM   LFR Technologies

NovaSoft Systems   Oracle, Inc.   SAP, AG

ISO 9000 Software

3C Technologies   American Software   Business Challenge, Inc.   CFM Inc.

DP Solutions, Inc.   Interspan, Inc.   IQS Inc.   JBL Systems

Micromagic   National Instruments   Pilgrim Software   PowerWay, Inc.

Prosys   Q Soft Solutions   Quality Resource   Reality Interactive

Readers should note that there are also several technical newsletters in the marketplace that review several of the systems listed and highlighted previously.

CONCLUSION

As we have seen from the information provided, environmental, health, and safety considerations are a complex set of rules, regulations, requirements, and data points that are constantly in a state of flux. The days of manual input and management are no more, and given additional external and internal factors that constantly barrage EH&S professionals in their day-to-day activities, there is not enough time in a day to accomplish all these goals. Turning to software systems, can enhance the compilation and accumulation of data to address regulatory issues and other management concerns.

The software packages highlighted in our case studies and the magazines we reference provide just a glimpse into what is currently available in the marketplace for specialty software solutions. The key to picking the right software is for a prospective end-user to understand their needs and understand which software package can provide the most benefit to their specific needs.

For example, Foxboro®, a long-time provider to the environmental safety markets, embodies this general concept in their Dynamic Performance Monitor. Con™ (software) product brochure[19] that lists among their product's attributes: (1) increased production throughput; (2) decreased consumable materials; (3) decreased power consumption; (4) decreased off-spec production; and (5) bottom line savings.

This software product is based on a patented approach of The Foxboro Company that entails the object-based, real-time modeling of key performance metrics using process instrumentation. This software package utilizes the information generated by these live, real-time models to provide the operators with the information required making effective high level decisions about the process. Much of this data is presented in economic terms rather than pure engineering terms, but good environmental management decisions should have clear economic impact, and the decision makers should understand what that impact is.

EH&S software, in combination with the right platform and other compatible software, such as ERP,[20] EDM and real-time intelligent systems, can accomplish the five tasks listed in The Foxboro Company's brochure and more—it's a matter of choosing the right package, customizing it as necessary, and ensuring that all the bugs are out of it so that it functions for you as designed.[21]

Good luck in your EH&S software related endeavors.

ENDNOTES

[1] Provided with permission from the Anheuser-Busch Companies, © 1997 by Anheuser-Busch Companies, Inc. All rights reserved.

[2] One of the first documents that described audit protocols and a how-to for environmental auditing was a work commissioned by the Edison Electric Institute titled, *Environmental Auditing Workbook*, Edison Electric Institute, 2nd ed., 1983, and an EPA document, *Environmental Audit Protocol for EPA Facilities*, prepared by Arthur D. Little, Inc., November 1986.

[3] The International Standard, ISO 14012, Guidelines for environmental auditing—Qualification criteria for environmental auditors, © International Organization for Standardization, 1996.

[4] Readers are encouraged to contact the various publishers themselves for additional back issues dealing with EH&S software applications.

[5] The author first presented a summary of his research into this area at Chemputers-Northeast, sponsored by *Chemical Engineering*, September 1996 in King of Prussia, Pennsylvania; and a revised version at the IPC Printed Circuits Expo, April 1998, in Long Beach, California.

[6] B. Nelson and A. Sziklai, "The KEY to Sustainable Development," *Environmental Protection*, June 1998, p. 46.

[7]One of the pioneers in the DfE software design arena is Ecobalance (of the Ecobilan group), in collaboration with IBM, Alcatel, Groupe Schneider and Thompson. For additional information readers may want to contact Ecobalance directly.

[8]A. Holynshi and S. Wood, "Developing an ESH Auditing System Using Lotus Notes," presented at an IBC Conference on environmental auditing, October 15–17, 1997, Philadelphia, PA.

[9]Provided with the permission of Ocean Spray Cranberries, Inc. The author first presented this case study as part of his presentation, *The Next Generation of Environmental Management* at CARE INNOVATION '96, November 1996 in Frankfurt, Germany.

[10]Reprinted with the permission of International Business Machines (IBM).

[11]Reprinted with the permission of Northern Telecom.

[12]Nortel is among the original signatories of the PERI Guidelines, and includes among its signatories representatives from Amoco, DOW, DuPont, IBM, British Petroleum, and Phillips Petroleum. We provide additional information about PERI in Chapter 10 and in Appendix D.

[13]Provided with the permission of Peregrine, Inc.

[14]Some conference providers: IBC Group, *Chemical Engineering* and Kalthoff International.

[15]As of the time this manuscript was being prepared, EPA was working with one contractor that had developed a P2 software tool that was going through beta testing.

[16]One such product is "G2" developed by Gensym Corporation.

[17]Among the conferences that showcase and exhibit a number of these specialty software providers is the Chemical Engineering Exposition & Conference, hosted by *Chemical Engineering.* The most recent conference was held June 3–4, 1998 in Houston, Texas.

[18]Among other conference providers that showcase and exhibit a number of these specialty software providers is Kalthoff International. The most recent, Kalthoff Spring 98, was held March 23–26, 1998 in San Diego, California.

[19]Reprinted with permission from The Foxboro Company, © The Foxboro Company, 1996.

[20]For additional insight, readers are referred to the article, "Running on Information" by E. M. Kirschner, *Chemical & Engineering News,* September 15, 1997, p. 15.

[21]For additional reading on the subject of information management and the computer's role in this area, readers should refer to the book, *What Will Be: How the New World of Information Will Change Our Lives,* by Michael Dertouzos, New York: HarperCollins Publishers, 1997.

REFERENCES

General

Dertouzos, M. *What Will Be: How the New World of Information Will Change Our Lives,* New York, NY: HarperCollins Publishers, 1997.

Environmental Auditing Workbook, Washington, DC: Edison Electric Institute, 2nd Ed., 1983.

Environmental Audit Protocol for EPA Facilities, Final Draft. Cambridge, MA: Arthur D. Little, 1986.

The International Standard, ISO 14012, Guidelines for environmental auditing—Qualification criteria for environmental auditors, International Organization for Standardization, 1996, Geneva, Switzerland.

Software References

Chemical Processing—July 1995; September 1995; July 1996; December 1996; and March 1998

Chemical Engineering—September 1995 through April 1998, inclusive

Control—July 1995 through April 1998, inclusive

Environmental Protection—February 1994; September 1994; December 1994; March 1995; September 1995; December 1996; April 1998; and June 1998

Occupational Health & Safety—June 1998

Pollution engineering—June 1995; January 1996; January 1997; and January 1998

Quality Digest—June 1996

Software Strategy—May 1998

Today's Chemist at Work—April 1998

PRESENTATIONS

Bamburger, A. "The Information Systems Perspective: Bridging the Gap Between EM and IS," EH&S MIS, Washington, D.C., March 11–12, 1997.

Crognale, G. "Next Generation Environmental Management Systems," Chemputers IV, King of Prussia, PA, 1996.

Crognale, G. "The Next Generation of Environmental Management", CARE INNOVATION '96, Frankfurt, a. M., Germany, November 18–20, 1996.

Crognale, G. "Utilizing Software to Streamline EH&S Data Management," IPC Printed Circuits Expo '98, Long Beach, CA, April 26–30, 1998.

Holynski, A. and S. Wood. "Developing an ESH Auditing System Using Lotus Notes," presented at an IBC Environmental Auditing Conference, Philadelphia, PA, October 15–17, 1997.

PROMOTIONAL LITERATURE FROM:

IBM—Process Solutions: Enterprise Document Management Solutions

Caribou Systems—The Caribou System Case Study—Nortel; Perigrine Inc.

Ecobalance—Life Cycle Assessment: Expanding the View beyond facility boundaries

The Foxboro Company—Dynamic Performance Measures: Your passport to world-class manufacturing in the '90s

CHAPTER 14

SOURCES OF ENVIRONMENTAL INFORMATION

Gabriele Crognale, P.E. with contribution from Roland W. Schumann III[1]

> *"Knowledge is of two kinds. We know a subject ourselves, or we know where we can find information upon it."*
>
> *Samuel Johnson*

APPRECIATING THE ENORMITY OF KEEPING CURRENT

The number of environmental, health, and safety (EH&S) regulations is growing at a phenomenal rate. At the same time, existing regulations are being expanded and toughened at such a rapid pace that it can be almost overwhelming to ensure that your company is truly in compliance with the spirit and the letter of those laws and regulations. At this point, there are more than twelve thousand pages of regulations in Title 40 CFR (plus an additional twenty-six hundred pages in Titles 29 and 49). Given the historical rate of regulatory expansion, there could be more than twenty thousand pages of EH&S regulations by the end of the decade! In addition to staying current on the laws and regulations, those companies that are following them must also keep up with the myriad of internal directives, compliance enforcement guidance documents, and the letters of interpretation that are frequently issued by each of the regulating agencies.

When faced with the prospect of monitoring and complying with the regulations, the frequent changes and the new directives, it's logical to also examine the costs and benefits associated with non-compliance. Indeed, outside the obvious point that complying with the laws and regulations is "the right thing to do," there is the issue of individual and corporate liability. Not too long ago, the likelihood of being caught and prosecuted for an environmental violation was fairly remote. There were far more businesses that were engaged in operations than there were enforcement officials to check-up on and investigate violators. Then along came stepped up enforcement in the form of increased hiring of inspection personnel at all levels and more sophisticated electronic tracking of potential vio-

388

lators. Additionally, environmental violations were no longer being handled exclusively by the civil courts. Rather, a new trend had emerged to identify those violators.

At the same time that the cost of committing environmental violations had become much more significant, a means of incorporating leniency into the penalty assessment process had been established. The net result is that fines may be reduced if the violating company can prove that it has an *effective* program in place to prevent and detect violations of the law. The operative word there is *effective*. It is not sufficient to merely have a written program. Implicit in the guideline is the understanding that organizations must provide current information to their employees and also establish oversight responsibility to specific individuals within the higher-level ranks.

WHAT ARE THE ODDS YOU WILL BE CAUGHT?

As the degree of enforcement has been ratcheted up several notches, so too has the pace of technological advances in the area of damage assessment. Not long ago, sampling methods detected contaminants in parts-per-million (PPM), whereas now they look at parts-per-billion (or PPB). Thus, since pollutants that once escaped detection are now more easily found, a growing number of organizations are being identified as "polluters" and now must deal with the costs of remediation. In turn, it is now more important than ever that everyone involved in activities which can potentially cause environmental damage fully understand what the various laws and regulations require them to do to stay in compliance.

THE RELATIONSHIP BETWEEN LAWS AND REGULATIONS

The cornerstones of environmental compliance are the *United States Code* (or "U.S.C.") and the *Code of Federal Regulations* (or "CFRs"). The U.S.C. contains the actual laws or statutes that were signed by the president. Once a law is created, the appropriate government agencies (EPA, DOT, OSHA, etc.) are empowered to write regulations intended to enforce those laws. There are more than twenty books containing regulations that fall under the broad umbrella of EH&S regulations.

The CFRs are published by the federal government annually, but complying with the regulations is not simply a matter of buying the most recent book and thumbing through it. After all, there are plenty of new regulations which are promulgated between editions and which must be observed by businesses. A key challenge to the EH&S professional is to not only learn the myriad of existing regulations, but to also stay abreast of changes that affect them as well as any new regulations that come along.

One mechanism the government uses to get the information out is through a booklet it publishes each business day entitled the *Federal Register*. It contains a mixture of information that ranges from the mundane to the very significant. The trick is to separate the wheat from the chaff. Although it contains information on every activity within the government, there are three main types of information that regularly appear and should be of particular importance to EH&S professionals. First, there are "notices of proposed rulemaking." These are notices which give the public an idea of new regulations that are on the horizon.

The second category of important information consists of temporary regulations that are also referred to as "interim rules." They stay in effect until the regulations become finalized. While the first two categories are important, it is the third major category which contains new regulations that will eventually be included in the CFRs. These regulations are called "final rules." Unfortunately, given the other clutter which appears in the Federal Register, it can be overwhelming to rely solely upon either the Federal Register or the US Code as sources of environmental information.[2]

Thankfully, there are other options. Throughout this chapter, you will encounter a variety of suggestions regarding where you should look for your compliance information. In many instances, it will be suggested that you turn to your local public library. This suggestion is necessary because many of the directories and catalogs you will be using to locate appropriate newsletters, journals, magazines, and so forth will be of little value to you after your initial search for information. Thus, it doesn't make sense to simply provide you with information about how to order those directories if what you really need is a one-time read-through of the information that is contained in them. At this point it is essential to understand the two major types of information.[3]

WHAT ARE "PRIMARY" AND "SECONDARY" SOURCES OF INFORMATION?

Primary sources of information are those items which were just discussed. They are the actual laws & regulations. As might be expected, they are frequently hard to decipher or place into practical context. After all, most of them were written to cover a wide range of industries and business activities. While there's really no way around having to maintain a set of the current CFRs or statutes on-hand, it doesn't mean they have to be relied upon exclusively to learn what your legal obligations are under the various legal and regulatory programs. That's where the secondary sources of information come in handy.

By their nature, secondary sources of information take the often wordy and confusing laws and regulations and provide non-legalese analyses and interpretations. Depending upon the orientation you are seeking, you can often find books, newsletters, and online services which will take a complex subject and tailor it to your particular interests. Later in this chapter we will discuss online services in much greater detail, but now for the sake of illustration, we'll go over a few no-cost (free!) online services offered by the USEPA.

Since EPA has separate program offices for each of the major environmental subsets, it's not surprising that there are also many niche-online options. To say that EPA has quite a few free electronic bulletin board systems (BBSs) oriented to specific environmental media is an understatement. For example, there are several BBSs which provide only air or water information and there are others which are terrific sources of hazardous waste and Superfund-type information. Keep in mind that because environmental compliance information changes so rapidly, it's a good idea to develop several good sources of information drawing from each of the various platforms—books, journals, newsletters, courses, online services, etc.

A word of caution . . . It is essential that you check the credentials of the authors and speakers who you have chosen for your information. With the boom in environmental enforcement activities and the subsequent realization that there is money to be made in environmental education, many people with questionable, if not dubious, knowledge and expertise are jumping on the "self-proclaimed expert" bandwagon.

ACQUIRING A GOOD BASE OF COMPLIANCE KNOWLEDGE

There is no shortage of specialty companies to provide you with the information you require. While some of these organizations focus solely on providing information via books or videotapes, there are others which can provide information in a variety of methods—a sort of *"one-stop shopping"* approach to compliance information.

Books are the first and most logical place to begin your search for compliance information. However, given the semipermanent nature of books, they can't be relied upon exclusively to provide the most current information (e.g., *What is the latest change to the RQ, or "Reportable Quantity", for a spill of trichlorethylene?*). Simply put, the span of time between when a book is written and when it's published prevents most books from serving as chief sources of technical compliance information. Furthermore, since publishers endeavor to keep their books on the shelves for at least two or three years between editions, there is a tendency to keep the information just general enough to allow the books to stay somewhat current until the inventory is depleted and the new edition is ready. Thus, areas of the law which see frequent and dramatic changes (e.g., Clean Air Act and RCRA) are also more likely to have subject matter which causes books to become outdated.

USING VIDEOTAPES AS A SOURCE OF EH&S INFORMATION

Videotapes can be a great way to gain a quick understanding of a complex subject, but professional-level videos are also notoriously expensive and of short duration. Prices for programs in this category range from the low hundreds to more than $1000 and most don't last any longer than fifteen or twenty minutes. Given their high prices, videos are best-suited for training *groups* of people rather than individuals.

If you elect to incorporate videos into your information acquisition program, it's a good idea to turn to one of the larger videotape distribution companies. These organizations can save you a considerable amount of time. In their quest to offer you the "best of the best," they screen lots of videotapes that were produced by multiple video production companies before they choose the programs that seem the most appropriate for their clientele. The net result is that the overall quality of the programs they offer tends to be higher than if you stick with a single company. Additionally, quite often you will have a larger pool of choices if you go with a distribution company since they can focus on, and expand, their offerings on subjects they know are of interest to their customers. In contrast, videotape production companies are generally forced try to try a melting-pot approach wherein they offer a range of programs that will, presumably, sell to a variety of audiences. As such, they often don't have the luxury of producing more than one or two tapes on a given topic. Indeed, after one video has been made, it's time to move on to another topic. Conversely, looking through the catalogs of videotape distribution companies will quickly show that, instead of one or two tapes per topic, there are many more tapes to choose from on the more popular topics.

CONSIDERING LOOSELEAF SERVICES FOR ACCESS TO CHANGING INFORMATION

A growing number of EH&S publishers are learning that looseleaf subscriptions are veritable *cash cows* when it comes to attracting customers and hooking them on the monthly, quarterly, or "as needed" updates they offer. Looseleaf service are typically much more expensive than standard bound books—not unlike the price of professional videos, the prices of these looseleaf service typically start in the low hundreds and can surpass $1,000 for a one-year subscription. The cost usually depends upon the frequency of the updates as well as the complexity of the topic. A service that informs its readers about narrowly focused changes to a high-level topic, such as the legal implication of trading air emissions points, would cost significantly more than a broadly-focused service that simply provides general tidbits of information.

In most cases, looseleaf services contain several useful components such as a durable binder with tabs containing a base document, a monthly newsletter, and periodic updates that are intended to be inserted into the base document. There are two key elements to consider when considering a looseleaf subscription: The *length of time the service has been around* (looseleaf subscriptions are always coming and going), and *the credentials of the author and editors.* That last point is of paramount importance in light of the looseleaf management practices that are common in the publishing industry. Usually, one or more prestigious authors will sign a contract with a publisher to draft a portion of the initial materials. However, once the base document has been completed, those authors will drop out of sight and the publisher's own editors and writers will take on the task of maintaining the service. The original author(s) will still be consulted occasionally for advice and editorial guidance, but the bulk of the writing will be done by others. That explains why it is important to evaluate who the author and editors of a service are before deciding whether or not to subscribe.

In similar fashion, the environmental compliance arena has generated a healthy business in the area of professional conferences, seminars and courses that are offered in various parts of the United States and overseas by different promoters. Depending upon the timeleness of the subject matter and the invited faculty at these offerings, they can provide relevant and timely information on a particular subject matter that can be of interest to the environmental professional. In addition, many colleges and universities have developed continuing education programs in environmental management and compliance, health and safety, and OSHA-required courses, as vehicles to help many working EH&S professionals who cannot attend some of these offerings due to budget or work constraints.

As our research found, whether a particular course, conference or seminar producer can provide its audience timely, appropriate, and relevant information, it is absolutely essential that a prospective attendee evaluate several elements in helping to make an appropriate decision.[4] Taking the time to do so can help determine whether or not one should attend, and whether attending can enhance their EH&S skills.

From our experience these elements include:

- What value will the conference, seminar, or course provide that attendees can bring back to the facility to share with their colleagues?
- Are the conference, seminar, or course providers well-known in the industry, and what is their track record?
- Will the subject matter cover new ground, or it is a marketing "spin" of a known subject area?
- Who are the guest faculty or course instructors? What experience and credentials do they bring to qualify them as subject matter experts?
- What feedback, if any, was received from previous offerings?
- Or, any other considerations the prospective attendee may wish to explore further to make an informed decision.

NEWSLETTERS CAN PROVIDE TIMELY INFORMATION AT A REASONABLE PRICE

With the proliferation of personal computers, feature-laden word processors and low-cost laser printers, desktop publishing has evolved to the point where anyone with access to WordPerfect® or Microsoft Word® can publish a professional-looking newsletter. In fact, so many newsletters abound that merely selecting one to subscribe to can be a significant challenge. A recent check of Oxbridge's newsletter directory yielded page after page of EH&S newsletters. Since that reference book is oriented to covering commercially-produced (for profit) publications, it doesn't attempt to list the even greater number of free newsletters that are available for the asking from many law firms and engineering firms across the country. Many of those publications are of equal or better quality than their often expensive commercial counterparts.

A good place to get started is to look through the Oxbridge directory at your local library and make a list of those products which sound most interesting. The entries in the directory contain enough information to put you in contact with the various publishers. A short phone call to those who publish the newsletters should produce file folders full of sample issues that can then be evaluated side-by-side, feature-for-feature. A similar approach with *Martindale-Hubell's Law Firm Directory* (also available in the reference section of your public library) can produce similar results with law firm newsletters. Look up the phone numbers to a handful of the national law firms—those with offices in several states—and give them a call. In most cases they will be delighted to place you on their mailing lists; after all, in their eyes, you are a potential client and sending you a copy of their newsletter is an inexpensive way to demonstrate their knowledge of complex and rapidly changing topics.

FINDING OUT ABOUT ENVIRONMENTAL TRADE MAGAZINES AND JOURNALS

There are two main categories of professional magazines: standard *fee-based* publications like *Forbes* or *Fortune,* and controlled-circulation or *"nonpaid"* subscriptions. There is a surprising number of excellent-quality periodicals that fit into the latter category. Indeed, there are so many good free magazines these days that it is difficult to justify paying for a magazine subscription. Here is a brief explanation of how some magazines can be given away free while others charge hefty fees for each issue. Those magazines which require paid subscriptions, are financed both by the subscribers and by the advertisers. Anyone who is willing to pay the fee can receive the subscription. With publications that are in business for the long haul, the money earned through sales of subscriptions and advertisements covers, presumably, the cost of producing and distributing the magazines.

To the contrary, the controlled-circulation magazines derive their income almost exclusively from the pool of advertisers that take out large blocks of space for their advertisements and who also rent the magazine's mailing lists. These publications use a detailed subscription questionnaire to screen prospective subscribers. By selectively determining who will get a free subscription, the publishers can build a finely-tuned database of individuals who fit the profiles that are most often sought by the advertisers; such as, compliance officers at manufacturing facilities which employ more than one hundred workers at multiple locations. To help ensure the integrity of their mailing list claims, these publishers are audited on an annual basis. Random sample selections of the mailing lists are taken and those people are interviewed over the telephone. The purpose is to verify that those people have, indeed, signed up on their own and haven't simply had their names pulled from one magazine's mailing list and plugged into another's.

Given that, there is still the challenge to identify those magazines, free or otherwise, which will best suit your needs. One easy, although hardly encompassing, way to start your search is to survey others in your field, either at your company or at professional seminars and conferences, to learn where they get their information. Using that approach can save time but you run the risk of inadvertently overlooking some important magazines. Another more systematic approach is to review the entries listed in the *environment publications* section of *Standard Rates & Data Services* (or "SRDS"). This is a huge, phonebook-like reference publication that contains, among other things, the information that direct mail marketing professionals need to accurately assess a magazine's mailing list for direct mail purposes. You can use the same information to determine how widely read a particular publication really is and how many people have paid to get it. Each entry in SRDS indicates the total number of "paid" and "nonpaid" subscriptions that are mailed out each month. The statistical data in SRDS for publications like *Time* or *Newsweek,* shows that nearly all subscribers have paid for their copies.

However, a quick look at the information provided for the leading EH&S magazines like *Pollution Engineering, Environmental Protection, Industrial Safety & Health News* shows that the vast majority of readers who receive copies in the mail each month have gotten them for free. This doesn't mean the publishers don't try to sell their magazines, it just means that they are also willing to give them away. Take this into consideration when you complete the publisher's subscription questionnaire. If the answers you provide paints a picture of someone who matches the magazine's "profile" of the ideal subscriber, you are more likely to be given a complimentary subscription.[5]

LOOKING MORE CLOSELY AT LIBRARIES

BUSINESS LIBRARIES

As hackneyed as it may sound, don't overlook the benefit of visiting your local library. Although the corner library may, indeed, have what you are looking for, the best resources are usually found in another type of library. In addition to the standard public library, most counties operate at least one designated "business library"—often somewhere near the county seat or courthouse. These business libraries can yield a bonanza of worthwhile EH&S information. Not only are they generally better-supplied with a broader range of expensive professional journals, the business libraries are also more likely to have state-of-the-art computer systems equipped with software such as OSHA's CD-ROM subscription service or other commercially produced electronic products. An example of an exceptional CD-ROM service that many libraries now have on-hand is called "Infotrac" and it serves the same function as the old *Reader's Guide To Periodic Literature.* Infotrac is actually a two-part service. While the main attribute is the CD-ROM reference service, it is tied closely to a microfilm service. Literally tens of thousands of article summaries (a.k.a. abstracts) can be reviewed on the monitor nearly instantly. In many cases, the entries are much longer than traditional one-paragraph abstracts. Where the brevity provided by an abstract is insufficient, it's usually a snap to locate the actual article since infotrac uses a unique coding system that allows the user to easily find the complete text of most articles "off-line" in microfilm format.

Most business libraries also have some access to excellent online services such as Dialog, Dunn & Bradstreet, and/or the Internet. Another system making its way into many public libraries is known as CARL, which is an acronym for the Colorado Alliance of Research Libraries. This is a massive online service that, despite its name, has information that appeals to many people well outside the borders of Colorado. Detailed company profiles and the full text of countless journal articles are just two of the useful resources found in CARL. There are instances where an article is listed in the database, but isn't available online. In many of those cases, the system does offer a reasonably-priced document delivery capability.

COLLEGE LIBRARIES

Even the smaller junior colleges typically have business-oriented libraries that will contain information useful to the EH&S professional. Like the business libraries mentioned above, on-campus libraries can yield a wealth of otherwise hard-to-find compliance information. While usage procedures vary widely among private institutions, the libraries at most state colleges have policies which permit non-students to access, if not check-out, books and other information sources.

COUNTY OR STATE LAW LIBRARIES

Just like many of the business libraries, county and state law libraries are often located somewhere near the county seat or the state capital. As the name implies, these libraries contain books, periodicals, and online services that are legally oriented. Unfortunately, like other players in the compliance picture, these libraries have also felt the sting of budget cuts and fat-trimming; in turn, they often don't have the most recent version of a specific book on hand. Thus, the book that sits on the shelves can be woefully outdated.[6]

MOVING AHEAD WITH ELECTRONIC INFORMATION

HOW MUCH DO YOU NEED?

Only a few years ago, electronic reference materials were something that could only be found at the library or within a government agency like EPA or DOT. Now, however, the opportunity to use electronic sources of information in the comfort of your home or office is a realistic and cost-effective option. There are three basic considerations that face the EH&S professional who is considering investing in electronic products:

1. Do you need access to electronic information often enough to justify buying your own software package or will a trip to the local library be sufficient? (If you choose to rely on the library, remember that although they may have some sort of electronic information available, it probably won't be the same software package you would've selected);

2. If the decision to make a purchase has been made, how much have you budgeted to spend (prices start in the hundreds and rapidly climb into the thousands); and,

3. How important are frequent updates to you?

How Often Do You Need It?

Some software products are updated annually while others are updated quarterly or monthly. There are even a few that offer weekly CD-ROM updates. It generally holds true that the more often you need something updated, the more expensive it will be. If your job responsibilities require you to stay current on just the "big picture" items, a quarterly update service supplemented by a good collection of monthly trade magazines will probably suffice. On the other hand, if you have the responsibility of keeping an entire company informed about even the slightest changes to the already complex compliance landscape, it's undoubtedly in your best interest to obtain everything you can that will help you pass along that information.

What are the Main Differences Between Floppy Disks & CD-ROM?

Before delving into the various sources of electronic EH&S information, it may be worthwhile to go over the major platforms—or formats—commonly available in the EH&S electronic information marketplace. Data can either be obtained from an information provider on some sort of storage media like CD-ROM, floppy disk, and magnetic tape, or it can be accessed online using a PC equipped with a modem.

First, we'll look at the disk/CD options. As software programs grow in complexity and as more PCs are being outfitted with CD-ROM hardware, fewer new software programs are being released on floppy disks. Here's why: A high-density floppy disk can hold approximately 1.4 megabytes of information. Even with compression (a software program that allows more information to be squeezed onto the disk), the maximum amount of information that can fit onto the disk is less than 3 megabytes. This presents a problem since the size of software programs keeps growing as more of them are written for a Microsoft Windows™ environment (Windows™ programs require lots of files and gobble up the storage space.) Depending upon the quantity of disks the software company buys at a time, the disks can cost the software publisher around $1 a piece. On the other hand, CD-ROMs can hold up to 680 megabytes of information and they cost just over a dollar each to duplicate.[7]

What this means in real terms is that a relatively small database, such as a 70-megabyte program[8] containing the EH&S CFRs, would cost the publisher $35 to $40 in materials cost to produce the thirty-five floppy disks that would be required. The same program loaded onto a single CD-ROM would cost the publisher around $1.25. Equally important to the end-user, information on a CD-ROM can either remain on the disk or it can be installed on a computer's hard disk in one or two straightforward steps. A floppy-disk based version of the same software package would have to be installed and, with thirty-five disks, it would take literally hours of inserting and removing disks. Furthermore, the same CD-ROM described before still has enough room to hold nearly 180 more CFR volumes and yet the cost of the disk would remain the same! You can see why publishers are leaning towards CD-ROM for information delivery.

WHAT MAKES ONLINE SERVICES BETTER THAN DISKS OR CDS FOR OBTAINING INFORMATION?

The main downfall of any disk or CD-based update service is the lag time encountered between when the information was initially released by the government and when it actually appears on a new disk release. Unlike looseleaf publishers who can get away with saying they will publish updates on an "as needed" basis, most disk and CD publishers have adopted schedules, or cut-off dates, after which they will produce updates to their products. If something new occurs after the cut-off date, it is held until the next issue is compiled. That can mean a delay of weeks or months, before you see the change on your computer screen.

On the other hand, online services generally update their information as changes occur, so there can often be less than a one or two day delay between when a change was made to the law or regulation and when the change appears online. It's that near real-time aspect of online research that makes it so appealing to people who absolutely require that degree of timeliness. Of course, that ongoing effort to maintain the currency of an online database often comes with a hefty price tag.

WHAT IS THE DIFFERENCE BETWEEN A FREE ONLINE SERVICE AND ONE THAT COSTS MONEY?

Not unlike the earlier discussion regarding the merits of free magazines versus fee-based subscriptions, many of the same considerations apply here. Additionally, although there are only a handful of fee-based systems that are oriented to the EH&S professional, there are an estimated forty thousand independent electronic bulletin board systems (BBSs) installed across the country. Many of those BBSs contain worthwhile EH&S information that can assist in keeping you up-to-date on recent changes. Unfortunately, identifying those BBSs which have an EH&S slant can be formidable challenge. There are a few commercially-produced directories which attempt to make that weeding-out process a bit more manageable. A point emphasized in *Eco-Data: Using Your PC to Obtain Free Environmental Information*[9] is worth considering here: The vast majority of those forty thousand-plus BBSs can be categorized as "general interest" systems. In other words, they contain information covering a broad range of topics, from things like human rights—to radical environmentalism—to photography. Likewise, there are also countless BBSs which specialize on very narrowly-focused topics like water quality in Lake Michigan.

It is surprising how much useful material you can find on even the general interest BBSs. Many of them have online conferences (discussion groups) that can get pretty focused. The discussions often get very animated, with people taking sides and drawing lines in the proverbial sand. Don't automatically assume that only zealous environmentalists call those BBSs. Indeed, many callers who have significant expertise often participate in those lively online discussions and you may be surprised at the level of detail that is covered. After all, many of the callers who frequent those online conferences either work for one of the regulating agencies or with private companies which must deal with the many of the same compliance dilemmas that you are also facing. Thus, it's worth spending a little time visiting a variety of BBSs to see which ones contain information that may be of use to you.

For all their recent notoriety as a cheap means of providing electronic information to the masses, BBSs are still something of a pop culture. Many of them are run on old PCs in the back room or basement of somebody with an axe to grind. While many of them do contain good information and thought-provoking discussions, few of them can really claim to contain the latest compliance information. For that, you will need to work with either an established *commercial* online information provider or the Internet.

Somewhat like the narrow-focus versus general interest BBS discussion above, the same considerations come up when considering commercial online services. There are general interest services (e.g., America Online, Compuserve, Prodigy, etc.) which appeal to a broad range of interests, but which contain relatively little depth to their topics. Then there are several business-oriented services (e.g., Dialog, Lexis/Nexis, Legislate) which don't contain the mass-appeal information, but which do have much greater depth to their subjects. The general interest services typically do contain EH&S forums, or interest areas, like America Online's (AOL's) "Network Earth." Those forums can be used to locate timely articles on related topics.

The downside is that most articles are oriented to the consumer/nonprofessional reader. For example, an article on the logging industry's role in regional deforestation might approach the topic from the perspective that the solution lies in more involvement by environmental activists, rather than emphasizing the importance of a more forward-thinking strategy on the part of the logging companies. On the positive side, those services also offer a bonanza of interesting entertainment-oriented services as well as a gateway to the Internet—more about that in a moment.

For access to the purely business side of environmental compliance, a subscription to one of the business-oriented services will be necessary. Long-time industry leader Lexis (owned by *Mead Data General* of Dayton, Ohio) has been providing legal/regulatory information to law firms and professional service firms for years. A relatively recent addition to the highly competitive field of online EH&S information providers is a firm named Legislate. It is a subsidiary of the *Washington Post Group*; as such, Legislate has the resources necessary to maintain an electronic version of the CFRs with updates posted on a *daily* basis. A *final rule* that appears in the *Federal Register* one day, is incorporated into the CFRs by noon the following business day. Thus, it's safe to say that people using Legislate or Lexis, have more current information at their fingertips than do the inspectors or enforcement personnel at the regulatory agencies. Of course, that kind of firepower comes with a daunting price tag. Obtaining access to Legislate's basic CFR service will cost you more than $8000 per year.

WHAT IS THE INTERNET AND HOW CAN IT BE USED TO OBTAIN EH&S INFORMATION?

The Internet is essentially a "network of networks" that has an estimated thirty-plus million users worldwide. It is the starting point (or on-ramp!) for the now infamous information superhighway. Originally conceived to be an indestructible means of communication between members of the military and key research institutions in the event of nuclear war, the Internet (or "Net") has recently expanded its presence into the commercial sector. At once it is both hopelessly chaotic and incredibly well-organized. In the Net's earlier days, users had to be expert computer wizards knowledgeable of complex Unix commands. Thankfully, there is now an abundance of much more user-friendly interfaces to the Net which make it significantly easier for nonexpert users to get in and benefit from the incredible wealth of information that the Net has to offer.

It's worth mentioning that until fairly recently, only college students researchers, and members of the military could roam the electronic corridors of the Net. Then, word began to spread, presumably first among frustrated students who graduated from college and who no longer had access to the Net, that the Internet was a wonderful way to make contact with other people who shared common interests and to stay in touch with them essentially for free. The beauty of the Net is that there are so many computers connected to it that connections from one point on the Net to another are almost always local calls. So, even if you are in New York and you want to send a message to a colleague in California, it can be sent as if it were only traveling across town—not across the country. The computers knows how to route the message from one computer system to another, always taking advantage of local rates. This becomes important once you learn how to harness the power of the Net.

GAINING ACCESS TO THE NET

As mentioned earlier, the private sector has been diligently working to provide user-friendly entry-points to the Net. Typical of other competitive situations in the communications field (i.e., long distance telephone companies trying to get you to switch services, cellular phone carriers selling phones for 1¢ when you purchase an "activation", the corner video store offering two rentals for the price of one, etc.), the demand for commercial access to the Net has prompted many new companies to enter the picture and set up competing online packages. The biggest difference between the various Internet access providers has to do with the degree of access that is offered. Some services provide only electronic mail (or "e-mail") transfers, while others offer complete Internet access with file-transfer options. In most cases where the Net is going to be used to obtain EH&S information, having a good e-mail capability is the most important aspect to look for.

As such, if you decide that you want to explore the Net, the easiest way to get on it is to use the service provided by one of the big commercial providers such as America Online. Although there is a monthly charge for access ($9.95)[10] and an hourly charge of $3.50 (after you use up the five free hours you receive each month,) their software makes it so incredibly easy to send and receive e-mail messages that the money you spend will make the experience much more enjoyable. Additionally, most of the commercial providers have some sort of introductory package that will give you a chance to try out the service before you make a commitment. For example, America Online offers one free month and ten hours of free time to explore the capabilities of the service. As mentioned earlier, access to the Net is just one of many services that the commercial providers offer to their subscribers.

WHAT ARE "LISTSERVS" AND NEWSGROUPS, AND WHAT'S WRONG WITH GETTING "FLAMED"?

LISTSERVS

Now that you've found your way onto the Net, where do you go from here? Looking for information on the Internet is referred to as "surfing the Net." The easiest way to plug into the flow of EH&S information is to join related Listservs. These are like the online conferences or discussion groups that were described in the section of this chapter that addressed BBSs. Users send messages to the Listserv moderator and he/she screens them before sending copies to the other subscribers who have signed up to the Listserv. Each time a message is posted to a Listserv, a copy of it is sent to everyone else on the distribution list. While it *is* possible to send private messages to other people on the Net, these Listservs give you a chance to ask questions of a large group and then to benefit from their collective knowledge.

For example, say that you have a pressing question on reportability of a particular kind of spill. You could research the answer yourself and, quite possibly, spend hours or days in the process. Or, you could go into a large conference hall (Listserv) filled with EH&S professionals, step up to the podium, state your question to the group, and sit back and wait while the answers poured in. There are literally thousands of Listservs covering a number of various topics. The intent of this discussion is to make you aware that Listservs exist and that they can be great sources of information. Finding out more about specific Listservs which may appeal to you will be easier if you pick up one of the commercially-produced Listserv directories that are available at bookstores which normally carry computer books.

NEWSGROUPS

These are very similar to Listservs in two ways: They are gathering places on the Net for people with common interests and there are Newsgroups on almost every topic conceivable. There are two significant differences, though. First, with Listservs, copies of every public message are sent to your e-mail mailbox. This can quickly cause mail to pile up in your account. With Newsgroups, the messages are not actually sent to you, only the "subject" lines from the messages will appear on your Newsgroup directory (much like a table of contents). When you see a topic that looks interesting to you, you can select just that message and moments later it is on your screen. This keeps your mailbox from becoming engorged overnight. Another major difference between Listservs and Newsgroups is that, unlike the Listservs, Newsgroups are *not* moderated so the type of questions and answers which appear can vary in quality and appropriateness.

The typical scenario is to find ones that interest you (again, there are commercially published directories that make this easy) and then add them to your profile. Each time you log on to your Internet access provider's computer, it's a simple matter of checking your Newsgroups to see how many new messages (subjects) they have. With America Online, the software automatically keeps track of the total number of messages in the Newsgroup as well as the number of those messages that have not been read yet. It was mentioned earlier that Newsgroups are not moderated and, as such, there is no single person to ensure that everything which is posted is appropriate for the Newsgroup audience. To deal with this, a sort of frontier justice has evolved on the Net where users decide for themselves which messages are acceptable and they deal swiftly with those who violate the unwritten code of ethics (also known as Internet etiquette or "Netiquette.") This takes us to the subject of "flaming."

FLAMING

When you first venture onto the Internet, it's a good idea to spend your first week or two reading the messages that other people are sending to each other. This is referred to as "lurking" and it will give you a better sense of what the decorum is for that particular group of people. Each Newsgroup and Listserv has its own personality. Behavior that is acceptable on one is not necessarily tolerated on another. What constitutes unacceptable behavior? While different groups have their own quirks, one way that people get into trouble is when they join a Newsgroup or Listserv and immediately begin trying to suck information out of the other people without first getting a feel for the protocol and scope of material that is discussed online. For example, if the question you posted was just asked and answered the previous day, you can expect to have people give you a hard time.

This usually takes the form of an inflammatory ("flaming") comment or two being sent to you . . . or actually, the comment is *about* you and is most often sent to the Listserv or Newsgroup for everyone else to see. That way, everyone gets a chance to read about your mistake. If your "violation" was deemed severe enough, you can receive literally thousands of pieces of "hate mail" in your mailbox. However, this should not discourage you from using the resources contained on the Net. Our recommendation? Just use the Net wisely and be aware of its constraints.

RISING TO THE CHALLENGE

As new regulations are promulgated and as existing rules become more stringent, it remains imperative that EH&S professionals be aware of constantly changing legal responsibilities. To obtain the latest information, you must be resourceful and imaginative. It can be catastrophic to believe that studying something once will ensure your competence in the future. With the possible exception of the medical field, no other career or regulatory area requires as extensive training and retraining as does the environmental regulatory compliance field.

In addition, with the increase in popularity of the Internet as an information source, many private and public organizations are constantly updating the information they can provide the global public, especially environmental information. Notably, EPA is one public organization that has risen to the demand to satisfy the public's demand for environmental information, and constantly updates its web site with pertinent information.[11]

In addition to the suggestions contained throughout this chapter, attending "update" courses on a regular basis (and more often as regulations are promulgated) is an absolute necessity. As an additional source of information to help navigate and "mine" the net's resources, a section of Appendix G is devoted to homepages of various organizations (that can be of benefit to readers, including: federal and state agencies, professional and trade organizations, and a sampling of regulated companies and public interest groups. These sites and their links to dated sites can help provide pertinent information related to current EH&S issues.

At the same time, you should begin to build and maintain (or learn where to find) a library of current publications. To start you on your journey, various trade publications are listed in Appendix B to supplement the information you can obtain from organizations listed in Appendix G. As you endeavor to obtain the best resources available, remember that new sources of environmental information are constantly emerging and the existing ones are frequently updated and improved. It may seem that the wealth of EH&S information sources of today (newsletters, books, videos, and electronic platforms such as the Internet and CD-ROMs) is overwhelming. But bear in mind that, to a large degree, those products are just different means of conveying the same information depending upon your level of comfort with the various technologies involved. It's up to you to decide how you can best learn the information that will keep you and your company in compliance with both the spirit and the letter of the law. Good luck!

ENDNOTES

[1]Disclaimer: The information and opinions expressed in this chapter are the contributing author's and do not necessarily reflect the ideals or policies of the contributing author's employer. The material developed for this chapter is provided with permission by Roland Schumann, the contributing author, formerly with TASC, Inc.

[2]*The Federal Register* notices of some noteworthy EPA programs and policies were described in the previous chapters of the book.

[3]For added reader benefit, Appendix B contains a partial listing of professional and trade organization, specialty publishers, and other listings that the reader may find useful in pursuit of additional information the author describes.

[4]For additional insight, refer to G. Crognale, "ISO Audit Training—Choosing the Right Course Provider" *Quality Management,* Waterford, CT: Bureau of Business Practice, pp. 4–6), September 25, 1998.

[5]Another disclaimer: The contributing author is not advocating that false or misleading information be provided; rather, it is suggested that potential subscribers answer all questions truthfully.

[6]A recent visit to a highly-regarded county law library in a suburb of Washington, D.C. (on a search for *timely* information on RCRA) yielded abysmal results: A copy of Ridgeway Hall's RCRA *Hazardous Waste Handbook* (published by Government Institutes) was on the shelves. The problem? Although the book is currently in its 10th edition, the copy on the shelves was the 2nd edition.

[7]Like almost any purchasing environment, there are economies of scale. Software publishers that develop programs which sell in the thousands get a much better deal when they buy blank disks or press CD-ROMs than those companies which sell only a few hundred copies of a program.

[8]Government Institutes, Inc. has produced a CFR subscription service on CD-ROM that contains all of title 40, the H&S volumes in title 29, and the HAZMAT volumes in title 49. A few useful OSHA documents like the Field Operations Manual have also been thrown in and the total size of the electronic files is slightly more than 70 megabytes.

[9]*Eco-Data: Using Your PC to Obtain Free Environmental Information,* Schumann, R., Rockville, MD: Government Institutes, Inc., 1994.

[10]Disclaimer: At the time this chapter went to press, the rates offered by the various Net providers such as AOL and Prodigy, have been moving steadily downward and are difficult to estimate. The costs shown are for informational purposes only, and do not reflect current prices.

[11]The entire issue of the EPA Newsletter, *New Directions: A Report on Regulatory Reinvention,* EPA 100-R-98-04, August 1998, is devoted to examples of the type of information EPA provides the public daily. This newsletter can be accessed via www.epa.gov/reinvent.

REFERENCES

Crognale, G. "Choosing the Right Seminar," *Leadership For The Front Lines,* Waterford, CT: A Bureau of Business Practice Newsletter, November 10, 1996, p. 1.

Crognale, G. "ISO Audit Training—Choosing the Right Course Provider," *Quality Management,* Waterford, CT: Bureau of Business Practice, September 25, 1998, pp. 4–6.

Schumann, R. *Eco-Data: Using Your PC to Obtain Free Environmental Information,* Rockville, MD: Government Institutes, Inc., 1994.

EPA Newsletter, *New Directions: A Report on Regulating Reinvention,* EPA 00-R-98-04, August 1998.

Erickson, C. and T. Murphy, *Environmental Guide to the Internet,* 3rd ed., Rockville, MD: Government Institutes, Inc., 1997.

CHAPTER 15

WHAT THE FUTURE MAY LOOK LIKE IN ENVIRONMENTAL MANAGEMENT: A PERSPECTIVE

Gabriele Crognale, P.E., with contribution from Daniel McDonnell and Diana Bendz, IBM

> *"Even if you're on the right track, you'll get run over if you just sit there"*
>
> *Will Rogers*

INTRODUCTION

When we first conceptualized this book back in 1992, the environmental management arena, as we know it today, was slowly coming into being. Looking back briefly to that period, the EMAS standards had been finalized (then known as the eco-audit scheme), while ISO 9000 had begun its slow evolution to usher in ISO 14000 that is slowly gaining momentum as it strives to attain critical mass within the EPA-regulated community in the U.S. and other regulated organizations around the world. Our belief now, as it was then, is that the field of environmental management (at regulated companies) needs to shed its old image of playing regulatory compliance catch-up as it puts out regulatory "forest fires" as well as embrace cutting-edge and innovative ideas as it coexists with other business units within its own company.

As such, our focus in this chapter is to look at where we are at present in terms of environmental management matters, describe in greater detail some of the drivers that led to this crossroads, and take a look at some companies as they re-tool or re-shift their focus to stay competitive, the "living company" concept that Arie De Geus described in his book, *The Living Company.* The overriding theme to environmental management as we proceed into the next century, in our estimation, is that it will be business driven, and how well business can respond to the many underlying and dynamic forces shaping decisions each and every day, could play a pivotal role in the success of future environmental management programs, as well as businesses themselves. We also share with readers our insight into some emerging considerations we see on the radar screen. These cutting-edge considerations include a sampling of business and strategy books and related publications, national business newspapers, and topics from emerging environmental and business strategy conferences.

From a historical perspective, some of these emerging considerations were first brought on by a general paradigm shift in company management philosophies and objectives in the early 1990s, such that many companies flocked to embrace trendy management techniques. These "novel" techniques included the ubiquitous re-engineering and TQM concepts, which almost exclusively focused on cost-cutting measures and staff cuts. These, in turn, were the catalysts for full-scale "downsizing" and "rightsizing" that led to massive employee layoffs[1] as one way to achieve the corporate desired cost-cutting objectives.

Let's pause for a moment to digest this information. As readers may recall, the earlier part of the 1990s brought on a wide-sweeping change by many businesses to widely and perhaps blindly embrace the more trendy business management techniques of the time as described in the new items previously referenced. Such widespread dramatic changes of fundamental core business values and principles were not always as effective and beneficial as first perceived. The bottom line was that some companies were trying whatever they could to squeeze a profit—sometimes at the expense of workers. The backlash effect to this strategy however, is that it can sometimes lead to worker anxiety and disloyalty.[2]

From another viewpoint, James C. Collins states, ". . . reengineering and other prevailing management fads that urge dramatic change and fundamental transformation on all fronts are not only wrong, they are dangerous. Any great and enduring human institution must have an underpinning of core values and sense of timeless purpose that should never change. The same lesson applies to corporations."[3] Among the companies he refers to having successfully adapted to a changing world without losing their core values over the decades include Hewlett-Packard (H-P), Disney, Boeing, and IBM. By coincidence, we chose three of the four companies listed as case studies in our book because of our admiration of them as companies to watch. One interesting point Collins makes refers to H-P founders Bill Hewlett and David Packard, who made respect for the individual employee a core value at H-P. This respect was not developed for strategic advantage, but as they believed, it is the morally right way to manage. They are not alone in this belief.[4]

As in other business units within organizations, the EH&S units also felt the company ax as a result of management cost-cutting measures. In some organizations, this cost-cutting may have come at the expense of EH&S staff reductions from layoffs or attrition. To sustain these ongoing efforts, some positions may have also been incorporated into other remaining EH&S positions. For example, at some companies, management began to combine health and safety responsibilities with environmental responsibilities, so that environmental managers who may have previously been dealing only with environmental issues were now being saddled with the added responsibility of health & safety, making them true EH&S managers. Combining two former positions into one was one sure-fire way to decrease overhead costs. The downside to such combined positions can be that environmental managers and personnel can be taxed beyond their limit and quality of overall worker performance may suffer as a result. At other companies, managers began to experiment with outsourcing all or a portion of their EH&S

responsibilities to outside contractors to help pick up the slack from the void of laid-off EH&S employees. This, too, was viewed by management as a means to decrease costs without sacrificing quality or performance. Outsourcing also ushered in a whole new business enterprise for some service providers.

Another business area that seems to have become a venue for some organizations to help control spiraling costs is the joint venture in which competitors enter into agreements to work with each other to create newer products at a lower cost.[5] This trend toward joint ventures created a new term, "co-opetition." Adam M. Brandenburger and Barry J. Nalebuff, the co-authors of the book *Co-opetition*,[6] describe throughout their book how some companies have learned not to destroy their competition to win in business, and have instead opted to cooperate when a market area is being created, and to compete when that market is to be divided. In their book, the authors provide case studies of a number of high-profile companies and how the reader can learn from their real-life experiences. There are lessons that can be learned in this reference book for astute EH&S managers as well.

Rounding out other areas that may also have an influence on EH&S managers and staff as we proceed into the future include: business creativity to sustain a business unit, the creation of intelligent business alliances to spur additional business advances, and following through with strategies that lead to action. Among other noteworthy books in the marketplace that focus on these business strategies include: *Intelligent Business Alliances*,[7] *The Balanced Scorecard*,[8] and *Jamming*.[9] Each book focuses on a different facet of business strategies that EH&S managers can incorporate as their own in addressing EH&S and core business issues as a means to advance in their objectives and also dispel any preconceived notions that EH&S concerns are a stand-alone and revenue-draining cost center within a regulated company.

For example, in *The Balanced Scorecard*,[10] the authors refer regularly to various building-block diagrams that can translate strategy into operational terms, broken down as: (1) vision and strategy that points to (2) *financial*; (3) *internal business process*; (4) *learning and growth*; and (5) *customer*—objectives, measures, targets and initiatives at each step. Such diagram exercises could also be very useful as metrics tools for EH&S managers in calculating various EH&S activities that relate to other business units.

In another example, John Kao, the author of *Jamming*, extols creativity in the workplace as an important worker tool key to the creation of new products, provide opportunities to improve the company's bottom line, and help the company increase in value. This is a general theme of his throughout his book. In his estimation,[11] there are eight reasons why business creativity is so critical in today's business landscape. Among them include:

1. Computer-generated information technology allows more people to be creative (*as we described in Chapter 13*);
2. Companies realize that knowledge or intellectual property can be more important that physical assets (*as we described in further detail in our chapter about benchmarking*);

3. To attain growth, companies constantly need to reinvent themselves (as Arie De Geus described in his book, *The Living Company, highlighted in Chapter 1*);

4. Many workers in the 1990s business climate want to do more creative work (*as we note in The Wall Street Journal feature stories described herein at Endnote No. 12 and 13*);

5. End users are placing a high priority on how products they buy look (*as seen by consumer electronics companies as one example*);

6. This is a customer/end user market (*among the drivers for ISO 9001 and 14001*);

7. Commerce has become global, and as such, competition provides for ideas to flow cross-borders, and the final reason,

8. There is a growing paradigm shift in the managerial mindset that fosters creativity versus stifling it as before.

Interestingly, there are a number of nuggets of creative ideas from this list of eight business reasons that can also be applied to EH&S issues. One poignant example that refers back to the main topic of this chapter deals with how EH&S management has had to change to meet market demands. In the heyday of regulatory compliance requirements, EH&S was a strong driving force for having companies toe the line. As companies and EH&S programs matured, the regulatory driver was became less of a major force and companies could afford to perform some cost cutting and downsizing of EH&S staff and still maintain compliance. Where absolutely necessary, some companies went to outside vendors for outsourcing their services. What had happened to the EH&S group as a result? Internal customers—other business units and top management—had begun to see the EH&S group as a cost without tangible benefits and had made strategic decisions to cut staff wherever practical. Had EH&S groups initiated various creative and innovative measures to satisfy their existing (internal) customers and plan for future consideration, some might still be at full staff. As readers may recall, we provided several examples in the earlier chapters for fostering creativity within EH&S ranks that could be useful in particular situations where job security could come into play.

Interestingly though, while all this business and workplace upheaval has been going on these past few years, the pendulum may have just begun to swing in the opposite direction with respect to worker issues, and business may have just undergone another paradigm shift back to fundamental basics. For example, two story features that regularly appear in *The Wall Street Journal,* "Workplace"[12] and "Managing Your Career"[13] noted a number of companies that are beginning to shift their focus—again!—but this time it has a softer touch. In the first example, the author refers to an employee at Foldcraft Corporation, a maker of restaurant furniture, who was chosen to take an educational tour of El Salvador and Guatemala. In another case, Silicon Graphics, a California computer-maker, gives employees awards to employees who express their ideas in poems that deal with themes such as, "encouraging creativity" and other beneficial themes. In addition, Lotus Development Corporation, a unit of IBM, was highlighted for forming committees to improve employee morale.

The underlying theme in each of these examples, the author points out, is that some companies are vying for a new corporate goal: helping employees realize their values and find real meaning on the job. Does this also reflect back to increased empowerment for more employees at the lower level of the corporate food chain? Perhaps, but like most other solutions, this so-called self-actualization is not a "one-size-fits-all" solution. What works for one person may not work for another, and a delicate balance may need to be struck, viewed on a case-by-case basis.

Taking this concept of 'meaningless' work into the EH&S area, there may be situations where plant employees may feel the same way about certain repetitive work related to EH&S regulatory requirements. Finding ways to dispel such misgivings may prove beneficial all around. For example, this may be a worthwhile exercise for the reader to bring up for discussion at the next environmental or health & safety monthly meeting, or as a punch-list item to be discussed at a business managers' roundtable discussion.

In the second example cited, the author notes that companies that once bragged about their re-engineered processes and quality metrics are now looking more closely at the importance of workers themselves. At issue are the number of companies that are attempting to rebuild their cultures uprooted by mergers and acquisitions and relentless cost cutting measures that are realizing real labor shortages. What it all boils down to is that employers and employees still want the same things: employers want committed workers who can stand up to challenging problems, and employees want job security and stability, interesting work, a boss they can look up to, and good pay and benefits in exchange. It would seem that we may be coming back full circle to the work ethic of the 50s and 60s.

Environmental, health & safety management, like other business units within a company, are no different than the rest, in the sense that employees have the same aspirations, wants and needs as other company employees, and may have also been affected by business upheavals of the past several years. More so than employees in other business units or cost centers, though, EH&S employees have in recent years faced such additional obstacles as management reluctance to fund additional expenditures for regulatory considerations, experiencing the MEGO (*my-eyes-glaze-over*) syndrome while being perceived as extolling regulatory doom and gloom to top managers, and other obstacles befitting the "green wall" that EH&S managers face from time to time.

In retaliation of sorts, a number of EH&S managers began to apply business measures to counter the effects of the "green wall" and help get their message across to senior management. They began introducing facts and figures in terms non-EH&S managers could readily understand with a clear delineation to monies and materials saved[14] in relation to EH&S expenditures to show a one-to-one relationship. This practice has begun to win over a number of top managers who like what they see, and would like their company to strive for more. Herein lies the catch: to do more, essential EH&S staff cannot be cut. Cost cutting cannot extend to the elimination of a certain number of positions, otherwise the effort that may be required to monitor, evaluate and assess the areas that can generate real cost savings may be too much for overtaxed staff to handle, and the full effect of cost savings programs may not be realized.

WHAT SOME COMPANIES ARE DOING

OUTSOURCING CORE EH&S RESPONSIBILITIES

In the previous section we provided a brief insight into some of the business issues and considerations that organizations may be facing today, including those that may also affect environmental and health & safety core competencies. The most compelling of these forces most often relates to financial issues and how these may affect decisions placed upon a company's environmental manager by senior management. In any organization in which compliance with applicable environmental and health and safety regulations is an integral part of the business environment, the ability of the top environmental person to do his or her best also hinges upon a core staff of professionals and other specialists to perform their respective duties. How well the organization is staffed by competent employees at each of its facilities is gauged by a combination of several factors, including: the ability to employ and maintain qualified staff, work conditions, the company culture, and the amount of allocated corporate funds available to sustain and improve EH&S programs as necessary, to name a few.

As we pointed out in the introduction, several factors that have come in play over the past several years may have had an adverse effect on a number of organizations, including scores of layoffs of EH&S personnel and the combining of EH&S responsibilities of some individuals to compensate for staff shortfalls. Much of this was accomplished to improve bottom line numbers at affected companies as one way to introduce cost cutting measures. Along the way, some organizations may have also taken a long, hard look at occurrences within their facilities with respect to EH&S responsibilities, and may have begun to evaluate various alternatives that could best maintain core EH&S responsibilities with existing EH&S staff and still meet budget constraints. One of the most frequent alternatives management arrived at was outsourcing, and a new era began for many contracted service providers, including consultants and outside laboratories, to name a few.

Outsourcing of certain environmental services has been around for some time, and has most commonly been used in areas such as third-party environmental audits, cleanup activities as part of EPA-led Superfund or RCRA corrective action activities, due diligence, and site assessments as part of a site acquisition or merger, and some permitting activities, among others.

More recently, as more organizations began to feel the belt tightening of downsizing and other cost cutting corporate measures, their managers began to look at outsourcing more and more EH&S functions, perceived by some business managers as being non-core business functions. Of course, with each piece of work that is transferred to an outside contractor, additional responsibilities are placed onto facility EH&S staff to monitor, review and approve their work, that can take away valuable time from their other core EH&S functions.

As Mark Posson, the manager of environmental protection programs at Lockheed Martin Missiles and Space, noted in his interview[15] of six companies that had outsourced environmental management, the decision to effectively outsource some part of EH&S responsibilities hinges upon a number of key factors in selecting an outsourcing partner. These include: (1) take the time to define the scope of work—*ensure each point is clear,* (2) establish a sliding compensation rate for poor performance—*not unlike the contract awards provisions in EPA contract work;* and *near and dear to us,* (3) meet the individuals who will be doing the work—*past experiences have shown that some contractors list specific individuals in their proposals who are then replaced by others once a contract is awarded.*

By his account of the companies he interviewed, all six determined that "... outsourcing of environmental programs should be the exception rather than the rule. Specific tasks within programs are therefore selectively outsourced based on the factors discussed.[16] These included the four primary areas that encompassed the survey results: (1) outsourcing compliance programs; (2) other outsourcing, such as, due diligence, site assessment and cleanup; (3) outsourcing influences, such as, "lifetime employment" philosophies; and (4) outsourcing decreased compliance.

In conclusion, outsourcing of specific EH&S responsibilities is still a market area for a good number of environmental consulting firms. Just what that relationship is for each of their industrial clients is gauged on a case-by-case basis. Some companies, like the survey respondents, have found that certain core EH&S competencies are best left to internal EH&S staff as the most effective and efficient way to perform these tasks. If we also take into account some of the points raised by the previous business strategy book authors, top decision makers at regulated organizations could gain some insight from their recommendations and apply it to their EH&S staff. In this fashion, their organizations could gain financial benefits without unnecessary cost cutting (layoff)[17] measures and could achieve increased efficiencies by their EH&S staff. After all, with respect to core EH&S responsibilities integral to the company's sustaining and exceeding EH&S compliance, internal EH&S specialists know the organization best, and their experience and expertise cannot easily be matched by outsourcing companies, unless these companies have on staff key senior people with extensive experience from that or a similar field. For example, a number of senior environmental professionals in the chemical or petroleum sector usually take some form of early retirement to pursue financially rewarding and stimulating consulting practices in their fields, sometimes with their former employers. And why not? They know this specialized area and can continue to provide their specialized expertise while pursuing other, less strenuous ventures, like playing a few rounds of golf.

With respect to how some regulated companies view their EH&S staff at present, a number of their business area managers view their internal staff as valuable resources. Will this trend continue? We think so. Will outsourcing continue to flourish in this environment? We also think so, but how effective outsourcing will become may be dependent on a number of considerations. This may include how well some contractors perform as the environmental services market continues to consolidate, and their ability to aggressively pursue this market with innovative and creative ideas to garner the attention of regulated companies. All too often, many vendors seem to provide the "same-old same-old" to their clients. In this fast-changing market, traditional thinking and risk-adverse marketing is giving way to a new order to pursue customers. Vendors need to be quick, risk-taking, out-of-the-box thinkers with innovative ideas that can translate to value to their customers. In other words, their customers may just need to be "wowed" to retain their services.

Given such a wide open field where nimbleness and innovative thinking rules, there just may be a small percentage of service providers that can fit this bill. Even so, since competition is so keen, a buyers market situation exists so that the customers, EPA-regulated companies, can use this knowledge to their strategic advantage and procure consultant services at very competitive rates to address their specific needs.

SOME COMPANIES RELY ON BOUTIQUE FIRMS: STRATEGIC "OUTSOURCING"

In some regulated organizations, the need for full-scale outsourcing is not as critical, especially if these organizations maintain a strong EH&S corporate group supported by key EH&S managers and staff at each of their facility locations. In cases such as these, corporate and plant EH&S managers may look to specialty consultant "boutique" firms or independent consultants who understand the business area in which the organization may have a specific need. The example previously alluded to in the petrochemical field is one venue for former retirees, now independent consultants, who can provide strategic consulting to their former employers. In addition, some specialty consulting firms in this field also retain such individuals to provide them contract services as required.

Another area where regulated organizations can tap into specialty expertise is in the computer software arena, such as we discussed in Chapter 13, where more and more organizations are looking at software system vendors, consultants and systems integrators to help them choose a specialty EMIS software program, help them develop an in-house tailor-made version, or a combination of both. As companies evolve and generate more and more data that becomes assimilated by its various business units, much of this data can also have links to EH&S requirements and responsibilities. As such, many companies are turning to specialized software programs to manage and "mine" such information overload. As EMIS becomes more prominent, and very powerful ERP software systems and intelligent systems come more into the business mainstream, such software will become very key to business decision-making, and specialists in this field will be very much in demand. As an example, specialized EMIS and intelligent "real-time" software applications are widely used in a number of regulated industries, such as the petrochemical, chemical, pharmaceutical and other feedstock industries. Many of these organizations have been pleased with the results such software provides.

Another area that has developed into a market niche for specialized boutique service providers is the specialized training and consulting arena brought on by the widespread approval of the International Organization for Standardization (ISO) standards in quality management (the ISO 9000 series) and environmental management (the ISO 14000 series).

The advent of these standards has generated a specialized service area for the support services market of ISO 9000 and 14000 trainers, course providers, consultants, auditors and registrars who specialize in these fields and are recognized as being accredited to practice in these specialty niche areas. For example, as we previously described in Chapter 12 regarding ISO 14001, organizations that desire to become certified to this ISO standard need to follow certain steps to achieve that goal. First, their facilities in question must be audited by an individual who has attained accreditation by the Registrar Accreditation Board (RAB), a U.S. organization that is part of the American Society for Quality (ASQ) headquartered in Milwaukee, Wisconsin. Second, depending upon the outcome of this audit, the facility can either declare self-certification, or retain the service of an accredited registrar to certify that facility to ISO 14001. If readers are interested in obtaining additional information about the RAB, their web site is www.rab.org.

As a result of the ISO thrust, market forces have provided a niche-services area for a number of organizations that have received accreditation to teach the required ISO 14001 EMS Lead Auditor course and provide related ISO 14001 services. This is the first step a prospective EMS Lead Auditor needs to accomplish as part of his or her accreditation process stipulated by the RAB. Regulated organizations that desire to have the in-house capability to audit their own facilities, or assist in third-party EMS audits, will need to procure the services of one of these course providers at some point.[18]

As a supplement to the required ISO training, regulated organizations that do not wish to self-declare their conformance to ISO 14001 are also relying upon third party accredited ISO 14001 registrars to certify their facilities to ISO 14001. Those organizations that can provide these specialized services to the U.S. regulated community will be in various stages of demand by their clients based on how well the U.S. regulated community embraces ISO 14001 as a valuable business tool.

From our perspective, many regulated companies are still looking to EPA for guidance on how the agency will view ISO 14001 certification as an incentive for decreased regulatory compliance inspectors. As of this writing, EPA's latest view on ISO 14001[19] is as follows: "... EPA supports and will help promote the development and use of EMSs, including those based on the ISO 14001 standard, that help an organization achieve its environmental obligations and broader environmental performance goals." Compliance is still a big draw for EPA, and many organizations know this all too well. Should EPA give its go-ahead to ISO 14001 certification, that push will be a driver that is very hard for companies to resist for jumping onto the ISO 14001 certification bandwagon.

As a closing consideration to compliance, enforcement and the potential benefits of thorough environmental management systems (that can lead to ISO 14001 certified facilities), those of us in the environmental community, meanwhile, should not lose sight of EPA's continued focus on enforcement and compliance assurance, as succinctly put by Steven A. Herman,[20] of EPA's Office of Enforcement and Compliance Assurance: "... the Office of Enforcement and Compliance Assurance's mission is to assure compliance with our environmental laws, deter noncompliance, punish polluters and other violators, ensure that violators do not gain an economic advantage, stop illegal behavior, and correct any damage to the environment."

Rather than run for cover hearing such wide-sweeping statements, regulated entities should view such EPA statements as the "carrot" as opposed to the "stick" with respect to what the future may hold for some regulated organizations should they decide to flaunt the regulators' primary goal. Specialty service providers that are strategically poised to help such organizations move closer in this direction can provide a valuable service to companies and help increase their bottom line as well. Their combined efforts may take some time and dedication to accomplish stated company goals, but the efforts may just net them positive results, as our research found from a few readily available case studies.[21]

STRATEGIC ALLIANCES: ANOTHER ALTERNATIVE

As one picks up a copy of *The Wall Street Journal* or other business publication, there is usually a story about one blockbuster corporate merger or another. Similar mergers also abound in the environmental services field, where one firm decides to merge or acquire another firm for a number of business reasons, whether it is to enter into a whole new market area, grow larger to pursue bigger clients, or just to survive in this very competitive field.

While mergers and acquisitions have garnered front page news in business circles for some time, another force has emerged as an alternative to straight mergers to survive—the strategic alliance, or as also referred to, "co-opetition." While the primary focus of strategic alliances is business-oriented, that is, to compete more effectively in the marketplace, there is also room for organizations to look at strategic alliances as one way to help address and conquer environmental and health and safety issues.

In the environmental services area, some firms do procure from time to time the services of smaller firms to assist them in areas in which they have little to no expertise, or in client markets in which they do not have a foothold. They have thus formed a strategic alliance. The benefit of such an alliance is that collectively both firms can pursue a client that individually they might not have successfully been able to pursue. As a result, both firms benefit. A symbiotic relationship in the simplest, purest form.

Taking strategic alliances one step further, Larraine Segil, the author of *Intelligent Business Alliances*,[22] refers to business alliances as one of the most important keys to competitive advantage in business today. When planned *intelligently*, strategic alliances can provide companies access to a number of new venues: lucrative markets, breakthrough technologies, and core competencies of skilled workers at other firms.

Key to embarking on an alliance with another company, Segil stresses that you should start by analyzing your own company first, then choose the right partner and project, and with those objectives accomplished, you should develop a plan to implement the alliance. Setting the alliance in motion, both parties should be clear on several guidelines, addressed as a series of eight questions,[2] which we revised as:

1. How will the alliance process begin?
2. Who will perform what function within each partner company? When? How?
3. Who will compile, track and assimilate any information obtained as a result of this process?
4. What can the partners agree on as criteria for attaining their objectives?
5. How will each milestone attained be communicated to each partner?
6. What will each partner gain as a result of this alliance?
7. What incentive programs are appropriate?
8. How will the partnership fit with the existing relationships of both companies?

While the primary thrust of *Intelligent Business Alliances* is a business enhancement primer, many of the author's points can also be relevant to EH&S considerations. As such, we revised five of the eight questions suggested by Segil in a way that could be adapted to an environmental study, or a benchmarking study that focuses on specific environmental issues that could be of interest or concern to one or both partner organizations.

Also, within the environmental consciousness sector, a number of organizations have been formed that have initiated strategic alliances with private sector (industrial) companies as one way to promote favorable changes, such as: reducing waste, preventing pollution, conserving raw materials and other natural resources, while providing an outlet for these industrial organizations to be seen as role models for other businesses to follow. Among organizations that have forged strategic alliances with private sector companies include:

- The H. John Heinz III Center for Science, Economics, and the Environment;
- Management Institute for Environment and Business; and
- Business for Social Responsibility Education Fund.

Among the companies that have entered into strategic alliances with these organizations include: AT&T, the Gap, Johnson & Johnson, Polaroid, Bristol-Myers Squibb and DuPont. Their joint focus has been to facilitate collaborative efforts to research and implement innovative solutions to environmental challenges.[24]

In addition, watchdog-type organizations such as, the Council of Economic Priorities (CEP), the Investor Responsibility Research Center (IRRC), the Coalition of Environmentally Responsible Economies (CERES), keep track of and follow the activities of a number of high profile publicly held companies. Their in-depth research can complement the work of the alliances through the research they conduct on behalf of various stakeholders, including the public.

For example, the CEP issued a report[25] in March of 1998 that focused on fifteen of the largest petroleum refining companies in the US that were the subject of a comprehensive environmental performance analysis encompassing 1997 and 1998. As the executive summary of the report notes, each company's overall performance rank was obtained by a weighing of three main factors: (1) environmental impact (60 percent); (2) environmental management systems (30 percent); and (3) corporate environmental reporting (10 percent).

In other matters, the CEP continues its on-going efforts to improve global ethics, and presented its Eleventh Annual Corporate Conscience Award in 1997,[26] where it recognized twelve companies for their *community involvement, employee relations, environmental stewardship, global ethics* and *pioneer awards in global ethics.* Among the 1997 award recipients included: the Kellogg Company for community involvement, and Novo Nordisk A/S (Denmark), J Sainsbury plc (UK) and Wilkhahn, Wilkening + Hahne GmbH + Co (Germany) in Environmental Stewardship.

What can such strategic alliances and conscience awards really do for regulated organizations? If such alliances and related activities are seen by organizations as tools that can help them overcome certain obstacles, they can provide opportunities for regulated companies to showcase their progressive efforts through third-party channels, such as the non-profit organizations, while exhibiting a non-biased approach. While styles and individuals may vary, these vehicles can bring to the table an eclectic collection of stakeholders that collectively work toward a common goal. In a similar way, experiences gained by participants of EPA's CSI program can provide parallels in how various stakeholders come together in developing a consensus approach to achieve a common goal, probably much like the alliances described herein. We are still in the learning curve in this area, and there is still a good amount of trial and error to overcome as we progress forward. The key is to continue forward, not backward.

OK then, where might alliances be headed in the future as they produce a greater following? We see such alliances being valuable tools that could be used by companies as part of their participation in any voluntary EPA pilot programs down the road, or help to highlight their pollution prevention, cost-cutting an innovation programs as case studies for successes achieved in conferences or as noteworthy trade journal articles. In turn, these successes could be linked to achievements in their environmental management program, their EMIS software programs, their established EH&S auditing programs, or their implementation of the ISO standards, among other considerations, and highlighted through appropriate external communication channels as a benefit.

WHAT THIS PORTENDS FOR NEXT-GENERATION ENVIRONMENTAL MANAGEMENT

Outsourcing, strategic alliances, specialty boutique services ... what are these items saying about the environmental management arena, about business in general? Any way you look at it, environmental management, as we know it today has changed from what it used to be in the early days of "command-and-control." In order to stay a step ahead and be considered a viable business entity in the face of all this constant upheaval, environmental management, whether as an internal function of a company, or as an outsourced service provided, may need the assistance of nimble-footed individuals at the helm who are not averse to new opportunities and suggestions, and perhaps most importantly, may need to be somewhat un-conventional. We need look no further for innovative ideas than the commercials of two European car makers: Saab and Volkswagen. Finding your own road may just be what it's all about.

Where can organizations of all types start in this evolutionary process? Where it makes the most sense—at the beginning—the front line workers, and then move your way up through the ranks even as policies and directives from the top are filtered down through the ranks.

What we are really talking about here is re-shaping the organization's internal culture to allow employees to think outside the box to expand their horizons and to give new meaning to their day-to-day activities, to generate a new feeling of self. This is also about building bridges between one another in a work environment to instill in employees that each one makes a contribution in their own way, and with the proper feeling of empowerment, employees can take on an added dimension to their work responsibility that allows them to feel they are actually accomplishing something: whether it's a better product made, wasting less raw materials or off-spec products, maintaining compliance with applicable environmental regulations, or other feelings of accomplishment. All of these individual factors can add up to helping to build up trust and loyalty with workers, maintaining exceptional and well-trained workers, and having fewer employees leave to pursue other jobs.

In the following example, provided courtesy of Daniel McDonnell and Diana Bendz of IBM, the authors provide a case study of what IBM is doing to lessen the environmental impact of electronic products as part of its commitment to the environment. This is one example depicting an innovative approach to address the impacts computers and other electronic products can have upon the environment and what can be done to mitigate those impacts.

THE IBM CASE STUDY[27]

As more computers are being built and updated to accommodate user demands, certain environmental issues are also created as a result. One of the greatest issues centers around the disposal of used and obsolete computers and other electronic products, which by McDonnell's estimate, are 300,000 tons/year that enter into landfills, with a projected estimate of 2 million tons/year by the year 2000. A breakdown of current practices show that computers and other electronic products are segregated into: 3 percent recycled; 7.1 percent sold; 14.1 percent landfilled; and a whopping 75.8 percent still in storage in some fashion.

For IBM's part, the company's Corporate Policy Statement 139A provides that "... IBM is committed to environmental affairs leadership in all of its business activities." Furthermore, as part of the company's Corporate Product Stewardship, "... IBM recognizes its responsibility to develop and manufacture products that are energy efficient, protective of the environment, and can be recycled and disposed of safely." IBM, as a business entity, also recognizes that the reuse and recycling of valuable machines and parts is a business opportunity. What does this translate to? Two things: It's safe for the planet; and Waste = $ off the bottom line.

Among the key tenets of IBM's Corporate Product Stewardship, the company focuses on several strategic initiatives, including: (1) environmentally conscious products (ECP), such as Design for Environment (DfE); (2) product end-of-life management (PELM); and (3) central reutilization in its Endicott facility, encompassing: reutilization, used parts return, asset protection competency center and technology center.

The Endicott Reutilization Center utilizes a three-step approach to process used machines from throughout the US from its various plant locations, IBM's customers, and IBM's US sales and service locations. The process involves reutilization for receipt, sorting and disposition of the machines. Following processing, machines are designated for: *Reuse*—parts in field service programs, and machines used internally after reconditioning; *Resell*—as industry standard parts and reconditioned machines; and *Recycle*—by material content.

Among statistical figures, used parts average 32,000 pieces/week, which translates to 650,000 to 700,000 pounds/week being reutilized, or 15 to 17.5 million tons/year. The output generated by the Reutilization Center, expressed as commodities distribution by percentage breaks down as: Displays—16.3 percent; Ferrous—23.8 percent; Other—9.2 percent; Precious Metals—4 percent; Trash—7 percent; Plastics—5.6 percent; Non-ferrous—22 percent; and Reuse—12.1 percent. In gross totals, this program has allowed 2+ million pounds (over 1000 tons) of machine equipment to avoid landfills altogether. While that may be a small percentage of the total computers that are disposed of annually in landfills, it shows determination on the part of IBM to do its part and more to manage the issue of computers going to landfills.

OTHER CONSIDERATIONS

In addition, organizations such as the Conference Board (New York, New York) continue to promote conferences and symposia geared to business considerations, among them: (1) developing community relations between organizations and local communities; (2) environmental, health, and safety excellence (as we described in greater detail in Chapter 7); (3) business and education reform, such as the "Fourth Wave"; (4) information management conferences dealing with information technology as today's competitive weapon; and (5) a leadership conference on global corporate citizenships. Groups such as the Conference Board are well respected for their insight and forward-thinking, key to keeping abreast of emerging considerations that may affect a company's business groups. Another of their more recent brochures highlights outsourcing strategies that can be applicable in EH&S business considerations. This conference is titled "The 1999 Strategic Outsourcing Conference: Optimizing the Value of Outsourcing."

In the final analysis, after companies have had an opportunity to assess the full impact of various management initiatives, such as re-engineering, downsizing & rightsizing (a fancy name for lay-off), and how these initiatives may have adversely affected business units within a company, including the EH&S units, their top managers may come around full circle in their thoughts regarding the contributions employees can actually make. Today's rapid-paced business envi ronment has forever changed the work scene from many workers. However, in some companies, as the founders of Hewlett-Packard noted, respect for the individual employee is key to a company's well-being. In addition, if companies opt to build trust with employees and instill a feeling of loyalty, there is a good likelihood that workers will respond in kind by feeling good about their job, feel a sense of gratification in any empowerment bestowed upon them as part of their work duties, and may minimize situations where some employees may want to job shop as a means for improvement.

In closing, what may be in store for environmental, health, and safety management, whether from within a regulated organization, or the legion of professional service providers, such as consultants, trainers and course/conference providers, or the regulators themselves, is not so much dealing with more complex issues—whether more rigorous regulations, enforcement of these regulations, or better products to market—but rather a different way of looking at existing conditions to attract more common sense solutions.

If we stop to look at the bigger picture, we will realize that products will still continue to be made, raw materials will still be used, and some wastes will still be generated—*the key is and will continue to be*—what products can be useful and beneficial, how can their components be reutilized, and what raw materials can be substituted to make their production less wasteful and harmful for the benefit of workers and the environment. As more companies police themselves to ensure these tenets are maintained regularly, perhaps rigorous and sustained regulatory compliance and enforcement may become a thing of the past as more companies follow the lead of high-profile environmental stewards and emulate them to become more mature with respect to their EH&S responsibilities, as depicted in the IBM example we provide.

Among the innovative and effective tools that can be useful in this regard, that can apply in a number of business units, whether EH&S, purchasing, research and development, manufacturing, distribution, or human resources, are those tools that instill new energy into employees. This energy can translate to opportunities for workers to allow them to showcase their achievements that can be seen and measured.

As another example, one of the companies renowned for encouraging creativity among its workers is the 3M Company.[28] The company identified six "drivers of innovation" as keys to its success, according to 3M's senior vice president of research and development, William Coyne. These are: (1) "Vision"—3M's vision is to be the most innovative enterprise in the world; (2) "Foresight"—is the market ready for a new approach?; (3) "Stretch objectives"—pushing the employee envelope of creativity and innovation; (4) "Empowerment"—at 3M, the company practices "say what you do and do what you say" regarding employee empowerment; (5) "Communication and networking"—this is strongly encouraged companywide among the various business units where employees can combine their collective experiences to come up with innovative ideas—similar to our approach regarding various business units keeping open lines of communication with the EH&S groups; and (6) "Peer recognition"—innovation at 3M is recognized in one of three ways: the Technical Circle of Excellence Award, the Innovator Award, and membership in the Carelton society.

In light of all these aspects, top managers at organizations of all types, including allied service providers, need to ask themselves ... "Are we ready to commit to whatever it takes to make beneficial changes happen, regardless of the carrots or sticks that may or may not be imposed by regulators, watchdog groups, customers and other influential stakeholders and work from within our own ranks to execute these changes, or do we need an external "prod" to jump start our activities?"

Only the regulated organizations, their customers, distributors, suppliers, service providers, and regulators themselves can best answer questions such as these, and how their answers may beneficially effect environmental, health, and safety matters as well as other related business considerations, and the applicable and relevant regulations.

Collectively, we hope that such innovative thinking as described in these fifteen chapters can make a difference in making the world in which we live a better place to be for all of us.

ENDNOTES

[1] For an historical perspective, the following articles that appeared in *The Wall Street Journal* are worth reading regarding re-engineering and other management ideas. These include A. Ehrbar, "Price of Progress: 'Re-Engineering' Gives Firms New Efficiency, Workers the Pink Slip," March 16, 1993, p. A1; A. K. Naj, "Shifting Gears: Some Manufacturers Drop Efforts to Adopt Japanese Techniques," May 7, 1993, p. A1; F. Bleakley, "Many Companies Try Management Fads, Only To See Them Flop," July 6, 1993, p. A1; and G. Fuchsberg, "Small Firms Struggle With Latest Management Trends," August 26, 1993, p. B2.

[2] See Murray, "Thanks, Goodbye: Amid Record Profits, Companies Continue to Lay Off Employees," *The Wall Street Journal*, May 4, 1995, p. A1.

[3] For additional information, see J. Collins, "Change is Good—But First, Know What Should Never Change," *Fortune*, May 29, 1995, pp. 65–70. Collins is the co-author of *Built to Last: Successful Habits of Visionary Companies.*

[4] A similar story can be found in the *San Diego News Tribune*, M. Kinsman, "Generous employers are likely to be repaid," January 19, 1996, p. C-1.

[5] One of the first stories about joint ventures was featured in *The Wall Street Journal* by N. Templeton, "Strange Bedfellows: More and More Firms Enter Joint Ventures with Big Competitors," November 1, 1995, p. A1.

[6] Adam M. Brandenburger and Barry J. Nalebuff, *Co-opetition.* New York: Bantam Doubleday Dell Publishing Group, Inc., 1996.

[7] Larraine Segil, *Intelligent Business Alliances: How to Profit Using Today's Most Important Strategic Tool.* Random House, Inc., 1996.

[8] Robert S. Kaplan and David P. Norton, *The Balanced Scorecard: Translating Strategy Into Action.* Boston, MA: Harvard Business School Press, 1996.

[9] John Kao, *Jamming: The Art and Discipline of Business Creativity.* New York, NY: Harper Collins Publishers, Inc., 1996.

[10] R. S. Kaplan and D. P. Norton "Using the Balanced Scorecard as a Strategic Management System," *Harvard Business Review,* January–February 1996, p. 76.

[11]At No. (9), pp. 4–17.

[12]G. P. Zachary, "The New Search for Meaning in 'Meaningless' Work," *The Wall Street Journal,* January 9, 1997, p. B1.

[13]H. Lancaster, "Hiring a Full Staff May Be the Next Fad in Management," *The Wall Street Journal,* April 28, 1998, p. B1.

[14]One of the first companies to implement and track with metrics a cost savings program tied to pollution prevention efforts was the 3M Corporation with its highly successful, corporate-wide 3P Program—Pollution Prevention Pays—initiated back in late-1980s. 3M's strategy is based on source reduction and the reclamation and reuse of process waste. In addition, 3M had developed two videos in 1991 that highlighted the company's pollution prevention achievements and overall strategy. These videos are titled, *3M's Pollution Prevention Pays Program* and *Challenge to Innovation,* listed in EPA's booklet, *Pollution Prevention Training Opportunities in 1991.* Raytheon is another company that developed information in this area.

[15]See M. Posson, "The Risks and Benefits of Outsourcing Environmental Management," *Corporate Environmental Strategy,* v3n3, Spring 1996, pp. 5–11.

[16]Ibid., p. 11.

[17]See J. Moynihan, "Outsourcing: Boom or Bane? It Depends on Whom You Talk To," *Today's Chemist At Work,* June 1997, p. 17. Ms. Moynihan notes that outsourcing is seen as a threat by some employees whose jobs are taken over by "outsources" and slashed from the work force.

[18]For additional insight into how to choose an ISO course provider, see G. Crognale, "Auditor Training—Choosing The Right Course Provider," *Quality Management,* September 25, 1998, pp. 4–6.

[19]From EPA's position statement on EMSs and ISO 14001 in the *Federal Register,* Vol. 63, No. 48, March 12, 1998, pp. 12094–12096.

[20]S. Herman, "EPA"s 1998 Enforcement and Compliance Assurance Priorities," *National Environmental Enforcement Journal,* Vol. 13, No. 1, February 1998, p. 3.

[21]See G. Crognale, "Tallying the EMS Bottom Line: How To Sell Real Cost Benefits," *ISO 14000 & EMS Advisor,* Issue # 116, May–June 1998, pp. 2–4.

[22]At No. (7), pp. 184–196.

[23]At n. 7, p. 185.

[24]See "Grand Alliances," *The GreenBusiness Letter,* February 1996, p. 1.

[25]"Petroleum Refining Industry Report, Campaign for Cleaner Corporations," March 1998, the Council on Economic Priorities, R. I. Chin, V. L. Ganek and A. D. Muska. Council on Economic Priorities, New York, NY. At the time the report was issued, Philips Petroleum was concerned about its ranking due to unconfirmed data, and upon being notified, CEP corrected the company's ranking.

[26]Reprinted with permission from the Council on Economic Priorities (CEP) from the CEP's published report, *Council on Economic Priorities Eleventh Annual Corporate Conscience Awards,* New York: Council on Economic Priorities, 1997.

[27]This case study depicting IBM's Endicott facility was first presented by Diana Bendz at CARE INNOVATION '96 in Frankfurt, Germany, November 18, 1996, and later by Daniel McDonnell at an Inverse Manufacturing Symposium in Tokyo, Japan, March 10, 1997.

[28]Excerpted from an article by B. Filipczak, "It Takes All Kinds: Creativity in the Workforce," *Training,* May 1997, p. 36. Follow-on articles that provide additional depth

to this subject area include: (1) J. Murphy, "Results First, Change Second," *Training,* May 1997, pp. 59–67; (2) T. Esque and T. Gilbert, "Making Competencies Pay Off," *Training,* January 1995, pp. 44–50; and (3) C. Griffith, "Building a Resilient Work Force," *Training,* January 1998, pp. 54–58.

REFERENCES

ARTICLES

Bleakley, F. "Many Companies Try Management Fads, Only To See Them Flop," *The Wall Street Journal,* July 6, 1993, p. A1.

Crognale, G. "Tallying the EMS Bottom Line: How To Sell Real Cost Benefits," *ISO 14000 & EMS Advisor,* Issue # 116, Madison, CT: Business & Legal Reports, May–June, 1998, pp. 2–4.

Crognale, G. "Auditor Training—Choosing The Right Course Provider," *Quality Management,* Issue 2118, Waterford, CT: Bureau of Business Practice, September 25 1998, pp. 4–6.

Collins, J. "Change is Good—But First, Know What Should Never Change," *Fortune,* May 29, 1995, pp. 65–70.

Ehrbar, A. "Price of Progress: 'Re-Engineering' Gives Firms New Efficiency, Workers the Pink Slip," *The Wall Street Journal,* March 16, 1993, p. A1.

Esque, T. and T. Gilbert. "Making Competencies Pay Off," January 1995, *Training,* January 1995, pp. 44–50.

Flipczak, G. "It Takes All Kinds: Creativity in the Workforce," *Training,* May 1997, p. 36.

Fuchsberg, G. "Small Firms Struggle With Latest Management Trends," *The Wall Street Journal,* August 26, 1993, p. B7.

Griffith, C. "Building a Resilient Work Force," *Training,* January 1998, pp. 54–58.

"Grand Alliances: A New Network of Nonprofits Is Helping Environmental Partnerships Grow and Prosper," author not listed, *The GreenBusiness Letter,* February 1996, p. 1.

Herman, S. (1998). "EPA's 1998 Enforcement and Compliance Assurance Priorities," *National Environmental Enforcement Journal,* Vol. 13, No. 1, February 1998, pp. 3–14.

Kaplan, R. and D. Norton, (1996). "Using the Balanced Scorecard as a Strategic Management System," *Harvard Business Review,* January–February 1996, p. 70.

Kinsman, M. "Generous employers are likely to be repaid," *San Diego News Tribune,* January 19, 1996, p. C-1.

Lancaster, H. "Hiring a Full Staff May Be the Next Fad in Management," *The Wall Street Journal,* April 28, 1998, p. B1.

Moynihan, J. "Outsourcing: Boom or Bane? It Depends on Whom You Talk To," *Today's Chemist At Work,* Washington, DC: American Chemical Society, 1997, p. 17.

Murray, M. "Thanks, Goodbye: Amid Record Profits, Companies Continue to Lay Off Employees," *The Wall Street Journal,* May 4, 1995, p. A1.

Naj, A. "Shifting Gears: Some Manufacturers Drop Efforts To Adopt Japanese Techniques," *The Wall Street Journal,* May 7, 1993, p. A1.

Posson, M. "The Risks and Benefits of Outsourcing Environmental Management," *Corporate Environmental Strategy,* v3n3, Spring 1996, pp. 5–11.

Templeton, N. "Strange Bedfellows: More and More Firms Enter Joint Ventures With Big Competitors," *The Wall Street Journal,* November 1, 1995, p. A1.

Zachary, G. "The New Search for Meaning in 'Meaningless' Work," *The Wall Street Journal,* January 9, 1997, p. B1.

BOOKS

Brandenburger A. M. & B. J. Nalebuff. *Co-opetition.* New York: Bantam Doubleday Dell Publishing, 1996.

De Geus, A. *The Living Company: Habits for survival in a turbulent business environment,* Boston, MA: Harvard Business School Press, 1997.

Segil, L. *Intelligent Business Alliances: How To Profit Using Today's Most Important Strategic Tool,* New York: Random House, 1996.

Kaplin R. S. & D. P. Norton. *The Balanced Scorecard: Translating Strategy Into Action,* Boston, MA: Harvard Business School Press, 1996.

Kao, J. *Jamming: The Art and Discipline of Business Creativity.* New York, NY: Harper Collins Publishers, 1996.

EPA DOCUMENTS

Pollution Prevention Training Opportunities in 1991

Progress on Reducing Industrial Pollutants, EPA 2HP-3003, October 1991

Federal Register, Vol. 63, No. 48, March 12, 1998, pp. 12094–12096.

OTHER RELATED DOCUMENTS

Chin, R., Ganek, V., and A. Muska. "Petroleum Refining Industry Report, Campaign for Cleaner Corporations" New York, NY: Council on Economic Priorities, 1998.

Council on Economic Priorities Eleventh Annual Corporate Conscience Awards, New York, NY: Council on Economic Priorities, 1997.

APPENDIX A

SAMPLE EPA INSPECTION CHECKLIST

This appendix contains a sample of an EPA Multi-Media Inspection checklist. We hope the information will be useful to readers for providing them additional insight into what Federal & State inspectors can look for during typical inspections.

MULTI-MEDIA CHECKLIST

GENERAL INFORMATION Inspector: _____ Date: _____

Facility Name: _____ Contact: _____

Address: _____
       (STREET)   (CITY)   (STATE)   (ZIP)

Phone No.: (___) ___ – ___ SIC Code: _____ No. Employees: _____

Products mfgd. and description of facility: _____

_____

_____

Air: Stationary Source Compliance

1.O   Did you observe opaque smoke emitted from a smokestack (dark enough to obscure anything behind the plume)? _____ —If yes—Which process line (be specific, i.e., boiler No. 4)? _____ —Air pollution control equipment out of service? _____ —If yes—When will it be back on line? _____

2.OP Did you smell any strong odors? _____ If yes, from what process? _____ What chemicals (i.e., solvents) were causing the odors? _____ Is the process controlled by air pollution control equipment?

425

3.1   Has the facility added any processes or expanded any preexisting processes in the last two years which emits air pollutants? _____ —If yes, what type of process was added? _____ Did the facility obtain a state air permit?

4.1   a) Has the facility undergone any demolitions within the last 18 months? _____

      b) Has the facility removed any asbestos from any facility components (pipes, boilers, ducts, etc.) within the past 18 months?_____ —If yes— Approx. how many square feet or linear feet? [Units should be expressed in terms of feet for pipe (length) and square feet for all other facility components (area)]. _____

      If either 4.a or 4.b were "yes"—answer the following:

      c) Was notification for the project provided to EPA or any other regulatory agency? _____ Name and address of any contractors involved:

_____

EPCRA (Chem. Inventory)

1.1   Does the facility have on-site at any time during the calendar year threshold quantities of any Extremely Hazardous Substances (EHS) or any other hazardous materials requiring submission of Tier II chemical inventory forms? _____ —If yes—Have Tier II forms been filed with local and state planning authorities? _____

EPCRA Section 313 (Toxic Release Inventory)

1.1   Does the facility manufacture, process, or use any toxic chemicals in a quantity greater than 10,000 lbs per year?

2.1   Has the facility submitted any toxic chemical release forms (Form R) to EPA?

FIFRA

1.1   Does the facility manufacture, distribute, repackage, relabel, store or use pesticides? (Product which would be considered pesticides include disinfectants, sterilizers, germicides, algicides, virucides, swimming pool compounds, insecticides, fungicides, herbicides, etc.) _____

2.1   If yes, does the label bear an EPA registration and establishment number? _____

RCRA

1.1   Does the facility generate or otherwise handle hazardous waste? If so, describe the types of hazardous waste generated/handled, and state whether it is generated on-site or received from off-site.

_____

_____

_____

2.OP Do you see any waste stored in containers, drums, tanks, pails, or dumpsters? Note the approximate quantity of waste, and its location.

_____

_____

3.OP Are there any containers or tanks of hazardous waste which are open or in poor condition (leaking, corroded, etc)? If so, describe waste (ie., liquid, sludge, etc.), indicate markings on containers/tanks and the container/tank location(s).

_____

_____

4.OP Is there any evidence of spills or leaks or dumping to the ground, pits or lagoons? If so, note location and extent of release.

_____

_____

5.I Does the facility operate a boiler or industrial furnace? Has there been any incineration of hazardous waste on-site? If so, what type of hazardous waste, and is this an ongoing operation?

_____

_____

_____

SPCC

1.I How many gallons of oil does the facility store above and below ground?

_____

—If the facility stores more than 660 gallons in a single tank or more than 1320 gallons in a number of tanks above ground _or_ more than 42,000 gallons below ground—Does the facility have a certified SPCC (Spill Prevention, Control, and Countermeasure) plan signed by a P.E.?

_____

TSCA PCB

1.OP Is there any evidence of spills or leaks from transformers, capacitors, or other liquid-filled electrical equipment that may contain PCBs? Yes: _____ No: _____
If "yes", describe type of equipment and spill or leak:

_____

_____

2.IP If the above equipment has leaks or spills, is it considered to contain PCBs:
based on—marking with "Large PCB Mark"? Yes: _____ No: _____
based on—equipment Nameplate? Yes: _____ No: _____
based on—information from facility rep.? Yes: _____ No: _____

3.OP   Are there any PCB items (equipment, drums of waste or other containers) in storage for disposal? Yes: _____ No: _____

4.I   Where are these items being stored and what is their condition? _____

_____

_____

_____

TSCA Core

1.I   Does the facility manufacture (synthesize anew) any chemical substances in any amount? _____ If so, in simple terms, what chemical(s) do they make? _____

2.I   Does the facility import any chemical substances into the United States? (Company is "Importer of Record") _____

UST

1.I   Does the facility store in USTs motor fuels, waste oils, and/or hazardous substances? _____ YES _____ NO

(Note: USTs containing heating fuels for on-site heating purposes are exempted from RCRA UST.)

If Yes, ask:

2.I   Are the USTs registered with the state? _____ YES _____ NO

(Each state keeps notification data for USTs)

3.I   Is some form of leak detection in use for the USTs (system is tank and associated piping)? _____ YES _____ NO

4.I   Are records available showing registration and monthly leak detection along with the yearly UST system tightness test? _____ YES _____ NO

WATER

A. DIRECT (NPDES) & INDIRECT (PRETREATMENT) DISCHARGERS

1.P   Does the facility use water in its manufacturing processes? _____ If yes—

I   a) Does the facility discharge wastewater (process, sanitary, cooling, etc.) into a surface water, municipal sewer system, or a subsurface system? ___

I   b) Are all of the discharges covered by a permit? _____

2.P   Does the facility have floor drains? _____ If yes—

a) Are materials stored in a manner that leaks or spills could enter the floor drains? _____

b) Are materials dumped down the floor drains? _____

I   c) Where do the floor drains discharge (1) treatment facility, 2) municipal sewer, 3) subsurface system, or 4) surface water)? _____

3.   Does the facility treat its process and/or sanitary wastewater prior to discharge? _____ If yes—

OP   a) Is the treatment equipment operational, clean, and well maintained? _____

OP   b) Is the discharge free of solids, color and odor? _____

B. STORM WATER

1.0   Are there catch basins, drains, culverts, ditches, etc. on the property in-tended to convey storm water? _____ If yes—
a) Is the storm water conveyed to a (1) treatment facility, (2) combined sewer, (3) separate storm sewer, or (4) surface water? _____

2.1   Are the storm water discharges covered by a permit or has the discharger applied for a permit? _____

3.0   Are materials stored outside? _____ If yes—
a) Are materials (1) stored in sealed containers, under tarps or roofs, or (2) are they open to contact with precipitation? _____
b) Are outside material handling/storage areas clean and kept in a man-ner to prevent contamination of runoff? _____

UIC

1.1   Does this facility *discharge* any *fluids to* drains, plumbing, or drainage sys-tems connected to *UIC wells* that are designed for the subsurface em-placement of fluids *INTO the ground?* (See UIC well types and fluid types below).
yes _____.   no _____.

2.1   If yes, indicate *fluid type(s)* and *UIC well type(s)* below.
_____.

3.1   Does this facility have a federal, state, or local permit authorizing this (these) underground injection system(s).
yes _____.   no _____.

4.1   If treated sewage effluent is discharged below ground, are other fluid types disposed of in the sewage waste stream?
yes _____.   no _____.
If yes, indicate other fluid types disposed.

WETLANDS

1.0   Within view, are there a) streams, ponds or other water bodies; b) vege-tated areas with standing water; or c) areas with mucky, peaty or satu-rated (squishy) soils? If yes, have any of these areas been disturbed by waste/refuse disposal, storage of materials, ditching, or filling? If yes, briefly describe: _____
_____
_____

2.1   If yes to both "Observable" questions, does facility have a federal CWA section 404 permit, or any state or local permit authorizing the activity (ies) observed?
_____
_____

APPENDIX B

PUBLICATIONS, PROFESSIONAL & INDUSTRY ASSOCIATIONS, PUBLIC INTEREST GROUPS AND GOVERNMENTAL AGENCIES

This appendix contains a listing of contacts the reader may utilize to obtain additional sources of information that could also serve as a reference guide. Additional information, such as fax numbers, e-mail, and web pages, are included as available. A more complete listing of web sites is found in Appendix G.

Publications and Newsletters

Journal of Corp. Env. Strategy
AHC Group
1223 Peoples Avenue
Troy, NY 12080-3590
518-276-2669
(fax)518-276-2051

Environmental Business Journal
P.O. Box 371769
San Diego, CA 92137

Today's Chemist at Work
American Chemical Society
1155 16th Street, NW
Washington, DC 20036

Chemical Week
888 Seventh Avenue
New York, NY 10106
212-621-4900
(fax)212-621-4950

Journal of Business & Strategy
Faulkner & Gray, Inc.
11 Penn Plaza
New York, NY 10001
800-535-8403

Quality Digest
QCI International
40 Declaration Drive
Chico, CA 95973
916-893-4095
(fax)916-893-0395

Environmental Protection Magazine
  and *Occupational Health & Safety*
Stevens Publishing Corp
5151 Beltline Road, Suite 1010
Dallas, TX 75240
972-687-6700

Occupational Hazards Magazine
1350 Connecticut Ave., NW
Washington, DC 20036

Pollution Engineering
1350 E. Touhy Avenue
Des Plains, IL 60018
847-390-2615
(fax)847-390-2636
PolEngineering@cahners.com

Chemical Engineering and
    Engineering News-Record
McGraw-Hill, Inc.
1221 Ave. of the Americas
New York, NY 10020
212-512-2000

Chemical Processing; Control
Putnam Publishing Company
555 W. Pierce Road, Suite 301
Itasca, IL 60143
630-467-1300
(fax)630-467-1124

Environment
Heldief Publications
1319 Eighteenth Street, NW
Washington, DC 20036; 202-276-6267

Strategy & Business
Booz-Allen & Hamilton
101 Park Ave.
New York, NY 10178

Leader to Leader
The Drucker Foundation
320 Park Avenue, 3rd Floor
New York, NY 10022

Chemical & Engineering News and
    Environmental Science & Technology
American Chemical Society
1156 16th Street, NW
Washington, DC 20036
202-872-4501 (*C&EN*)
202-872-4582 (*ES&T*)
(fax)202-872-4403
est@acs.org

National Environmental Enforcement Journal
National Association of Attorneys General
750 First Street, NE
Suite 1100
Washington, DC 20002
202-326-6044
(fax)202-408-6982

Oil & Gas Journal
PennWell Publishing Company
1421 S. Sheridan Road
Box 1260
Tulsa, OK 74101
http://www.ogjonline.com

Safety Compliance Letter
    Safety Management and
    Quality Management
Bureau of Business Practice
24 Rope Ferry Road
Waterford, CT 06386
800-243-0876
(fax)860-437-3150
http://www.bbpnews.com

Environmental Management Report
    Quality Report
McGraw-Hill, Inc.
Fairfax, VA 22032
703-591-9008
(fax)703-591-0971

Business Ethics
Mavis Publications, Inc.
PO Box 8439
Minneapolis, MN 55408
612-879-0695
(fax)612-879-0699

Business in the Environment
Business in the Community Initiative
5 Cleveland Place
London SW1V 6JJ
+44-71-321-6430
(fax)+44-71-321-6410

Environmental Claims Journal and
  Environmental Quality Management
John Wiley & Sons, Inc.
605 Third Avenue
New York, NY 10158
212-850-6479
SUBINFO@wiley.com

Greener Management International
Greenleaf Publishing
Broom Hall
8-10 Broomhall Road
Sheffield, S10 2 DR, UK
+44-114-266-3789
(fax)+44-114-267-9403

*International Environmental Systems
  Update & Integrated Management
  Systems Update*
CEEM, Inc.
10521 Braddock Road
Fairfax, VA 22032;
800-745-5565
(fax)703-250-5313

*Environmental Compliance & Litigation
  Strategy*
345 Park Avenue South
New York, NY 10010
800-888-8300

Harvard Management Update
Harvard Business School Publishing
60 Harvard Way
Boston, MA 02163
800-988-0886

Environmental Business Journal
Environmental Business International, Inc.
4452 Park Blvd. #301
San Diego, CA 92116
619-295-7685
(fax)619-295-5743
ebi@ebiusa.com
http://www.wbiuse.com

Business and the Environment
Cutter Information Corp.
37 Broadway
Arlington, MA 02174

Tom Peters Fast Forward
PO Box 652
Mt. Morris, IL 61054
800-827-3095

Imprimis
Hillsdale College
Hillsdale, MI 49242
800-334-8904

Professional & Industry Associations

Chemical Manufacturers' Assoc.
1300 Wilson Blvd.
Arlington, VA 22209
703-741-5000
(fax)703-741-6000

Amer. Assoc. of Eng. Societies
1111 19th Street, NW
Washington, DC 20036
202-296-2237

Nat'l Assoc. of Environ. Mgmt.
2025 Eye Street, Suite 1126
Washington, DC 20006
202-986-6616; 800-391-NAEM
(fax)202-530-4408

Air & Waste Mgmt. Assoc.
3 Gateway Center
Pittsburg, PA 15330
412-232-3444

American Petroleum Institute
1220 L Street, NW
Washington, DC 20005
202-682-8000
(fax)202-682-8232

American Productivity & Quality Center
123 North Post Oak Lane, 3rd Floor
Houston, TX 77024
713-685-4642
(fax)713-681-8575
http:\\www.apqc.org

Environmental Auditing Roundtable
35888 Midland Avenue
North Ridgefield, OH 44039
216-327-6605
(fax)216-327-6609
cc004250@interramp.com

Global Environmental Management
    Initiative (GEMI)
Law Companies Group, Inc.
2000 L Street, NW, Suite 710
Washington, DC 20036
202-296-7449

American Academy of Environmental
    Engineers
130 Holiday Court
Annapolis, MD 21401
301-266-3311

American Chemical Society
1155 16th Street, NW
Washington, DC 20036
202-872-4600/4582
(fax)202-872-4403

American Inst. of Chem. Engineers
345 E. 47th Street
New York, NY 10017
212-705-7338

National Association of Env. Professionals
6524 Ramoth Drive
Jacksonville, FL 32226

Pharmaceutical Manufacturers Assoc.
1100 Fifteenth Street, NW
Washington, DC 20005
202-835-3400

National Center for Manufacturing
    Sciences
3025 Boardwalk
Ann Arbor, MI 48108
313-995-0300

Edison Electric Institute
201 Pennsylvania Avenue, NW
Washington, DC 20004
202-508-5000
http://www.eei.org

American Plastics Council
1801 K Street, NW
Washington, DC 20006
202-974-5400

American Bar Association
Natural Resources, Energy & Environment
750 No. Lake Shore Drive
Chicago, IL 60611
312-988-5724
(fax)312-988-5572

American Forest & Paper Association
1111 19th Street, NW
Washington, DC 20036
202-463-2455
(fax)202-463-2785

American Society of Civil Engineers
1801 Alexander Bell Drive
Reston, VA 20191
703-295-6000; 800-548-2723

National Association of Manufacturers
1331 Pennsylvania Avenue, NW
Washington, DC 20004
202-637-3000

Interconnecting Packaging Circuits
2215 Sanders Road
Northbrook, IL 60062
847-509-9700
(fax)847-509-9798

National Association of Chemical
    Distributors
1525 Wilson Blvd., Suite 750
Arlington, VA 22209
703-527-6223
(fax)703-527-7747
http://www.ncad.com

National Environmental Training
    Association
3020 East Camelback Road, Suite 399
Phoenix, AZ 85106
(602)956-6099
(fax)602-956-6399

Public Interest Groups

Greenpeace USA
1436 U Street, NW
Washington, DC 20009
202-462-1177

Council on Economic Priorities
30 Irving Place
New York, NY 10003
212-420-1133
(fax)212-420-0988

Earthjustice Legal Defense Fund
180 Montgomery Street, Suite 1400
San Francisco, CA 94104
415-627-6700
(fax)415-627-6740

Friends of the Earth
218 D Street, SE
Washington, DC 20003

Ecological Management Foundation
Prinsengracht 240
Amsterdam, Holland 1017 YM

World Wildlife Fund
1250 24th Street, NW
Washington, DC 20037
202-293-4800
(fax)202-293-9211

Coalition for Environmentally Responsible
   Economies (CERES)
711 Atlantic Avenue
Boston, MA 02111
617-451-0927
(fax)617-482-2028

The National Environmental Education
   & Training Foundation
*Institute for Corporate Environmental
   Mentoring (ICEM)*
734 Fifteenth Street, NW, Suite 420
Washington, DC 20005
202-628-8200
(fax)202-628-8204
NEETF@NEETF.org

Environmental Defense Fund
257 Park Avenue South
New York, NY 10010
212-686-4191

The Sierra Club
85 Second Street, 2nd Floor
San Francisco, CA 94105
415-977-5500

Business Council for Sustainable
   Development
World Trade Center Building
Route de Aeroport 10
CH 1215 Geneva 15, Switzerland
+41-22-788-3202

Investor Responsibility Research Center
1350 Connecticut Ave, NW, Suite 700
Washington, DC 20036
202-833-0700
(fax)202-833-3555

Earth Share
3400 International Drive, NW
Washington, DC 20008
202-537-7100

Renew America
1400 16th Street, NW
Suite 710
Washington, DC 20036
202-232-2252
(fax)202-232-2617
www.crest.org/renewamerica

Government & Related Agencies

US Environmental Protection Agency
401 M Street, SW
Washington, DC 20460
202-260-4700

EPA Region I
One Congress Street
Boston, MA 02203
617-565-3420

EPA Region II
290 Broadway
New York, NY 10007
212-262-2400

EPA Region III
841 Chestnut Building
Philadelphia, PA 19107
215-566-5565

EPA Region IV
61 Forsythe St, SW
Atlanta, GA 30365
404-562-8327

EPA Region V
77 W. Jackson Blvd.
Chicago, IL 60604
312-353-2000

EPA Region VI
1455 Ross Avenue
Dallas, TX 75202
214-655-6444

EPA Region VII
726 Minnesota Ave.
Kansas City, KS 66101
913-551-7000

EPA Region VIII
999 18th Street
Denver, CO 80202
303-312-6312

EPA Region IX
75 Hawthorne Street
San Francisco, CA 94105
415-744-1305

EPA Region X
1200 Sixth Avenue
Seattle, WA 98101
206-553-1200

National Enforcement Investigations
    Center (NEIC)
US EPA
Building 53
PO Box 25227-DFC
Denver, CO 80255
303-236-3636/5123

National Environmental Training Institute
    (NETI)
12345 West Almeda Parkway
Lakewood, CO 80228
303-969-5815

Department of Commerce
Herber C. Hoover Building
14th Street & Constitution Ave.
Washington, DC 20230
202-482-2000

Department of the Interior
Washington, DC 20240
202-208-3100

PPIC (EPA)
202-260-1023
(fax)202-260-4659
ppic@epamail.epa.gov

Northeastern Governors Association
400 N. Capitol Street, NE
Washington, DC 20001
202-624-8450

US Government Printing Office
Superintendent of Documents
North Capitol and H Streets, NW
Washington, DC 20401
202-512-1800

Department of Defense
The Pentagon
Washington, DC 20301
703-695-6639

US Dept of Labor, OSHA
Voluntary Protection Programs,
  Room N3700
200 Constitution Ave., NW
Washington, DC 20210
202-219-7266
(fax)202-219-8783

US Dept. of Energy
James Forrestal Building
1000 Independence Avenue
Washington, DC 20585
202-586-5000

Department of Justice
Environment & Nat'l Resources
10th St. & Constitution Ave, NW
Washington, DC 20530
202-514-2701

US Food and Drug Administration
SB 8, Room 3807 (food)
200 C Street, SW
Washington, DC 20204
202-205-4144
(fax)202-205-5004

Park Lane Building (drugs)
Room 15A-11
5600 Fishers Lane
Rockville, MD 20857
301-827-4573
301-827-3990 (medical devices)

Federal Emergency Management Agency
Federal Center Plaza
500 C Street, SW
Washington, DC 20472
202-646-3923

Commission of the European Community
34 Rue Belliard
1049 Brussels, Belgium

United Nations Environment Program
Tour Mirabeau
39-43 Quai Andre Citroen
75739 Paris CEDEX 15, France
+33-1-4058-8850

US Chamber of Commerce
1615 H Street, NW
Washington, DC 20062
202-659-6000

Interagency Environmental Technologies
  Office
955 L'Enfant Plaza North, SW, Suite 5322
Washington, DC 20024
etstrategy@gnet.org

Department of Transportation
400 Seventh Street, SW
Washington, DC 20590
202-366-4000

US Securities & Exchange Commission
450 Fifth Street, NW
Washington, DC 20549
202-942-4150
http://www.sec.gov

United Nations Development Program
336 East 45th Street
New York, NY 10017
212-906-3683

International Chamber of Commerce
38 Cours Albert 1er
75008 Paris, France
+33-1-4953-2828

National Institute for Occupational Safety
  & Health
Centers for Disease Control
1600 Clifton Road, NE
Atlanta, GA 30333
404-639-3534
http://www.cdc.gov/niosh/homepage.
  html

US Council for Int'l Business
1212 Ave. of the Americas
New York, NY 10036
212-354-4480
(fax)212-575-0327
http://www.uscib.org

Key ISO Environmental Standards Groups

American National Standards Institute
(ANSI)
11 West 42nd Street
New York, NY 10036
212-642-4900
(fax)212-398-0023
http://www.ansi.org
info@ansi.org

American Society for Quality (ASQ)
611 East Wisconsin Ave.
PO Box 3005
Milwaukee, WI 53201
800-248-1946; 414-272-8575
(fax)1-414-272-1734

International Organization for
Standardization (ISO)
1, Rue de Varembe
Case Postal 50
CH-1211 Geneva 20, Switzerland
+41-22-749-0111
(fax)+41-22-733-3430
http://www.iso.ch
central@isocs.iso.ch

British Standards Institute (BSI)
389 Chiswick High Road
London W4 4L, UK
+44-181-996-9000
(fax)+44-181-996-7400
info@bsi.org.uk
http://www.bsi.org.uk

Standards Australia (SSA)
PO Box 1055
Strathfield, NSW 2135, Australia
+61-2-97-46-4700
(fax)+61-2-97-46-8450
intsect@standards.com.au
http://www.standards.com.au

Association francais de normalisation
(AFNOR)
Tour Europe
92049 Paris La Defense Cedex, France
+33-1-42-91-5555
(fax)+33-1-42-91-5656
international @email.afnor.fr
http://www.afnor.fr/

Canadian Standards Association (CSA)
178 Rexdale Boulevard
Rexdale, Toronto, Ontario
Canada M9W 1R3
416-747-4155
(fax)416-747-4149

American Society for Testing & Materials
100 Bar Harbor Drive
West Conshhohocken, PA 19428
610-832-9500
(fax)610-832-9555/9666
service @local.astm.org

Netherlands Standardization Institute
(NNI)
Kalfjeslaan 2, PO Box 5059
2600 GB Delft
Netherlands
+31-15-269-0390
(fax)+31-15-269-0190
info@nni.nl
http://www.nni.nl

Norges Standardiseringsforbund (NSF)
Drammensveien 145A
Postboks 353 Skoyen
0212 Oslo, Norway
+47-220-49200
(fax)+47-220-49211
firmapost@nsf.telemax.no
http://www.standard.no/nsf

Deutches Institut fur Normung (DIN)
Burggrafenstrasse 6
10787 Berlin, Germany
+49-30-26-010
(fax)+49-30-26-011231
postmaster@din.de
http://www.din.de/

APPENDIX C

LIST OF ACRONYMS

Appendix C lists most or all of the acronyms referred to in the book.

ANSI	American National Standards Institute
API	American Petroleum Institute
APQC	American Productivity Quality Center
ASQ	American Society for Quality
ASTM	American Society for Testing and Materials
CAA	Clean Air Act
CAS No.	Chemical Abstract Service Number
CERES	Coalition for Environmentally Responsible Economies
CMA	Chemical Manufacturers Association
CSI	Common Sense Initiative (EPA voluntary program)
CWA	Clean Water Act
DfE	Design for the Environment
DOE	Department of Energy
DOJ	Department of Justice
DOT	Department of Transportation
EARA	Environmental Auditors Registration Accreditation
EDM	Electronic Document Management (software)
EH&S	Environmental, Health & Safety
ELP	Environmental Leadership Program (EPA voluntary initiative)
EMIS	Environmental Management Information System (software)
EMS	Environmental Management System
EPA	(US) Environmental Protection Agency
EPCRA	Emergency Planning and Community Right-to-Know Act
EPE	Environmental Performance Evaluation (part of ISO 14000)
FDA	Food and Drug Administration
GAO	General Accounting Office
HSWA	Hazardous & Solid Waste Amendments of RCRA
ICC	International Commerce Commission

IDEA	Integrated Data for Enforcement Analysis (EPA tracking tool)
IRRC	Investor Responsibility Research Center
ISO	International Organization for Standardization
ISO/TC 207	Technical Committee 207 (charged with drafting the ISO 14000s)
LCA	Life Cycle Assessment (part of ISO 14000)
MFP	(EPA's) Model Facility Program (part of ELP)
MMI	Man-machine Interface (a type of control software)
MOU	Memorandum of Understanding
MSV	Management System Verification (Part of Responsible Care®)
NAEM	National Association for Environmental Management
NCP	New Chemicals Program (TSCA)
NEIC	National Enforcement Investigations Center (part of EPA's OECA)
NETI	National Environmental Training Institute (part of EPA)
NOV	Notice of Violation
OECA	(EPA's) Office of Enforcement & Compliance Assurance
OSHA	Occupational, Safety & Health Administration
OSWER	Office of Solid Waste & Emergency Response (EPA)
PDCA	plan-do-check-act (part of the Demming process in quality circles)
PERI	Public Environmental Reporting Initiative
PP&E	(Chevron's policy) Protecting People & the Environment
Project XL	Project for eXcellence and Leadership (EPA voluntary program)
PSM	Process Safety Management—an OSHA rule (see also RMP)
RCRA	Resource Conservation & Recovery Act
RCRIS	RCRA Compliance Reporting Information System
RMP	Risk Management Program—an EPA rule (see also PSM)
ROI	Return on Investment
SAGE	Strategic Advisory Group on the Environment
SAR	Supplied Air Respirators
SARA	Superfund Amendments & Reauthorization Act
SCADA	Supervisory Control and Data Acquisition (control software)
SDWA	Safe Drinking Water Act
SEC	(US) Securities & Exchange Commission
SEP	Supplemental Environmental Project (in EPA enforcement settlements)
SIC	Standard Industrial Classification (industry code)
STEP	(API's) Strategies for Today's Environmental Partnership
TMB	Technical Management Board (establishes committees for ISO)
TQEM	Total Quality Environmental Management
TRI	Toxic Release Inventory
TSCA	Toxic Substances Control Act
TSD	Treatment, Storage, & Disposal facility (also TSDF)
UKAS	United Kingdom Accreditation Service
UNCED	United Nations Conference on Environment & Development
UNEP	United Nations Environment Programme
USTAG	US Technical Advisory Group (to TC207)
VOCs	Volatile Organic Compounds
WRAP	(Dow's) Waste Reduction Always Pays (company program)

ENVIRONMENTAL GUIDING PRINCIPLES & CODES OF ETHICS

This appendix contains the guiding principles and codes of ethics of a representative sample of trade and environmental organizations. We chose these organizations to highlight each group's contribution to protecting human health and the environment and other considerations, and to do so in a prudent fashion while conducting business. These organizations also provide environmental professionals and other stakeholders a forum with which to exchange ideas and network with others with similar ideals and objectives.

If readers are interested in learning more about the organizations represented, they should refer to Appendix B for contact information.

The guiding principles are:

- Responsible Care®
- The Public Environmental Reporting Initiative (drafted and agreed to by an ad hoc group of companies)
- The CERES Principles

RESPONSIBLE CARE® GUIDING PRINCIPLES

The Guiding Principles are the foundation of the Responsible Care® ethic. They outline each member and Partner commitment to environmental, health and safety responsibility in managing chemicals. CMA members and Partners pledge to manage their businesses according to these principles:

- To recognize and respond to community concerns about chemicals and our operations.
- To develop and produce chemicals that can be manufactured, transported, used and disposed of safely.
- To make health, safety and environmental considerations a priority in our planning for all existing and new products and processes.

- To report promptly to officials, employees, customers and the public, information on chemical-related health or environmental hazards and to recommend protective measures.
- To counsel customers on the safe use, transportation and disposal of chemical products.
- To operate our plants and facilities in a manner that protects the environment and the health and safety of our employees and the public.
- To extend knowledge by conducting or supporting research on the health, safety and environmental effects of our products, processes and waste materials.
- To work with others to resolve problems created by past handling and disposal of hazardous substances.
- To participate with government and others in creating responsible laws, regulations and standards to safeguard the community, workplace and environment.
- To promote the principles and practices of Responsible Care® by sharing experiences and offering assistance to others who produce, handle, use, transport or dispose of chemicals.

1. GUIDING PRINCIPLES

Statements that outline each member company and Partner's commitment to environmental, health and safety responsibility in managing chemicals. CMA members and Partners pledge to manage their businesses according to these principles.

2. CODES OF MANAGEMENT PRACTICES

At the heart of the Responsible Care® initiative are the six Codes of Management Practices. The Codes focus on management practices in specific areas of chemical manufacturing, transportation and handling. Members and Partners must make good-faith efforts to attain the goals of each Code.

Community Awareness and Emergency Response (CAER) Code The CAER Code brings together the chemical industry and local communities through communication and cooperative emergency planning. The CAER Code emphasizes dialogue and interaction with many audiences and requires companies to establish facility outreach programs. The Code also requires facilities to develop and annually test an emergency plan in coordination with the community.

Pollution Prevention Code The Pollution Prevention Code is designed to achieve ongoing reduction in the amount of contaminants and pollutants released to the air, water, and land from member and Partner facilities.

Process Safety Code The Process Safety Code is designed to prevent fires, explosions and accidental chemical releases. The Code identifies areas where members and Partners can improve their safety performance, from process design through continued operation and routine maintenance.

Distribution Code The Distribution Code assists member and Partner companies in reducing the risk of harm posed by the distribution of chemicals to the general public; to carrier, distributor, contractor and chemical industry employees; and to the environment.

Employee Health and Safety Code The Employee Health and Safety Code protects and promotes the health and safety of people working at or visiting CMA member or Partner company sites.

Product Stewardship Code The Product Stewardship Code is designed to promote the safe handling of chemicals at all stages—from initial manufacture to distribution, sale and ultimate disposal.

3. PUBLIC ADVISORY PANEL

This panel, a group of environmental, health and safety thought leaders, has been in existence since the adoption of Responsible Care® in 1988 and has provided insight into all aspects of the initiative.

4. MEMBER SELF-EVALUATIONS

Member companies and Partners submit reports annually to CMA on their progress in implementing the six Codes. These self-evaluations measure progress and are a valuable management tool for CMA and its individual members and Partners.

5. MANAGEMENT SYSTEMS VERIFICATION

In 1996, the CMA Board adopted a management systems verification (MSV) process to assist members and Partners in continuously improving their management and implementation of Responsible Care®. The process gives those companies who participate an outside view of their company's progress, and helps demonstrate the integrity of the initiative to key audiences, including employees, local communities, public officials and others.

6. MEASURES OF PERFORMANCE

CMA members and Partners have developed credible external performance measures to demonstrate the progress being made through Responsible Care®.

7. EXECUTIVE LEADERSHIP GROUPS (ELGs)

Senior level support for Responsible Care® continues to be an essential ingredient to the initiative's success. The regional ELGs provide a forum for member and Partner executive contacts to regularly address their progress in implementing Responsible Care®.

8. MUTUAL ASSISTANCE

Member-to-member mutual assistance has surfaced as one of the most effective methods for advancing Responsible Care®. Through the mutual assistance network, members and Partners at the executive contact, Responsible Care® Coordinator and practitioner levels regularly come together share information vital to the successful implementation of the initiative.

9. PARTNERSHIP PROGRAM

The Partnership Program allows eligible, non-CMA companies and state or national trade associations to participate directly in Responsible Care®.

10. OBLIGATION OF MEMBERSHIP

CMA bylaws obligate member companies and Partners to participate in Responsible Care® as defined by the CMA Board of Directors. The Board currently defines each member's obligation to Responsible Care® by requiring CMA members and Partners to subscribe to the Guiding Principles, to participate in the development of the initiative and to make good faith efforts to implement the program elements of the initiative.

PERI GUIDELINES

GUIDELINE COMPONENTS

Each reporting organization may decide how, when, and to what extent to present the PERI reporting components listed below. No specific order of presentation is mandatory or encouraged. The recommended content to be included is as follows:

1. ORGANIZATIONAL PROFILE

Provide information about the organization that will allow the environmental data to be interpreted in context:

- Size of the organization (e.g., revenue, employees)
- Number of locations
- Countries in which the organization operates
- Major lines of activity, and
- The nature of environmental impacts of the organization's operations.

Provide a contact name in the organization for information regarding environmental management.

2. ENVIRONMENTAL POLICY

Provide information on the organization's environmental policy(ies), (e.g., scope and applicability, content, goals and date of introduction or revision, if relevant).

3. ENVIRONMENTAL MANAGEMENT

Summarize the level of organizational accountability for environmental policies and programmes and the environmental management structure (e.g., corporate environmental staff and/or organizational relationships). Indicate how policies are implemented throughout the organization and comment on such items as:

- Board involvement and commitment to environmental matters
- Accountability of other functional units of the organization
- Environmental management systems in place (if desired, include references or registration under—or consistency with—any relevant national or international standards).
- Total Quality Management (TQM), Continuous Improvement or other organization-wide programmes that may embrace environmental performance.
- Identify and quantify the resources committed to environmental activity (e.g., management, compliance, performance, operations, auditing).
- Describe any educational/training programmes in place that keep environmental staff and management current on their professions and responsibilities.

Summarize overall environmental objectives, targets and goals, covering the entire environmental management programme.

4. ENVIRONMENTAL RELEASES

Environmental releases are one indicator of an organization's impact on the environment. Provide information that quantifies the amount of emissions, effluents or wastes released to the environment.

Information should be based on the global activity of the organization, with detail provided for smaller geographic regions, if desired.

Provide the baseline data against which the organization measures itself each year to determine its progress, and quantify, to the extent possible, the following—including historical information (e.g., last three years, where available) to illustrate trends:

- Emissions to the atmosphere, with specific reference to any:
  - Chemical-based emissions (include those listed in any national reportable inventories, e.g., TRI in the U.S., NPRI in Canada, SEDESOL's Emissions Inventory in Mexico)
  - Use and emissions of ozone-depleting substances
  - Greenhouse gas emissions, e.g., carbon dioxide, methane, nitrous oxide, and halocarbons.
- Discharges to water (include those considered to be a priority for your organization).
- Hazardous waste, as defined by national legislation. Indicate the percentage of hazardous waste that was recycled, treated, incinerated, deep-well injected, or otherwise handled, either on- or off-site. Comment on how hazardous waste disposal contractors (storers, transporters, recyclers, or handlers of waste) are monitored or investigated by the organization.
- Waste discharges to land. Include information on toxic/hazardous wastes, as well as solid waste discharges from facilities, manufacturing processes, or operations.
- Objectives, targets, and progress made regarding the above-listed items, including any information on other voluntary programme activity (e.g., U.S. EPA 33/50 programme).

- Identify the extent to which the organization uses recommended practices or voluntary standards developed by other organizations, such as the International Chamber of Commerce, the International Standards Organizations, CMA, API, CEFIC, U.S. EPA, Environment Canada, MITI Guidelines, etc.

5. RESOURCE CONSERVATION

- **Materials conservation** Describe the organization's commitment to the conservation and recycling of materials and the use and purchase of recycled materials. Include efforts to reduce, minimize, reuse, or recycle packaging.
- **Energy conservation** Describe the organization's activity and approach to energy conservation: commitment made to reduce energy consumption, or to use renewable or more environmentally benign energy sources, energy efficiency programme activities, reductions achieved in energy consumption and the resulting reductions achieved in VOCs, NOX, air toxics and greenhouse gas emissions.
- **Water conservation** Describe the organization's efforts in reducing its use of water or in recycling of water.
- **Forest, land and habitat conservation** Describe the organization's activities to conserve or reduce/minimize its impact on natural resources such as forest, lands and habitats.

6. ENVIRONMENTAL RISK MANAGEMENT

Describe the following:

- Environmental audit programmes and their frequency, scope, number completed over the past two years—as well as extent of coverage. Indicate whether the audits are conducted by internal or external personnel or organizations, and to whom and to which management levels the audit findings are reported. Describe follow-up efforts included in the programme to ensure improved performance.
- Remediation programmes in place or being planned, indicating type and scope of activity.
- Environmental emergency response programmes, including the nature of training at local levels, frequency, and the extent of the programme. Indicate the degree and method of communications extended to local communities and other local organizations regarding mutual aid procedures and evacuation plans in case of an emergency.
- Workplace hazards. Indicate the approach taken to minimize health and safety risks in the organization's operations, and describe any formal policies or management practices to reduce these risks (e.g., employee and contractor safety training and supervision, statistical reporting).

7. ENVIRONMENTAL COMPLIANCE

Provide information regarding the organization's record of compliance with laws and regulations. Summary history for the last three years should be given. Additional detail should be provided for any significant incidents of non-compliance since the last report, including:

- Significant fines or penalties incurred (define in accordance with local situation, e.g., over $25,000 in the U.S.) and the jurisdiction in which it applied

- The nature of the non-compliance issues (e.g., reportable, uncontrolled releases, including oil and chemical spills at both manufacturing and distribution operations)
- The scope and magnitude of any environmental impact
- The programmes implemented to correct or alleviate the situation.

8. PRODUCT STEWARDSHIP

This component defines "product" as the outcome of the organization's activity and is applicable whether an organization manufactures, provides services, advocates, governs, etc. In addition, the section is intended to focus on both the organization's activities in producing its products or services not addressed elsewhere in the guidelines and any activities associated with the "end-of-life" of products or services.

Provide information that indicates the degree to which the organization is committed to evaluating the environmental impact of its products, processes and/or services.

Describe any programme activity, procedure, methodology, or standard that may be in place to support the organization's commitment to reduce the environmental impacts of its products and services. For example:

- Discuss technical research or design: (e.g., new products, services or practices, redesign of existing products or services, practices implemented or discontinued for environmental reasons, design for recyclability or disassembly, or redesign of accounting practices).
- Provide information on waste reduction/pollution prevention programmes from the organization's products, processes or services, including conservation and reuse of materials, and the use of recycled materials.
- Describe the organization's efforts to make its products, processes and services more energy efficient.
- Describe post-consumer materials management, or end-of-life programmes, such as product take-back.
- Detail customer cooperative or partnership programmes and their development: (e.g., used oil collection and energy efficiency services).
- Describe supplier programmes and cooperative or partnership activities designed to reduce environmental impacts or add environmental value to the design or redesign of products and services.
- Include information regarding selection criteria for environmentally responsible suppliers and standards to which they must adhere.
- Identify the scope of the supplier certification process (e.g., all suppliers, major suppliers, or those in specific sectors).

Other components:

- Specify product stewardship targets and goals, and comment on established procedures to monitor and measure company performance.
- Provide any baseline data against which the organization can measure its progress.

9. EMPLOYEE RECOGNITION

Include information regarding employee recognition and reward programmes that encourage environmental excellence. Comment on other education and information programmes that motivate employees to engage in sound environmental practices.

10. STAKEHOLDER INVOLVEMENT

Describe the organization's efforts to involve other stakeholders in its environmental initiatives.

Indicate any significant work undertaken with research or academic organizations, policy groups, non-governmental organizations, and/or industry associations on environmental issues—including cooperative efforts in environmentally preferable technologies.

Describe how the organization relates to the communities in which it operates, and provide a description of its activities. For example, indicate the degree to which the organization shares pertinent facility-specific environmental information with the communities in which it has facilities.

THE CERES PRINCIPLES

By adopting these Principles, we publicly affirm our belief that corporations have a responsibility for the environment, and must conduct all aspects of their business as responsible stewards of the environment by operating in a manner that protects the Earth. We believe that corporations must not compromise the ability of future generations to sustain themselves.

We will update our practices constantly in light of advances in technology and new understandings in health and environmental science. In collaboration with CERES, we will promote a dynamic process to ensure that the Principles are interpreted in a way that accommodates changing technologies and environmental realities. We intend to make consistent, measurable progress in implementing these Principles and to apply them to all aspects of our operations throughout the world.

PROTECTION OF THE BIOSPHERE

We will reduce and make continual progress toward eliminating the release of any substance that may cause environmental damage to the air, water, or the earth or its inhabitants. We will safeguard all habitats affected by our operations and will protect open spaces and wildnerness, while preserving biodiversity.

SUSTAINABLE USE OF NATURAL RESOURCES

We will make sustainable use of renewable natural resources, such as water, soils and forests. We will conserve non-renewable natural resources through efficient use and careful planning.

REDUCTION AND DISPOSAL OF WASTES

We will reduce and where possible eliminate waste through source reduction and recycling. All waste will be handled and disposed of through safe and responsible methods.

ENERGY CONSERVATION

We will conserve energy and improve the energy efficiency of our internal operations and of the goods and services we sell. We will make every effort to use environmentally safe and sustainable energy sources.

RISK REDUCTION

We will strive to minimize the environmental, health and safety risks to our employees and the communities in which we operate through safe technologies, facilities and operating procedures, and by being prepared for emergencies.

SAFE PRODUCTS AND SERVICES

We will reduce and where possible eliminate the use, manufacture or sale of products and services that cause environmental damage or health or safety hazards. We will inform our customers of the environmental impacts of our products or services and try to correct unsafe use.

ENVIRONMENTAL RESTORATION

We will promptly and responsibly correct conditions we have caused that endanger health, safety or the environment. To the extent feasible, we will redress injuries we have caused to persons or damage we have caused to the environment and will restore the environment.

INFORMING THE PUBLIC

We will inform in a timely manner everyone who may be affected by conditions caused by our company that might endanger health, safety or the environment. We will regularly seek advice and counsel through dialogue with persons in communities near our facilities. We will not take any action against employees for reporting dangerous incidents or conditions to management or to appropriate authorities.

MANAGEMENT COMMITMENT

We will implement these Principles and sustain a process that ensures that the Board of Directors and Chief Executive Officer are fully informed about pertinent environmental issues and are fully responsible for environmental policy. In selecting our Board of Directors, we will consider demonstrated environmental commitment as a factor.

AUDITS AND REPORTS

We will conduct an annual self-evaluation of our progress in implementing these Principles. We will support the timely creation of generally accepted environmental audit procedures. We will annually complete the CERES Report, which will be made available to the public.

Disclaimer
These Principles establish an environmental ethic with criteria by which investors and others can assess the environmental performance of companies. Companies that endorse these Principles pledge to go voluntarily beyond the requirements of the law. The terms may and might in Principles one and eight are not meant to encompass every imaginable consequence, no matter how remote. Rather, these Principles obligate endorsers to behave as prudent persons who are not governed by conflicting interests and who possess a strong commitment to environmental excellence and to human health and safety. These Principles are not intended to create new legal liabilities, expand existing rights or obligations, waive legal defenses, or otherwise affect the legal position of any endorsing company, and are not intended to be used against an endorser in any legal proceeding for any purpose.

APPENDIX E

ORGANIZATIONS' ENVIRONMENTAL POLICIES, MISSION STATEMENTS, AND OTHER SUPPORTIVE DOCUMENTS

This appendix lists examples of environmental policies and other supportive documents from a sampling of regulated organizations. The documents are representative samples depicting the efforts of companies committed to protecting the environment.

The companies highlighted are:

- Chevron Corporation
- Hewlett-Packard Corporation

CHEVRON: PROTECTING PEOPLE AND THE ENVIRONMENT

ENVIRONMENTAL POLICY

We are committed to protecting the safety and health of people and the environment. We will conduct our business in a socially responsible and ethical manner. Our goal is to be the industry leader in safety and health performance, and to be recognized worldwide for environmental excellence.

We will achieve this goal through:

Safety	Safety is everyone's responsibility. Design, operate and maintain our facilities to prevent injury, illness and accidents.
Compliance	Establish processes to ensure that all of us understand our roles and all operations are in compliance.

Pollution Prevention	Continually improve our processes to minimize pollution and waste.
Community Outreach	Communicate openly with the public regarding the possible impact of our business on them or the environment.
Product Stewardship	Manage potential risks of our products with everyone involved through the product's life cycle.
Conservation	Conserve company and natural resources by continually improving our processes and measuring our progress.
Advocacy	Work cooperatively with public representatives to base laws and regulations on sound risk management and cost-benefit principles.
Property Transfer	Assess and manage environmental liabilities prior to any property transaction.
Transportation	Work with our carriers and distributors to ensure safe distribution of our products.
Emergency Response	Be prepared for any emergency and mitigate any incident quickly.

Graphical Examples of the Chevron Corporation's Environmental Efforts

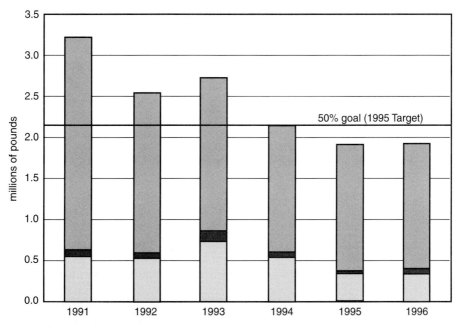

EPA 33/50 Program Emissions
17 Target Chemicals, Ongoing Operations

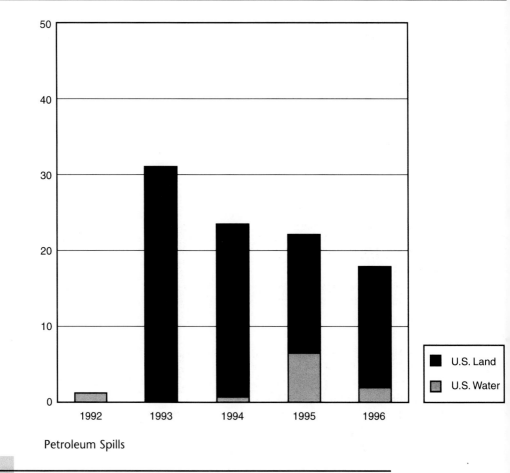

Petroleum Spills

HEWLETT-PACKARD COMPANY: HP ENVIRONMENTAL MANAGEMENT POLICY

Hewlett-Packard Company is committed to conducting its business in an ethical and socially responsible manner. An aggressive approach to environmental management, which includes occupational health, industrial hygiene, safety management and ecological protection, is consistent with the spirit and intent of our established corporate objectives and cultural values.

We recognize that our company, together with other business organizations, has a major responsibility for protecting the ecology, and the health and safety of our employees, our customers and the communities in which we operate worldwide.

The Hewlett-Packard environmental management objective is:

To provide products and services that are environmentally sound throughout their life cycle and to conduct our operations worldwide in an environmentally responsible manner.

The following principles guide us in achieving this objective:

- Recognize that excellence in environmental performance is consistent with our corporate objectives and essential to our continued business success.
- Ensure that environmental policies, programs and performance standards are an integral part of our planning and decision-making process.
- Regard sound environmental management as an integral part of our total quality commitment and to apply the principles and practice of continuous improvement accordingly.
- Be open and responsive to the environmental expectations and concerns of our employees, customers, government agencies and the public by providing clear and candid information about the environmental impact of our products, services and operations.
- Design and construct our facilities to minimize waste generation and promote energy efficiency and ecosystem protection.
- Design our products and services and their associated manufacturing and distribution processes to be safe in their operation; minimize use of hazardous materials; make efficient use of energy and other resources; and to enable recycling and reuse.
- Pursue a strategy that substantially reduces or eliminates the generation of chemical and solid waste.
- Proactively address environmental contamination resulting from any HP operation.
- Foster environmental responsibility among our employees and encourage their initiative and involvement.
- Contribute constructively to the shaping of public policy based on sound business and scientific principles.
- Ensure our suppliers support our Environmental Management Policy and to encourage them to adopt similar principles.

APPENDIX F

ADDITIONAL READINGS

This appendix lists additional sources of information the author reviewed and did not specifically reference in the book. These sources can provide additional reference information and cross-referencing for the reader.

BOOKS

P. Block, *Stewardship: Choosing Service over Self-Interest,* San Francisco: Berrett-Koehler Publishers, 1993.

E. Callenbach, F. Capra, L. Goldman, et al., *EcoManagement: The Elmwood Guide to Ecological Auditing and Sustainable Business,* San Francisco: Berrett-Koehler Publishers, 1993.

F. Cairncross, *Costing the Earth,* Boston, MA: Harvard Business School Press, 1991.

G. Durnill, *The Making of a Conservative Environmentalist,* Bloomington, IN: Indiana University Press, 1995.

B. Gentry, *Private Investment and the Environment,* Office of Development Studies, New York: United Nations Development Programme, 1996.

C. Glassic, M. Huber, and G. Maeroff, *Scholarship Assessed: Evaluation of the Professoriate,* San Francisco: Jossey-Bass Publishers, 1997.

F. Hesselbein, M. Goldsmith, and R. Beckhard, (Eds.), *The Organization of the Future,* San Francisco: Jossey-Bass Publishers, 1997.

J. Hradesky, *Total Quality Management Handbook,* New York: McGraw-Hill, Inc., 1995.

E. B. Harrison, *Going Green: How to Communicate Your Company's Environmental Commitment,* Burr Ridge, IL: Irwin Professional Publishing, 1993.

R. Livernash, W. Faries, W. Lusigi et al., *Valuing the Global Environment: Actions & Investments for a 21st Century,* Washington, DC: Global Environmental Facility, 1998.

Managing Planet Earth (a compilation of works from *Scientific American*), New York: W. H. Freeman & Co., 1990.

H. B. Maynard & S. E. Mehrtens, *The Fourth Wave: Business in the 21st Century,* San Francisco: Berrett-Koehler Publishers, 1993.

ARTICLES

Crognale, G. "From Permits to the Press: Responsibilities of an Environmental Manager," *Safety Management,* November 1997, p. 6.

DeCrane, A. Jr., Chairman and Chief Executive Officer, Texaco, Inc., "The Ecological Imperative for Business," *Directors & Boards,* Vol. 18, No. 1, Fall 1993, pages not listed.

Gentry B. and L. Fernandez, *Valuing the Environment: How Fortune 500 CFOs and Analysts Measure Corporate Performance,* United Nations Development Programme (UNDP), Bureau for Development Policy, New York, NY: Office of Development Studies, Briefing Paper, Fall 1997.

Hasselbein, F. "Barriers to Leadership," *Leader to Leader,* San Francisco, CA: Jossey-Bass Publishers, No. 3, Winter 1997, pp. 4–5.

Hill, M. "A Trend Toward 'Smart' Regulations," *Environmental Solutions,* January 1997, pp. 14–16.

"The Evolving Professional," *Tom Peters: Fast Forward,* Palo Alto, CA: TPG Communications, January 1997, p. 1.

Roe, D. Dr. W. Pease, K. Florini, et al., "Toxic Ignorance," New York, NY: Environmental Defense Fund, Summer 1997.

EPA DOCUMENTS

1996 Toxics Release Inventory: Public Data Release—Ten Years of Right-to-Know, Office of Pollution Prevention and Toxics, EPA 745-R-98-005, May 1998.

EPA's 33/50 Program Second Progress Report: Reducing Risks Through Voluntary Action, February 1992.

Managing for Better Environmental Results, Office of Administrator, EPA 100-R-97-004, March 1997.

Multi-Media Investigation Manual, US EPA Office of Enforcement, Government Institutes, Inc., Rockville, MD.

New Directions: A Report on Regulatory Reinvention, EPA 100-R-98-04, May 1998.

Partnerships in Preventing Pollution, Office of the Administrator, EPA 100-B-96-001, Spring 1996.

Pollution Prevention Success Stories, Office of Pollution Prevention and Toxics, EPA/742/96/002, April 1996.

Principles of Environmental Enforcement, Office of Enforcement, February 1992.

The Changing Nature of Environmental and Public Health Protection: An Annual Report on Reinvention, Office of Reinvention, EPA 100-R-98-003, March 1998.

GAO DOCUMENTS

EPA AND THE STATES: Environmental Challenges Require a Better Working Relationship, GAO/RCED-95-64, April 1995.

ENVIRONMENTAL PROTECTION: Key Management Issues Facing EPA, B-279755, April 23, 1998.

ENVIRONMENTAL PROTECTION: EPA's and the States' Efforts to "Reinvent" Environmental Regulation, GAO/T-RCED-98-33, November 1997.

PEER REVIEW: EPA's Implementation Resources Uneven, GAO/RCED-96-236, September 1996.

REGULATORY BURDEN: Measurement Challenges and Concerns Raised by Selected Companies, GAO/GGD-97-2, November 1996.

RESOURCE CONSERVATION AND RECOVERY ACT: Inspections of Facilities Treating and Using Hazardous Waste Fuels Show Some Noncompliance, GAO/RCED-96-211, August 1996.

Testimony on environmental crimes, *Issues Related to Justice's Criminal Prosecution of Environmental Offenses,* November 3, 1993, GAO/T-GGD-94-33.

MISCELLANEOUS DOCUMENTS & REPORTS

Bridge to a Sustainable Future, National Environmental Technology Strategy, Washington, DC: Interagency Environmental Technologies Office, April 1995.

Funding Cost—Effective Pollution Prevention Initiatives: Incorporating Environmental Costs into Business Decision Making, a primer, Washington, DC: Global Environmental Management Initiative, 1994.

Hold the Applause! A Case Study of Corporate Environmentalism as Practiced at DuPont, Washington, DC: Friends of The Earth, 1991.

Leader to Leader, San Francisco, CA: Jossey-Bass Publishers, No. 4, Spring 1997.

1996 Governor's Awards for Environmental Excellence, Commonwealth of Pennsylvania, Department of Environmental Protection, 7000-BK-DEP2045, Harrisburg, PA.

1998 Handbook of Business Strategy, New York, NY: Faulkner & Gray, 1997.

The National Report Card on Environmental Knowledge, Attitudes and Behaviors, Washington, DC: National Environmental Education and Training Foundation (NEETF), November 1997.

Petroleum Refining Industry Report, Campaign for Cleaner Corporations, March New York, NY: The Council on Economic Priorities, 1998.

Partnerships to Progress: The Report of the President's Commission on Environmental Quality (PCEQ), Washington, DC, January 1993.

Progress on the Environmental Challenge: A Survey of Corporate America's Environmental Accounting and Management, Price Waterhouse LLP, 1994.

United Nations Environment Programme, Industry and Environment Programme Activity Centre, Technical Report No 11, *From Regulations to Industry Compliance: Building Institutional Capabilities*, Paris, France: UNEP, 1992.

APPENDIX G

SUPPLEMENTAL INFORMATION

This appendix lists a representative sample of additional and relevant information to supplement the material referenced in the book. This information includes:

1. Some useful websites of representative federal, state, and international agencies, regulated organizations, and trade and special interest groups. A number of these sites provide links to related groups and organizations that should provide additional information.
2. Some related and useful items of interest from a sampling of regulated organizations.
   - Texaco
   - Boeing

1. WEB SITES

(1) US FEDERAL AGENCIES

Environmental Protection Agency—www.epa.gov (main)

Office of Air and Radiation—www.epa.gov/oarhome.html

Office of Enforcement and Compliance Assurance—
www.epa.gov/oecaerth/index.html

Office of Prevention, Pesticides, and Toxic Substances—www.epa.gov/internet/oppts/

Office of Reinvention—www.wpa.gov/reinvent

Office of Solid Waste and Emergency Response—www.epa.gov/swerrims/

Office of Water—www.epa.gov/OW/

458

Brownfields—www.epa.gov/brownfields/

Common Sense Initiative—www.epa.gov/commonsense/

DfE Program—www.epa.gov/dfe

Environmental Appeals Board—www.epa.gov/boarddec/index.html

Envirofacts—www.epa.gov/envior/index_java.html

Environmental Leadership Program—www.epa.gov/reinvent/elp.htm

One Stop Reporting—www.epa.gov/reinvent/onestop/

Partnership Programs—www.epa.gov/partners/

Project XL—www.epa.gov/ProjectXL/

33/50 Program—www.epa.gov/reinvent/3350.htm

Chemical Safety Board—www.chemsafety.gov

Food and Drug Administration—www.fda.gov

National Aeronautics and Space Administration—www.nasa.gov

National Institute of Environmental Health Sciences—www.niehs.nih.gov

National Institute for Occupational Safety and Health—www.cdc.gov/niosh/homepage.html

National Oceanic and Atmospheric Administration—www.noaa.gov

Nuclear Regulatory Commission—www.nrc.gov

Securities and Exchange Commission—www.sec.gov

US Army Corps of Engineers—www.usace.army.mil

US Department of Agriculture—www.usda.gov

US Department of Commerce—www.doc.gov

    Office of Policy & Strategy Planning—www.osec.doc.gov/opsp/

    National Oceanic & Atmospheric Administration—www.noaa.gov

    National Institutes of Standards and Technology—www.atp.nist.gov/

    National Technical Information Services—www.ntis.gov

US Department of Education—www.ed.gov

US Department of Energy—www.doe.gov

    Federal Energy Regulatory Commission—www.fedworld.gov/ferc

US Department of the Interior—www.doi.gov

US Department of Justice—www.usdoj.gov

US Department of Labor—www.dol.gov

    Occupational Safety and Health Administration (OSHA)—www.osha.gov

    US Department of Transportation—www.dot.gov

    US General Accounting Office—www.gao.gov

(2) STATE ORGANIZATIONS

CA WMB—www.ciwmb.ca.gov/

IN DEM—www.state.in.us/idem

LA DEQ—www.deq.state.la.us

MA DEP—www.state.ma.us

MI DNR—www.dnr.state.mi.us/

NY DEC—www.dec.state.ny.us/

Ohio EPA—www.epa.ohio.gov/oepa.html

PA DEP—www.dep.state.pa.us

TX NRCC—www.tnrcc.state.tx.us

WA DOE—www.wa.gov/ecology/

(3) PROFESSIONAL AND TRADE ORGANIZATIONS

American Chemical Society—www.acs.org, chemCenter.org

American Electroplaters and Surface Finishers Society—www.aesf.org

American Forest and Paper Association—www.afandpa.org/home

American Petroleum Institute—www.api.org

American Plastics Council—www.plastics.org

American Productivity and Quality Center—www.apqc.org

Chemical Manufacturers Association—www.cmahq.com/

Clean Washington Center (Pacific Northwest Economic Region)—www.cwc.org/

Edison Electric Institute—www.eei.org/enviro/

Electric Power Research Institute—www.epri.com/

Global Environmental Management Initiative—www.gemi.org

International Institute for Management Development—www.imd.ch/

National Association of Chemical Distributors—www.nacd.com

National Association for Environmental Management—www.naem.org.com

National Association of Manufacturers—www.nam.org

National Petroleum Council—www.npc.org

National Petroleum Refiners Association—www.npradc.org/

Society of Automotive Engineers—www.sae.org/

Society of Environmental Journalists—www.sej.org/

Society of Petroleum Engineers—www.spe.org

US Chamber of Commerce—www.uschamber.org/

> Nations' Business (a publication of the Chamber)—www.nationsbusiness.org

World Petroleum Congress—www.world-petroleum.org/

(4) REGULATED ORGANIZATIONS

Acushnet Rubber—www.acushnet.com

AlliedSignal—www.alliedsignal.com

Allen-Bradley—www.ab.com

Amoco—www.amocochem.com

AMP Incorporated—www.amp.com

Anheuser-Busch—www.anheuser-busch.com

AT&T—www.att.com

Ashland Chemical—www.ashchem.com

Atlantic Richfield—www.arco.com/

BASF Corporation—www.basf.com

BP America—www.bp.com

Boeing—www.boeing.com

Chevron—www.chevron.com

Colgate-Palmolive Company—www.colgate.com

Conoco—www.conoco.com/

Dow Chemical—www.dow.com

Eastman Kodak—www.kodak.com

Elf Aquitaine—www.elf.fr/us/

Exxon—www.exxon.com

Foxboro—www.foxboro.com

Ford—www.ford.com/

Fujitsu—www.fujitsu.com

General Dynamics—www.gdls.com

Gensym—www.gensym.com

General Motors—www.gm.com

Gillette—www.gillette.com

Goodyear Tire & Rubber Company—www.goodyear.com

Hallmark Cards—www.hallmark.com

Hewlett-Packard—www.hp.com

Hitachi—www.hitachi.com/

Honda—www.honda.com

IBM—www.ibm.com

> IBM's ISO 14001 letter—
> www.ibm.com/procurement/html/suppliers/html

International Paper—www.ipaper.com

Johnson & Johnson—www.johnson.com

Kraft General Foods—www.kraftfoods.com

Lockheed Martin Corporation—www.lmco.com

Lucent Technologies—www.lucent.com

Mars—www.mars.com

McDonald's—www.mcworld.mcdonalds.com

Millipore—www.millipore.com

Mobil—www.mobil.com/

Monsanto—www.monsanto.com

Motorola—www.mot.com

Nabisco—www.nabisco.com

Nalco Chemical—www.nalco.com

Nestle USA—www.nestle.com

Nortel—www.nortel.com

Northrop Grumman—www.northgrum.com

Pfizer—www.pfizer.com

Perigrine—www.peregrineinc.com

Philips Petroleum—www.philips66.com

Polaroid—www.polaroid.com

Praxair—www.praxair.com

Proctor & Gamble—www.pg.com

Raytheon—www.raytheon.com

Rockwell International—www.rockwell.com

Seagram's—www.seagram.com

Shell Oil—www.shell.com

Siemens—www.siemens.com

Sony Electronics—www.sony.com

Texaco—www.texaco.com

Texas Instruments—www.ti.com

Textron—www.textron.com

Toyota—www.toyota.com

Union Camp—www.unioncamp.com

USX—www.usx.com

Walt Disney Company—www.disney.com

Xerox—www.xerox.com

3M—3m.com

(5) NATIONAL ORGANIZATIONS AND SPECIAL INTEREST GROUPS

Americans for the Environment—www.AforE.org

American National Standards Institute—www.ansi.org

American Society for Quality—www.asq.org

    Registrar Accreditation Board—www.rab.org

American Standards for Testing and Materials—www.astm.org

Association francais de normalization—www.afnor.fr/

British Standards Institute—www.bsi.org.uk/

Business for Social Responsibility—www.bsr.org

Coalition for Environmentally Responsible Economies—www.ceres.org

Commission for Environmental Cooperation—www.cec.org

Committee for the National Institute for the Environment—www.enie.org

Council on Economic Priorities—www.cepnyc.org

Deutches Institut fur Normung—www.din.de/

Earth Island Institute—www.earthisland.org

Environmental and Energy Study Institute—www.eesi.org

Environmental Legal Defense Fund—www.wldf.org

Environment Canada—www.doe.ca

European Environmental Agency—www.eea.dk

    European Environmental Law—www.eel.nl/

    Joint Research Center of the EC—www.ei.jrc.it/

Earth Share—www.earthshare.org

Environmental Change Network—www.nmw.ac.uk/ecn/index.html

Environmental Defense Fund—www.edf.org

    "Toxic Ignorance"—www.edf.org/pubs/Reports/ToxicIgnorance

    Scorecard—www.scorecard.org

Foundation for Clean Air Progress—www.cleanairprogress.org/

Friends of the Earth—www.foe.org

Global Change Data and Information System—www.gcdis.usgcrp.gov/

Global Change Research Information Office—www.gcrio.org

Global Climate Coalition—www.worldcorp.com/dc-on-line/gcc/index.html

Global Climate Information Project—www.climatefacts.org/

Great Lakes Information Network—www.great-lakes.net

Greenpeace—www.greenpeace.org

Global Futures—www.globalff.org/

Investor Research Responsibility Center—www.irrc.org

International Organization for Standardization—www.iso.ch

National Library for the Environment—www.cnie.org/nle/

National Resources Defense Council—www.nrdc.org

National Wildlife Federation—www.nwf.org

Netherlands Normalization Institute—www.nni.nl

Norges Standardiseringsforbund—www.standard.no/nsf

Organization for Economic Cooperation and Development—www.oecd.org

Pennsylvania Resources Council—www.prc.org

Renew America—www.crest.org/renewamerica

Rocky Mountain Institute—www.rmi.org

Standards Australia—www.standards.com.au

United Nations Educational, Scientific & Cultural Organization—www.unesco.org/

United Nations Environment Programme—www.unep.ch

US Council for International Business—www.uscib.org

World Health Organization—www.who.int/

World Resources Institute—www.wri.org

World Wildlife Fund—www.wwf.org

(6) OTHER ORGANIZATIONS

Best Manufacturing Practices Center of Excellence—www.bmpoe.org/

> Environmental Best Manufacturing Practices—www.bmpoe/ebmp/index.html

Canadian Center for Pollution Prevention (C2P2)—www.c2p2.sarnia.com/

Chicago Tribune—www.chicago.tribune.com

Council for Agricultural Science and Technology—www.cast-science.org/

ENEA Ambiente (Italy)—www.wwwamb.casaccia.enea.it/

> ANDREA—Archieve for Environmental Education—www.psicoped.rm.cnr.it/   .

Earth Network—www.ecouncil.ac.cr/

Envirolink Network—www.envirolink.org

Environmental News Network—www.een.com

Fertilizer Institute, The—www.tfi.org/

GreenBusiness Letter, The—www.greenbiz.com

Gov Exec (Government executive magazine)—www.govexec.com/features/

Japan Marine Science & Technology Center—www.jamstec.go.ip/

> Japan Printed Circuit Association—www.japca

National Coalition Against the Misuse of Pesticides—www.ncamp.org/

Nature Conservancy, The—www.tnc.org

Organization for Economic Cooperation and Development—www.oced.org

> Environmental Issues link—www.oced/env/

Pace University, School of Law—www.law.pace.edu/env/

Pesticides Action Network—www.panna.org/panna/

Public Interest Research Group—www.pirg.org

Rete Ambiente (Italy)—www.reteambiente.it/

Valdez Society (Japan)—(fax) 81-3-3263-9463

Wall Street Journal, The—www.wsj.com

Waste Prevention Association "3R"—www.rec.org/poland/wpa/wpa.htm

News From TEXACO
Public Relations
2000 Westchester Avenue
White Plains, NY 10650
(914) 253-4177

TEXACO CHAIRMAN MAPS AN APPROACH TO SUSTAINABLE
GROWTH THAT EMBRACES A CLEAN ENVIRONMENT AND A
VIBRANT ECONOMY

FOR RELEASE: TUESDAY, OCTOBER 26, 1993.

WHITE PLAINS, N.Y., Oct. 26—Speaking at the Marketing Conference of The
Conference Board today, Texaco Chairman of the Board and Chief Executive Officer Alfred C. DeCrane, Jr., said the United States should accept that a concept of
sustainable growth, which embraces both a clean environment and a vibrant
economy, driven by reliable, readily available, cost competitive, environmentally
compatible fuels is the best way to insure an improving standard of living for
today and the future.

To promote such sustainable growth, DeCrane called on American business
to demonstrate leadership in making sure that future laws and regulations balance both environmental and economic issues.

"Rather than looking for hypothetically cleaner, less polluting, as-yet-
undetermined and undeveloped alternatives, government should complement
the reformulated gasolines the industry is putting into the market with other
cost-effective, practical environmental efforts," DeCrane said.

Among those practical environmental efforts are enhanced automobile inspection and maintenance, the scrapping of old, high-emitting cars and better
control of stationary sources of nitrous oxide, such as factory and utility smokestacks, DeCrane said.

"These ideas are do-able; they're affordable and they work," he said. "We
can put them into effect right now."

The need for a common-sense approach to regulatory matters is acute, De-
Crane said. Despite several concrete gains in clean air in recent years, the U.S. continues to hamstring its economic competitiveness and to smother job creation with an
unhealthy addiction to excessive command-and-control environmental regulations.

"Too often command-and-control regulations are not based on science, nor
on facts, nor on cost-effectiveness," DeCrane said. "Rather, they are political responses to popular panics and pressures. The result: an excessive burden of incidents like acid rain legislation, the dioxin and alar scares, which damaged the
U.S. economy and competitiveness by restrictions that were not scientifically supported."

Today, the U.S. spends more than $140 billion a year on environmental regulations, only part of the total regulatory cost to this nation of $400 billion to $500 billion a year, which represents a "hidden tax" of roughly $4,000 to $5,000 a year on the average American family. The oil industry alone is slated to spend an estimated $37 billion in the 1990s to meet environmental requirements on refineries and their products, which amounts to a figure greater than the $31 billion value of the refineries themselves.

Sustainable growth seeks to balance a nation's natural, human, and material resources, DeCrane said. Moreover, it does not accept the concept that developed or advanced societies must be forced to shift into neutral, or reverse, so developing societies can be subsidized or granted special "rights" to advance.

"So, one of our key marketing messages must be to not follow the siren song of those who demand, in the name of environmental protection, that we stall economic growth and turn away from resources critical to maintaining and improving economic well-being," DeCrane noted.

One such example is petroleum—oil and natural gas—a resource that DeCrane called "the most accessible, reliable, transportable, affordable and highly performing energy source in the history of this planet." The challenge for the petroleum industry is to provide the facts that overcome society's skepticism and confirm that petroleum and a clean environment are not only compatible, but equally important components of sustainable growth.

Meeting these challenges can, in part, be accomplished by questioning the costs and the scientific bases of the flood of environmental regulations, DeCrane said. They also can be met by communicating, clearly and persuasively, everything that the petroleum and other industries are doing to assure that society can enjoy both growth and a clean environment.

Some examples he noted of recent steps by industry are:

- The oil industry has been working with the auto industry in a $40 million, four-year effort to develop the facts necessary to support environmentally cleaner cars and cleaner fuels.
- Working within the limited flexibility provided by the Clean Air Act, the oil industry, the Environmental Protection Agency and environmental groups in 1992 collaborated in arriving at a new reformulated gasoline that meets the Clean Air Act emission standards for 1995 at an increased, but affordable, manufacturing cost.
- The industry set up a national marine spill response cooperative to deal with major oil spills in U.S. waters.
- The industry has developed safer ways to reduce, recycle and dispose of wastes produced while manufacturing oil-based products.
- The chemical industry is substantially reducing air emissions from plant sites—reductions that outpace legislated mandates.

"President Clinton has rightly said that the time for making policy on the basis of hyperbole, scare tactics and intolerance is over. Instead, he says, we need a balanced policy to preserve jobs and protect the environment," said DeCrane.

"If we are to have sustainable growth," DeCrane concluded, "we must spend judiciously and effectively—and maintain the right balance for our social priorities."

CONTACTS: Dave Dickson (914) 253-4128
                     Jim Reisler   (914) 253-4389

OPERATIONS TECHNOLOGY PROCEDURE	O.T.P. 122
BOEING COMMERCIAL AIRPLANE GROUP	February 29, 1996

GENERAL

BACKGROUND

In 1991, an Operations Development procedure (ODP 005) was published to define the cross-functional assignment process. Since that time, Operations Technology has improved the process and provided a framework by which the process continued.

Shadowing was approved by the Leadership Council because it provides exposure to the challenges faced by related groups and organizations in executing their duties. The program is to be used for career development, gaining a broader view of Operations Technology and The Boeing Company, and increasing knowledge of and performance with customers or suppliers.

PARTICIPATION

Internal cross-functional and shadowing assignments may be arranged in any Operations Technology group(s). External organizations that might be considered for participation in these processes include, but are not limited to, the following:

- Computing
- Customer Services
- Engineering
- Facilities
- Finance
- Human Resources
- Industrial Engineering
- Manufacturing
- Manufacturing Engineering
- Materiel
- Program Management
- Quality Assurance
- Sales and Marketing
- Tooling

ORGANIZATION BENEFITS

Benefits to the organization include:

- Increased knowledge of our customer/supplier requirements.
- Improved working relationships and teamwork.

- Transfer of production process knowledge to process owners and support organizations.
- Exchange of knowledge within the organization to support strategy and process improvements.

EMPLOYEE BENEFITS

Employee benefits include:

- Improved performance in his/her present or future job.
- Preparation for business, managerial, or technical responsibilities.

CROSS-FUNCTIONAL PARTICIPANT SELECTION CRITERIA

ASSESSMENT

Participants are selected to best achieve the purposes of the cross-functional assignment. Assessment shall be based on:

1. Work history.
2. Strong career growth potential.
3. Significant technical and/or business expertise.

ORGANIZATION GUIDELINES

1. Maintain ongoing participation for internal and external assignments. The target level of participation will be a percentage of total headcount plus or minus "x" percent as determined by the director.
2. Maintain a variety of assignments (host organizations, pay codes, work groups).
3. Consider project assignments that are familiar to the participant, as well as those that differ from the participant's past experience.
4. Consider some management positions as developmental opportunities and use these to rotate managers.
5. Ensure that current job assignments are not jeopardized by participation on cross-functional assignments.
6. Look for assignments that produce long-term payoffs for the organization and the participant.

Index

469

Professional and Industry Associations, 427–428
  specific web sites, 455
Proctor & Gamble, 92
Public Environmental Reporting Initiative
  (PERI)
  description of, 274–275
  discussion of, 7, 92, 158, 171, 253
  guidelines, 439–442
Public interest groups, 429
  specific web sites, 457–459
  other organizations' web sites, 459–460
Public Service Electric and Gas, 373

R

Registrar Accreditation Board (RAB), 408
Registrars, 169, 328–329
Renew America, 265
Risk management
  examples of, 293–306
  relative to a company's financial state, 309
  steps to develop, 308
Ritchie, Ingrid, 314
Rockwell International, 274, 366
Rogers, Will, 400
Royal Dutch/Shell, 281

S

SAP AG, 371
S&P 500, 175
SGS-Thompson, 126, 331–347
Saab, 412
Seagal, Steven, 188
Segil, Lorraine, 409–410
Silicon Graphics, 370, 403
Shell Oil, 132
Sheldon, Christopher, 185, 186
Shelton, Robert, 140
Six Sigma, 330
Software
  auditing, see also Audits
  capabilities, 377
  environmental, health & safety, 370–373
  environmental management systems, see also
    Audits
  EDM, 136–137
  ERP, 136–137, 366
  MMI, 366
  real-time, 377
  SCADA, 366–370
SPECTRUM, 259–260
  Statistical process control (SPC), 345
Strategic Group on the Environment (SAGE), 20,
  317–318
Strategic sourcing, 165
Standard Rates and Data Services, 389
Sullivan, J. Kirk, 146

Sun Microsystems, 370
Swiss Bank Corp, 174

T

TQEM, 330
TQM, 330, 401
Texaco
  discussion of, 86, 186
  environmental report, 1996, 128,
    281–283
  Press Release, 461–462
Three Mile Island, 304
Toshiba, 173
Trade publications, listings, 425–427
Training
  discussion of, 27, 215
  EMS auditor, 169
  EMS awareness, 26
  requirements, 201, 340–341
Trending, 345
Truman, Harry S., 115
Turning Point, 293–294

U

US Chamber of Commerce, 50, 163
US Department of Energy, 196, 316
US Department of Justice, 98
US Department of Transportation, 27
US Environmental Protection Agency
  civil monetary penalties, 99
    discussion of, 139, 316, 385
  command-and-control, 33
  compliance, 35
  enforcement
    administrative orders (AOs), 35
    databases
      Enviorfacts, 118, 164
      IDEA, 101
    notices of violation (NOVs), 35, 161, 193
    offices
      National Enforcement Investigations Center
        (NEIC), 35, 269–270
      Office of Criminal Enforcement (OCE), 35
      Office of Enforcement and Compliance Assur-
        ance (OECA), 34, 195
    policies, 204
    sample enforcement actions, 110, 132, 139
  environmental auditing, see also audits
    self-policing and disclosure policy, 7, 34, 85, 214,
      221
  guidance documents
    Exercise of Investigative Discretion, 35
    FY 1997 Enforcement and Corrective Action Pri-
      orities, 37–39
    Five-Year Plan, 36
    Four-Year Plan, 36

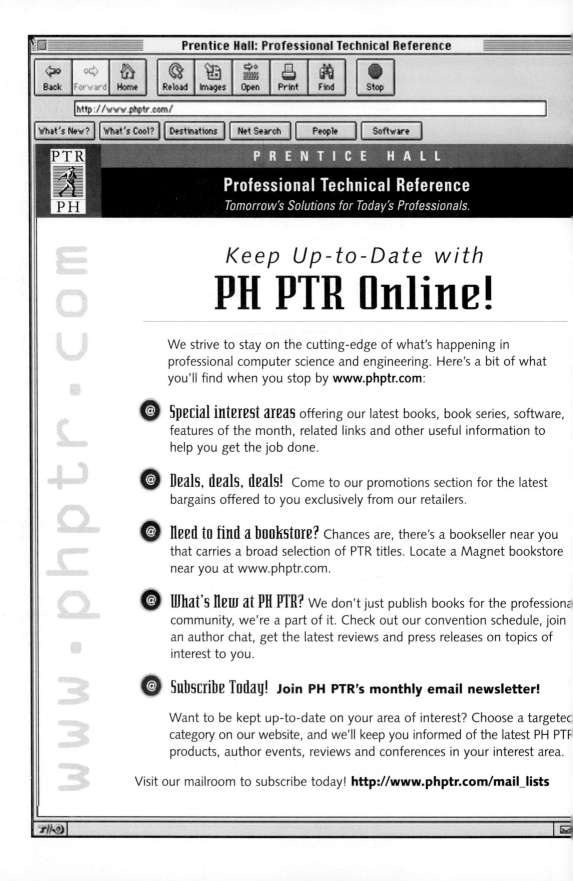